Progress in Mathematics
Volume 301

Michael Ruzhansky
Mitsuru Sugimoto
Jens Wirth
Editors

Evolution Equations of Hyperbolic and Schrödinger Type

Asymptotics, Estimates and Nonlinearities

Editors

Michael Ruzhansky
Department of Mathematics
Imperial College London
London
United Kingdom

Mitsuru Sugimoto
Graduate School of Mathematics
Nagoya University
Nagoya
Japan

Jens Wirth
Fachbereich Mathematik
Universität Stuttgart
Stuttgart
Germany

ISBN 978-3-0348-0453-0   ISBN 978-3-0348-0454-7 (eBook)
DOI 10.1007/978-3-0348-0454-7
Springer Basel Heidelberg New York Dordrecht London

Library of Congress Control Number: 2012945184

Mathematics Subject Classification (2010): 35-06, 35Axx, 35Lxx, 35Pxx, 35Qxx, 35Sxx, 35B34, 42B35, 47A10, 49K20, 58J45, 74B20, 78A46, 82C40

Printed on acid-free paper

Springer Basel AG is part of Springer Science+Business Media (www.birkhauser-science.com)

Contents

vi   Contents

Preface

Asymptotic constructions and large-time asymptotic estimates for solutions to evolution equations of hyperbolic or more general p-evolution type are an active field of current research. We took this as an incentive to organise an international workshop on

Asymptotic Properties of Solutions to Hyperbolic Equations

at Imperial College London. It took place in March 2011 and included 32 speakers presenting current results of their work. We are grateful to EPSRC supporting this meeting within the Pathways to Impact Award scheme.

The papers collected in this volume are authored by participants of that meeting. They focus on different aspects of current research and are, in particular, centred around

- symbolic and other parametrix constructions;
- energy estimates in various function spaces;
- asymptotic behaviour of solutions to the Cauchy problem;
- microlocal analysis and Fourier integral operators;
- problems for strictly and non-strictly hyperbolic equations and systems and their non-linear aspects;
- hyperbolic constructions in different settings;
- applications: elasticity, general relativity, etc.

The aim of this volume is two-fold. On one hand it shall give an overview on the great variety of ongoing current research in the field and, therefore, allow researchers as well as students to grasp new aspects and broaden their understanding of the area. We put a particular emphasis on detailed proofs of results and completeness of presentations. On the other hand, all contained papers are full research papers presenting new results. This allows experts in the field to describe deeper inside views and will hopefully lead to further collaborative work in the area.

The papers are in alphabetical order.

The speakers at the above-mentioned workshop were Piero d'Ancona (Rome), Matania Ben-Artzi (Jerusalem), Nikolaos Bournaveas (Edinburgh), Massimo Cicognani (Bologna), Ferruccio Colombini (Pisa), Mihalis Dafermos (Cambridge), Claudia Garetto (London), Vladimir Georgiev (Pisa), Todor Gramchev (Cagliari), Günther Hörmann (Vienna), Maarten de Hoop (Purdue), Tynysbek Kalmenov (Almaty), Sergiu Klainerman (Princeton), Hideo Kubo (Tohoku), Tokio Matsuyama

(Tokyo), Michael Oberguggenberger (Innsbruck), Cesare Parenti (Bologna), Alberto Parmeggiani (Bologna), Luigi Rodino (Torino), Michael Ruzhansky (London), Daniele Del Santo (Trieste), Semyon Serovaisky (Almaty), Kanat Shakenov (Almaty), Sergio Spagnolo (Pisa), Mitsuru Sugimoto (Nagoya), Hiroyuki Takamura (Hakodate), Mirko Tarulli (London), Igor Trooshin (Tohoku), Gunther Uhlmann (Washington), Baoxiang Wang (Beijing), Jens Wirth (Stuttgart), and Ingo Witt (Göttingen). We thank all participants for making the meeting a success.

We point out that, in fact, the conference was also the 5th meeting in a series of international meetings devoted to "Function Spaces and Partial Differential Equations", initiated by the first two editors of this volume, where the four previous ones were held at:

- Osaka University, Japan, February 18–20, 2008;
- Imperial College London, UK, December 3–5, 2008;
- Nagoya University, Japan, September 28–October 1, 2009;
- University of Göttingen, Germany, June 14–17, 2010;

The following 6th meeting will take place at the Aalto University, Helsinki, Finland, in June 2012, and will be devoted to "Fourier analysis and pseudo-differential operators".

We would also like to thank other members of the organising committee of the conference and their contributions in different ways, in particular, Claudia Garetto for doing excellent organisational work, as well as PhD students Donal Connolly, David Rottensteiner, and Mirko Tarulli.

Michael Ruzhansky, Mitsuru Sugimoto, and Jens Wirth

Progress in Mathematics, Vol. 301, 1–40

Divergence-type Operators:
Spectral Theory and Spacetime Estimates

Matania Ben-Artzi

Abstract. The paper is concerned with various aspects of the spectral structure of the operator

$$H = -\sum_{j,k=1}^{n} \partial_{x_j} a_{j,k}(x) \partial_{x_k}.$$

It is assumed to be formally self-adjoint in $L^2(\mathbb{R}^n)$, $n \geq 2$. The real coefficients $a_{j,k}(x) = a_{k,j}(x)$ are assumed to be bounded and H is assumed to be uniformly elliptic and to coincide with $-\Delta$ outside of a ball. A Limiting Absorption Principle (LAP) is proved in the framework of weighted Sobolev spaces. It is then used for (i) A general eigenfunction expansion theorem and (ii) Global spacetime estimates for the associated (inhomogeneous) generalized wave equation.

Mathematics Subject Classification. Primary 47A10; Secondary 35P05, 35L15, 47F05.

Keywords. Spectral derivative, divergence-type operator, limiting absorption principle, eigenfunction expansion, spacetime estimates.

1. Introduction

Let H be a self-adjoint (bounded or unbounded) operator in a Hilbert space $\mathcal{H}$. The classical spectral theorem [59] gives a representation of H,

$$H = \int_{\mathbb{R}} \lambda \, dE(\lambda),$$

in terms of its (uniquely determined) spectral family (of projections) $\{E(\lambda)\}$.

The knowledge of $\{E(\lambda)\}$ yields valuable information on the spectral structure of H; the location of its singular or absolutely continuous spectrum, as well as its eigenvalues.

On the other hand, there are important issues (typically related to partial differential operators) that cannot be resolved simply on the basis of the spectral theorem. We pick here one important topic and expound it in more detail, so as to illustrate the point at hand.

Assuming that $\{E(\lambda)\}$ is (strongly) continuous from the left, one might think of $E(\lambda + 0) - E(\lambda)$ as a projection on the *eigenspace* associated with λ. However, if λ is not an eigenvalue, this projection clearly vanishes. On the other hand, the mathematical foundation of quantum mechanics has turned the *expansion by generalized eigenfunctions* (such as the Fourier transform with respect to the Laplacian) into a basic tool of the theory (see, e.g., [86] for an early treatment). So the question is how (if at all possible) to incorporate such an expansion into the abstract framework of the spectral theorem. We shall address this question in Section 5, where we show how the basic premise of this review, namely, the *smoothness* concept of the spectral family, leads to an eigenfunction expansion theorem for the class of *divergence-type* operators.

Using a formal point of view we can say that the bridge between the spectral theorem and the aforementioned *eigenfunction expansion theorem* is obtained by replacing the above difference $E(\lambda + 0) - E(\lambda)$ by its *scaled version*, the (formal, at this stage) derivative $\frac{d}{d\lambda}E(\lambda)$. In fact, this derivative is the cornerstone of the present review.

Certainly, this derivative is far from being a *new object*. In the physical literature it is known as the *density of states* [29, Chapter XIII]. It has appeared implicitly in many mathematical studies of quantum mechanics.

After introducing our basic notational conventions and functional spaces in Section 2, we present the basic abstract setting in Section 3. This structure was first established in a joint work with the late A. Devinatz [15]. It relies on the fundamental hypothesis that the spectral derivative is Hölder continuous in a suitable functional setting. The primary aim is to establish a *Limiting Absorption Principle* (LAP), namely, that the resolvents (from either side of the spectrum) remain continuous up to the (absolutely continuous) spectrum in this setting. Once established for an operator H, we show in Subsection 3.2 that it persists to functions $f(H)$, for a wide family of functions f, with interesting results for operators of mathematical physics, such as the *relativistic Schrödinger* operator. It is pointed out that without the smoothness assumption, the validity of the LAP for H does not necessarily imply its validity even for H^2.

The next three sections are devoted to the main application considered in this review, namely, a detailed study of the operator

$$H = -\sum_{j,k=1}^{n} \frac{\partial}{\partial x_j} a_{j,k}(x) \frac{\partial}{\partial x_k},$$

which is assumed to be formally self-adjoint in $L^2(\mathbb{R}^n)$, $n \geq 2$. The real coefficients $a_{j,k}(x) = a_{k,j}(x)$ are assumed to be bounded and H is assumed to be uniformly elliptic and to coincide with $-\Delta$ outside of a ball. In particular, the coefficients can be discontinuous. It is readily seen that these assumptions imply that $\sigma(H)$, the spectrum of H, is the half-axis $[0, \infty)$, and is entirely continuous. The *threshold* $z =$

0 plays a special role in this setting. The absolute continuity of the spectrum was established in [11]. It is a straightforward result of the LAP, which is established in Section 4. In particular, we show that the limiting values of the resolvent remain continuous across the threshold (which is therefore not a resonance).

Since its appearance in the classical works of Eidus [40] and Agmon [1], the LAP has proven to be a fundamental tool in the study of spectral and scattering theory. The method of Eidus (for second-order elliptic operators) relied on careful elliptic estimates while the method of Agmon used Fourier analysis (division by symbols with simple zeros), followed by a perturbative ("bootstrap") argument to deal with lower-order terms. This latter method, extended to simply characteristic operators of any order, is expounded in [49, Chapter 14]. The method of Mourre (also known as the "conjugate operator method") [68] paved the way to the breakthrough in the study of the (quantum) N-body problem [70]. We refer to the monographs [4, 36] for the presentation of Mourre's method in an abstract framework. We also refer to the recent paper [41], where the LAP is proved by using a combination of Mourre's method and energy estimates.

The LAP for the divergence-type operator H introduced above cannot be obtained by a straightforward application of either one of these methods. Firstly, the presence of the non-constant coefficients $a_{j,k}(x)$ means that H is not a relatively compact perturbation of the Laplacian, and the perturbation method of Agmon cannot be applied. Secondly, if one insists (as we do here) on assuming only boundedness (and not smoothness) of these coefficients, the method of Mourre, as used in the semiclassical literature [76], cannot be applied (the conjugate operator is related to a generator of the corresponding flow that, in turn, relies on smoothness). In contrast, our approach to the LAP enables us to obtain resolvent estimates for the Laplacian *beyond* the L^2 setting, by using $H^{-1,s}$ weighted Sobolev spaces (see Subsection 4.1). In this context the operator H can be handled as a perturbation of the Laplacian.

We note in addition that both Agmon's and Mourre's methods cannot be applied across the threshold at $z = 0$. Here we obtain continuity of the limiting values of the resolvent across the threshold, at the expense of using a more restrictive weight function. This fact is essential in the treatment of global spacetime estimates in Section 6.

A more detailed discussion of the relevant literature is given in Section 4.

Section 5 is devoted to the eigenfunction expansion theorem (by generalized eigenfunctions) associated with the operator H. We have already touched upon this topic above, illustrating the differences between the general (abstract) spectral theorem and the detailed *Fourier-type* expansion needed in applications. We expand on this issue in the section.

A global spacetime estimate for the associated (inhomogeneous) generalized wave equation is proved in Section 6. We chose to bring this example (instead of the simpler Schrödinger-type equation) in order to stress the various possibilities available with the tool of the spectral derivative. In doing so we need to restrict much further our class of coefficient matrices. In fact, in order to obtain good control on the behavior of the limiting values of the spectral derivative at *high energy*, we need to use geometric assumptions (*non-trapping trajectories*), which are common in semiclassical theory.

2. Functional spaces and notation

We collect here some basic notations and functional spaces to be used throughout this paper.

The closure of a set Ω (either in the real line $\mathbb{R}$ or in the complex plane $\mathbb{C}$) is denoted by $\overline{\Omega}$.

For any two normed spaces X, Y, we denote by $B(X, Y)$ the space of bounded linear operators from X to Y, equipped with the operator norm $\| \ \|_{B(X,Y)}$ topology (to which we refer as the *uniform operator topology*). In the case $X = Y$ we simplify to $B(X)$.

The following weighted L^2 and Sobolev spaces will appear frequently. First, for $s \in \mathbb{R}$ and m a non-negative integer, we define

$$L^{2,s}(\mathbb{R}^n) := \left\{ u(x) \mid \|u\|_{0,s}^2 = \int_{\mathbb{R}^n} (1 + |x|^2)^s |u(x)|^2 dx < \infty \right\}$$

$$H^{m,s}(\mathbb{R}^n) := \left\{ u(x) \mid D^\alpha u \in L^{2,s}, \ |\alpha| \le m, \ \|u\|_{m,s}^2 = \sum_{|\alpha| \le m} \|D^\alpha u\|_{0,s}^2 \right\}$$

(we write L^2 for $L^{2,0}$ and $\|u\|_0 = \|u\|_{0,0}$). More generally, for any $\sigma \in \mathbb{R}$, let $H^\sigma \equiv H^{\sigma,0}$ be the Sobolev space of order σ, namely,

$$H^\sigma = \{ \hat{u} \mid u \in L^{2,\sigma} \},$$

$\|\hat{u}\|_{\sigma,0} = \|u\|_{0,\sigma}$, where the Fourier transform is defined as usual by

$$\hat{u}(\xi) = (2\pi)^{-\frac{n}{2}} \int_{\mathbb{R}^n} u(x) \exp(-i\xi x) dx. \tag{2.1}$$

For negative indices, we denote by $\{H^{-m,s}, \| \cdot \|_{-m,s}\}$ the dual space of $H^{m,-s}$. In particular, observe that any function $f \in H^{-1,s}$ can be represented (not uniquely) as

$$f = f_0 + \sum_{k=1}^n i^{-1} \frac{\partial}{\partial x_k} f_k, \quad f_k \in L^{2,s}, \ 0 \le k \le n. \tag{2.2}$$

In the case $n = 2$ and $s > 1$, we define

$$L_0^{2,s}(\mathbb{R}^2) = \{ u \in L^{2,s}(\mathbb{R}^2) \mid \hat{u}(0) = 0 \},$$

and set $H_0^{-1,s}(\mathbb{R}^2)$ to be the space of functions $f \in H^{-1,s}(\mathbb{R}^2)$ which have a representation (2.2), where $f_k \subset L_0^{2,s}$, $k = 0, 1, 2$.

3. The basic abstract structure

Let $\mathcal{H}$ be a Hilbert space over $\mathbb{C}$ (the complex numbers), whose scalar product and norm we denote, respectively, by $(\,,\,)$ and $\|\;\|$.

Let $\mathcal{X}$ be another Hilbert space such that $\mathcal{X} \subseteq \mathcal{H}$, where the embedding is dense and continuous. In other words, $\mathcal{X}$ can be considered as a dense subspace of $\mathcal{H}$, equipped with a stronger norm. Then, of course, $\mathcal{X} \hookrightarrow \mathcal{H} \hookrightarrow \mathcal{X}^*$, where $\mathcal{X}^*$ is the anti-dual of $\mathcal{X}$, i.e., the continuous additive functionals l on $\mathcal{X}$, such that $l(\alpha v) = \overline{\alpha}\, l(v)$, $\alpha \in \mathbb{C}$. The (linear) embedding $h \in \mathcal{H} \hookrightarrow x^* \in \mathcal{X}^*$ is obtained as usual by the scalar product (in $\mathcal{H}$), $x^*(x) = (h, x)$.

We use $\|x\|_{\mathcal{X}}$, $\|x^*\|_{\mathcal{X}^*}$ for the norms in $\mathcal{X}$, $\mathcal{X}^*$, respectively, and designate by $\langle\,,\,\rangle$ the $(\mathcal{X}^*, \mathcal{X})$ pairing.

Let H be a self-adjoint (in general unbounded) operator on $\mathcal{H}$ and let $\{E(\lambda)\}$ be its spectral family. Let

$$R(z) = (H - z)^{-1}, \quad z \in \mathbb{C}^{\pm} = \{z \mid \pm \operatorname{Im} z > 0\},$$

be the associated resolvent operator. We denote by $\sigma(H) \subseteq \mathbb{R}$ the spectrum of H.

Clearly, if $\lambda \in \sigma(H)$, then $R(z)$ cannot converge to a limit in the uniform operator topology of $B(\mathcal{H})$ as $z \to \lambda$. However, a basic notion in our treatment is the fact that such *continuity up to the spectrum* of the resolvent can be achieved in a weaker topology. We begin with the following definition.

Definition 3.1. Let $[\kappa_1, \kappa_2] \subseteq \mathbb{R}$. We say that H satisfies the *Limiting Absorption Principle* (LAP) in $[\kappa_1, \kappa_2]$ if $R(z)$, $z \in \mathbb{C}^{\pm}$, can be extended continuously to $\operatorname{Im} z = 0$, $\operatorname{Re} z \in [\kappa_1, \kappa_2]$, in the uniform operator topology of $B(\mathcal{X}, \mathcal{X}^*)$. In this case we denote the limiting values by $R^{\pm}(\lambda)$, $\kappa_1 \leq \lambda \leq \kappa_2$.

Remark 3.2. By the well-known Stieltjes formula [59], for all $x \in \mathcal{X}$,

$$((E(\delta) - E(\gamma))x, x) = \frac{1}{2\pi i} \int_{\gamma}^{\delta} \left\langle \left(R^+(\lambda) - R^-(\lambda)\right)x, x \right\rangle d\lambda, \quad [\gamma, \delta] \subseteq [\kappa_1, \kappa_2],$$

it follows that H is absolutely continuous in $[\kappa_1, \kappa_2]$.

Remark that our assumptions readily imply that the uniform operator topology of $B(\mathcal{X}, \mathcal{X}^*)$ is weaker than that of $B(\mathcal{H})$. Also note that the limiting values $R^-(\lambda)$ are, generally speaking, different from $R^+(\lambda)$.

For reasons to become clear later, we introduce still another Hilbert space $\mathcal{X}_H^*$, which is a dense subspace of $\mathcal{X}^*$, equipped with a stronger norm (so that the embedding $\mathcal{X}_H^* \hookrightarrow \mathcal{X}^*$ is continuous). However, we do not require that $\mathcal{H}$ be embedded in $\mathcal{X}_H^*$. As indicated by the notation, $\mathcal{X}_H^*$ may depend on H (see Example 3.5 below). A typical case would be when H can be extended as a densely defined operator in $\mathcal{X}^*$ and $\mathcal{X}_H^*$ would be its domain there, equipped with the graph norm. This will be the case in Theorem 3.11 below.

Let $\{E(\lambda)\}$ be the spectral family of H. When there is no risk of confusion, we also use $E(B)$ to denote the spectral projection on any Borel set B (so that $E(\lambda) = E(-\infty, \lambda)$).

Definition 3.3. Let $U \subseteq \mathbb{R}$ be open and let $0 < \alpha \leq 1$. Assume that U is of *full spectral measure*, namely, $E(\mathbb{R} \setminus U) = 0$. Then H is said to be of *type* $(\mathcal{X}, \mathcal{X}_H^*, \alpha, U)$ if the following conditions are satisfied:

1. The operator-valued function

$$\lambda \to E(\lambda) \in B(\mathcal{X}, \mathcal{X}^*), \quad \lambda \in U,$$

   is weakly differentiable with a locally Hölder continuous derivative in $B(\mathcal{X}, \mathcal{X}_H^*)$; that is, there exists an operator-valued function

$$\lambda \to A(\lambda) \in B(\mathcal{X}, \mathcal{X}_H^*), \quad \lambda \in U,$$

   so that (recall that $(\, , \,)$ is the scalar product in $\mathcal{H}$ while $\langle \, , \, \rangle$ is the $(\mathcal{X}^*, \mathcal{X})$ pairing)

$$\frac{d}{d\lambda}(E(\lambda)x, y) = \langle A(\lambda)x, y \rangle, \quad x, y \in \mathcal{X}, \lambda \in U,$$

   and such that for every compact interval $K \subseteq U$, there exists an $M_K > 0$ satisfying

$$\|A(\lambda) - A(\mu)\|_{B(\mathcal{X}, \mathcal{X}_H^*)} \leq M_K |\lambda - \mu|^\alpha, \quad \lambda, \mu \in K.$$

2. For every bounded open set $J \subseteq U$ and for every compact interval $K \subseteq J$, the operator-valued function (defined in the weak sense)

$$z \to \int_{U \setminus J} \frac{A(\lambda)}{\lambda - z} \, d\lambda, \quad z \in \mathbb{C}, \operatorname{Re} z \in K, |\operatorname{Im} z| \leq 1,$$

   takes values and is Hölder continuous in the uniform operator topology of $B(\mathcal{X}, \mathcal{X}_H^*)$, with exponent α.

Remark 3.4. We could *localize* this definition and, in particular, relax the assumption that $E(\mathbb{R} \setminus U) = 0$. However, this is not needed for the operators discussed in this review, typically perturbations of operators with absolutely continuous spectrum (see the following example below).

Example 3.5 ($H_0 = -\Delta$). (This example will be continued in Subsection 4.1.)

We take the operator H_0 to be the unique self-adjoint extension of the restriction of $-\Delta$ to smooth compactly supported functions [59]. Let $\{E_0(\lambda)\}$ be the spectral family associated with H_0 so that, using the Fourier notation introduced in Section 2,

$$(E_0(\lambda)h, h) = \int_{|\xi|^2 \leq \lambda} |\hat{h}|^2 \, d\xi, \quad \lambda \geq 0, \ h \in L^2(\mathbb{R}^n). \tag{3.1}$$

We refer to Section 2 for definitions of the weighted L^2 and Sobolev spaces involved in the sequel. Recall that by the standard trace lemma, we have

$$\int_{|\xi|^2 = \lambda} |\hat{h}|^2 \, d\tau \leq C \|\hat{h}\|_{H^s}^2, \quad s > \frac{1}{2}, \lambda > 0, \tag{3.2}$$

where $C > 0$ is independent of λ and $d\tau$ is the restriction of the Lebesgue measure (see [15] for the argument that it can be used for the full half-axis, not just compact intervals).

We conclude that the weak derivative $A_0(\lambda) = \frac{d}{d\lambda}E_0(\lambda)$ exists in the space $B(L^{2,s}, L^{2,-s})$ for any $s > \frac{1}{2}$ and $\lambda > 0$ and satisfies

$$\langle A_0(\lambda)h, k \rangle = (2\sqrt{\lambda})^{-1} \int_{|\xi|^2 = \lambda} \hat{h}\overline{\hat{k}}\, d\tau, \quad h, k \in L^{2,s}, \tag{3.3}$$

where $\langle\ ,\ \rangle$ is the $(L^{2,-s}, L^{2,s})$ pairing (conjugate linear with respect to the second term) and $d\tau$ is the Lebesgue surface measure (we write $L^{2,s}$ for $L^{2,s}(\mathbb{R}^n)$).

Furthermore, by taking s large in (3.2) (it suffices to take $s > \frac{n}{2}+2$) and using the Sobolev imbedding theorem we infer that $A_0(\lambda)$ is locally Lipschitz continuous in the uniform operator topology, so that by interpolation it is locally Hölder continuous in the uniform operator topology of $B(L^{2,s}, L^{2,-s})$ for any $s > \frac{1}{2}$.

Finally, since the (distributional) Fourier transform of $A_0(\lambda)h$ is the surface density $(2\sqrt{\lambda})^{-1}\delta_{|\xi|^2 - \lambda}\hat{h}(\xi)\, d\tau$, we conclude that actually $A_0(\lambda)h \in H^{m,-s}$, $s > \frac{1}{2}$, for any $m > 0$, and $A_0(\lambda)$ is locally Hölder continuous in the uniform operator topology of $B(L^{2,s}, H^{m,-s})$ for any $s > \frac{1}{2}$.

Thus, all the requirements of Definition 3.3 are satisfied with $\mathcal{X} = L^{2,s}(\mathbb{R}^n)$, $\mathcal{X}^*_{H_0} = H^{2,-s}(\mathbb{R}^n)$, $s > \frac{1}{2}$.

3.1. The limiting absorption prinicple – LAP

Recall first the classical Privaloff-Korn theorem (see [31] for a proof).

Theorem. *Let $f\colon \mathbb{R} \to \mathbb{C}$ be a compactly supported Hölder continuous function so that, for some $N > 0$ and $0 < \alpha < 1$,*

$$|f(\lambda_2) - f(\lambda_1)| \leq N\,|\lambda_2 - \lambda_1|^\alpha, \quad \lambda_2, \lambda_1 \in \mathbb{R}.$$

Let

$$F^\pm(z) = \int_\mathbb{R} \frac{f(\lambda)}{\lambda - z}\, d\lambda, \quad z \in \mathbb{C}^\pm.$$

Then, for every $\mu \in \mathbb{R}$, the limits

$$F^\pm(\mu) = \lim F(z) = \pm i\pi f(\mu) + \mathrm{P.\,V.}\int_\mathbb{R} \frac{f(\lambda)}{\lambda - \mu}d\lambda \quad \text{as } z \to \mu,\ \pm \operatorname{Im} z > 0,$$

exist and moreover, for every compact $K \subseteq \overline{\mathbb{C}^+}$ (or $K \subseteq \overline{\mathbb{C}^-}$), there exists a constant M_K so that

$$|F^\pm(z_2) - F^\pm(z_1)| \leq N M_K\,|z_2 - z_1|^\alpha, \quad z_1, z_2 \in K.$$

We can now state our basic theorem, concerning the LAP in the abstract setting. We remark that a slightly different version will appear in Subsection 4.3.

Theorem 3.6. *Let H be of type $(\mathcal{X}, \mathcal{X}^*_H, \alpha, U)$ (where $U \subseteq \mathbb{R}$ is open and $0 < \alpha \leq 1$). Then H satisfies the LAP in U. More explicitly, the limits*

$$R^\pm(\lambda) = \lim_{\epsilon \downarrow 0} R(\lambda \pm i\epsilon), \quad \lambda \in U,$$

exist in the uniform operator topology of $B(\mathcal{X}, \mathcal{X}_H^*)$ and the extended operator-valued function

$$R(z) = \begin{cases} R(z), & z \in \mathbb{C}^+, \\ R^+(z), & z \in U, \end{cases}$$

is locally Hölder continuous in the same topology (with exponent α).

A similar statement applies when $\mathbb{C}^+$ is replaced by $\mathbb{C}^-$, but note that the limiting values $R^\pm(\lambda)$ are in general different.

Proof. Let $J \subseteq U$ be a bounded open set such that $\overline{J} \subseteq U$ and $K \subseteq J$ be a compact interval. Let $\varphi \in C_0^\infty(U)$ be a cutoff function with $\varphi \equiv 1$ on J. Taking $x, y \in \mathcal{X}$, we have, for $\operatorname{Re} z \in K$, $\operatorname{Im} z \neq 0$,

$$(R(z)x, y) = \int_U \frac{\varphi(\mu)\langle A(\mu)x, y\rangle}{\mu - z}\, d\mu + \int_{U \setminus J} \frac{(1 - \varphi(\mu))\langle A(\mu)x, y\rangle}{\mu - z}\, d\mu$$
$$= (R_1(z)x, y) + (R_2(z)x, y).$$

By hypothesis (see Definition 3.3) the operator-valued function

$$R_2(z) = \int_{U \setminus J} \frac{(1 - \varphi(\mu))A(\mu)}{\mu - z}\, d\mu,$$

belongs to $B(\mathcal{X}, \mathcal{X}_H^*)$, and it is locally Hölder continuous for $\operatorname{Re} z \in K$. Thus, we are reduced to considering R_1.

Observe that the integral

$$R_1(z) = \int_{U \setminus J} \frac{\varphi(\mu)A(\mu)}{\mu - z}\, d\mu,$$

is well defined as a Riemann integral, since the integrand is continuous in the uniform norm topology of $B(\mathcal{X}, \mathcal{X}_H^*)$. Thus $R_1(z) \in B(\mathcal{X}, \mathcal{X}_H^*)$. It remains to prove the assertion concerning its Hölder continuity.

Note that the embeddings $\mathcal{X} \hookrightarrow \mathcal{H} \hookrightarrow \mathcal{X}^*$ and $\mathcal{X}_H^* \hookrightarrow \mathcal{X}^* \hookrightarrow \mathcal{X}_H^{**}$ are dense and continuous. Thus, we can view $\mathcal{X}$ as embedded in $\mathcal{X}_H^{**}$, so that the pairing $\langle A(\mu)x, y\rangle$ can be regarded as an $(\mathcal{X}_H^*, \mathcal{X}_H^{**})$ pairing.

Suppose now that $\operatorname{Im} z_i > 0$, $\operatorname{Re} z_i \in K$, $i = 1, 2$, so that the Privaloff-Korn theorem yields, for $x, y \in \mathcal{X}$,

$$|([R_1(z_2) - R_1(z_1)]x, y)|$$
$$\leq M_K \sup_{\mu_1 \neq \mu_2} \frac{|\langle [\varphi(\mu_2)A(\mu_2) - \varphi(\mu_2)A(\mu_2)]x, y\rangle|}{|\mu_2 - \mu_1|^\alpha} |z_2 - z_1|^\alpha,$$

and as observed above

$$|\langle [\varphi(\mu_2)A(\mu_2) - \varphi(\mu_2)A(\mu_2)]x, y\rangle|$$
$$\leq \left\| [\varphi(\mu_2)A(\mu_2) - \varphi(\mu_2)A(\mu_2)]x \right\|_{\mathcal{X}_H^*} \|y\|_{\mathcal{X}_H^{**}}$$
$$\leq \left\| \varphi(\mu_2)A(\mu_2) - \varphi(\mu_2)A(\mu_2) \right\|_{B(\mathcal{X}, \mathcal{X}_H^*)} \|x\|_{\mathcal{X}} \|y\|_{\mathcal{X}_H^{**}}.$$

Thus,
$$\left|([R_1(z_2) - R_1(z_1)]x, y)\right| \le NM_K |z_2 - z_1|^\alpha \, \|x\|_{\mathcal{X}} \, \|y\|_{\mathcal{X}_H^{**}},$$
where
$$N = \sup_{\mu_1 \ne \mu_2} \frac{\|\varphi(\mu_2)A(\mu_2) - \varphi(\mu_2)A(\mu_2)\|_{B(\mathcal{X}, \mathcal{X}_H^*)}}{|\mu_2 - \mu_1|^\alpha}.$$
Since $\mathcal{X}$ is dense in $\mathcal{X}_H^{**}$, the last estimate yields
$$\|R_1(z_2) - R_1(z_1)\|_{B(\mathcal{X}, \mathcal{X}_H^*)} \le NM_K |z_2 - z_1|^\alpha,$$
and the proof is complete. □

Corollary 3.7. *In view of the Stieltjes formula* (*see* Remark 3.2 *above*) *we have*
$$A(\lambda) = \frac{1}{2\pi i} \left(R^+(\lambda) - R^-(\lambda)\right), \quad \lambda \in U.$$
In particular, H is absolutely continuous in U and $R^+(\lambda) - R^-(\lambda)$ cannot vanish on a subset of $\sigma(H) \cap U$ of positive (Lebesgue) measure.

Remark 3.8. The operator $A(\lambda)$, $\lambda \in [0, \infty)$, is known in the physical literature as the *density of states* [29, Chapter XIII].

Also, combining the theorem with the observations in Example 3.5 we obtain the following corollary, which is Agmon's classical LAP theorem [1].

Corollary 3.9. *Let $H_0 = -\Delta$ and set $R_0(z) = (H_0 - z)^{-1}$, $\operatorname{Im} z \ne 0$. Then the limits*
$$R_0^\pm(\lambda) = \lim_{\epsilon \downarrow 0} R_0(\lambda \pm i\epsilon), \quad \lambda \in (0, \infty),$$
exist in the uniform operator topology of $B(L^{2,s}, H^{2,-s})$, $s > \frac{1}{2}$. Furthermore, these limiting values are Hölder continuous in this topology.

Remark 3.10. The considerations of Example 3.5, based on trace estimates, can be applied to a wide range of constant coefficient partial differential operators (so-called *simply characteristic* operators, including all principal-type operators). Hence, a suitable LAP can be established for such operators. We shall not pursue this direction further in this review, but refer the reader to [15].

In general, it is easier to verify the conditions of Definition 3.3 for the operator space $B(\mathcal{X}, \mathcal{X}^*)$ than for $B(\mathcal{X}, \mathcal{X}_H^*)$. However, in some circumstances it is enough to establish the conditions in the latter space. This is expressed in the following theorem.

Theorem 3.11. *Let H be densely defined and closable in $\mathcal{X}^*$, with closure $\overline{H}$. Take $\mathcal{X}_H^* = D(\overline{H})$ (its domain), equipped with the graph norm*
$$\|x\|_{\mathcal{X}_H^*}^2 = \|x\|_{\mathcal{X}^*}^2 + \|\overline{H}x\|_{\mathcal{X}^*}^2.$$
Suppose that H is of type $(\mathcal{X}, \mathcal{X}^*, \alpha, U)$ (see Definition 3.3). Then in fact H is of type $(\mathcal{X}, \mathcal{X}_H^*, \alpha, U)$.

Proof. In view of Theorem 3.6 (where all assumptions hold in $B(\mathcal{X}, \mathcal{X}^*)$) we know that the limits

$$R^{\pm}(\lambda) = \lim_{\epsilon \downarrow 0} R(\lambda \pm i\epsilon), \quad \lambda \in U,$$

exist in the uniform operator topology, are locally Hölder continuous and, furthermore, for all $x \in \mathcal{X}$,

$$\lim_{\epsilon \downarrow 0} \overline{H} R(\lambda \pm i\epsilon)\, x = x + \lambda R^{\pm}(\lambda)x, \quad \lambda \in U.$$

Since $\overline{H}$ is closed in $\mathcal{X}^*$, we obtain

$$\overline{H} R^{\pm}(\lambda)\, x = x + \lambda R^{\pm}(\lambda)x \in \mathcal{X}^*,$$

so that $R^{\pm}(\lambda)x \in \mathcal{X}_H^*$. From the definition of the graph norm topology we see that $R^{\pm}(\lambda)$ is locally Hölder continuous in $B(\mathcal{X}, \mathcal{X}_H^*)$. Thus, using Eq. (3.7), we conclude that the same is true for $A(\lambda)$, so that the first condition in Definition 3.3 is satisfied.

To establish the second condition, let $J \subseteq U$ be an open set and $K \subseteq J$ compact. Let $z \in \mathbb{C}$ with $\operatorname{Re} z \in K$, and let $F(\lambda; z) = \frac{\chi_{U \setminus J}(\lambda)}{\lambda - z}$ (as usual, χ is the characteristic function of the indicated set). By the standard spectral calculus

$$HF(H; z) = \int_U \lambda F(\lambda; z) \, dE(\lambda) = \int_{U \setminus J} \frac{\lambda A(\lambda)}{\lambda - z} \, d\lambda,$$

so that both $F(H; z) = \int_{U \setminus J} \frac{A(\lambda)}{\lambda - z} d\lambda$ and $\overline{H} F(H; z)$ are in $B(\mathcal{X}, \mathcal{X}^*)$ and are, in fact, locally Lipschitz continuous in the uniform operator topology. Thus $z \to F(H; z)$ is locally Lipschitz continuous in $B(\mathcal{X}, \mathcal{X}_H^*)$, which concludes the proof. $\square$

3.2. Persistence of smoothness under functional operations

For a wide class of functions $f \colon \mathbb{R} \to \mathbb{R}$ the (self-adjoint) operator $f(H)$ is defined via the calculus associated with the spectral theorem [59], namely,

$$f(H) = \int_{\mathbb{R}} f(\lambda) \, dE(\lambda),$$

where $\{E(\lambda)\}$ is the spectral family of H.

Various spectral properties of $f(H)$ (whose spectrum is $\operatorname{Ran} f_{\sigma(H)}$) can be read off from the structure of f. (We use the notation $\operatorname{Ran} f_W$ for the image of $W \subseteq \mathbb{R}$ under f.)

However, one important aspect which is missing is the fact that if H satisfies the Limiting Absorption Principle in U, there is no guarantee that $f(H)$ satisfies the same principle in $\operatorname{Ran} f_U$ or any part thereof. This remains true even if f is very smooth, monotone, etc.

In contrast, if H is of type $(\mathcal{X}, \mathcal{X}_H^*, \alpha, U)$, then also $f(H)$ is of that type (with U replaced by $\operatorname{Ran} f_U$ and perhaps a different Hölder exponent), for a rather broad family of functions. This is the content of the next theorem. In particular, in view of Theorem 3.6, also $f(H)$ satisfies the LAP.

We do not attempt to make the most general statement, but instead refer the reader to [20] for further details.

Theorem 3.12. *Let H be of type $(\mathcal{X}, \mathcal{X}_H^*, \alpha, U)$ (where $U \subseteq \mathbb{R}$ is open and $0 < \alpha \leq 1$). Let $f \colon \mathbb{R} \to \mathbb{R}$ be a locally Hölder continuous function. Assume, in addition, that the restriction of f to U is continuously differentiable, and that its derivative f' is positive and locally Hölder continuous on U.*

Then the operator $f(H)$ is of type $(\mathcal{X}, \mathcal{X}_H^*, \alpha', \operatorname{Ran} f_U)$, for some $0 < \alpha' \leq 1$.

Proof. Let $\{F(\lambda)\}$ be the spectral family of $f(H)$. If $B \subseteq \mathbb{R}$ is a Borel set, the spectral theorem yields

$$F(B) = E(f^{-1}(B)),$$

and since $E(\mathbb{R} \setminus U) = 0$ (see Definition 3.3), we can further write

$$F(B) = E(f^{-1}(B) \cap U).$$

Since $f' > 0$ in U, an easy calculation gives for the (weak) derivative

$$\frac{d}{d\mu} F(\mu) = f'(\lambda)^{-1} \frac{d}{d\lambda} E(\lambda), \quad \lambda = f(\mu) \in U.$$

The assertion of the theorem follows directly from this formula. $\square$

In view of Theorem 3.6 we infer that $f(H)$ satisfies the LAP in $\operatorname{Ran} f_U$.

Remark 3.13. It should be remarked that if H satisfies the LAP in the sense of Definition 3.1 (including all the functional setting mentioned there), there is no guarantee that H^2 satisfies the LAP in $\{\mu = \lambda^2 \mid \lambda \in U\}$. For this to be false, however, one needs to find an example where the limiting values of the resolvent are not Hölder continuous.

Continuing Corollary 3.9 and taking $f(\lambda) = \sqrt{|\lambda|} + 1$, we obtain a LAP for the *relativistic Schrödinger operator* [20].

Corollary 3.14. *Let $L = \sqrt{-\Delta} + I$ and set $P(z) = (L - z)^{-1}$, $\operatorname{Im} z \neq 0$. The spectrum of L is $\sigma(L) = [1, \infty)$ and is absolutely continuous. The limits*

$$P^{\pm}(\lambda) = \lim_{\epsilon \downarrow 0} P(\lambda \pm i\epsilon), \quad \lambda \in (1, \infty),$$

exist in the uniform operator topology of $B(L^{2,s}, H^{2,-s})$, $s > \frac{1}{2}$. Furthermore, these limiting values are Hölder continuous in this topology.

4. The limiting absorption principle for second-order divergence-type operators

In the following sections we consider *divergence-type* second-order operators. As perturbations of the Laplacian they do not belong to any of the above categories; the difference between such an operator and the Laplacian is not even compact. However, our aim is to show that we can still deal with such operators, starting from the smoothness properties of (the spectral derivative of) the Laplacian.

Let $H = -\sum_{j,k=1}^{n} \partial_j a_{j,k}(x)\partial_k$, where $a_{j,k}(x) = a_{k,j}(x)$, be a formally self-adjoint operator in $L^2(\mathbb{R}^n)$, $n \geq 2$. The notation $\partial_j = \frac{\partial}{\partial x_j}$ is used throughout the following sections.

We assume that the real measurable matrix function $a(x) = \{a_{j,k}(x)\}_{1 \leq j,k \leq n}$ satisfies, with some positive constants $a_1 > a_0 > 0$, $\Lambda_0 > 0$,

$$a_0 I \leq a(x) \leq a_1 I, \qquad x \in \mathbb{R}^n, \qquad\qquad (4.1)$$

$$a(x) = I, \qquad\qquad |x| > \Lambda_0. \qquad\qquad (4.2)$$

In what follows we shall use the notation $H = -\nabla \cdot a(x)\nabla$.

We retain the notation H for the self-adjoint (Friedrichs) extension associated with the form $(a(x)\nabla\varphi, \nabla\psi)$, where $(\,,\,)$ is the scalar product in $L^2(\mathbb{R}^n)$. When $a(x) \equiv I$, we get $H = H_0 = -\Delta$.

We refer to Section 2 for definitions of the various functional spaces that will appear in what follows.

Let

$$R_0(z) = (H_0 - z)^{-1}, \quad R(z) = (H - z)^{-1}, \quad z \in \mathcal{C}^{\pm} = \{z \mid \pm \operatorname{Im} z > 0\},$$

be the associated resolvent operators.

We note that the operator H can be extended in an obvious way (retaining the same notation) as a bounded operator $H \colon H^1_{\mathrm{loc}} \to H^{-1}_{\mathrm{loc}}$. In particular, $H \colon H^{1,-s} \to H^{-1,-s}$, for all $s \geq 0$. Furthermore, the graph norm of H in $H^{-1,-s}$ is equivalent to the norm of $H^{1,-s}$.

Similarly, we can consider the resolvent $R(z)$ as defined on $L^{2,s}$, $s \geq 0$, where $L^{2,s}$ is densely and continuously embedded in $H^{-1,s}$.

The fundamental result presented in this section is that H satisfies the LAP over the *whole real axis*. The exact formulation is as follows:

Theorem 4.1. *Suppose that $a(x)$ satisfies (4.1), (4.2). Assume further that the operator H satisfies the unique continuation property. Then the operator H satisfies the LAP in $\mathbb{R}$. More precisely, let $s > 1$ and consider the resolvent $R(z) = (H - z)^{-1}$, $\operatorname{Im} z \neq 0$, as a bounded operator from $L^{2,s}(\mathbb{R}^n)$ to $H^{1,-s}(\mathbb{R}^n)$.*
Then:

(a) *$R(z)$ is bounded with respect to the $H^{-1,s}(\mathbb{R}^n)$ norm. Using the density of $L^{2,s}$ in $H^{-1,s}$, we can therefore view $R(z)$ as a bounded operator from $H^{-1,s}(\mathbb{R}^n)$ to $H^{1,-s}(\mathbb{R}^n)$.*

(b) *The operator-valued functions, defined respectively in the lower and upper half-planes,*

$$z \to R(z) \in B(H^{-1,s}(\mathbb{R}^n), H^{1,-s}(\mathbb{R}^n)), \quad s > 1, \pm \operatorname{Im} z > 0, \qquad (4.3)$$

can be extended continuously from $\mathcal{C}^{\pm} = \{z \mid \pm \operatorname{Im} z > 0\}$ to $\overline{\mathcal{C}^{\pm}} = \mathcal{C}^{\pm} \bigcup \mathbb{R}$ (with respect to the uniform operator topology of $B(H^{-1,s}(\mathbb{R}^n), H^{1,-s}(\mathbb{R}^n))$). In the case $n = 2$ replace $H^{-1,s}$ by $H_0^{-1,s}$.

We denote the limiting values of the resolvent on the real axis by

$$R^{\pm}(\lambda) = \lim_{z \to \lambda, \, \pm \operatorname{Im} z > 0} R(z). \tag{4.4}$$

Remark 4.2. Since $L^{2,s}$ (resp. $H^{1,-s}$) is densely and continuously embedded in $H^{-1,s}$ (resp. $L^{2,-s}$), we conclude that the resolvents $R_0(z)$, $R(z)$ can be extended continuously to $\overline{C^{\pm}}$ in the $B(L^{2,s}(\mathbb{R}^n), L^{2,-s}(\mathbb{R}^n))$ uniform operator topology.

The spectrum of H is therefore entirely absolutely continuous. In particular, it follows that the limiting values $R^{\pm}(\lambda)$ are continuous at $\lambda = 0$ and H has no resonance there.

The study of the resolvent near the threshold $\lambda = 0$ is sometimes referred to as *low energy estimates*. Following the proof of the theorem, at the end of Subsection 4.2, we review some of the existing literature concerning such estimates, as well as some other results pertaining to the LAP in *non short-range* settings.

Before proceeding to the proof of the theorem, we need to obtain more information on the resolvent of the Laplacian.

4.1. The operator $H_0 = -\Delta$ – revisited

The basic properties of this operator have already been discussed in Example 3.5 and Corollary 3.9. In particular, the explicit form of $\{E_0(\lambda)\}$, its spectral family, is given in Eq. (3.1), and the spectral derivative A_0 is given explicitly in Eq. (3.3).

The weighted L^2 estimates for A_0 were obtained by using the trace estimate (3.2).

However, we can refine this estimate near $\lambda = 0$ as follows.

Proposition 4.3. *Let* $\frac{1}{2} < s < \frac{3}{2}$, $h \in L^{2,s}$. *For* $n = 2$ *assume further that* $s > 1$ *and* $h \in L_0^{2,s}$. *Then*

$$\int_{|\xi|^2 = \lambda} |\hat{h}|^2 \, d\tau \leq C \, \min\{\lambda^{\gamma}, 1\} \|\hat{h}\|_{H^s}^2, \tag{4.5}$$

where

$$\begin{aligned} 0 < \gamma = s - \frac{1}{2}, \quad n \geq 3, \\ 0 < \gamma < s - \frac{1}{2}, \quad n = 2, \end{aligned} \tag{4.6}$$

and $C = C(s, \gamma, n)$.

Proof. If $n \geq 3$, the proof follows as in [19, Appendix], when we take into account the fact (*generalized Hardy inequality*) that multiplication by $|\xi|^{-s}$ is bounded from H^s into L^2 [45] (see also [64, Section 9.4]).

If $n = 2$ and $1 < s < \frac{3}{2}$ we have, for $h \in L_0^{2,s}$,

$$|\hat{h}(\xi)| = |\hat{h}(\xi) - \hat{h}(0)| \leq C_{s,\delta} |\xi|^{\delta} \|\hat{h}\|_{H^s},$$

for any $0 < \delta < \min\{1, s-1\}$. Using this estimate in the integral in the right-hand side of (4.5), the claim follows also in this case. $\qquad\square$

Combining Eqs. (3.3), (3.2) and (4.5), we conclude that

$$
\begin{aligned}
|\langle A_0(\lambda)f, g\rangle| &\le \langle A_0(\lambda)f, f\rangle^{\frac{1}{2}} \langle A_0(\lambda)g, g\rangle^{\frac{1}{2}} \\
&\le C \min\left\{\lambda^{-\frac{1}{2}}, \lambda^n\right\} \|f\|_{0,s} \|g\|_{0,\sigma}, \quad f \in L^{2,s}, g \in L^{2,\sigma},
\end{aligned}
\tag{4.7}
$$

where

(i) $n \ge 3$, $\quad \dfrac{1}{2} < s, \sigma < \dfrac{3}{2}$, $\quad s + \sigma > 2$ and $\quad 0 < 2\eta = s + \sigma - 2$,

or $\hspace{10cm}$ (4.8)

(ii) $n = 2$, $\quad 1 < s < \dfrac{3}{2}$, $\quad \dfrac{1}{2} < \sigma < \dfrac{3}{2}$, $\quad s + \sigma > 2$, $\quad 0 < 2\eta < s + \sigma - 2$

$$\text{and} \quad \hat{f}(0) = 0.$$

In both cases, $A_0(\lambda)$ is Hölder continuous and vanishes at $0, \infty$, so we obtain as in [15]:

Proposition 4.4. *The operator-valued function*

$$
z \to R_0(z) \in
\begin{cases}
B(L^{2,s}, L^{2,-\sigma}), & n \ge 3, \\
B(L_0^{2,s}, L^{2,-\sigma}), & n = 2,
\end{cases}
\tag{4.9}
$$

where s, σ *satisfy* (4.8), *can be extended continuously from* $\mathcal{C}^{\pm}$ *to* $\overline{\mathcal{C}^{\pm}}$, *in the respective uniform operator topologies.*

Remark 4.5. We note that the conditions (4.8) yield the continuity of $A_0(\lambda)$ *across the threshold* $\lambda = 0$ and hence the continuity property of the resolvent as in Proposition 4.4. However, for the local continuity at any $\lambda_0 > 0$, it suffices to take $s, \sigma > \frac{1}{2}$, as has been stated in Corollary 3.9, which is Agmon's original result [1].

This remark applies equally to the statements below, where the resolvent is considered in other functional settings.

We shall now extend this proposition to more general function spaces. Let $g \in H^{1,\sigma}$, where s, σ satisfy (4.8). Let $f \in H^{-1,s}$ have a representation of the form (2.2). Eq. (3.3) can be extended to yield an operator (for which we retain the same notation)

$$A_0(\lambda) \in B(H^{-1,s}, H^{-1,-\sigma}),$$

defined by (where now $\langle\,,\,\rangle$ is used for the $(H^{-1,s}, H^{1,\sigma})$ pairing),

$$
\left\langle A_0(\lambda)\left[f_0 + i^{-1}\sum_{k=1}^{n}\frac{\partial}{\partial x_k}f_k\right], g\right\rangle
$$

$$
= (2\sqrt{\lambda})^{-1} \int\limits_{|\xi|^2 = \lambda}\left[\hat{f}_0(\xi) + \sum_{k=1}^{n}\xi_k\hat{f}_k(\xi)\right]\overline{\hat{g}(\xi)}\,d\tau, \quad f \in H^{-1,s}, g \in H^{1,\sigma}.
$$

$$\tag{4.10}$$

(replace $H^{-1,s}$ by $H_0^{-1,s}$ if $n = 2$).

Observe that this definition makes good sense even though the representation (2.2) is not unique, since

$$f = f_0 + \sum_{k=1}^{n} i^{-1} \frac{\partial}{\partial x_k} f_k = \tilde{f}_0 + \sum_{k=1}^{n} i^{-1} \frac{\partial}{\partial x_k} \tilde{f}_k,$$

implies

$$\hat{f}_0(\xi) + \sum_{k=1}^{n} \xi_k \hat{f}_k(\xi) = \hat{\tilde{f}}_0(\xi) + \sum_{k=1}^{n} \xi_k \hat{\tilde{f}}_k(\xi)$$

(as tempered distributions).

To estimate the operator-norm of $A_0(\lambda)$ in this setting we use (4.10) and the considerations preceding Proposition 4.4, to obtain, instead of (4.7), for $k = 1, 2, \ldots, n$,

$$\left| \left\langle A_0(\lambda) \frac{\partial}{\partial x_k} f_k, g \right\rangle \right| \tag{4.11}$$
$$\leq C \min\{\lambda^{-\frac{1}{2}}, \lambda^n\} \|f\|_{-1,s} \|g\|_{1,\sigma}, \quad f \in H^{-1,s}, g \in H^{1,\sigma},$$

where s, σ satisfy (4.8) (replace $H^{-1,s}$ by $H_0^{-1,s}$ if $n = 2$).

We now define the extension of the resolvent operator by

$$R_0(z) = \int_0^\infty \frac{A_0(\lambda)}{\lambda - z} \, d\lambda, \quad \text{Im } z \neq 0. \tag{4.12}$$

The convergence of the integral (in the operator norm) follows from the estimate (4.11).

The LAP in this case is given in the following proposition.

Proposition 4.6. *The operator-valued function $R_0(z)$ is well defined (and analytic) for non-real z in the following functional setting.*

$$z \to R_0(z) \in \begin{cases} B(H^{-1,s}, H^{1,-\sigma}), & n \geq 3, \\ B(H_0^{-1,s}, H^{1,-\sigma}), & n = 2, \end{cases} \tag{4.13}$$

where s, σ satisfy (4.8). Furthermore, it can be extended continuously from $\mathcal{C}^{\pm}$ to $\overline{\mathcal{C}^{\pm}}$, in the respective uniform operator topologies. The limiting values are denoted by $R_0^{\pm}(\lambda)$.

The extended function satisfies

$$(H_0 - z) R_0(z) f = f, \quad f \in H^{-1,s}, z \in \overline{\mathcal{C}^{\pm}}, \tag{4.14}$$

where for $z = \lambda \in \mathbb{R}$, $R_0(z) = R_0^{\pm}(\lambda)$.

Proof. For simplicity we assume $n \geq 3$. By Definition (4.12) and estimate (4.11), we get readily $R_0(z) \in B(H^{-1,s}, H^{-1,-\sigma})$ if $\text{Im } z \neq 0$ as well as the analyticity of the map $z \to R_0(z)$, $\text{Im } z \neq 0$. Furthermore, the extension to $\text{Im } z = 0$ is carried out as in [15].

Eq. (4.14) is obvious if $\operatorname{Im} z \neq 0$ and $f \in L^{2,s}$. By the density of $L^{2,s}$ in $H^{-1,s}$, the continuity of $R_0(z)$ on $H^{-1,s}$ and the continuity of $H_0 - z$ (in the sense of distributions) we can extend it to all $f \in H^{-1,s}$.

As $z \to \lambda \pm i \cdot 0$ we have $R_0(z)f \to R_0^{\pm}(\lambda)f$ in $H^{-1,-\sigma}$. Applying the (constant coefficient) operator $H_0 - z$ yields, in the sense of distributions, $f = (H_0 - z) R_0(z)f \to (H_0 - \lambda) R_0^{\pm}(\lambda)f$ which establishes (4.14) also for $\operatorname{Im} z = 0$.

Finally, the established continuity of $z \to R_0(z) \in B(H^{-1,s}, H^{-1,-\sigma})$ (up to the real boundary) and Eq. (4.14) imply the continuity of the map $z \to H_0 R_0(z) \in B(H^{-1,s}, H^{-1,-\sigma})$.

The stronger continuity claim (4.13) follows, since the norm of $H^{1,-\sigma}$ is equivalent to the graph norm of H_0 as a map of $H^{-1,-\sigma}$ to itself. $\qquad\square$

Remark 4.7. The main point here is the fact that the limiting values can be extended continuously to the threshold at $\lambda = 0$.

In the neighborhood of any $\lambda > 0$ this proposition follows from [79, Theorem 2.3], where a very different proof is used. In fact, using the terminology there, the limit functions $R_0^{\pm}(\lambda)f$ are the unique (on either side of the positive real axis) radiative functions and they satisfy a suitable *Sommerfeld radiation condition*. We recall it here for the sake of completeness, since we will need it in the next section.

Let $z = k^2 \in \mathcal{C} \setminus \{0\}$, $\operatorname{Im} k \geq 0$. For $f \in H^{-1,s}$ let $u = R_0(z)f \in H^{1,-\sigma}$ be as defined above. Then

$$\mathcal{R}u = \int_{|x|>\Lambda_0} \left| r^{-\frac{n-1}{2}} \frac{\partial}{\partial r} (r^{\frac{n-1}{2}} u) - iku \right|^2 \, dx < \infty, \tag{4.15}$$

where $r = |x|$. We shall refer to $\mathcal{R}u$ as the radiative norm of u.

Furthermore, we can take $s, \sigma > \frac{1}{2}$, as in Remark 4.5.

4.2. Proof of the LAP for the operator H

We start with some considerations regarding the behavior of the resolvent near the spectrum.

Fix $[\alpha, \beta] \subset \mathbb{R}$ and let

$$\Omega = \{z \in \mathcal{C}^+ \mid \alpha < \operatorname{Re} z < \beta, \, 0 < \operatorname{Im} z < 1\}. \tag{4.16}$$

Let $z = \mu + i\varepsilon \in \Omega$ and consider the equation

$$(H - z) u = f \in H^{-1,s}, \quad u \in H^{1,-\sigma} \ (f \in H_0^{-1,s} \text{ if } n = 2). \tag{4.17}$$

(Observe that in the case $n = 2$ also $u \in L_0^{2,\sigma}$.)

With Λ_0 as in (4.2), let $\chi(x) \in C^{\infty}(\mathbb{R}^n)$ be such that

$$\chi(x) = \begin{cases} 0, & |x| < \Lambda_0 + 1, \\ 1, & |x| > \Lambda_0 + 2. \end{cases} \tag{4.18}$$

Eq. (4.17) can be written as

$$(H_0 - z)(\chi u) = \chi f - 2\nabla\chi \cdot \nabla u - u\Delta\chi. \tag{4.19}$$

Letting $\psi(x) = 1 - \chi(\frac{x}{2}) \in C_0^\infty(\mathbb{R}^n)$ and using Proposition 4.6 and standard elliptic estimates, we obtain from (4.19)

$$\|u\|_{1,-\sigma} \leq C \left[\|f\|_{-1,s} + \|\psi u\|_{0,-s} \right], \tag{4.20}$$

where s, σ satisfy (4.8), $\sigma' > \sigma$ and $C > 0$ depends only on Λ_0, σ, s, n.

We note that, since ψ is compactly supported, the term $\|\psi u\|_{0,-s}$ can be replaced by $\|\psi u\|_{0,-s'}$ for any real s'.

In fact, the second term in the right-hand side can be dispensed with, as is demonstrated in the following proposition.

Proposition 4.8. *The solution to* (4.17) *satisfies,*

$$\|u\|_{1,-\sigma} \leq C \|f\|_{-1,s}, \tag{4.21}$$

where s, σ satisfy (4.8) *and $C > 0$ depends only on σ, s, n, Λ_0.*

Proof. In view of (4.20) we only need to show that

$$\|\psi u\|_{0,-s} \leq C \|f\|_{-1,s}. \tag{4.22}$$

Since $L^{2,s}(\mathbb{R}^n)$ is dense in $H^{-1,s}(\mathbb{R}^n)$, it suffices to prove this inequality for $f \in L^{2,s}(\mathbb{R}^n) \cap H^{-1,s}(\mathbb{R}^n)$ (using the norm of $H^{-1,s}$).

We argue by contradiction. Let

$$\{z_k\}_{k=1}^\infty \subseteq \Omega, \quad \{f_k\}_{k=1}^\infty \subseteq L^{2,s}(\mathbb{R}^n) \cap H^{-1,s}(\mathbb{R}^n)$$

(with $\hat{f}_k(0) = 0$ if $n = 2$) and

$$\{u_k = R(z_k)f_k\}_{k=1}^\infty \subseteq H^{1,-\sigma}(\mathbb{R}^n)$$

be such that

$$\|\psi u_k\|_{0,-s} = 1, \quad \|f_k\|_{-1,s} \leq k^{-1}, \quad k = 1, 2, \ldots,$$
$$z_k \to z_0 \in \overline{\Omega} \quad \text{as } k \to \infty. \tag{4.23}$$

By (4.20), $\{u_k\}_{k=1}^\infty$ is bounded in $H^{1,-\sigma}$. Replacing the sequence by a suitable subsequence (without changing notation) and using the Rellich compactness theorem we may assume that there exists a function $u \in L^{2,-\sigma'}$, $\sigma' > \sigma$, such that

$$u_k \to u \quad \text{in } L^{2,-\sigma'} \text{ as } k \to \infty. \tag{4.24}$$

Furthermore, by weak compactness we actually have (restricting again to a subsequence if needed)

$$u_k \xrightarrow{w} u \quad \text{in } H^{1,-\sigma} \text{ as } k \to \infty. \tag{4.25}$$

Since H maps continuously $H^{1,-\sigma}$ into $H^{-1,-\sigma}$, we have

$$H u_k \xrightarrow{w} H u \quad \text{in } H^{-1,-\sigma} \text{ as } k \to \infty,$$

so that from $(H - z_k) u_k = f_k$ we infer that

$$(H - z_0) u = 0. \tag{4.26}$$

In view of (4.19) and Remark 4.7 the functions χu_k are *radiative functions*. Since they are uniformly bounded in $H^{1,-\sigma}$, their *radiative norms* (4.15) are uniformly bounded.

Suppose first that $z_0 \neq 0$. In view of Remark 4.7 we can take $s, \sigma > \frac{1}{2}$. Then the limit function u is a radiative solution to $(H_0 - z_0) u = 0$ in $|x| > \Lambda_0 + 2$ and hence must vanish there (see [79]). By the unique continuation property of solutions to (4.26) we conclude that $u \equiv 0$. Thus by (4.24) we get $\|\psi u_k\|_{0,-\sigma'} \to 0$ as $k \to \infty$, which contradicts (4.23).

We are therefore left with the case $z_0 = 0$. In this case $u \in H^{1,-\sigma}$ satisfies the equation

$$\nabla \cdot (a(x)\nabla u) = 0. \tag{4.27}$$

In particular, $\Delta u = 0$ in $|x| > \Lambda_0$ and

$$\int_{\Lambda_0}^{\infty} \int_{|x|=r} r^{-2\sigma} \left(|u|^2 + |\frac{\partial u}{\partial r}|^2 \right) d\tau dr < \infty. \tag{4.28}$$

Consider first the case $n \geq 3$. We may then use the representation of u by spherical harmonics so that, with $x = r\omega$, $\omega \in S^{n-1}$,

$$u(x) = r^{-\frac{n-1}{2}} \left\{ \sum_{j=0}^{\infty} b_j r^{\mu_j} h_j(\omega) + \sum_{j=0}^{\infty} c_j r^{-\nu_j} h_j(\omega) \right\}, \quad r > \Lambda_0, \tag{4.29}$$

where

$$\mu_j(\mu_j - 1) = \nu_j(\nu_j + 1) = \lambda_j + \frac{(n-1)(n-3)}{4}, \tag{4.30}$$
$$0 = \lambda_0 < \lambda_1 \leq \lambda_2 \leq \cdots$$

being the eigenvalues of the Laplace-Beltrami operator on S^{n-1}, and $h_j(\omega)$ the corresponding spherical harmonics. Since $\lambda_1 = n - 1$, it follows that

$$\mu_0 = \frac{n-1}{2}, \quad \mu_0 + 1 \leq \mu_1 \leq \mu_2 \leq \cdots, \quad \frac{n-3}{2} = \nu_0 < \nu_1 \leq \nu_2 \leq \cdots \tag{4.31}$$

We now observe that (4.28) forces

$$b_0 = b_1 = \cdots = 0.$$

Also, by (4.29)

$$\int_{|x|=r} \frac{\partial u}{\partial r} d\tau = -(n-2)|S^{n-1}| c_0, \quad r > \Lambda_0, \tag{4.32}$$

($|S^{n-1}|$ is the surface measure of S^{n-1}), while integrating (4.27) we get

$$\int_{|x|=r} \frac{\partial u}{\partial r} d\tau = 0, \quad r > \Lambda_0. \tag{4.33}$$

Thus $c_0 = 0$. It now follows from (4.29) that, for $r > \Lambda_0$,

$$\int_{|x|=r} \left(|u|^2 + \left| \frac{\partial u}{\partial r} \right|^2 \right) d\tau \leq \left(\frac{r}{\Lambda_0} \right)^{-2\nu_1} \int_{|x|=\Lambda_0} \left(|u|^2 + \left| \frac{\partial u}{\partial r} \right|^2 \right) d\tau. \qquad (4.34)$$

Multiplying (4.27) by $\bar{u}$ and integrating by parts over the ball $|x| \leq r$, we infer from (4.34) that the boundary term vanishes as $r \to \infty$. Thus $\nabla u \equiv 0$, in contradiction to (4.23)–(4.24).

It remains to deal with the case $n = 2$. Instead of (4.29) we now have

$$u(x) = r^{-\frac{1}{2}} \left\{ \widetilde{b_0} r^{\frac{1}{2}} \log r + \sum_{j=0}^{\infty} b_j r^{\mu_j} h_j(\omega) + \sum_{j=1}^{\infty} c_j r^{-\nu_j} h_j(\omega) \right\}, \quad r > \Lambda_0, \quad (4.35)$$

where $\mu_0 = \frac{1}{2}$, $\mu_1 = \frac{3}{2}$, $\nu_1 = \frac{1}{2}$. As in the derivation above, the condition (4.28) yields $b_0 = b_1 = \cdots = 0$. Also, we get $\widetilde{b_0} = 0$ in view of (4.33). It now follows that

$$\int_{|x|=r} \bar{u} \frac{\partial u}{\partial r} d\tau = -2\pi \sum_{j=1}^{\infty} \left(\nu_j + \frac{1}{2} \right) |c_j|^2 r^{-2\nu_j - 1}, \quad r \geq \Lambda_0, \qquad (4.36)$$

from which, as in the argument following (4.34), we deduce that $u \equiv 0$, again in contradiction to (4.23)–(4.24). $\qquad \square$

Proof of Theorem 4.1. Part (a) of the theorem is actually covered by Proposition 4.8. Moreover, the proposition implies that the operator-valued function

$$z \to R(z) \in B(H^{-1,s}(\mathbb{R}^n), H^{1,-\sigma}(\mathbb{R}^n)), \quad s > 1, z \in \Omega,$$

is uniformly bounded, where s, σ satisfy (4.8). Here and below replace $H^{-1,s}$ by $H_0^{-1,s}$ if $n = 2$.

We next show that the function $z \to R(z)$ can be continuously extended to $\overline{\Omega}$ in the *weak toplogoy* of $B(H^{-1,s}(\mathbb{R}^n), H^{1,-\sigma}(\mathbb{R}^n))$. To this end, we take $f \in H^{-1,s}(\mathbb{R}^n)$ and $g \in H^{-1,\sigma}(\mathbb{R}^n)$ and consider the function

$$z \to \langle g, R(z)f \rangle, \quad z \in \Omega,$$

where $\langle \, , \, \rangle$ is the $(H^{-1,\sigma}, H^{1,-\sigma})$ pairing. We need to show that it can be extended continuously to $\overline{\Omega}$.

In view of the uniform boundedness established in Proposition 4.8, we can take f, g in dense sets (of the respective spaces). In particular, we can take $f \in L^{2,s}(\mathbb{R}^n)$ and $g \in L^{2,\sigma}(\mathbb{R}^n)$, so that the continuity property in Ω is obvious.

Consider therefore a sequence $\{z_k\}_{k=1}^{\infty} \subseteq \Omega$ such that $z_k \xrightarrow[k \to \infty]{} z_0 \in [\alpha, \beta]$. The sequence $\{u_k = R(z_k)f\}_{k=1}^{\infty}$ is bounded in $H^{1,-\sigma}(\mathbb{R}^n)$. Therefore there exists a subsequence $\{u_{k_j}\}_{j=1}^{\infty}$ which converges to a function $u \in L^{2,-\sigma'}$, $\sigma' > \sigma$.

We can further assume that $u_{k_j} \xrightarrow[j \to \infty]{w} u$ in $H^{1,-\sigma}$. It follows that

$$\langle g, u_{k_j} \rangle \xrightarrow[j \to \infty]{} \langle g, u \rangle.$$

Passing to the limit in $(H - z_{k_j}) u_{k_j} = f$ we see that the limit function satisfies

$$(H - z_0) u = f.$$

We now repeat the argument employed in the proof of Proposition 4.8. If $z_0 \neq 0$ we note that the functions $\{\chi u_k\}_{k=1}^{\infty}$ are radiative functions with uniformly bounded *radiative norms* (4.15) in $|x| > \Lambda_0 + 2$. The same is therefore true for the limit function u.

If $z_0 = 0$, then the function $u \in H^{1,-\sigma}$ solves $Hu = f$.

In both cases this function is unique and we get the convergence

$$\langle g, R(z_k)f \rangle = \langle g, u_k \rangle \xrightarrow[k \to \infty]{} \langle g, u \rangle.$$

We can now define

$$R^+(z_0)f = u, \tag{4.37}$$

with an analogous definition for $R^-(z_0)$.

At this point we can readily deduce the following extension of the resolvent $R(z)$ as the inverse of $H - z$.

$$(H - z) R(z)f = f, \quad f \in H^{-1,s}, \; z \in \overline{\mathcal{C}^{\pm}}, \tag{4.38}$$

where $R(z) = R^{\pm}(\lambda)$ when $z = \lambda \in \mathbb{R}$.

Indeed, observe that if $\operatorname{Im} z \neq 0$ then $(H - z)R(z)f = f$ for $f \in L^{2,s}(\mathbb{R}^n)$ and $(H - z)R(z) \in B(H^{-1,s}, H^{-1,-\sigma})$, so the assertion follows from the density of $L^{2,s}(\mathbb{R}^n)$ in $H^{-1,s}(\mathbb{R}^n)$. For $z = \lambda \in \mathbb{R}$ we use the (just established) weak continuity of the map $z \mapsto (H - z)R(z)$ from $H^{-1,s}$ into $H^{-1,-\sigma}$ in $\overline{\mathcal{C}^{\pm}}$.

The passage *from weak to uniform continuity* (in the operator topology) is a classical argument due to Agmon [1]. In [9] we have applied it in the case $n = 1$. Here we outline the proof in the case $n > 1$.

We establish first the continuity of the operator-valued function $z \to R(z)$, $\overline{\Omega}$, in the *uniform operator topology* of $B(H^{-1,s}(\mathbb{R}^n), L^{2,-\sigma}(\mathbb{R}^n))$.

Let $\{z_k\}_{k=1}^{\infty} \subseteq \overline{\Omega}$ and $\{f_k\}_{k=1}^{\infty} \subseteq H^{-1,s}(\mathbb{R}^n)$ be sequences such that $z_k \xrightarrow[k \to \infty]{}$ $z \in \overline{\Omega}$ and f_k converges weakly to f in $H^{-1,s}(\mathbb{R}^n)$. It suffices to prove that the sequence $u_k = R(z_k)f_k$, which is bounded in $H^{1,-\sigma}(\mathbb{R}^n)$, converges strongly in $L^{2,-\sigma}(\mathbb{R}^n)$. Since this is clear if $\operatorname{Im} z \neq 0$, we can take $z \in [\alpha, \beta]$.

Note first that we can take $\frac{1}{2} < \sigma' < \sigma$ so that s, σ' satisfy (4.8). Then the sequence $\{u_k\}_{k=1}^{\infty}$ is bounded in $H^{1,-\sigma'}(\mathbb{R}^n)$ and there exists a subsequence $\{u_{k_j}\}_{j=1}^{\infty}$ which converges to a function $u \in L^{2,-\sigma}$.

We can further assume that $u_{k_j} \xrightarrow[j \to \infty]{w} u$ in $H^{1,-\sigma}$.

It follows that the limit function satisfies (see Eq. (4.38))

$$(H - z) u = f.$$

Once again we consider separately the cases $z \neq 0$ and $z = 0$.

In the first case, in view of (4.38) and Remark 4.7, the functions χu_k are *radiative functions*. Since they are uniformly bounded in $H^{1,-\sigma}$ their *radiative norms* (4.15) are uniformly bounded, and we conclude that also $\mathcal{R}u < \infty$.

In the second case, we simply note that $u \in H^{1,-\sigma}$ solves $Hu = f$.

As in the proof of Proposition 4.8 we conclude that in both cases the limit is unique, so that the whole sequence $\{u_k\}_{k=1}^{\infty}$ converges to u in $L^{2,-\sigma}(\mathbb{R}^n)$. Thus, the continuity in the uniform operator topology of $B(H^{-1,s}(\mathbb{R}^n), L^{2,-\sigma}(\mathbb{R}^n))$ is proved.

Finally, we claim that the operator-valued function $z \to R(z)$ is continuous in the *uniform operator topology* of $B(H^{-1,s}(\mathbb{R}^n), H^{1,-\sigma}(\mathbb{R}^n))$. Indeed, if we invoke Eq. (4.38), we get that also $z \to HR(z)$ is continuous in the uniform operator topology of $B(H^{-1,s}(\mathbb{R}^n), H^{-1,-\sigma}(\mathbb{R}^n))$.

Since the domain of H in $H^{-1,-\sigma}(\mathbb{R}^n)$ is $H^{1,-\sigma}(\mathbb{R}^n)$, the claim follows. The conclusion of the theorem follows by taking $\sigma = s$. $\qquad\square$

Remark 4.9. In view of (4.19) and Remark 4.7 it follows that for $\lambda > 0$ the functions $R^{\pm}(\lambda)f$, $f \in H^{-1,s}$, are *radiative*, i.e., satisfy a Sommerfeld radiation condition.

Remark 4.10. The assumption of unique continuation property satisfied by H forces us to assume that the coeffcients are Lipschitz continuous. If we do not make this assumption, it is readily seen from the proof of the theorem that a discrete set of eigenvalues, with no finite accumulation point, can still be embedded in the continuous spectrum. However, there is no singular continuous spectrum and zero is neither an eigenvalue nor a resonance.

I am grateful to Professor Hendrik Vogt for pointing it out to me.

The fact that the limiting values of the resolvent are continuous across the threshold at $\lambda = 0$ has been established in the case $H = H_0$ [14, Appendix A], and in the one-dimensional case ($n = 1$) in [9, 12, 30]. The paper [74] deals with the two-dimensional ($n = 2$) case, but the resolvent $R(z)$ is restricted to continuous compactly supported functions f, thus enabling the use of pointwise decay estimates of $R(z)f$ at infinity. In the case of the closely related *acoustic propagator*, where the matrix $a(x) = b(x_1)I$ is scalar and dependent on a single coordinate, there are in general countably many thresholds embedded in the continuous spectrum. Any study of the LAP must therefore deal with this difficulty. We mention here the papers [12, 24, 23, 39, 32, 34, 57, 58, 63, 85], as well as the *anisotropic* case where $b(x_1)$ is a general positive matrix [13].

We mention next some related studies concerning the LAP where, however, the threshold has been avoided. Our discussion is restricted, however, to operators that can be characterized as "perturbations of the Laplacian". The extensive literature concerning the N-body operators is omitted, apart from the monographs [4, 36] that have already been mentioned in the Introduction in connection with Mourre's approach to the LAP.

The pioneering works of Eidus and Agmon have already been mentioned in the Introduction. Under assumptions close to ours here (but also assuming that $a(x)$ is continuously differentiable) a weaker version (roughly, *strong* instead of *uniform* convergence of the resolvents) was obtained by Eidus [40, Theorem 4 and

Remark 1]. For $H = H_0$ the LAP has been established by Agmon [1]. Indeed, it was established for operators of the type $H_0 + V$, where V is a short-range perturbation. The short-range potential V was later replaced by a long-range or Stark-like potential [53, 6], a potential in $L^p(\mathbb{R}^n)$ [44, 55], a potential depending only on direction $x/|x|$ [46] and a perturbation of such a potential [71, 72]. In these latter cases the condition $\alpha > 0$ is replaced by $\alpha > \limsup_{|x| \to \infty} V(x)$.

We refer to [20] for the LAP for operators of the type $f(-\Delta) + V$ for a certain class of functions f.

We refer to [76] and references therein for the case of perturbations of the Laplace-Beltrami operator Δ_g on noncompact manifolds. The LAP (still in $(0, \infty)$) holds under the assumption that g is a smooth metric on $\mathbb{R}^n$ that vanishes at infinity. We make use of this result in the proof of Theorem 6.1 (see Section 6).

The LAP for the periodic case (namely, $a(x)$ is symmetric and periodic) has recently been established in [69]. Note that in this case the spectrum is absolutely continuous and consists of a union of intervals (*bands*).

4.3. An application: Existence and completeness of the wave operators $W_\pm(H, H_0)$

A nice consequence of Theorem 4.1 is the existence and completeness of the wave operators. We recall first the definition [59, Chapter X].

Consider the family of unitary operators

$$W(t) = \exp(itH)\exp(-itH_0), \quad -\infty < t < \infty.$$

The strong limits $W_\pm(H, H_0) = s\text{-}\lim_{t \to \pm\infty} W(t)$, if they exist, are called the *wave operators* (relating H, H_0). They are clearly isometries. If their ranges are equal, we say that they are *complete*.

Using a well-known theorem of Kato and Kuroda [61], we have the following corollary.

Corollary 4.11. *The wave operators $W_\pm(H, H_0)$ exist and are complete.*

Indeed, all that is needed is that H, H_0 satisfy the LAP in $\mathbb{R}$, with respect to the same operator topologies.

We refer to [54], where the existence and completeness of the wave operators $W_\pm(H, H_0)$ is established under suitable smoothness assumptions on $a(x)$. (However, $a(x) - I$ is not assumed to be compactly supported and H can include also magnetic and electric potentials.)

5. An eigenfunction expansion theorem

In the Introduction we mentioned the connection (as well as the *gap*) between the spectral theorem (for self-adjoint operators) in its functional-analytic formulation and the *generalized eigenfunction theorem*, a fundamental tool in the study of partial differential operators (and scattering theory). It was mentioned there that these theorems should be connected through the Limiting Absorption Principle. This is indeed the purpose of this section.

We derive an eigenfunction expansion theorem for a divergence-type operator H, the operator considered in Section 4.

Let $\{E(\lambda), \lambda \in \mathbb{R}\}$ be the spectral family associated with H and $A(\lambda) = \frac{d}{d\lambda} E(\lambda)$ be its weak derivative. We use the formula (3.7),

$$A(\lambda) = \frac{1}{2\pi i} \lim_{\epsilon \downarrow 0} \left(R(\lambda + i\epsilon) - R(\lambda - i\epsilon) \right) = \frac{1}{2\pi i} \left(R^+(\lambda) - R^-(\lambda) \right).$$

By Theorem 4.1 we know that $A(\lambda) \in B(L^{2,s}(\mathbb{R}^n), L^{2,-s}(\mathbb{R}^n))$, for values of s as given in the theorem.

The formal relation $(H - \lambda) A(\lambda) = 0$ can be given a rigorous meaning if, for example, we can find a bounded operator T such that $T^* A(\lambda) T$ is bounded in $L^2(\mathbb{R}^n)$ and has a complete set (necessarily at most countable) of eigenvectors. These will serve as *generalized eigenvectors* for H. We refer to [22, Chapters V, VI] and [25] for a development of this approach for self-adjoint elliptic operators. Note that by this approach we have at most a countable number of such generalized eigenvectors for any fixed λ. In the case of $H_0 = -\Delta$ they correspond to

$$|x|^{-\frac{n-3}{2}} J_{\sqrt{\kappa_j}}\left(\sqrt{\lambda}\,|x|\right) \psi_j(\omega),$$

where $\kappa_j = \lambda_j + \frac{(n-1)(n-3)}{4}$, λ_j being the jth eigenvalue of the Laplace-Beltrami operator on the unit sphere S^{n-1}, ψ_j the corresponding eigenfunction and J_ν is the Bessel function of order ν.

On the other hand, the inverse Fourier transform

$$g(x) = (2\pi)^{-\frac{n}{2}} \int_{\mathbb{R}^n} \hat{g}(\xi) e^{i\xi x} \, d\xi, \tag{5.1}$$

can be viewed as expressing a function in terms of the *generalized eigenfunctions* $\exp(i\xi x)$ of H_0. Observe that now there is a continuum of such functions corresponding to $\lambda > 0$, namely, $|\xi|^2 = \lambda$.

From the physical point-of-view this expansion in terms of *plane waves* proves to be more useful for many applications. In particular, replacing $-\Delta$ by the Schrödinger operator $-\Delta + V(x)$ one can expect, under certain hypotheses on the potential V, a similar expansion in terms of *distorted plane waves*. This has been accomplished, in increasing order of generality (more specifically, decay assumptions on $V(x)$ as $|x| \to \infty$) in [73, 52, 1, 79, 2]. See also [87] for an eigenfunction expansion for relativistic Schrödinger operators.

Here we use the LAP result of Theorem 4.1 in order to derive a similar expansion for the operator H. In fact, our generalized eigenfunctions are given by the following definition.

Definition 5.1. For every $\xi \in \mathbb{R}^n$, let

$$\psi_\pm(x,\xi) = -R^\mp\left(|\xi|^2\right)((H - |\xi|^2)\exp(i\xi x))$$

$$= R^\mp(|\xi|^2)\left(\sum_{l,j=1}^n \partial_l(a_{l,j}(x) - \delta_{l,j})\partial_j\right)\exp(i\xi x). \tag{5.2}$$

The *generalized eigenfunctions* of H are defined by

$$\varphi_\pm(x,\xi) = \exp(i\xi x) + \psi_\pm(x,\xi). \tag{5.3}$$

We assume $n \geq 3$ in order to simplify the statement of the theorem. As we show below (see Proposition 5.3) the generalized eigenfunctions are (at least) continuous in x, so that the integral in the statement makes sense.

Theorem 5.2. *Suppose that $n \geq 3$ and that $a(x)$ satisfies (4.1), (4.2). For any compactly supported $f \in L^2(\mathbb{R}^n)$ define*

$$(\mathbb{F}_\pm f)(\xi) = (2\pi)^{-\frac{n}{2}}\int_{\mathbb{R}^n} f(x)\overline{\varphi_\pm(x,\xi)}\,dx, \quad \xi \in \mathbb{R}^n. \tag{5.4}$$

Then the transformations $\mathbb{F}_\pm$ can be extended as unitary transformations (for which we retain the same notation) of $L^2(\mathbb{R}^n)$ onto itself. Furthermore, these transformations diagonalize H in the following sense:

$f \in L^2(\mathbb{R}^n)$ is in the domain $D(H)$ if and only if $|\xi|^2(\mathbb{F}_\pm f)(\xi) \in L^2(\mathbb{R}^n)$ and

$$H = \mathbb{F}_\pm^* M_{|\xi|^2}\mathbb{F}_\pm, \tag{5.5}$$

where $M_{|\xi|^2}$ is the multiplication operator by $|\xi|^2$.

Before starting the proof of the theorem, we collect some basic properties of the generalized eigenfunctions in the following proposition.

Proposition 5.3. *The generalized eigenfunctions*

$$\varphi_\pm(x,\xi) = \exp(i\xi x) + \psi_\pm(x,\xi)$$

(see (5.3)) are in $H^1_{\mathrm{loc}}(\mathbb{R}^n)$ for each fixed $\xi \in \mathbb{R}^n$ and satisfy the equation

$$(H - |\xi|^2)\,\varphi_\pm(x,\xi) = 0. \tag{5.6}$$

In addition, these functions have the following properties:

(i) *The map*

$$\mathbb{R}^n \ni \xi \to \psi_\pm(\cdot,\xi) \in H^{1,-s}(\mathbb{R}^n), \quad s > 1,$$

is continuous.

(ii) *For any compact $K \subseteq \mathbb{R}^n$, the family of functions $\{\varphi_\pm(x,\xi) \mid \xi \in K\}$ is uniformly bounded and uniformly Hölder continuous in $x \in \mathbb{R}^n$.*

Proof. Since $(H - |\xi|^2) \exp(i\xi x) \in H^{-1,s}$, $s > 1$, Eq. (5.6) follows from the definition (5.2) in view of Eq. (4.38).

Furthermore, the map

$$\mathbb{R}^n \ni \xi \to (H - |\xi|^2) \exp(i\xi x) \in H^{-1,s}(\mathbb{R}^n), \quad s > 1,$$

is continuous, so the continuity assertion (i) follows from Theorem 4.1.

For $s > 1$, the set of functions $\{\psi_\pm(\cdot, \xi) \mid \xi \in K\}$ is uniformly bounded in $H^{1,-s}$. Thus, in view of (5.6), it follows from the De Giorgi-Nash-Moser Theorem [42, Chapter 8] that the set $\{\varphi_\pm(x, \xi) \mid \xi \in K\}$ is uniformly bounded and uniformly Hölder continuous in $\{|x| < R\}$ for every $R > 0$. In particular, we can take $R > \Lambda_0$ (see Eq. (4.2)). In the exterior domain $\{|x| > R\}$ the set $\{\psi_\pm(x, \xi) \mid \xi \in K\}$ is bounded in $H^{1,-s}$, $s > 1$, and we have $(H_0 - |\xi|^2) \psi_\pm(x, \xi) = 0$.

In addition, the boundary values $\{\psi_\pm(x, \xi) \mid |x| = R, \xi \in K\}$ are uniformly bounded. From well-known properties of solutions of the Helmholtz equation we conclude that this set is uniformly bounded and therefore, invoking once again the De Giorgi-Nash-Moser Theorem, uniformly Hölder continuous. □

Proof of Theorem 5.2. We use the LAP proved in Theorem 4.1, adapting the methodology of Agmon's proof [1] for the eigenfunction expansion in the case of Schrödinger operators with short-range potentials. To simplify notation, we prove for $\mathbb{F}_+$.

Let $u \in H^1$ be compactly supported. For any z such that $\operatorname{Im} z \neq 0$ we can write its Fourier transform as

$$\hat{u}(\xi) = (2\pi)^{-\frac{n}{2}} \int_{\mathbb{R}^n} u(x) \exp(-i\xi x) \, dx = \frac{(2\pi)^{-\frac{n}{2}}}{|\xi|^2 - z} \int_{\mathbb{R}^n} u(x) (H_0 - z) \exp(-i\xi x) \, dx.$$

Let $\theta \in C_0^\infty(\mathbb{R}^n)$ be a (real) cutoff function such that $\theta(x) = 1$ for x in a neighborhood of the support of u.

We can rewrite the above equality as

$$\hat{u}(\xi) = \frac{(2\pi)^{-\frac{n}{2}}}{|\xi|^2 - z} \langle (H_0 - z) u(x), \theta(x) \exp(i\xi x) \rangle,$$

where $\langle \, , \, \rangle$ is the $(H^{-1,s}, H^{1,-s})$-sesquilinear pairing (conjugate linear with respect to the second term).

We have therefore, with $f = (H - z) u$,

$$\hat{u}(\xi) = \frac{(2\pi)^{-\frac{n}{2}}}{|\xi|^2 - z} \left(\langle (H - z) u(x), \theta(x) \exp(i\xi x) \rangle + \overline{\langle (H_0 - H) \exp(i\xi x), u(x) \rangle} \right)$$

$$= \frac{(2\pi)^{-\frac{n}{2}}}{|\xi|^2 - z} \left(\langle f(x), \theta(x) \exp(i\xi x) \rangle + \langle f(x), R(\bar{z}) (H_0 - H) \exp(i\xi x) \rangle \right). \quad (5.7)$$

Introducing the function

$$\widetilde{f}(\xi, z) = \hat{f}(\xi) + (2\pi)^{-\frac{n}{2}} \langle f(x), R(\bar{z}) (H_0 - H) \exp(i\xi x) \rangle,$$

we have

$$\hat{u}(\xi) = \widehat{R(z)f}(\xi) = \frac{\widetilde{f}(\xi, z)}{|\xi|^2 - z}, \quad \mathrm{Im}\, z \neq 0. \tag{5.8}$$

We now claim that this equation is valid for all compactly supported $f \in H^{-1}$.

Indeed, let $u = R(z)f \in H^{1,-s}$, $s > 1$. Let $\psi(x) = 1 - \chi(x)$, where $\chi(x)$ is defined in (4.18). We set

$$u_k(x) = \psi(k^{-1}x)u(x), \quad f_k(x) = (H - z)\left(\psi(k^{-1}x)u(x)\right), \quad k = 1, 2, 3, \ldots$$

The equality (5.8) is satisfied with u, f replaced, respectively, by u_k, f_k. Since

$$\psi(k^{-1}x)u(x) \xrightarrow[k\to\infty]{} u(x)$$

in $H^{1,-s}$, we have

$$(H - z)\left(\psi(k^{-1}x)u(x)\right) \xrightarrow[k\to\infty]{} (H - z)\, u = f(x)$$

in $H^{-1,-s}$, where in the last step we have used Eq. (4.38).

In addition, since $(H_0 - H)\exp(i\xi x)$ is compactly supported

$$\langle f_k(x), R(\bar{z})(H_0 - H)\exp(i\xi x)\rangle = \overline{\langle (H_0 - H)\exp(i\xi x), R(z)f_k(x)\rangle}$$
$$\xrightarrow[k\to\infty]{} \overline{\langle (H_0 - H)\exp(i\xi x), R(z)f\rangle} = \langle f, R(\bar{z})(H_0 - H)\exp(i\xi x)\rangle.$$

Combining these considerations with the continuity of the Fourier transform (on tempered distributions) we establish that (5.8) is valid for all compactly supported $f \in H^{-1}$.

Let $\{E(\lambda), \lambda \in \mathbb{R}\}$ be the spectral family associated with H. Let $A(\lambda) = \frac{d}{d\lambda}E(\lambda)$ be its weak derivative. More precisely, we use the relation (3.7), to get (using Theorem 4.1), for any $f \in H^{-1,s}$, $s > 1$,

$$\langle f, A(\lambda)f \rangle = \frac{1}{2\pi i}\langle f, \left(R^+(\lambda) - R^-(\lambda)\right)f\rangle.$$

We now take $f \in L^2$ and compactly supported. From the resolvent equation we infer

$$R(\lambda + i\epsilon) - R(\lambda - i\epsilon) = 2i\epsilon R(\lambda + i\epsilon)R(\lambda - i\epsilon), \quad \epsilon > 0,$$

so that

$$\langle f, A(\lambda)f \rangle = \lim_{\epsilon \downarrow 0} \frac{\epsilon}{\pi}\|R(\lambda + i\epsilon)f\|_0^2, \quad \epsilon > 0.$$

Using Eq. (5.8) and Parseval's theorem, we therefore have

$$\langle f, A(\lambda)f \rangle = \lim_{\epsilon \to 0+} \frac{\epsilon}{\pi}\left\||\xi|^2 - (\lambda + i\epsilon))^{-1}\widetilde{f}(\xi, \lambda + i\epsilon)\right\|_0^2, \quad \epsilon > 0. \tag{5.9}$$

Note that $\widetilde{f}(\xi, z)$ can be extended continuously as $z \to \lambda + i \cdot 0$ by

$$\widetilde{f}(\xi, \lambda) = \hat{f}(\xi) + (2\pi)^{-\frac{n}{2}}\langle f(x), R^-(\lambda)(H_0 - H)\exp(i\xi x)\rangle. \tag{5.10}$$

In order to study properties of $\tilde{f}(\xi, z)$ as a function of ξ we compute

$$\tilde{f}(\xi, z) = \hat{f}(\xi) + (2\pi)^{-\frac{n}{2}} \overline{\left\langle \left(\sum_{l,j=1}^{n} \partial_l(a_{l,j}(x) - \delta_{l,j}) \partial_j \right) \exp(i\xi x), R(z)f(x) \right\rangle} \quad (5.11)$$

$$= \hat{f}(\xi) + (2\pi)^{-\frac{n}{2}} i \sum_{l,j=1}^{n} \xi_j \int_{\mathbb{R}^n} (a_{l,j}(x) - \delta_{l,j}) \, \partial_l(R(z)f(x)) \exp(-i\xi x) \, dx,$$

where in the last step we have used that both $\partial_l(R(z)f(x))$ and $(a_{l,j}(x) - \delta_{l,j})$ $\exp(-i\xi x)$ are in L^2.

Consider now the integral

$$g(\xi, z) = \int_{\mathbb{R}^n} (a_{l,j}(x) - \delta_{l,j}) \, \partial_l(R(z)f(x)) \exp(-i\xi x) \, dx, \quad z \in \Omega,$$

where Ω is as in (4.16).

In view of Theorem 4.1, the family $\{\partial_l R(z)f(x)\}_{z \in \Omega}$ is uniformly bounded in $L^{2,-s}$, $s > 1$, so by Parseval's theorem we get

$$\|g(\cdot, z)\|_0 < C, \quad z \in \Omega,$$

where C only depends on f.

This estimate and (5.11) imply that, if $f \in L^2$ is compactly supported,

(i) The function

$$\mathbb{R}^n \times \overline{\Omega} \ni (\xi, z) \to \tilde{f}(\xi, z)$$

is continuous. For real z it is given by (5.10).

(ii)
$$\lim_{k \to \infty} \int_{|\xi|>k} (|\xi|^2 - z)^{-1} |\tilde{f}(\xi, z)|^2 \, d\xi = 0,$$

uniformly in $z \in \Omega$.

As $z \to |\xi|^2 + i \cdot 0$, we have by Theorem 4.1 and Eq. (5.3),

$$\lim_{z \to |\xi|^2 + i \cdot 0} \tilde{f}(\xi, z) = (2\pi)^{-\frac{n}{2}} \int_{\mathbb{R}^n} f(x) \overline{\varphi_+(x, \xi)} \, dx = \mathbb{F}_+ f(\xi),$$

so that, taking (i) and (ii) into account, we obtain from (5.9), for any compactly supported $f \in L^2$,

$$\langle f, A(\lambda)f \rangle = \frac{1}{2\sqrt{\lambda}} \int_{|\xi|^2 = \lambda} |\mathbb{F}_+ f(\xi)|^2 \, d\sigma, \quad \lambda > 0, \quad (5.12)$$

where $d\sigma$ is the surface Lebesgue measure.

It follows that, for any $[\alpha, \beta] \subset (0, \infty)$,

$$((E(\beta) - E(\alpha))f, f) = \int_\alpha^\beta \langle f, A(\lambda)f \rangle d\lambda = \int_{\alpha \le |\xi|^2 \le \beta} |\mathbb{F}_+ f(\xi)|^2 \, d\xi. \quad (5.13)$$

Letting $\alpha \to 0$, $\beta \to \infty$, we get

$$\|f\|_0 = \|\mathbb{F}_+ f\|_0. \tag{5.14}$$

Thus $f \to \mathbb{F}_+ f \in L^2(\mathbb{R}^n)$ is an isometry for compactly supported functions, which can be extended by density to all $f \in L^2(\mathbb{R}^n)$.

Furthermore, since the spectrum of H is entirely absolutely continuous, it follows that for every $f \in L^2$, Eq. (5.12) holds for almost all $\lambda > 0$ (with respect to the Lebesgue measure).

Let $f \in D(H)$. By the spectral theorem

$$\langle Hf, A(\lambda)Hf \rangle = \lambda^2 \langle f, A(\lambda)f \rangle = \frac{1}{2\sqrt{\lambda}} \int_{|\xi|^2 = \lambda} \left| |\xi|^2 \, \mathbb{F}_+ f(\xi) \right|^2 d\sigma, \quad \lambda > 0.$$

In particular,

$$\|Hf\|_0^2 = \int_{\mathbb{R}^n} \left| |\xi|^2 \, \mathbb{F}_+ f(\xi) \right|^2 d\xi. \tag{5.15}$$

Conversely, if the right-hand side of (5.15) is finite, then $\int_0^\infty \lambda^2 \langle f, A(\lambda)f \rangle \, d\lambda < \infty$, so $f \in D(H)$.

The adjoint operator $\mathbb{F}_+^*$ is a partial isometry (on the range of $\mathbb{F}_+$). If $f(x) \in L^2(\mathbb{R}^n)$ is compactly supported and $g(\xi) \in L^2(\mathbb{R}^n)$ is likewise compactly supported, then

$$(\mathbb{F}_+ f, g) = (2\pi)^{-\frac{n}{2}} \int_{\mathbb{R}^n} \left(\int_{\mathbb{R}^n} f(x) \, \overline{\varphi_+(x, \xi)} \, dx \right) \overline{g(\xi)} \, d\xi$$

$$= (2\pi)^{-\frac{n}{2}} \int_{\mathbb{R}^n} f(x) \left(\int_{\mathbb{R}^n} \overline{g(\xi) \, \varphi_+(x, \xi)} \, d\xi \right) dx,$$

where in the change of order of integration Proposition 5.3 was taken into account.

It follows that, for a compactly supported $g(\xi) \in L^2(\mathbb{R}^n)$,

$$(\mathbb{F}_+^* g)(x) = (2\pi)^{-\frac{n}{2}} \int_{\mathbb{R}^n} g(\xi) \, \varphi_+(x, \xi) \, d\xi, \tag{5.16}$$

and the extension to all $g \in L^2(\mathbb{R}^n)$ is obtained by the fact that $\mathbb{F}_+^*$ is a partial isometry.

Now if $f \in D(H)$, $g \in L^2(\mathbb{R}^n)$, we have

$$(Hf, g) = \int_{\mathbb{R}^n} |\xi|^2 \, \mathbb{F}_+ f(\xi) \, \overline{\mathbb{F}_+ g(\xi)} \, d\xi = \int_{\mathbb{R}^n} \mathbb{F}_+^* \left(|\xi|^2 \, \mathbb{F}_+ f(\xi) \right) \overline{g(\xi)} \, d\xi,$$

which is the statement (5.5) of the theorem.

It follows from the spectral theorem that, for every interval $J = [\alpha, \beta] \subseteq [0, \infty)$ and for every $f \in L^2(\mathbb{R}^n)$, we have, with $E_J = E(\beta) - E(\alpha)$ and χ_J the characteristic function of J,

$$E_J f(x) = \mathbb{F}_+^* \left(\chi_J(|\xi|^2) \, \mathbb{F}_+ f(\xi) \right)$$

or

$$\mathbb{F}_+ E_J f(\xi) = \chi_J(|\xi|^2) \, \mathbb{F}_+ f(\xi).$$

It remains to prove that the isometry $\mathbb{F}_+$ is onto (and hence unitary). So, suppose to the contrary that, for some nonzero $g(\xi) \in L^2(\mathbb{R}^n)$,

$$(\mathbb{F}_+^* g)(x) = 0.$$

In particular, for any $f \in L^2(\mathbb{R}^n)$ and any interval J as above,

$$0 = \left(E_J f, \mathbb{F}_+^* g \right) = (\mathbb{F}_+ E_J f, g) = \left(\chi_J(|\xi|^2) \mathbb{F}_+ f(\xi), g(\xi) \right) = \left(\mathbb{F}_+ f(\xi), \chi_J(|\xi|^2) g(\xi) \right),$$

so that $\mathbb{F}_+^* \left(\chi_J(|\xi|^2) \, g(\xi) \right) = 0$.

By Eq. (5.16) we have, for any $0 \le \alpha < \beta$,

$$\int\limits_{\alpha < |\xi|^2 < \beta} g(\xi) \, \varphi_+(x, \xi) \, d\xi = 0$$

so that, in view of the continuity properties of $\varphi_+(x, \xi)$ (see Proposition 5.3), for a.e. $\lambda \in (0, \infty)$,

$$\int\limits_{|\xi|^2 = \lambda} g(\xi) \, \varphi_+(x, \xi) \, d\sigma = 0. \tag{5.17}$$

From the definition (5.3) we get

$$\int\limits_{|\xi|^2 = \lambda} g(\xi) \exp(i\xi x) \, d\sigma - \int\limits_{|\xi|^2 = \lambda} g(\xi) R^-(\lambda) \left((H - \lambda) \exp(i\xi x) \right) d\sigma = 0. \tag{5.18}$$

Since $(H - \lambda) \exp(i\xi x)$ is compactly supported (when $|\xi|^2 = \lambda$), the continuity property of $R^-(\lambda)$ enables us to write

$$\int\limits_{|\xi|^2 = \lambda} g(\xi) \, R^-(\lambda) \left((H - \lambda) \exp(i\xi x) \right) d\sigma = R^-(\lambda) \int\limits_{|\xi|^2 = \lambda} g(\xi) \, (H - \lambda) \exp(i\xi x) \, d\sigma,$$

which, by Remark 4.9, satisfies a Sommerfeld radiation condition. We conclude that the function

$$G(x) = \int\limits_{|\xi|^2 = \lambda} g(\xi) \exp(i\xi x) \, d\sigma \in H^{1, -s}, \quad s > \frac{1}{2},$$

is a radiative solution (see Remark 4.7) of $(-\Delta - \lambda) \, G = 0$ and hence must vanish. Since this holds for a.e. $\lambda > 0$, we get $\hat{g}(\xi) = 0$, hence $g = 0$. $\qquad\square$

6. Global spacetime estimates for a generalized wave equation

The Strichartz estimates [83] have become a fundamental ingredient in the study of nonlinear wave equations. They are L^p spacetime estimates that are derived for operators whose leading part has constant coefficients. We refer to the books [81, 82] and [5] for detailed accounts and further references.

Here we focus on spacetime estimates pertinent to the framework of this review, namely, weighted L^2 estimates.

We recall first some results related to the Cauchy problem for the classical wave equation,

$$\Box u = \frac{\partial^2 u}{\partial t^2} - \Delta u = 0, \tag{6.1}$$

subject to the initial data

$$u(x,0) = u_0(x), \quad \partial_t u(x,0) = v_0(x), \quad x \in \mathbb{R}^n. \tag{6.2}$$

The Morawetz estimate [66] yields

$$\int_{\mathbb{R}} \int_{\mathbb{R}^n} |x|^{-3} |u(x,t)|^2 \, dx dt \leq C \left(\|\nabla u_0\|_0^2 + \|v_0\|_0^2 \right), \quad n \geq 4,$$

while in [8] we gave the estimate

$$\int_{\mathbb{R}} \int_{\mathbb{R}^n} |x|^{-2\alpha-1} |u(x,t)|^2 \, dx dt \leq C_\alpha \left(\||\nabla|^\alpha u_0\|_0^2 + \||\nabla|^{\alpha-1} v_0\|_0^2 \right), \quad n \geq 3,$$

for every $\alpha \in (0,1)$.

Related results were obtained in [65] (allowing also dissipative terms), [50] (with some gain in regularity), [88] (with short-range potentials) and [47] for spherically symmetric solutions.

Here we consider the equation

$$\frac{\partial^2 u}{\partial t^2} + Hu = \frac{\partial^2 u}{\partial t^2} - \sum_{i,j=1}^n \partial_i a_{i,j}(x) \partial_j u = f(x,t), \tag{6.3}$$

subject to the initial data (6.2).

We first replace the assumptions (4.1), (4.2) by stronger ones as follows:

(H1) $$a(x) = g^{-1}(x) = \left(g^{i,j}(x) \right)_{1 \leq i,j \leq n}, \tag{6.4}$$

where $g(x) = (g_{i,j}(x))_{1 \leq i,j \leq n}$ is a smooth Riemannian metric on $\mathbb{R}^n$ such that

$$g(x) = I, \quad |x| > \Lambda_0.$$

(H2)   The Hamiltonian flow associated with $h(x,\xi) = (g(x)\xi, \xi)$
         is nontrapping for any (positive) value of h.   (6.5)

Recall that (H2) means that the flow associated with the Hamiltonian vectorfield $\mathcal{H} = \frac{\partial h}{\partial \xi} \frac{\partial}{\partial x} - \frac{\partial h}{\partial x} \frac{\partial}{\partial \xi}$ leaves any compact set in $\mathbb{R}_x^n$.

Identical hypotheses are imposed in the study of resolvent estimates in semi-classical theory [26, 27].

In our estimates we use *homogeneous Sobolev spaces* associated with the operator H.

We let $G = H^{\frac{1}{2}}$ which is a positive self-adjoint operator. Note that $\|G\theta\|_0$ is equivalent to the homogeneous Sobolev norm $\|\nabla\theta\|_0$.

Theorem 6.1. *Suppose that $n \geq 3$ and that $a(x)$ satisfies* Hypotheses (H1)–(H2). *Let $s > 1$.*

(a) (*local energy decay*) *There exists a constant $C_1 = C_1(s,n) > 0$ such that the solution to* (6.3), (6.2) *satisfies*

$$\int_{\mathbb{R}} \int_{\mathbb{R}^n} \left(1 + |x|^2\right)^{-s} \left[|Gu(x,t)|^2 + |u_t(x,t)|^2\right] dx\,dt$$

$$\leq C_1 \left\{ \|Gu_0\|_0^2 + \|v_0\|_0^2 + \int_{\mathbb{R}} \int_{\mathbb{R}^n} |f(x,t)|^2 \, dx\,dt \right\}. \quad (6.6)$$

(b) (*amplitude decay*) *Assume that $f = 0$. There exists a constant $C_2 = C_2(s,n) > 0$ such that the solution to* (6.3), (6.2) *satisfies,*

$$\int_{\mathbb{R}} \int_{\mathbb{R}^n} \left(1 + |x|^2\right)^{-s} |u(x,t)|^2 \, dx\,dt \leq C_2 \left[\|u_0\|_0^2 + \|G^{-1}v_0\|_0^2\right]. \quad (6.7)$$

This estimate generalizes similar estimates obtained for the classical ($g = I$) wave equation [8, 65].

Remark 6.2. The estimate (6.6) is an *energy decay estimate* for the wave equation (6.3). A localized (in space) version of the estimate has served to obtain global (small amplitude) existence theorems for the corresponding nonlinear equation [27, 48].

The weighted L^2 spacetime estimates for the *dispersive* equation

$$i^{-1}\frac{\partial}{\partial t} u = Lu,$$

have been extensively treated in recent years. In general, in this case there is also a gain of derivatives (so-called *smoothing*) in addition to the energy decay. For the Schrödinger operator $L = -\Delta + V(x)$, with various assumptions on the potential V, we refer to [3, 7, 8, 17, 19, 50, 62, 78, 80, 89] and references therein. In [33] the case of magnetic potentials is considered. The Schrödinger operator on a Riemannian manifold is treated in [26, 38]. For more general operators see [16, 20, 28, 51, 67, 77, 84] and references therein.

Proof of Theorem 6.1. (a) Define, with $G = H^{\frac{1}{2}}$,

$$u_{\pm} = \frac{1}{2}\left(Gu \pm iu_t\right).$$

Then

$$\partial_t u_\pm = \mp i\, G u_\pm \pm \frac{i}{2} f. \tag{6.8}$$

Defining

$$U(t) = \begin{pmatrix} u_+(t) \\ u_-(t) \end{pmatrix}, \tag{6.9}$$

we have

$$i^{-1} U'(t) = -KU + F, \tag{6.10}$$

where

$$K = \begin{pmatrix} G & 0 \\ 0 & -G \end{pmatrix}, \quad F(t) = \begin{pmatrix} \frac{1}{2} f(\cdot, t) \\ -\frac{1}{2} f(\cdot, t) \end{pmatrix}.$$

Note that, as is common when treating evolution equations, we write $U(t)$, $F(t)$, etc. for $U(x,t)$, $F(x,t)$, etc. when there is no risk of confusion.

The operator K is a self-adjoint operator on $\mathcal{D} = L^2(\mathbb{R}^n) \oplus L^2(\mathbb{R}^n)$. Its spectral family $E_K(\lambda)$ is given by $E_K(\lambda) = E_G(\lambda) \oplus (I - E_G(-\lambda))$, $\lambda \in \mathbb{R}$, where E_G is the spectral family of G.

Let $E(\lambda)$ be the spectral family of H, and let $A(\lambda) = \frac{d}{d\lambda} E(\lambda)$ be its weak derivative (3.7). By the definition of G we have

$$E_G(\lambda) = E(\lambda^2),$$

hence its weak derivative is given by

$$A_G(\lambda) = \frac{d}{d\lambda} E_G(\lambda) = 2\lambda\, A(\lambda^2), \quad \lambda > 0. \tag{6.11}$$

In view of the LAP (Theorem A) we therefore have that the operator-valued function

$$A_G(\lambda) \in B(L^{2,s}(\mathbb{R}^n), L^{2,-s}(\mathbb{R}^n))$$

is continuous for $\lambda \geq 0$.

Denoting $\mathcal{D}^s = L^{2,s}(\mathbb{R}^n) \oplus L^{2,s}(\mathbb{R}^n)$, it follows that

$$A_K(\lambda) = \frac{d}{d\lambda} E_K(\lambda) = A_G(\lambda) \oplus A_G(-\lambda), \quad \lambda \in \mathbb{R},$$

is continuous with values in $B(\mathcal{D}^s, \mathcal{D}^{-s})$ for $s > 1$.

Making use of Hypotheses (H1)–(H2), we invoke [76, Theorem 5.1] to conclude that $\limsup_{\mu \to \infty} \mu^{\frac{1}{2}} \|A(\mu)\|_{B(L^{2,s}, L^{2,-s})} < \infty$, so that by (6.11) there exists a constant $C > 0$ such that

$$\|A_G(\lambda)\|_{B(L^{2,s}, L^{2,-s})} < C, \quad \lambda \geq 0. \tag{6.12}$$

It follows that also

$$\|A_K(\lambda)\|_{B(\mathcal{D}^s, \mathcal{D}^{-s})} < C, \quad \lambda \in \mathbb{R}, \ s > 1, \ \lambda \in \mathbb{R}. \tag{6.13}$$

Let $\langle\ ,\ \rangle$ be the sesquilinear pairing between $\mathcal{D}^{-s}$ and $\mathcal{D}^s$ (conjugate linear with respect to the second term).

For any $\psi, \chi \in \mathcal{D}^s$ we have, in view of the fact that $A_K(\lambda)$ is a weak derivative of a spectral measure,

(i)
$$|\langle A_K(\lambda)\psi, \chi\rangle|^2 \le \langle A_K(\lambda)\psi, \psi\rangle \cdot \langle A_K(\lambda)\chi, \chi\rangle,$$

(ii)
$$\int_{-\infty}^{\infty} \langle A_K(\lambda)\psi, \psi\rangle d\lambda = \|\psi\|^2_{L^2(\mathbb{R}^n)\oplus L^2(\mathbb{R}^n)}. \tag{6.14}$$

We first treat the pure Cauchy problem, i.e., $f \equiv 0$.

To estimate $U(x,t) = e^{-itK}U(x,0)$ we use a duality argument. Some of the following computations will be rather formal, but they can easily be justified by a density argument, as in [8, 20]. We shall use $((\ ,\ ))$ for the scalar product in $L^2(\mathbb{R}^{n+1}) \oplus L^2(\mathbb{R}^{n+1})$.

Take $w(x,t) \in C_0^\infty(\mathbb{R}^{n+1}) \oplus C_0^\infty(\mathbb{R}^{n+1})$. Then,

$$((U,w)) = \int_{-\infty}^{\infty} e^{-itK}U(x,0) \cdot \overline{w(x,t)}\, dx dt = \int_{-\infty}^{\infty} \langle A_K(\lambda)U(x,0), \int_{-\infty}^{\infty} e^{it\lambda}w(\cdot,t)dt\rangle d\lambda$$

$$= (2\pi)^{1/2} \int_{-\infty}^{\infty} \langle A_K(\lambda)U(x,0), \tilde{w}(\cdot,\lambda)\rangle d\lambda,$$

where

$$\tilde{w}(x,\lambda) = (2\pi)^{-\frac{1}{2}}\int_{\mathbb{R}} w(x,t)e^{it\lambda}dt.$$

Noting (6.14), (6.13) and using the Cauchy-Schwarz inequality

$$|((U,w))| \le (2\pi)^{1/2} \|U(x,0)\|_0 \left(\int_{-\infty}^{\infty} \langle A_K(\lambda)\tilde{w}(\cdot,\lambda), \tilde{w}(\cdot,\lambda)\rangle d\lambda\right)^{1/2}$$

$$\le C\|U(x,0)\|_0 \left(\int_{-\infty}^{\infty} \|\tilde{w}(\cdot,\lambda)\|^2_{\mathcal{D}^s}\, d\lambda\right)^{\frac{1}{2}}.$$

It follows from the Plancherel theorem that

$$|((U,w))| \le C\|U(x,0)\|_0 \left(\int_{\mathbb{R}} \|w(\cdot,t)\|^2_{\mathcal{D}^s}\, dt\right)^{1/2}.$$

Let $\phi(x,t) \in C_0^\infty(\mathbb{R}^{n+1}) \oplus C_0^\infty(\mathbb{R}^{n+1})$ and take $w(x,t) = (1+|x|^2)^{-\frac{s}{2}}\phi(x,t)$ so that

$$\left|\left(\left((1+|x|^2)^{-\frac{s}{2}}U, \phi\right)\right)\right| \le C\|U(x,0)\|_0 \cdot \|\phi\|_{L^2(\mathbb{R}^{n+1})}.$$

This concludes the proof of the part involving the Cauchy data in (6.6), in view of (6.9).

To prove the part concerning the inhomogeneous equation, it suffices to take $u_0 = v_0 = 0$. In this case the Duhamel principle yields, for $t > 0$,

$$U(t) = \int_0^t e^{-i(t-\tau)K} F(\tau)d\tau,$$

where we have used the form (6.10) of the equation.

Integrating the inequality

$$\|U(t)\|_{\mathcal{D}^{-s}} \leq \int_0^t \left\| e^{-i(t-\tau)K} F(\tau) \right\|_{\mathcal{D}^{-s}} d\tau,$$

we get

$$\int_0^\infty \|U(t)\|_{\mathcal{D}^{-s}} \, dt \leq \int_0^\infty \int_\tau^\infty \left\| e^{-i(t-\tau)K} F(\tau) \right\|_{\mathcal{D}^{-s}} dt \, d\tau.$$

Invoking the first part of the proof we obtain

$$\int_0^\infty \|U(t)\|_{\mathcal{D}^{-s}} \, dt \leq C \int_0^\infty \|F(\tau)\|_0 \, d\tau,$$

which proves the part related to the inhomogeneous term in (6.6).

(b) Define

$$v_\pm(x,t) = \exp(\pm itG)\, \phi_\pm(x), \quad \text{where} \quad \phi_\pm(x) = \frac{1}{2}\left[u_0(x) \mp G^{-1}v_0(x) \right].$$

Then clearly

$$u(x,t) = v_+(x,t) + v_-(x,t).$$

We establish the estimate (6.7) for v_+.

Taking $w(x,t) \in C_0^\infty(\mathbb{R}^{n+1})$ we proceed as in the first part of the proof. Let $\langle \, , \rangle$ be the $(L^{2,-s}(\mathbb{R}^n), L^{2,s}(\mathbb{R}^n))$ pairing. Then

$$(v_+, w) = \int_{-\infty}^\infty e^{itG}\phi_+(x)\,\overline{w(x,t)}\, dxdt = \int_0^\infty \langle A_G(\lambda)\phi_+, \int_{-\infty}^\infty e^{-it\lambda}w(\cdot,t)dt\rangle d\lambda$$

$$= (2\pi)^{1/2} \int_0^\infty \langle A_G(\lambda)\phi_+, \tilde{w}(\cdot,\lambda)\rangle d\lambda,$$

where

$$\tilde{w}(x,\lambda) = (2\pi)^{-\frac{1}{2}} \int_{\mathbb{R}} w(x,t)e^{-it\lambda}\, dt.$$

Noting (6.12) as well as the inequalities (6.14) (with A_G replacing A_K) and using the Cauchy-Schwarz inequality

$$|(v_+, w)| \leq (2\pi)^{1/2} \|\phi_+\|_0 \left(\int_0^\infty < A_G(\lambda)\tilde{w}(\cdot, \lambda), \tilde{w}(\cdot, \lambda) > d\lambda \right)^{1/2}$$

$$\leq C \|\phi_+\|_0 \left(\int_0^\infty \|\tilde{w}(\cdot, \lambda)\|_{0,s}^2 \, d\lambda \right)^{\frac{1}{2}}.$$

The Plancherel theorem yields

$$|(v_+, w)| \leq C \|\phi_+\|_0 \left(\int_{\mathbb{R}} \|w(\cdot, t)\|_{0,s}^2 \, dt \right)^{1/2}.$$

Let $\omega \in C_0^\infty(\mathbb{R}^{n+1})$ and take $w(x,t) = \left(1 + |x|^2\right)^{-\frac{s}{2}} \omega(x,t)$ so that

$$\left| \left(\left(1 + |x|^2\right)^{-\frac{s}{2}} v_+, \omega \right) \right| \leq C \|\phi_+\|_0 \|\omega\|_{L^2(\mathbb{R}^{n+1})}.$$

This (with the similar estimate for v_-) concludes the proof of the estimate (6.7).

$\square$

Remark 6.3 (optimality of the requirement $s > 1$). A key point in the proof was the use of the uniform bound (6.13). In view of the relation (6.11), this is reduced to the uniform boundedness of $\lambda A(\lambda^2)$, $\lambda \geq 0$, in $B(L^{2,s}, L^{2,-s})$. By [76, Theorem 5.1] the boundedness at infinity, $\limsup_{\mu \to \infty} \mu^{\frac{1}{2}} \|A(\mu)\| < \infty$, holds already with $s > \frac{1}{2}$. Thus the further restriction $s > 1$ is needed in order to ensure the boundedness at $\lambda = 0$ (Theorem A).

Remark 6.4. Clearly we can take $[0, T]$ as the time interval, instead of $\mathbb{R}$, for any $T > 0$.

References

[1] S. Agmon, *Spectral properties of Schrödinger operators and scattering theory*, Ann. Sc. Norm. Super. Pisa **2** (1975), 151–218.

[2] S. Agmon, J. Cruz-Sampedro and I. Herbst, *Spectral properties of Schrödinger operators with potentials of order zero*, J. Funct. Anal. **167** (1999), 345–369.

[3] Y. Ameur and B. Walther, *Smoothing estimates for the Schrödinger equation with an inverse-square potential*, preprint (2007).

[4] W.O. Amrein, A. Boutet de Monvel and V. Georgescu, *C_0-Groups, Commutator Methods and Spectral Theory of N-Body Hamiltonians*, Birkhäuser, 1996.

[5] M. Beals and W. Strauss, *L^p estimates for the wave equation with a potential*, Comm. Partial Differential Equations **18** (1993), 1365–1397.

[6] M. Ben-Artzi, *Unitary equivalence and scattering theory for Stark-like Hamiltonians*, J. Math. Phys. **25** (1984), 951–964.

[7] M. Ben-Artzi, *Global estimates for the Schrödinger equation*, J. Funct. Anal. **107** (1992), 362–368.

[8] M. Ben-Artzi, *Regularity and smoothing for some equations of evolution*, in *Nonlinear Partial Differential Equations and Their Applications; Collège de France Seminar* (H. Brezis and J.L. Lions, eds.), vol. XI, Longman Scientific, pp. 1–12, 1994.

[9] M. Ben-Artzi, *On spectral properties of the acoustic propagator in a layered band*, J. Differential Equations **136** (1997), 115–135.

[10] M. Ben-Artzi, *Spectral theory for divergence-form operators* in *Spectral and Scattering Theory and Related Topics* (H. Ito, ed.), RIMS Kokyuroku 1607, pp. 77–84, 2008.

[11] M. Ben-Artzi, *Eigenfunction expansions and spacetime estimates for generators in divergence-form*, Reviews in Math. Physics **22** (2010), 1209–1240.

[12] M. Ben-Artzi, Y. Dermenjian and J.-C. Guillot, *Analyticity properties and estimates of resolvent kernels near thresholds*, Comm. Partial Differential Equations **25** (2000), 1753–1770.

[13] M. Ben-Artzi, Y. Dermenjian and A. Monsef, *Resolvent kernel estimates near thresholds*, Differential Integral Equations **19** (2006), 1–14.

[14] M. Ben-Artzi and A. Devinatz, *The limiting absorption principle for a sum of tensor products and applications to the spectral theory of differential operators*, J. Anal. Math. **43** (1983/84), 215–250.

[15] M. Ben-Artzi and A. Devinatz, *The Limiting Absorption Principle for Partial Differential Operators*, Mem. Amer. Math. Soc. **364** (1987).

[16] M. Ben-Artzi and A. Devinatz, *Local smoothing and convergence properties for Schrödinger-type equations*, J. Funct. Anal. **101** (1991), 231–254.

[17] M. Ben-Artzi and A. Devinatz, *Regularity and decay of solutions to the Stark evolution equation*, J. Funct. Anal. **154** (1998), 501–512.

[18] M. Ben-Artzi and A. Hardy, *On the expansion of an arbitrary field in terms of waveguide modes*, IEE Proc. Optoelectronics **141** (1994), 16–20.

[19] M. Ben-Artzi and S. Klainerman, *Decay and regularity for the Schrödinger equation*, J. Anal. Math. **58** (1992), 25–37.

[20] M. Ben-Artzi and J. Nemirovsky, *Remarks on relativistic Schrödinger operators and their extensions*, Ann. Inst. H. Poincaré Phys. Théor. **67** (1997), 29–39.

[21] M. Ben-Artzi and J. Nemirovsky, *Resolvent estimates for Schrödinger-type and Maxwell equations with applications*, in *Spectral and Scattering Theory* (A. Ramm, ed.), pp. 19–31, Plenum Publ., 1998.

[22] Ju.M. Berezanskii, *Expansion in Eigenfunctions of Selfadjoint Operators*, Transl. Math. Monogr., vol. 17, Amer. Math. Soc. 1968.

[23] T. Bouhennache, *Spectral analysis of an isotropic stratified elastic strip and applications*, Osaka J. Math. **37** (2000), 577–601.

[24] A. Boutet de Monvel-Berthier and D. Manda, *Spectral and scattering theory for wave propagation in perturbed stratified media*, J. Math. Anal. Appl. **191** (1995), 137–167.

[25] F.E. Browder, *The eigenfunction expansion theorem for the general self-adjoint singular elliptic partial differential operator. I. The analytical foundation* , Proc. Natl. Acad. Sci. USA **40** (1954), 454–459.

[26] N. Burq, *Semi-classical estimates for the resolvent in nontrapping geometries*, Int. Math. Res. Not. **5** (2002), 221–241.

[27] N. Burq, *Global Strichartz estimates for nontrapping geometries: About an article by H. Smith and C. Sogge*, Comm. Partial Differential Equations**28** (2003), 1675–1683.

[28] H. Chihara, *Smoothing effects of dispersive pseudodifferential equations*, Comm. Partial Differential Equations **27** (2002), 1953–2005.

[29] C. Cohen-Tannoudji, B. Diu and F. Laloë, *Quantum Mechanics*, John Wiley, 1977.

[30] A. Cohen and T. Kappeler, *Scattering and inverse scattering for steplike potentials in the Schrödinger equation*, Indiana Univ. Math. J. **34** (1985), 127–180.

[31] R. Courant and D. Hilbert, *Methods of Mathematical Physics II*, Interscience Publ. 1962.

[32] E. Croc and Y. Dermenjian, *Analyse spectrale d'une bande acoustique multistratifiée. Partie I: Principe d'absorption limite pour une stratification simple*, SIAM J. Math. Anal. **26** (1995), 880–924.

[33] P. D'Ancona and L. Fanelli, *Strichartz and smoothing estimates for dispersive equations with magnetic potentials*, Comm. Partial Differential Equations **33** (2008), 1082–1112.

[34] S. DeBièvre and W. Pravica, *Spectral analysis for optical fibers and stratified fluids I: The limiting absorption principle*, J. Funct. Anal. **98** (1991), 404–436.

[35] V.G. Deich, E.L. Korotayev and D.R. Yafaev, *Theory of potential scattering, taking into account spatial anisotropy*, J. Soviet Math. **34** (1986), 2040–2050.

[36] J. Derezinski and C. Gérard, *Scattering Theory of Classical and Quantum N-Particle Systems*, Springer-Verlag, Berlin, 1997.

[37] A. Devinatz, *Lectures on Spectral Calculus*, The Landau Center for Research in Mathematical Analysis, The Hebrew University of Jerusalem, vol. 19, 1992.

[38] S.-I. Doi, *Smoothing effects of Schrödinger evolution groups on Riemannian manifolds*, Duke Math. J. **82** (1996), 679–706.

[39] M. Durand, *Limiting absorption principle for stratified operators with thresholds: a general method*, Appl. Anal. **82**(2003), 821–838.

[40] D.M. Eidus, *The principle of limiting absorption*, Amer. Math. Soc. Transl. Ser. 2 **47** (1965), 157–192 (original in Russian: Mat. Sb. **57** (1962), 13–44).

[41] C. Gérard, *A proof of the abstract limiting absorption principle by energy estimates*, J. Funct. Anal. **254** (2008), 2707–2724.

[42] D. Gilbarg and N.S. Trudinger, *Elliptic Partial Differential Equations of Second Order*, Springer-Verlag 1977.

[43] J. Ginibre and M. Moulin, *Hilbert space approach to the quantum mechanical three body problem*, Ann. Inst. H. Poincaré Sect. A **21** (1974), 97–145.

[44] M. Goldberg and W. Schlag, *A limiting absorption principle for the three-dimensional Schrödinger equation with L^p potentials*, Int. Math. Res. Not. **75** (2004), 4049–4071.

[45] I. Herbst, *Spectral theory of the operator* $(p^2 + m^2)^{\frac{1}{2}} - Z\frac{e^2}{r}$, Comm. Math. Phys. **53** (1977), 285–294.

[46] I. Herbst, *Spectral and scattering theory for Schrödinger operators with potentials independent of* $|x|$, Amer. J. Math. **113** (1991), 509–565.

[47] K. Hidano, *Morawetz-Strichartz estimates for spherically symmetric solutions to wave equations and applications to semilinear Cauchy problems*, Differential Integral Equations **20** (2007), 735–754.

[48] K. Hidano, J. Metcalfe, H.F. Smith, C.D. Sogge and Y. Zhou, *On abstract Strichartz estimates and the Strauss conjecture for nontrapping obstacles*, Trans. Amer. Math. Soc. **362** (2010), 2789–2809.

[49] L. Hörmander, *The Analysis of Linear Partial Differential Operators II*, Springer-Verlag, 1983.

[50] T. Hoshiro, *On weighted L^2 estimates of solutions to wave equations*, J. Anal. Math. **72** (1997), 127–140.

[51] T. Hoshiro, *Decay and regularity for dispersive equations with constant coefficients*, J. Anal. Math. **91** (2003), 211–230.

[52] T. Ikebe, *Eigenfunction expansions associated with the Schrödinger operators and their application to scattering theory*, Arch. Rat. Mech. Anal. **5** (1960), 1–34.

[53] T. Ikebe and Y. Saito, *Limiting absorption method and absolute continuity for the Schrödinger operators*, J. Math. Kyoto Univ. Ser. A **7** (1972), 513–542.

[54] T. Ikebe and T. Tayoshi, *Wave and scattering operators for second-order elliptic operators in* $\mathbb{R}^n$, Publ. RIMS Kyoto Univ. Ser. A **4** (1968), 483–496.

[55] A.D. Ionescu and W. Schlag, *Agmon-Kato-Kuroda theorems for a large class of perturbations*, Duke Math. J. **131** (2006), 397–440.

[56] A. Jensen and T. Kato, *Spectral properties of Schrödinger operators and time-decay of the wave functions*, Duke Math. J. **46** (1979), 583–611.

[57] M. Kadowaki, *Low and high energy resolvent estimates for wave propagation in stratified media and their applications*, J. Differential Equations **179** (2002), 246–277.

[58] M. Kadowaki, *Resolvent estimates and scattering states for dissipative systems*, Publ. RIMS, Kyoto Univ. **38** (2002), 191–209.

[59] T. Kato, *Perturbation Theory for Linear Operators*, Springer-Verlag, 1966.

[60] T. Kato, *Smooth operators and commutators*, Studia Math. **31** (1968), 535–546.

[61] T. Kato and S.T. Kuroda, *The abstract theory of scattering*, Rocky Mountain J. Math. **1** (1971), 127–171.

[62] T. Kato and K. Yajima, *Some examples of smooth operators and the associated smoothing effect*, Rev. Math. Phys. **1** (1989), 481–496.

[63] K. Kikuchi and H. Tamura, *The limiting amplitude principle for acoustic propagators in perturbed stratified fluids*, J. Differential Equations **93** (1991), 260–282.

[64] V.G. Maz'ya and T.O. Shaposhnikova, *Theory of Sobolev Multipliers*, Springer-Verlag, 2008.

[65] K. Mochizuki, *Scattering theory for wave equations with dissipative terms*, Publ. RIMS, Kyoto Univ. **12** (1976), 383–390.

[66] C.S. Morawetz, *Time decay for the Klein-Gordon equation*, Proc. Roy. Soc. Ser. A **306** (1968), 291–296.

[67] K. Morii, *Time-global smoothing estimates for a class of dispersive equations with constant coefficients*, Ark. Mat. **46** (2008), 363–375.

[68] E. Mourre, *Absence of singular continuous spectrum for certain self-adjoint operators*, Comm. Math. Phys. **78** (1980/81), 391–408.

[69] M. Murata and T. Tsuchida, *Asymptotics of Green functions and the limiting absorption principle for elliptic operators with periodic coefficients*, J. Math. Kyoto Univ. **46** (2006), 713–754.

[70] P. Perry, I.M. Sigal and B. Simon, *Spectral analysis of N-body Schrödinger operators*, Ann. of Math. **114** (1981), 519–567.

[71] B. Perthame and L. Vega, *Morrey-Campanato estimates for Helmholtz equations*, J. Funct. Anal. **164** (1999), 340–355.

[72] B. Perthame and L. Vega, *Energy decay and Sommerfeld condition for Helmholtz equation with variable index at infinity*, preprint (2002).

[73] A.Ja. Povzner, *The expansion of arbitrary functions in terms of eigenfunctions of the operator* $-\Delta u + cu$, Amer. Math. Soc. Transl. Ser. 2 **60** (1966), 1–49 (original in Russian: Mat. Sb. **32** (1953), 109–156).

[74] A.G. Ramm, *Justification of the limiting absorption principle in* $\mathbb{R}^2$, in *Operator Theory and Applications* (A.G. Ramm, P.N. Shivakumar and A.V. Strauss, eds.), Fields Inst. Commun., vol. 25, Amer. Math. Soc., pp. 433–440, 2000.

[75] M. Reed and B. Simon, *Methods of Modern Mathematical Physics IV*, Academic Press, 1978.

[76] D. Robert, *Asymptotique de la phase de diffusion à haute énergie pour des perturbations du second ordre du laplacien*, Ann. Sci. École Norm. Sup. (4) **25** (1992), 107–134.

[77] M. Ruzhansky and M. Sugimoto, *Global L^2-boundedness theorems for a class of Fourier integral operators*, Comm. Partial Differential Equations **31** (2006), 547–569.

[78] M. Ruzhansky and M. Sugimoto, *A smoothing property of Schrödinger equations in the critical case*, Math. Ann. **335** (2006), 645–673.

[79] Y. Saito, *Spectral Representations for Schrödinger Operators with Long-Range Potentials*, Lecture Notes in Math., vol. 727, Springer-Verlag, 1979.

[80] B. Simon, *Best constants in some operator smoothness estimates*, J. Funct. Anal. **107** (1992), 66–71.

[81] C.D. Sogge, *Lectures on Non-Linear Wave Equations*, International Press, 2008.

[82] W.A. Strauss, *Nonlinear Wave Equations*, CBMS Reg. Conf. Ser. in Math., vol. 73, Amer. Math. Soc., 1989.

[83] R.S. Strichartz, *Restrictions of Fourier transforms to quadratic surfaces and decay of solutions of wave equations*, Duke Math. J. **44** (1977), 705–714.

[84] M. Sugimoto, *Global smoothing properties of generalized Schrödinger equations*, J. Anal. Math. **76** (1998), 191–204.

[85] H. Tamura, *Resolvent estimates at low frequencies and limiting amplitude principle for acoustic propagators*, J. Math. Soc. Japan **41** (1989), 549–575.

[86] E.C. Titchmarsh, *Eigenfunction expansions associated with second-order differential equations*, Clarendon Press, 1946.

[87] T. Umeda, *Generalized eigenfunctions of relativistic Schrödinger Operators I*, Electron. J. Differential Equations **127** (2006), 1–46.

[88] G. Vodev, *Local energy decay of solutions to the wave equation for short-range potentials*, Asymptot. Anal. **37** (2004), 175–187.

[89] B.G. Walther, *A sharp weighted L^2-estimate for the solution to the time-dependent Schrödinger equation*, Ark. Mat. **37** (1999), 381–393.

Matania Ben-Artzi
Institute of Mathematics
Hebrew University
Jerusalem 91904, Israel
e-mail: mbartzi@math.huji.ac.il

Progress in Mathematics, Vol. 301, 41–52

Kinetic Models of Chemotaxis

Nikolaos Bournaveas and Vincent Calvez

Abstract. We review some recent results on global existence and blow-up for kinetic models of chemotaxis and present new blow-up results.

Mathematics Subject Classification. Primary 35F10; Secondary 92C17, 82C40.

Keywords. Kinetic equations, chemotaxis, Strichartz estimates, blow-up.

1. Introduction

Chemotaxis is the directed motion of cells towards higher concentrations of chemo-attractants and plays a very important role in many biological processes. Experiments show that swimming bacteria such as *E.coli* undergo an erratic motion at the microscopic level and move in a series of 'run' and 'tumble'. A 'run' is motion in a straight line, and typically, a run in a favourable direction will be long and a run in an unfavourable direction will be short. At the end of a run the cell stops and 'tumbles' and this results in a reorientation and a new run in a new direction. Since the mean time for tumbling is much shorter than the mean time for running, we can model chemotaxis as a velocity jump process.

At the macroscopic level chemotaxis is described by systems of parabolic equations, the most famous of which is the Keller-Segel model. In its simplest form it is written as:

$$\partial_t n = \Delta n - \chi \nabla \cdot (n \nabla c), \tag{1.1a}$$

$$-\Delta c = n, \tag{1.1b}$$

where x varies in the whole space $\mathbb{R}^d$, $t \geq 0$, $n(t,x)$ denotes the density of the cells and $c(t,x)$ is the concentration of the chemoattractant. The first term on the right-hand side of (1.1a) represents the tendency of the cells to diffuse under their own Brownian motion and the second term their tendency to aggregate due to the presence of the chemoattractant. If $d = 1$ then diffusion is stronger than aggregation and solutions exist globally [21, 16]. If $d = 2$ then the two tendencies are evenly balanced. A typical result in this case is that we have global existence if the total mass of the cells is small and blow-up in finite time if it is large.

The critical value for the mass turns is $8\pi/\chi$ [2, 3, 14]. The proof of blow-up for $M > 8\pi/\chi$ relies on the identity

$$\frac{d}{dt}\int_{\mathbb{R}^2}|x|^2\,n(t,x)dx = 4M\left(1 - \frac{\chi}{8\pi}M\right). \tag{1.2}$$

The proof of global existence for $M < 8\pi/\chi$ uses the fact that the quantity

$$\mathcal{E}(t) = \int_{\mathbb{R}^2} n\,\log n\,dx + \frac{\chi}{4\pi}\int_{\mathbb{R}^2\times\mathbb{R}^2} n(t,x)\,n(t,y)\,\log|x-y|\,dxdy \tag{1.3}$$

is decreasing in time, since

$$\frac{d\mathcal{E}}{dt} = -\int_{\mathbb{R}^2} n\,|\nabla\log n - \chi\nabla S|^2\,dx, \tag{1.4}$$

together with the logarithmic Hardy-Littlewood-Sobolev inequality [1]

$$\frac{M}{2}\int_{\mathbb{R}^2} n\,\log n\,dx + \int_{\mathbb{R}^2\times\mathbb{R}^2} n(x)n(y)\,\log|x-y|\,dxdy \geq C(M). \tag{1.5}$$

We refer the reader to [17, 23] for more information on the Keller-Segel and other macroscopic models and to [8, 9] for the latest developments.

At the mesoscopic level chemotaxis is modelled by a kinetic equation for the density $f(t,x,v)$ of the cells coupled to a parabolic or elliptic equation for the concentration $S(t,x)$ of the chemoattractant. We focus our attention on the Othmer-Dunbar-Alt model [22]. In its simplest form it is written as:

$$\partial_t f + v\cdot\nabla_x f = \int_V T[S](t,x,v,v')f(t,x,v')dv' \tag{1.6a}$$

$$-\int_V T[S](t,x,v',v)f(t,x,v)dv', \tag{1.6b}$$

$$f(0,x,v) = f_0(x,v), \tag{1.6c}$$

$$-\Delta S = \rho := \int_V f(t,x,v)dv. \tag{1.6d}$$

Here $T[S](t,x,v,v') \geq 0$ is the turning kernel, a measure of the frequency of turning from velocity v' to velocity v at position x and time t. We'll discuss examples of interesting turning kernels below. For the sake of simplicity we take the velocity space V to be the unit ball. As in the case of the Keller-Segel system the total mass of the cells $M = \int_{\mathbb{R}^2\times V} f(t,x,v)dxdv$ is conserved, but there is no analogue of the energy $\mathcal{E}$ in (1.3).

The nonlinear IVP for (1.6) was first studied in [12]. The authors proved global existence of weak solutions in three dimensions under the hypothesis

$$0 \leq T[S](t,x,v,v') \leq C\left(1 + S(t,x+v) + S(t,x-v')\right)$$

and for initial data $0 \leq f_0 \in L^1 \cap L^\infty$. The proof uses the dispersive character of the equation for f and bootstraps higher $L^p_{x,v}$-norms starting from the base case $p = 1$ that corresponds to conservation of mass. The same method was used in [18] to prove global existence of weak solutions in two dimensions under the hypothesis

$$0 \leq T[S](t, x, v, v') \leq C(1 + S(t, x + v) + S(t, x - v')$$
$$+ |\nabla S(t, x + v)| + |\nabla S(t, x - v')|), \quad (1.7)$$

and in three dimensions under either of the hypotheses

$$0 \leq T[S](t, x, v, v') \leq C(1 + S(t, x + v) + |\nabla S(t, x + v)|)$$

or

$$0 \leq T[S](t, x, v, v') \leq C(1 + S(t, x - v') + |\nabla S(t, x - v')|).$$

As was pointed out in [18], the terms $S(t, x - v')$ and $S(t, x + v)$ require the use of different dispersion estimates with different integrability exponents. When both terms are present it is difficult to find one set of exponents that makes both dispersion estimates work.

2. Strichartz and dispersion estimates for the kinetic transport equation

Strichartz and dispersion estimates have proved to be very useful in the study of nonlinear equations. We refer the reader to [24, 27] and the references there in.

Strichartz and dispersion estimates for the kinetic transport equation were proved in [10] and the first application to a nonlinear kinetic model was given in [4]. They are as follows:

Proposition 2.1. (*Dispersion estimate,* [10]) *Let* $f_0 \in L^q(\mathbb{R}^d_x; L^p(\mathbb{R}^d_v))$ *where* $1 \leq q \leq p \leq \infty$, *and let* f *solve*

$$\partial_t f + v \cdot \nabla_x f = 0 \qquad (2.1)$$

with initial data $f(0, x, v) = f_0(x, v)$. *Then*

$$\|f(t)\|_{L^p(\mathbb{R}^d_x; L^q(\mathbb{R}^d_v))} \leq \frac{1}{|t|^{d\left(\frac{1}{q} - \frac{1}{p}\right)}} \|f_0\|_{L^q(\mathbb{R}^d_x; L^p(\mathbb{R}^d_v))}. \qquad (2.2)$$

Proposition 2.2. (*Strichartz estimates,* [10]) *Let* $d \geq 2$ *and let* $r, p, q, a \in [1, \infty]$ *satisfy the conditions*

$$p \geq q, \quad \frac{2}{r} = d\left(\frac{1}{q} - \frac{1}{p}\right) < 1, \quad a = \frac{2pq}{p+q} \leq 2. \qquad (2.3)$$

If $f(t, x, v)$ *solves*

$$\partial_t f + v \cdot \nabla_x f = g \ , \quad f(0, x, v) = 0, \qquad (2.4)$$

then

$$\|f\|_{L^r_t L^p_x L^q_v} \leq C\|g\|_{L^{r'}_t L^q_x L^p_v}. \qquad (2.5)$$

If $f(t, x, v)$ *solves*

$$\partial_t f + v \cdot \nabla_x f = 0 \ , \quad f(0, x, v) = f_0(x, v), \qquad (2.6)$$

then

$$\|f\|_{L^r_t L^p_x L^q_v} \leq C\|f_0\|_{L^a_{x,v}}. \qquad (2.7)$$

3. Global existence for some kinetic models of chemotaxis

We can use the estimates of Section 2 to improve the global existence results of [12] and extend to three dimensions the results of [18]. Using the dispersion estimate of Proposition 2.1 we can prove the following

Theorem 3.1. [4] *Let $d = 3$ and suppose that the (continuous) turning kernel $T[S]$ satisfies*

$$0 \le T[S](t, x, v, v') \le C\Big(1 + S(t, x + v) + S(t, x - v') + |\nabla S(t, x + v)|\Big). \quad (3.1)$$

Let $q \in (1, 3/2)$. Then there exists an exponent $p \in (3/2, 3)$ (depending on q) such that if the initial data $0 \le f_0 \in L^1(\mathbb{R}^6)$ is such that the norm $\|f_0(x - tv, v)\|_{L^p(\mathbb{R}^3_x; L^q(\mathbb{R}^3_v))}$ is finite for all $t > 0$, then the Cauchy problem (1.6) has a global weak solution f with $f(t) \in L^1(\mathbb{R}^6) \cap L^p(\mathbb{R}^3_x; L^q(V))$ for all $t \ge 0$.

The delocalization introduced by $x + v$ and $x - v'$ plays a fundamental role in the proof of this result. Observe that hypothesis (3.1) does not allow putting together the two gradients $\nabla S(t, x + v)$ and $\nabla S(t, x - v')$. This is still an open problem in three dimensions for large data. However, if we add the assumption that the critical $L^{3/2}_{x,v}$-norm of the initial data is sufficiently small then we can prove the following

Theorem 3.2. [4] *Let $d = 3$. Consider nonnegative initial data $f_0 \in L^1 \cap L^a$, where $\frac{3}{2} \le a \le 2$, and assume that $\|f_0\|_{L^a(\mathbb{R}^6)}$ is sufficiently small. Assume that the (continuous) turning kernel $T[S]$ satisfies the condition*

$$0 \le T[S](t, x, v, v') \le C \sum_{\pm} \Big[|S(t, x \pm v)| + |S(t, x \pm v')| +$$

$$|\nabla S(t, x \pm v)| + |\nabla S(t, x \pm v')|\Big]. \quad (3.2)$$

Then (1.6) has a global weak solution $f \in L^3_t\big([0, \infty); L^p(\mathbb{R}^3_x; L^q(V))\big)$, where $\frac{1}{p} = \frac{1}{a} - \frac{1}{9}$ and $\frac{1}{q} = \frac{1}{a} + \frac{1}{9}$. This result also holds if hypothesis (3.2) is replaced by the weaker: for all $p_1, p_2, p_3 \in [1, \infty]$ with $p_1 \ge \max(p_2, p_3)$, it holds

$$\|T[S](t, x, v, v')\|_{L^{p_1}_x L^{p_2}_v L^{p_3}_{v'}} \le C(|V|, p_2, p_3) \left[\|S(t, \cdot)\|_{L^{p_1}} + \|\nabla S(t, \cdot)\|_{L^{p_1}}\right]. \quad (3.3)$$

4. Critical nonlinearities

In this section we remove the delocalization that was so helpful in proving Theorem 3.1 and replace the right-hand side in hypothesis (3.1) by L^∞_x norms of S or ∇S. In other words, we are now considering models without direct memory effects. This leads to turning kernels that the Strichartz and dispersion estimates barely to handle. It turns out that these nonlinearities are critical.

In view of the results of Theorem 3.1 it is natural to ask whether global existence still holds in three dimensions if we assume that

$$0 \leq T[S](t, x, v, v') \leq C \left[1 + \|S(t, \cdot)\|_{L^\infty(\mathbb{R}^3)}^\alpha\right] \tag{4.1}$$

for some positive exponent α. In this case we can prove the following

Theorem 4.1. [5] *Let $d = 3$ and suppose that the turning kernel satisfies (4.1). Let the initial data satisfy $0 \leq f_0 \in L^1 \cap L^\infty$. Then:*

  i. *If $\alpha < 1$, then (1.6) has a global weak solution.*
  ii. *If $\alpha = 1$ and the critical $\|f_0\|_{L_{x,v}^{3/2}}$ is sufficiently small, then (1.6) has a global weak solution.*

The case of $\alpha = 1$ and large initial data in three dimensions is critical and remains open. In two dimensions having a term like $\|S(t, \cdot)\|_{L^\infty(\mathbb{R}^3)}^\alpha$ presents no problems.

Theorem 4.2. [5] *Let $d = 2$ and suppose that the turning kernel satisfies (4.1) with any $\alpha > 0$. Let the initial data satisfy $0 \leq f_0 \in L^1 \cap L^\infty$. Then (1.6) has a global weak solution.*

We can actually prove global existence even if we assume exponential growth of the nonlinearity in the sense

$$0 \leq T[S](t, x, v, v') \leq C \left(1 + \exp\left[\|S(t, \cdot)\|_{L^\infty(\mathbb{R}^2)}^\alpha\right]\right)$$

where $0 \leq \alpha < 1$. The case $\alpha = 1$ is critical and we can prove global existence if we assume in addition that the mass is small (the proof in [5] requires $M < \pi$). These results suggest that there is room for adding a gradient term to the turning kernel.

Theorem 4.3. [5] *Let $d = 2$ and suppose that the turning kernel satisfies*

$$0 \leq T[S](t, x, v, v') \leq C \left[1 + \|S(t, \cdot)\|_{L^\infty(\mathbb{R}^3)}^\alpha + \|\nabla S(t, \cdot)\|_{L^\infty(\mathbb{R}^3)}^\beta\right] \tag{4.2}$$

with any $\alpha > 0$ and $0 \leq \beta < 1$. Let the initial data satisfy $0 \leq f_0 \in L^1 \cap L^\infty$. Then (1.6) has a global weak solution.

The critical quantity $\|\nabla S(t, \cdot)\|_{L^\infty}$ in two dimensions behaves exactly as the critical quantity $\|S(t, \cdot)\|_{L^\infty}$ in three dimensions. Although we still don't know whether global existence for large data holds in this critical case in three dimensions, we do know that in two dimensions it doesn't, and that there exist critical turning kernels for which the solution blows up in finite time. We provide more details in the next section.

5. Blow-up in two dimensions

In this section we first recall some of the results of [6] and then prove some new blow-up results.

In view of the macroscopic theory discussed in Section 1 we expect that in dimension $d = 2$ solutions of (1.6) exist globally if the total mass of the cells is small, and blow-up in finite time if the total mass is large. Of course the behavior of the solution of the kinetic model depends not only on the total mass but also on the turning kernel. For example, for the turning kernel in (1.7), we have global existence even for large mass [18]. For simplicity of exposition we will always think of smooth compactly supported initial data and of solutions that decay sufficiently fast at infinity. We refer the reader to [4]–[6] for more precise assumptions.

The first blow-up result for a kinetic model of chemotaxis was given in [6]. Define

$$T_1[S](t, x, v, v') = (v \cdot \nabla S)_+ . \tag{5.1a}$$

Notice that T_1 vanishes when the velocity v after turning is more than 90 degrees away from ∇S and is positive when it is within 90 degrees from ∇S. We may say that under T_1 the cells always choose 'good directions'. It was shown in [6] that in the spherically symmetric case, if the total mass is sufficiently large then the solution can not exist globally.

We will show here that the same is true for the turning kernel

$$T_2[S](t, x, v, v') = |\nabla S| + v \cdot \nabla S. \tag{5.1b}$$

Notice that T_2 is small for velocities v near the direction of $-\nabla S$ and large for velocities near the direction of ∇S (and vanishes only in the exceptional case $v = -\frac{\nabla S}{|\nabla S|}$).

Clearly, equation (1.6d) doesn't determine S uniquely, so we need to clarify which solution we pick. Due to the choices in (5.1a) and (5.1b), only ∇S comes up in the equation for f, not S itself, so it is enough to specify ∇S. We make the obvious choice

$$\nabla S(t, x) = -\frac{1}{2\pi} \int_{\mathbb{R}^2} \rho(t, y) \frac{x - y}{|x - y|^2} dy. \tag{5.2}$$

We are now ready to state our first result.

Theorem 5.1. *A spherically symmetric solution of the kinetic model* (1.6) *with turning kernel* $T = T_1$ *or* T_2, *and sufficiently large mass* M, *cannot exist globally in time.*

Proof. Consider the positive quantity

$$I(t) = \int_{\mathbb{R}^2 \times V} |x|^2 f(t, x, v) dx dv = \int_{\mathbb{R}^2} |x|^2 \rho(t, x) dx.$$

Using the equation for f and the fact that the integral of the right-hand side with respect to v is zero, we have

$$\frac{dI}{dt} = 2 \int x \cdot v f dx dv.$$

Differentiating again we obtain (see [6] for more details)

$$\frac{d^2I}{dt^2} = 2 \int \int (x \cdot v)(-v \cdot \nabla f) dx dv \tag{5.3a}$$

$$+ 2 \int \int (x \cdot v) \int T[S](t, x, v, v') f(t, x, v') dv' dx dv \tag{5.3b}$$

$$- 2 \int \int (x \cdot v) \int T[S](t, x, v', v) f(t, x, v) dv' dx dv. \tag{5.3c}$$

For the term in (5.3a) we integrate by parts in x:

$$2 \int \int (x \cdot v)(-v \cdot \nabla f) dx dv = 2 \int \int |v|^2 f dx dv \le 2 \int \int f dx dv = 2M.$$

For the term in (5.3b) observe that $T[S](t, x, v, v')$ is independent of the velocity v', therefore

$$\text{term in } (5.3b) = 2 \int \int (x \cdot v) T[S](t, x, v, v') \rho(t, x) dx dv.$$

For $T = T_1$ we have

$$\text{term in } (5.3b) = 2 \int \rho(t, x) \int (x \cdot v) \, (v \cdot \nabla S(t, x))_+ dv \, dx = \frac{\pi}{4} \int \rho(t, x) \, x \cdot \nabla S(t, x) dx$$

and for $T = T_2$,

$$\text{term in } (5.3b) = 2 \int \rho(t, x) |\nabla S(t, x)| \int (x \cdot v) \, dv \, dx \tag{5.4}$$

$$+ 2 \int \rho(t, x) \int (x \cdot v) \, (v \cdot \nabla S(t, x)) \, dv \, dx \tag{5.5}$$

$$= \frac{\pi}{2} \int \rho(t, x) \, x \cdot \nabla S(t, x) dx. \tag{5.6}$$

Using $\nabla S(t, x) = -\frac{1}{2\pi} \int \rho(t, y) \frac{x-y}{|x-y|^2} dy$ and a simple argument familiar from the parabolic theory we find that $\int \rho x \cdot \nabla S dx = -\frac{1}{4\pi} M^2$, therefore

$$\text{term in } (5.3b) = -cM^2,$$

where $c = 1/16$ for $T = T_1$ and $c = 1/8$ for $T = T_2$. We see that the contribution of the first two terms in (5.3) is of the form $2M - cM^2$.

It remains to deal with the term in (5.3c). We have

$$\int T[S](t, x, v', v) f(t, x, v) dv' = c' |\nabla S(t, x)| f(t, x, v)$$

where $c' = 2/3$ if $T = T_1$ and $c' = \pi$ if $T = T_2$. Therefore,

$$\text{term in } (5.3c) = -2c' \int \int (x \cdot v) |\nabla S(t, x)| f(t, x, v) dx dv \tag{5.7}$$

$$= -2c' \int (x \cdot j(t, x)) |\nabla S(t, x)| dx, \tag{5.8}$$

where $j(t, x) = \int_V v f(t, x, v) dv$. Define $m(t, r) = \int_r^\infty \lambda \rho(t, \lambda) d\lambda$. Up to a constant, $m(t, r)$ is the mass outside a ball of radius r centered at the origin. It is easy to check that in the spherically symmetric case

$$|\nabla S(t, x)| = \frac{M/2\pi - m(t, r)}{r}, \quad x \cdot j(t, x) = \partial_t m(t, r)$$

therefore

$$\text{term in } (5.3c) = -4\pi c' \int_0^\infty (M/2\pi - m(t, r)) \partial_t m(t, r) dr \tag{5.9}$$

$$= -\frac{dP}{dt} \tag{5.10}$$

where

$$P(t) = -4\pi c' \int_0^\infty \left(\frac{M}{2\pi} m(t, r) - \frac{1}{2} m(t, r)^2 \right) dr. \tag{5.11}$$

Since $m(t, r) \leq \frac{M}{2\pi}$, we have $P \geq 0$. Putting everything together we have

$$\frac{d^2 I}{dt^2} \leq 2M - cM^2 - \frac{dP}{dt} \tag{5.12}$$

and integrating once we find

$$\frac{dI}{dt}(t) \leq \frac{dI}{dt}(0) + (2M - cM^2)t - P(t) + P(0). \tag{5.13}$$

Since $P(t) \geq 0$, it can be dropped from the rhs of (5.13). We get

$$\frac{dI}{dt}(t) \leq C_0 + (2M - cM^2)t$$

where $C_0 = \frac{dI}{dt}(0) + P(0)$. Integrating again we find

$$I(t) \leq I(0) + C_0 t + (2M - cM^2)\frac{t^2}{2}. \tag{5.14}$$

If $M > 32$ in the case of T_1, and $M > 16$ in the case of T_2, then (5.14) implies that $I(t)$ is negative for large t. It follows that the solution can not exist globally in time. $\qquad \square$

In the theorem above we have used the unit ball $\{|v| \leq 1\}$ as the velocity space. However, similar arguments can be used to show blow-up for T_1 and T_2 when the velocity space is the unit sphere $\{|v| = 1\}$. This is probably a more realistic choice from the biological point of view as cells move with almost constant speed. Developing Strichartz and dispersion estimates for the sphere is an interesting open problem.

In order to better understand the behavior of solutions that blow-up, we would like to have a more tractable model that retains the main features of the problem. One option is to replace ∇S in (5.1a) and (5.1b) by $-\frac{M}{2\pi} \frac{x}{|x|^2}$, which is what we would obtain from (5.2) if we replaced $\rho(y)$ by $\delta_0(y)$. (We could express this by saying that the signal thinks that all the mass is at the origin.) Then T_1 and T_2 become

$$\widetilde{T_1}(t,x,v,v') = \left(v \cdot \frac{-Mx}{2\pi|x|^2} \right)_+ \quad, \quad \widetilde{T_2}(t,x,v,v') = \frac{M}{2\pi|x|} \left(1 - \frac{v \cdot x}{|x|} \right). \quad (5.15)$$

(Although they are independent of t and v', we retain the notation (t,x,v,v') for the sake of clarity.) Our model now has a single equation, namely

$$\partial_t f + v \cdot \nabla_x f = \int_V T(t,x,v,v')f(t,x,v')dv' - \int_V T(t,x,v',v)f(t,x,v)dv' \quad (5.16)$$

where we choose again $V = \{|v| \le 1\}$.

Theorem 5.2. *A spherically symmetric solution of the kinetic model* (5.16) *with turning kernel* $T = \widetilde{T_1}$ *or* $\widetilde{T_2}$, *and sufficiently large mass* M, *cannot exist globally in time.*

Proof. We work as in the last theorem. The terms in (5.3a) and (5.3b) are treated in the same way and give $2M$ and $-cM^2$ respectively. For the term in (5.3c), if $T = \widetilde{T_1}$ we get

$$\int T(t,x,v',v)f(t,x,v)dv' = \frac{M}{3\pi|x|}f(t,x,v)$$

therefore

$$\text{term in } (5.3c) = -2\frac{M}{3\pi} \int\int \left(\frac{x}{|x|} \cdot v \right) f(t,x,v)dxdv$$
$$= -\frac{2M}{3\pi} \int \frac{x}{|x|} \cdot j(t,x)dx, \quad (5.17)$$

and using

$$x \cdot j(x) = rj(r) = \partial_t m \quad, \quad m(t,r) = \int_r^\infty \lambda\rho(t,\lambda)d\lambda, \quad (5.18)$$

we get

$$\text{term in } (5.3c) = -\frac{4M}{3} \int_0^\infty \frac{1}{r}\partial_t m(t,r)\, rdr = -\frac{dP}{dt}$$

where $P(t) = \frac{4M}{3} \int_0^\infty m(t,r)dr \ge 0$. The proof for $T = \widetilde{T_2}$ is similar. $\qquad\square$

We remark again that the ball $\{|v| \le 1\}$ can be replaced by the sphere $\{|v| = 1\}$. Several other variants are possible: the velocity v in (5.1a) and (5.1b) can be replaced by $\frac{v}{|v|}$, the operator $\partial_t + v \cdot \nabla_x$ can be replaced by $\partial_t + \frac{v}{|v|} \cdot \nabla_x$, etc.

6. Perspectives and open problems

The discussion in the previous sections raises many interesting questions.

In two dimensions we would like to know whether an optimal mass exists for the turning kernels T_1 and T_2, i.e., a constant M^* with the property that $M < M^*$ leads to global existence and $M > M^*$ leads to blow up. In the case of the Keller-Segel model the energy in (1.3) was used to find the critical mass, but no such quantity is known for the kinetic model.

In dimension three we would like to know whether blow-up is possible for critical turning kernels, i.e., for $T \simeq S$ in some sense (Theorem 4.1 says essentially that $T \simeq S^{1-\epsilon}$ leads to global existence). What quantity should be large to guarantee blow-up? What smallness conditions guarantee global existence? One expects $\|f_0\|_{L^{3/2}}$ to play a role here.

In all cases where blow-up occurs we would like to have a sharp estimate for the life span of the solution and to know what the solution looks like near the blow-up time. One expects Dirac masses to form but there is numerical evidence to suggest that the blow-up patterns may be much more interesting for kinetic models [28].

Kinetic models have been used to describe travelling pulses of bacteria. We refer the reader to [25, 26] for further information.

Acknowledgment: The authors acknowledge with pleasure several helpful discussions with Benoit Perthame.

The results in this paper were presented by the first author at the conference on *Asymptotic properties of solutions to hyperbolic equations* at Imperial College in March 2011. NB would like to thank the organisers for the invitation and financial support as well as for creating a stimulating environment for the exchange of ideas on hyperbolic PDEs.

References

[1] W. Beckner, Sharp Sobolev inequalities on the sphere and the Moser-Trudinger inequality. Ann. of Math. (2) 138 (1993), no. 1, 213–242.

[2] A. Blanchet, J. Dolbeault and B. Perthame, Two-dimensional Keller-Segel model: optimal critical mass and qualitative properties of the solutions, Electron. J. Differential Equations **44**, 32 pp. (electronic) (2006).

[3] A. Blanchet, J.A. Carrillo and N. Masmoudi, Infinite time aggregation for the critical Patlak-Keller-Segel model in $\mathbb{R}^2$. Comm. Pure Appl. Math. 61 (2008), no. 10, 1449–1481.

[4] N. Bournaveas, V. Calvez, S. Gutiérrez and B. Perthame, Global existence for a kinetic model of chemotaxis via dispersion and Strichartz estimates, Comm. Partial Differential Equations **33**, 79–95 (2008).

[5] N. Bournaveas and V. Calvez, Global existence for the kinetic chemotaxis model without pointwise memory effects, and including internal variables, Kinetic and Related Models **1**, 29–48 (2008).

[6] N. Bournaveas and V. Calvez, Critical mass phenomenon for a chemotaxis kinetic model with spherically symmetric initial data. Ann. Inst. H. Poincaré Anal. Non Linéaire 26 (2009), no. 5, 1871–1895.

[7] V. Calvez and L. Corrias, The parabolic-parabolic Keller-Segel model in $\mathbb{R}^2$, to appear in Comm. Math. Sci. (2008).

[8] V. Calvez, L. Corrias and M.A. Ebde, Blow-up, concentration phenomenon and global existence for the Keller-Segel model in high dimension, Communications on Partial Differential Equations, Volume 37, Issue 4, April 2012, 561–584.

[9] V. Calvez and J.A. Carrillo, Refined Asymptotics for the subcritical Keller-Segel system and Related Functional Inequalities, arXiv:1007.2837

[10] F. Castella and B. Perthame, Estimations de Strichartz pour les équations de transport cinétique, C. R. Acad. Sci. Paris Sér. I Math. **322**, 535–540 (1996).

[11] F.A.C.C. Chalub, Y. Dolak-Struss, P.A. Markowich, D. Oelz, C. Schmeiser and A. Soreff, Model hierarchies for cell aggregation by chemotaxis, Math. Models Methods Appl. Sci. **16**, 1173–1197 (2006).

[12] F.A.C.C. Chalub, P.A. Markowich, B. Perthame and C. Schmeiser, Kinetic models for chemotaxis and their drift-diffusion limits, Monatsh. Math. **142**, 123–141 (2004).

[13] Y. Dolak and C. Schmeiser, Kinetic models for chemotaxis: hydrodynamic limits and spatio-temporal mechanisms, J. Math. Biol. **51**, 595–615 (2005).

[14] J. Dolbeault and B. Perthame, Optimal critical mass in the two-dimensional Keller-Segel model in $\mathbb{R}^2$. C. R. Math. Acad. Sci. Paris 339 (2004), no. 9, 611–616.

[15] F. Filbet, Ph. Laurençot and B. Perthame, Derivation of hyperbolic models for chemosensitive movement, J. Math. Biol. **50**, 189–207 (2005).

[16] T. Hillen and A. Potapov, The one-dimensional chemotaxis model: global existence and asymptotic profile. Math. Methods Appl. Sci. 27 (2004), no. 15, 1783–1801.

[17] T. Hillen and K.J. Painter, A user's guide to PDE models for chemotaxis. J. Math. Biol. 58 (2009), no. 1-2, 183–217,

[18] H.J. Hwang, K. Kang and A. Stevens, Global solutions of nonlinear transport equations for chemosensitive movement, SIAM J. Math. Anal. **36**, 1177–1199 (2005).

[19] W. Jäger and S. Luckhaus, On explosions of solutions to a system of partial differential equations modelling chemotaxis, Trans. Amer. Math. Soc. **329**, 819–824 (1992).

[20] T. Nagai, Blow-up of radially symmetric solutions to a chemotaxis system, Adv. Math. Sci. Appl. **5**, 581–601 (1995).

[21] K. Osaki and A. Yagi, Finite-dimensional attractor for one-dimensional Keller-Segel equations. Funkcial. Ekvac. 44 (2001), no. 3, 441–469.

[22] H.G. Othmer, S.R. Dunbar and W. Alt, Models of dispersal in biological systems, J. Math. Biol. **26**, 263–298 (1988).

[23] B. Perthame, Transport equations in biology, Frontiers in Mathematics, Birkhäuser (2007).

[24] M. Ruzhansky and J. Smith, Dispersive and Strichartz estimates for hyperbolic equations with constant coefficients. MSJ Memoirs, 22. Mathematical Society of Japan, Tokyo, 2010. x+147 pp. ISBN: 978-4-931469-57-0.

[25] J. Saragosti, V. Calvez, N. Bournaveas, A. Buguin, P. Silberzan and B. Perthame, Mathematical description of bacterial traveling pulses. PLoS Comput. Biol. 6 (2010), no. 8, e1000890.

[26] J. Saragosti, V. Calvez, N. Bournaveas, B. Perthame, A. Buguin and P. Silberzan, Directional persistence of chemotactic bacteria in a traveling concentration wave, PNAS, 2011, 108 (39) 16235–16240.

[27] T. Tao, Nonlinear Dispersive Equations: Local and Global Analysis, AMS, 2006.

[28] N. Vauchelet, Numerical simulation of a kinetic model for chemotaxis, Kinetic and Related Models, 3 (2010), pp. 501–528.

Nikolaos Bournaveas
University of Edinburgh
School of Mathematics
JCMB, King's Buildings
Edinburgh, UK
e-mail: n.bournaveas@ed.ac.uk

Vincent Calvez
Unité de Mathématiques Pures et Appliquées
École Normale Supérieure de Lyon
UMR CNRS 5669
Lyon, France
e-mail: vincent.calvez@ens-lyon.fr

Progress in Mathematics, Vol. 301, 53–62

Modulus of Continuity and Decay at Infinity in Evolution Equations with Real Characteristics

Massimo Cicognani and Ferruccio Colombini

Abstract. In the hyperbolic Cauchy problem, the well-posedness in Sobolev spaces and the modulus of continuity of the coefficients are deeply connected. This holds true in the more general framework of evolution equations with real characteristics

$$D_t^2 u - \sum_{k=0}^{2p} a_k(t,x) D_x^k u = 0$$

($p = 1$ hyperbolic equations, $p = 2$ vibrating beam models,...) where a sharp scale of Hölder continuity, with respect to the time variable t, for the a_k's has been established.

We show that, for $p \geq 2$, a lack of regularity in t can be compensated by a decay as the space variable $x \to \infty$. This is not true in the hyperbolic case $p = 1$ because of the finite speed of propagation.

Mathematics Subject Classification. 35G10; 35L15.

Keywords. Evolution equations with real characteristics.

1. Introduction and main results

Let us consider the evolution operator

$$P := D_t^2 - \sum_{k=0}^{2p} a_k(t,x) D_x^k, \tag{1.1}$$

where $(t,x) \in [0,T] \times \mathbb{R}_x$, $D = \frac{1}{i}\partial$, and the positive integer p represents the evolution degree. The coefficients are at least such that

$$a_k \in C([0,T]; \mathcal{B}^\infty) \tag{1.2}$$

with $\mathcal{B}^\infty = \mathcal{B}^\infty(\mathbb{R}_x)$ the space of all bounded functions $a(x)$ together with all their derivatives.

We are interested in the Cauchy problem

$$\begin{cases} Pu = 0 \\ u(0,x) = u_0(x), \ D_t u(0,x) = u_1(x) \end{cases} \tag{1.3}$$

with Cauchy data in Sobolev spaces

$$u_0 \in H^s, \ u_1 \in H^{s-p}$$

and look for solutions which remain in Sobolev spaces at any time t. Precisely, we say that problem (1.3) is L^2 well posed if there exists a unique solution

$$u \in C([0,T];H^s) \cap C^1([0,T];H^{s-p})$$

and we say that it is well posed with a loss of derivatives if the unique solution is such that

$$u \in C([0,T];H^{s-\delta}) \cap C^1([0,T];H^{s-p-\delta})$$

for some positive δ.

From the Lax-Mizohata theorem, forward and backward well-posedness with Cauchy data at $t = t_0$ implies that the roots of

$$\tau^2 - a_{2p}(t,x)\xi^{2p}$$

are real at $t = t_0$. We assume

$$a_{2p}(t,x) \geq \lambda > 0 \tag{1.4}$$

in order to have real and distinct characteristics at any t. In particular, for $p = 1$ the operator P is strictly hyperbolic and $Pu = 0$ reduces to the wave equations

$$u_{tt} - a(t,x)u_{xx} + b(t,x)u_x + c(t,x)u = 0.$$

For $p = 2$ we have the beam models

$$u_{tt} + a(t,x)u_{xxxx} + b(t,x)u_{xxx} + c(t,x)u_{xx} + d(t,x)u_x + e(t,x)u = 0$$

and the related, by algebraic factorization of the symbol, first-order (in time) equation

$$\frac{1}{i}u_t + \alpha(t,x)u_{xx} + \beta(t,x)u_x + \gamma(t,x)u = 0$$

is a Schrödinger equation.

Starting from [3], we know that well-posedness and modulus of continuity of the coefficients are deeply connected in the hyperbolic Cauchy problem. For the wave equation

$$u_{tt} - a(t)u_{xx} = 0,$$

$a(t) \geq \lambda > 0$, the Lipschitz condition

$$|a(t+\tau) - a(t)| \leq L|\tau|$$

is sharp for L^2 well-posedness while the Log-Lipschitz regularity

$$|a(t+\tau) - a(t)| \leq L|\tau|\,|\log|\tau||$$

is the natural threshold for well-posedness with a loss of derivatives. Faster oscillations of the leading coefficient $a(t)$, allowed by any weaker continuity, may force the solution to get out of the distributions space. Coefficients in possible lower-order terms can be merely continuous and complex valued.

In [1], we have found the sharp relation between well-posedness and modulus of continuity, with respect to the time variable t, of the real parts of the coefficients in the case of a higher evolution degree $p > 1$.

Theorem 1.1 ([1]). *The Cauchy problem*

$$\begin{cases} D_t^2 u - \sum_{k=0}^{2p} a_k(t) D_x^k u = 0 \\ u(0, x) = u_0(x), \quad D_t u(0, x) = u_1(x), \end{cases} \tag{1.5}$$

is L^2 well posed assuming that the coefficients $a_k(t)$ with $k \geq p + 1$ are real, $a_{2p}(t) \geq \lambda > 0$, and are in the sharp scale

$$a_k \in C^{0,(k-p)/p}([0, T]; \mathbb{R}), \quad k = p + 1, \dots, 2p, \tag{1.6}$$

of Hölder continuity.
If the coefficients satisfy the Log-Hölder conditions

$$|a_k(t + \tau) - a_k(t)| \leq L|\tau|^{(k-p)/p} |\log|\tau||, \quad k = p + 1, \dots, 2p, \tag{1.7}$$

then the Cauchy problem (1.5) is well posed with a loss of derivatives.

The coefficients a_k with $k \leq p$ can be merely continuous and complex valued.

As far as general coefficients are concerned, $a_k = a_k(t, x)$ for $k \leq 2p$, $a_k(t, x) \in \mathbb{C}$ for $k \leq 2p - 1$, in the limit hyperbolic case $p = 1$ the L^2 well-posedness for the wave equation

$$D_t^2 u - a(t, x) D_x^2 u = 0$$

with $a \in C^{0,1}([0, T]; \mathcal{B}^\infty)$ is now quite standard. Of course, lower-order terms with coefficients in $C([0, T]; \mathcal{B}^\infty)$ can be considered also. For instance, one may apply the energy method to an equivalent first-order system

$$\partial_t U + \begin{pmatrix} i\sqrt{a(t, x)} D_x & 0 \\ 0 & -i\sqrt{a(t, x)} D_x \end{pmatrix} U + A(t, x, D_x) U = 0,$$

$A(t, x, \xi) \in L^\infty([0, T]; S^0)$, thanks to the boundedness in L^2 of the operators A and $i\sqrt{a(t, x)} D_x + (i\sqrt{a(t, x)} D_x)^* = -\frac{a_x}{2\sqrt{a}}$.

We refer to [4] for the case of a Log-Lipschitz coefficient $a(t, x)$. Also the isotropic case

$$|a(t + \tau, x + \eta) - a(t, x)| \leq L(|\tau||\log|\tau|| + |\eta||\log|\eta||)$$

is studied there with data and solutions in suitable Sobolev spaces.

Here we consider the case $p = 2$ in (1.1) taking a beam equation

$$D_t^2 u - a(t, x) D_x^4 u + b(t, x) D_x^3 u = 0. \tag{1.8}$$

Terms in $D_x^k u$ of order $k \leq 2$ can be added without any further assumption on them besides (1.2).

At the symbol level, we have the factorization

$$\tau^2 - a(t, x)\xi^4 + b(t, x)\xi^3 = \left(\tau - \sqrt{a}\xi^2 - \frac{b}{2\sqrt{a}}\xi\right)\left(\tau + \sqrt{a}\xi^2 + \frac{b}{2\sqrt{a}}\xi\right),$$

modulo terms of order 2 in the variable ξ, but now the operator

$$\left(i\sqrt{a}D_x^2 + i\frac{b}{2\sqrt{a}}D_x\right) + \left(i\sqrt{a}D_x^2 + i\frac{b}{2\sqrt{a}}D_x\right)^*$$

is of order 1 with principal part

$$-\frac{1}{\sqrt{a}}(a_x + \Im b)D_x.$$

In dealing with the elementary Schrödinger model equation

$$\frac{1}{i}u_t + u_{xx} + \beta(t,x)u_x = 0,$$

one finds that the boundedness of the integral function

$$\int_0^x \Re\beta(t,y)dy, \ x \in \mathbb{R},$$

is a necessary condition for L^2 well-posedness. If $\Re\beta$ is differentiable in t, then it is a sufficient condition too. In the case that $\Re\beta$ is merely continuous in t, we have the stronger sufficient decay condition

$$|\Re\beta(t,x)| \le g(\langle x \rangle)$$

with a decreasing function $g \in L^1(\mathbb{R}_+)$, $\langle x \rangle = \sqrt{1+x^2}$. We refer to [7], [8] for more general results.

Assuming first the sharp regularity in t for $a(t,x)$ and $\Re b(t,x)$ in (1.8), we can prove the following result:

Theorem 1.2. *The Cauchy problem*

$$\begin{cases} D_t^2 u - a(t,x)D_x^4 u + b(t,x)D_x^3 u = 0, \\ u(0,x) = u_0(x), \ D_t u(0,x) = u_1(x), \end{cases} \tag{1.9}$$

is L^2 well posed under the assumptions (1.4) for $a(t,x)$,

$$a \in C^{0,1}([0,T];\mathcal{B}^\infty), \ \Re b \in C^{0,1/2}([0,T];\mathcal{B}^\infty), \tag{1.10}$$

and

$$|a_x(t,x)| + |\Im b(t,x)| \le g(\langle x \rangle) \tag{1.11}$$

with a positive decreasing function g such that

$$g \in L^1(\mathbb{R}_+). \tag{1.12}$$

No further regularity in time is needed for $\Im b$ besides continuity. This suggests that a decay as $x \to \infty$ may compensate low regularity in t.

With $a(t,x)$ Log-Lipschitz and $\Re b(t,x)$ Log-Hölder functions in the variable t, we have well-posedness with a loss of derivatives.

In the case $a = a(t)$ and $b \in C^{0,1/2}([0,T];\mathcal{B}^\infty)$, Theorem 1.2 follows from Theorem 1.6 in [2], a complete proof will appear in a future work. Our aim in this paper is to show that a lack up to $1/2 = 1/p$ of Hölder regularity in time for $a(t,x)$

may be compensated by a, weaker than L^1, precise decay as $x \to \infty$. A typical example is given by

$$a(t, x) = 1 + \langle x \rangle^{-\delta} w(t),$$

$w \in C^{0,\alpha}$, $1/2 < \alpha < 1$, $|w(t)| < \lambda < 1$, that satisfies

$$|a(t + \tau, x) - a(t, x)| \leq L|\tau|^{\alpha} \langle x \rangle^{-\delta}$$

with a damping of the too fast oscillations in t by the amplitude $\langle x \rangle^{-\delta}$ as $x \to \infty$. The precise relation between the lack of Hölder regularity and the decay is given by

$$\delta = 2(1 - \alpha).$$

Here we prove the following:

Theorem 1.3. *The Cauchy problem*

$$\begin{cases} D_t^2 u - a(t, x) D_x^4 u = 0, \\ u(0, x) = u_0(x), \ D_t u(0, x) = u_1(x), \end{cases} \tag{1.13}$$

is L^2 well posed under the assumptions (1.4) for $a(t, x)$, (1.11) and (1.12) for $a_x(t, x)$, and with each C^{0,α_j} Hölder part a_{α_j} in

$$a(t, x) = \sum_{j=1}^{N} a_{\alpha_j}(t, x), \quad 1/2 < \alpha_j \leq 1, \tag{1.14}$$

that satisfies

$$|a_{\alpha_j}(t + \tau, x) - a_{\alpha_j}(t, x)| \leq L|\tau|^{\alpha_j} \langle x \rangle^{-2(1-\alpha_j)}, \ j = 1, \ldots, N. \tag{1.15}$$

In line with Theorem 1.2, the Lipschitz part $a_1(t, x)$ does not need to decay as $x \to \infty$.

In [1], the sharpness of Lipschitz regularity is proved by constructing a coefficient $w(t)$,

$$w \in \bigcap_{\alpha < 1} C^{0,\alpha}, \ |w(t)| \leq 1/2, \tag{1.16}$$

such that the Cauchy problem for the operator

$$D_t^2 - (1 + w(t)) D_x^4$$

is not well posed. Damping the oscillations of $w(t)$ by a factor $\langle x \rangle^{-\varepsilon}$ we now turn to L^2 well-posedness by Theorem 1.3.

The construction of $w(t)$ is inspired to the counter-examples of [3] and [6] for the hyperbolic operator

$$D_t^2 - (1 + w(t)) D_x^2.$$

Is it possible to get well-posedness assuming some, even strong, decay for the coefficient $g(x)w(t)$ in the model operator

$$D_t^2 - (1 + g(x)w(t)) D_x^2$$

to compensate the lack of Lipschitz regularity also in this case? The answer is no, because of the finite speed of propagation that characterizes hyperbolic equations

also with non-differentiable coefficients, see [5]. In fact, it is possible to construct
a coefficient $w(t)$ that satisfies (1.16) and a solution $u(t,x)$ of

$$D_t u^2 - (1 + w(t))D_x^2 u = 0$$

such that

$$u \in C^2(\mathbb{R}, (\gamma^\sigma)')$$

for every $\sigma > 1$, where $(\gamma^\sigma)'$ is the space of Gevrey ultradistributions of index σ,
and

$$u(0,x), u_t(0,x) \in \mathcal{B}^\infty(\mathbb{R}_x)$$

but $u(t,x)$ for $t > 0$ does not belong to any $H^s(I)$, with I a bounded neighborhood
of the origin $x = 0$. If we take a compactly supported cut-off function $\chi(x) \in \gamma^\sigma(\mathbb{R})$
for $\sigma > 1$, $\chi(x) = 1$ in $J \supset I$, then the solution $v(t,x)$ of

$$D_t^2 v - (1 + \chi(x)w(t))D_x^2 v = 0,$$

with the same Cauchy data, coincides with $u(t,x)$ for $x \in I$ and sufficiently small
t hence also $v(t,x)$ for $t > 0$ does not belong to $H^s(I)$ for any s.

In conclusion, it is possible to get advantage from the decay as the space
variable $x \to \infty$, and even to compensate a less regularity with respect to the time
variable t, only for $p > 1$ when the speed of propagation is not finite.

2. Proof of Theorem 1.3

We outline the main ideas in the case that the decomposition (1.14) reduces to a
single term $a(t,x)$ satisfying

$$|a(t + \tau, x) - a(t,x)| \le L|\tau|^\alpha \langle x \rangle^{-2(1-\alpha)}, \ 1/2 < \alpha \le 1. \tag{2.1}$$

The general case requires only a heavier notation.

First we regularize the non differentiable function $a(t,x)$ defining

$$\tilde{a}(t,x,\xi) = \int a(\tau,x)\varrho((t-\tau)\langle\xi\rangle^2))\langle\xi\rangle^2 d\tau, \tag{2.2}$$

with $\varrho \in C_0^\infty$, $0 \le \varrho(y) \le 1$, $\int_{-\infty}^{+\infty} \varrho(y)dy = 1$, $a(\tau,x) = a(0,x)$ for $\tau < 0$,
$a(\tau,x) = a(T,x)$ for $\tau > T$.
From (2.1), we have

$$|a(t,x) - \tilde{a}(t,x,\xi)| \le L\langle x \rangle^{-2(1-\alpha)}\langle\xi\rangle^{-2\alpha}, \tag{2.3}$$

$$|D_t\tilde{a}(t,x,\xi)| \le L\langle x \rangle^{-2(1-\alpha)}\langle\xi\rangle^{2(1-\alpha)}. \tag{2.4}$$

This gives the factorization

$$D_t^2 - a(t,x)D_x^4 = (D_t - \lambda(t,x,D_x))(D_t + \lambda(t,x,D_x)) + b(t,x,D_x)\langle D_x \rangle^2, \tag{2.5}$$

where the symbol

$$\lambda(t,x,\xi) = \sqrt{\tilde{a}(t,x,\xi)}\xi^2 \tag{2.6}$$

belongs to the Hörmander class $C([0,T]; S^2)$ and $b \in C([0,T]; S^{2(1-\alpha)})$ satisfies

$$|b(t,x,\xi)| \leq L\langle x \rangle^{-2(1-\alpha)}\langle \xi \rangle^{2(1-\alpha)}. \tag{2.7}$$

Defining $U =^t (v_0, v_1)$,

$$v_0 = \langle D_x \rangle^2 u, \quad v_1 = (D_t + \lambda(t,x,D_x))u,$$

and using again (2.3) after a standard diagonalization procedure of the principal part, the scalar equation $D_t^2 u - a(t,x)D_x^4 u = 0$ is equivalent to a 2×2 system $SU = 0$ with

$$S := \partial_t + \begin{pmatrix} i\sqrt{a(t,x)}D_x^2 & 0 \\ 0 & -i\sqrt{a(t,x)}D_x^2 \end{pmatrix} + B(t,x,D_x), \tag{2.8}$$

where the symbol full matrix $B \in C([0,T]; S^{2(1-\alpha)})$ is such that

$$|B(t,x,\xi)| \leq L\langle x \rangle^{-2(1-\alpha)}\langle \xi \rangle^{2(1-\alpha)}. \tag{2.9}$$

In order to obtain the L^2 well-posedness of the Cauchy problem for the system S with the energy method, our aim is to make

$$\begin{pmatrix} i\sqrt{a}D_x^2 & 0 \\ 0 & -i\sqrt{a}D_x^2 \end{pmatrix} + B$$

a bounded from below operator by means of a change of variable. In dealing with the operator

$$(\pm i\sqrt{a}D_x^2) + (\pm i\sqrt{a}D_x^2)^* = \mp\frac{a_x}{\sqrt{a}}D_x$$

we use (1.11) and (1.12) for $a_x(t,x)$.

For the full operator matrix B of order $2(1-\alpha) < 1$, we use the sharp Gårding inequality for systems and the decay condition (2.9). In particular, the part of $B(t,x,\xi)$ supported in a zone

$$\langle \xi \rangle_h \leq N\langle x \rangle, \quad \langle \xi \rangle_h = \sqrt{h^2 + \xi^2},$$

of the phase-space is of order 0. So, we need to make non-negative only the eigenvalues of

$$\chi\left(N\langle x \rangle/\langle \xi \rangle_h\right)(B(t,x,\xi) + B^*(t,x,\xi))$$

with $\chi(y)$ a cut-off function equal to 1 for $0 \leq y \leq 1$.

Correspondingly, in order to perform the change of variable, we define

$$\Lambda(x,\xi) = \Lambda_0(x,\xi) + \Lambda_1(x,\xi) \tag{2.10}$$

with

$$\Lambda_0(x,\xi) = K\omega(\xi/h)\int_0^x g(\langle y \rangle)dy, \tag{2.11}$$

$$\Lambda_1(x,\xi) = M\langle \xi \rangle^{-(2\alpha-1)}\omega(\xi/h)\int_0^x \langle y \rangle^{-2(1-\alpha)}\chi\left(N\langle y \rangle/\langle \xi \rangle_h\right)dy, \tag{2.12}$$

where $\omega(\eta)$ is a smooth function such that $\eta\omega(\eta) = |\eta|$ for $|\eta| \geq 1$, $g(\langle x \rangle)$ is the function in condition (1.11) for $a_x(t,x)$, the large parameters $K, h, M, N, M > K$, are to be fixed later. We can assume $g \in \mathcal{B}^\infty$ without any loss of generality.

For $|\xi| \geq h$, the symbols

$$\xi \partial_x \Lambda_0 = K|\xi|g(\langle x \rangle), \tag{2.13}$$

and

$$\xi \partial_x \Lambda_1 \geq \frac{M}{\sqrt{2}} \langle \xi \rangle^{2(1-\alpha)} \langle x \rangle^{-2(1-\alpha)} \chi \left(N \langle x \rangle / \langle \xi \rangle_h \right), \tag{2.14}$$

dominate the Hermitian parts of $i\sqrt{a}D_x^2$ and χB respectively.

Λ_0 is of order 0 thanks to $g \in L^1(\mathbb{R}_+)$, $\partial_\xi \Lambda_0$ is of order $-\infty$ because it is compactly supported in $|\xi| \leq h$. With $M \leq N^{2\alpha-1}$, Λ_1 is a family of uniformly bounded symbols of order 0. Taking the stronger inequality $MK \leq M^2 \leq N^{2\alpha-1}$, we have also uniform estimates

$$|\partial_\xi \Lambda_1 \partial_x \Lambda_0| \leq C \langle \xi \rangle_h^{-1} g(\langle x \rangle), \quad |\partial_\xi \Lambda_1 \partial_x \Lambda_1| \leq C \langle \xi \rangle_h^{-2\alpha} \langle x \rangle^{-2(1-\alpha)} \tag{2.15}$$

with a constant C independent of all parameters h, K, M, N.

The role of the parameter h is to provide invertibility for the operators $e^{\pm \Lambda}$. We have

$$e^{\Lambda} e^{-\Lambda} = I + R,$$

where the full symbol of the operator R satisfies

$$\left| \partial_\xi^\alpha \partial_x^\beta R(x, \xi) \right| \leq C(\alpha, \beta, K, M, N) \langle \xi \rangle_h^{-1-\alpha}$$

with constants $C(\alpha, \beta, K, M, N)$ independent of h. Thus, we can choose the parameter h in function of K, M, N in order to have an invertible operator $I + R$ by Neumann series.

With such a choice of h, taking (2.15) into account, the operators $e^{\pm \Lambda}$ are invertible with

$$(e^{\pm \Lambda})^{-1} = e^{\mp \Lambda}(I + R_{-1}^{\pm} + R_{-2\alpha}^{\pm} + R_{-2}^{\pm}), \tag{2.16}$$

where

$$|R_{-1}^{\pm}(x, \xi)| \leq C \langle \xi \rangle^{-1} g(\langle x \rangle), \quad |R_{-2\alpha}^{\pm}(x, \xi)| \leq C \langle \xi \rangle^{-2\alpha} \langle x \rangle^{-2(1-\alpha)}, \tag{2.17}$$

uniformly with respect to the large parameters K, M, N, and $R_{-2}^{\pm}$ is of order less than -2.

Taking also the uniform estimate

$$|\xi^2 \partial_\xi \Lambda_1 a_x| \leq C \langle \xi \rangle g(\langle x \rangle)$$

into account, from (2.16), (2.17) and (2.9), we can control the asymptotic expansion of the system operator

$$S_\Lambda := \begin{pmatrix} e^{\Lambda} & 0 \\ 0 & e^{-\Lambda} \end{pmatrix}^{-1} S \begin{pmatrix} e^{\Lambda} & 0 \\ 0 & e^{-\Lambda} \end{pmatrix} \tag{2.18}$$

after the change of variable in $SU = 0$.

Precisely, the symbol of

$$\begin{pmatrix} i\sqrt{a(t,x)}D_x^2 & 0 \\ 0 & -i\sqrt{a(t,x)}D_x^2 \end{pmatrix}_\Lambda + B_\Lambda$$

is given by

$$\begin{pmatrix} i\sqrt{a}\xi^2 + q_1 + 2\sqrt{a}\xi\partial_x\Lambda_0 & 0 \\ 0 & -i\sqrt{a}\xi^2 + q_1 + 2\sqrt{a}\xi\partial_x\Lambda_0 \end{pmatrix}$$

$$+ (B_1 + 2\sqrt{a}\xi\partial_x\Lambda_1 I) + B_0, \quad (2.19)$$

where $q_1 \in C([0,T]; S^1)$ is such that

$$|q_1(t, x, \xi)| \le C\langle\xi\rangle g(\langle x\rangle) \tag{2.20}$$

uniformly with respect to the parameters K, M, N.

The symbol of the full matrix B_1 belongs to $C([0,T]; S^{2(1-\alpha)})$ and has support included in that one of the function $\chi(N\langle x\rangle/\langle\xi\rangle_h)$ in the phase-space and satisfies

$$|B_1(t, x, \xi)| \le C(K)\langle\xi\rangle^{2(1-\alpha)}\langle x\rangle^{-2(1-\alpha)}\chi(N\langle x\rangle/\langle\xi\rangle_h) \tag{2.21}$$

with $C(K)$ depending on K but not on M, N provided that $M^2 \le N^{1-2\alpha}$. The symbol of the remainder B_0 belongs to $C([0,T]; S^0)$.

So far, it is sufficient to fix first K in (2.13) and then M in (2.14) sufficiently large in order to have positive, and of order less or equal to 1, Hermitian parts for

$$\pm i\sqrt{a}\xi^2 + q_1 + 2\sqrt{a}\xi\partial_x\Lambda_0$$

and

$$B_1 + 2\sqrt{a}\xi\partial_x\Lambda_1 I$$

in (2.19) for $|\xi| \ge h$. Thus, by the sharp Gårding inequality and the energy method, the Cauchy problem for the system S_Λ in (2.18) is L^2 well posed. The same holds true for the system S, hence for the scalar operator $D_t^2 - a(t,x)D_x^4$, because the operators $e^{\pm\Lambda}$ and $(e^{\pm\Lambda})^{-1}$ in the change of variable are of order 0.

Remark 2.1. Theorem 1.3 and its proof can be generalized to the equation

$$D_t^2 u - a(t,x)D_x^{2p}u = 0$$

with $p > 2$ assuming

$$a(t,x) = \sum_{j=1}^{N} a_{\alpha_j}(t,x), \quad (p-1)/p \le \alpha_j \le 1,$$

$$|a_{\alpha_j}(t+\tau, x) - a_{\alpha_j}(t,x)| \le L|\tau|^{\alpha_j}\langle x\rangle^{-p(1-\alpha_j)/(p-1)}, \quad j = 1, \dots, N.$$

We note that for $p > 2$ the limit case $\alpha_j = (p-1)/p$ can be included. For $p = 2$, in Theorem 1.3, we take $\alpha_j > 1/2$ to avoid the critical exponent -1 for $\langle x\rangle$ that can not be considered in our proof. We conjecture that $p = 2$ and $\alpha = 1/2$ lead to well-posedness with a loss of derivatives.

References

[1] Cicognani, M.; Colombini, F.: Sharp regularity of the coefficients in the Cauchy Problem for a class of evolution equations. *Differential and Integral Equations* **16** (2003), 1321–1344.

[2] Cicognani, M.; Colombini, F.: The Cauchy problem for p-evolution equations. *Transactions of the American Mathematical Society* **362** (2010), 4853–4869.

[3] Colombini, F.; De Giorgi, E.; Spagnolo, S.: Sur les équations hyperboliques avec des coefficients qui ne dépendent que du temps. *Ann. Sc. Norm. Sup. Pisa* **6** (1979), 511–559.

[4] Colombini, F.; Lerner, N.: Hyperbolic operators with non-Lipschitz coefficients. *Duke Math. J.* **77** (1995), 657–698.

[5] Colombini, F.; Métivier, G.: The Cauchy problem for wave equations with non Lipschitz coefficients; application to continuation of solutions of some nonlinear wave equations. *Ann. Sci. c. Norm. Supr.* **41** (2008), 177–220.

[6] Colombini F.; Spagnolo, S.: Some examples of hyperbolic equations without local solvability. *Ann. Sci. Ecole Norm. Sup.* **22** (1989), 109–125.

[7] Ichinose, W.: Some remarks on the Cauchy problem for Schrödinger type equations. *Osaka J. Math.* **21** (1984) 565–581.

[8] Kajitani, K.; Baba, A.: The Cauchy problem for Schrödinger type equations. *Bull. Sci. Math.* **119** (1995), 459–473.

Massimo Cicognani
Dipartimento di Matematica
Università di Bologna
Piazza di Porta San Donato, 5
I-40126 Bologna, Italy
e-mail: `cicognan@dm.unibo.it`

Ferruccio Colombini
Dipartimento di Matematica
Università di Pisa
Largo Bruno Pontecorvo, 5
I-56127 Pisa, Italy
e-mail: `colombin@dm.unipi.it`

Progress in Mathematics, Vol. 301, 63–85
© 2012 Springer Basel

Time-Frequency Analysis of Schrödinger Propagators

Elena Cordero, Fabio Nicola and Luigi Rodino

Abstract. We present a survey on recent results concerning applications of Time-Frequency Analysis to the study of Fourier Integral Operators (FIOs). In particular, we focus on Schrödinger-type FIOs, showing that Gabor frames provide optimally sparse representations of such operators. Using Maple software, new numerical examples for the Harmonic Oscillator are provided.

Mathematics Subject Classification. 35S30,47G30,42C15.

Keywords. Modulation spaces, localization operators, Fourier integral operators, Wigner distribution, short-time Fourier transform, Schatten class.

1. Introduction and definitions

At the basis of the Time-Frequency Analysis there are the linear operators of translation and modulation (so-called time-frequency shifts) given by

$$T_x f(t) = f(t - x) \quad \text{and} \quad M_\omega f(t) = e^{2\pi i \omega t} f(t). \tag{1}$$

These occur in the following time-frequency representation. Let g be a non-zero window function in the Schwartz class $\mathcal{S}(\mathbb{R}^d)$, then the short-time Fourier transform (STFT) of a signal $f \in L^2(\mathbb{R}^d)$ with respect to the window g is given by

$$V_g f(x, \omega) = \langle f, M_\omega T_x g \rangle = \int_{\mathbb{R}^d} f(t) \overline{g(t - x)} \, e^{-2\pi i \omega t} \, dt. \tag{2}$$

We have $V_g f \in L^2(\mathbb{R}^{2d})$. This definition can be extended to every pair of dual topological vector spaces, whose duality, denoted by $\langle \cdot, \cdot \rangle$, extends the inner product on $L^2(\mathbb{R}^d)$. For instance, it may be suited to the framework of tempered distributions.

We briefly explain the meaning of the previous "time-frequency" representation. If $f(t)$ represents a signal varying in time, its Fourier transform $\hat{f}(\omega)$ shows the distribution of its frequency ω, but its magnitude $|\hat{f}(\omega)|$ alone does not give information about "when" these frequencies appear. To overcome this problem, one may choose a non-negative window function g well localized around the origin. Then the information of the signal f at the instant x can be obtained by shifting the window g till the instant x under consideration, and by computing the Fourier transform of the product $f(x)g(t-x)$, that localizes f around the instant time x.

Once the analysis of the signal f is terminated, we can reconstruct the original signal f by a suitable inversion procedure. Namely, the reproducing formula related to the STFT, for every pairs of windows $\varphi_1, \varphi_2 \in \mathcal{S}(\mathbb{R}^d)$ with $\langle \varphi_1, \varphi_2 \rangle \neq 0$, reads as

$$\int_{\mathbb{R}^{2d}} V_{\varphi_1} f(x, \omega) M_\omega T_x \varphi_2 \, dx d\omega = \langle \varphi_2, \varphi_1 \rangle f. \tag{3}$$

In the present paper we shall mainly address to the discretized version of (2), namely the so-called Gabor frames. For $\alpha, \beta > 0$, $g \in L^2(\mathbb{R}^d)$, the set of time-frequency shifts $\mathcal{G}(g, \alpha, \beta) = \{g_{m,n} := M_n T_m g\}$, with $(m, n) \in \alpha \mathbb{Z}^d \times \beta \mathbb{Z}^d$, is a *Gabor frame* if there exist positive constants $A, B > 0$, such that

$$A\|f\|_{L^2} \leq \sum_{m,n} |\langle f, T_m M_n g \rangle|^2 \leq B\|f\|_{L^2}, \quad \forall f \in L^2(\mathbb{R}^d). \tag{4}$$

As a counterpart of (3), if condition (4) is satisfied, we may reconstruct the original signal from the Gabor coefficients $\langle f, g_{m,n} \rangle$ in terms of the so-called dual window γ. We refer to the next Section 2 for a somewhat more detailed presentation of the time-frequency methods and related references.

The objective of this paper is to apply Time-frequency Analysis to a certain class of Fourier Integral Operators (FIOs). FIOs are a mathematical tool to study a variety of problems arising in partial differential equations. Originally introduced by Lax [44] for the construction of parametrices in the Cauchy problem for hyperbolic equations, and set as a general mathematical theory by Hörmander [42], they have been widely employed in the framework of both pure and applied mathematics (see, e.g., the papers [22, 23], the books [43, 53, 57] and references therein). A particular class of FIOs was used by Helffer and Robert [40, 41] to study the spectral properties of a class of globally elliptic operators, generalizing the harmonic oscillator of the Quantum Mechanics.

The Fourier Integral operators we shall present possess a phase function similar to those of [40, 41]. Hence, the focus of applications are no longer hyperbolic problems, but Schrödinger's type propagators. A simple example is the resolvent of the Cauchy problem for the Schrödinger equation with a quadratic Hamiltonian. Precisely, consider the Cauchy problem

$$\begin{cases} i\dfrac{\partial u}{\partial t} + Hu = 0 \\ u(0, x) = u_0(x), \end{cases} \tag{5}$$

where H is the Weyl quantization of a quadratic form on $\mathbb{R}^d \times \mathbb{R}^d$ (see, e.g., [29]). Simple examples are $H = -\frac{1}{4\pi}\Delta + \pi|x|^2$, or $H = -\frac{1}{4\pi}\Delta - \pi|x|^2$ (see [6]). The solution to (5) is given by a one-parameter family of FIOs:

$$u(t, x) = e^{itH} u_0(x) = \int_{\mathbb{R}^d} e^{2\pi i \Phi(t, x, \eta)} \widehat{u}_0(\eta) \, d\eta$$

with a phase given by a quadratic form $\Phi(t, x, \eta)$ in the variables (x, η).

Let us just define the operators under consideration in the present paper and give a flavour of the results. For a given function f on $\mathbb{R}^d$ the *Fourier Integral Operator* (FIO) T with symbol σ and phase Φ on $\mathbb{R}^{2d}$ can be formally defined by

$$Tf(x) = \int_{\mathbb{R}^d} e^{2\pi i \Phi(x,\eta)} \sigma(x,\eta) \hat{f}(\eta) d\eta. \tag{6}$$

The phase function $\Phi(x,\eta)$ is smooth on $\mathbb{R}^{2d}$, fulfills the estimates

$$|\partial_z^\alpha \Phi(z)| \le C_\alpha, \quad |\alpha| \ge 2, \quad z = (x,\eta) \in \mathbb{R}^{2d}, \tag{7}$$

and the nondegeneracy condition

$$|\det \partial_{x,\eta}^2 \Phi(x,\eta)| \ge \delta > 0, \qquad (x,\eta) \in \mathbb{R}^{2d}. \tag{8}$$

The symbol σ on $\mathbb{R}^{2d}$ satisfies

$$|\partial_z^\alpha \sigma(z)| \le C_\alpha, \quad |\alpha| \le 2N, \quad \text{a.e. } z = (x,\eta) \in \mathbb{R}^{2d}, \tag{9}$$

for a fixed $N > 0$.

The action on L^2 and on the modulation spaces (see Definition in Section 2 below) of FIOs as above was studied in [18] (see also [1, 3, 8, 13, 14, 15, 16, 17, 19, 20, 50]) under even more general assumptions, namely for σ belonging to the Sjöstrand class $M^{\infty,1}(\mathbb{R}^{2d})$.

Time-frequency analysis can be used to rephrase the operator T as an infinite matrix. Similar ideas go back at least to [21] and were recently employed to study PDEs with not smooth coefficients, see, e.g., [52, 54].

As a basic result we have that the matrix representation of a FIO T with respect to a Gabor frame $g_{m,n} = M_n T_m g$ with $g \in \mathcal{S}(\mathbb{R}^d)$ is well organized, provided that the symbol σ satisfies the decay estimate for every $N > 0$ (see Theorem 3.1, cf. [18]):

Theorem 1.1. *For each $N > 0$, there exists a constant $C_N > 0$ such that*

$$|\langle Tg_{m,n}, g_{m',n'}\rangle| \le C_N \langle \chi(m,n) - (m',n')\rangle^{-2N}, \tag{10}$$

where χ is the canonical transformation generated by Φ.

In the special case of pseudodifferential operators such an almost diagonalization was already obtained in [49]; see also [34]. Indeed, pseudodifferential operators correspond to the phase $\Phi(x,\eta) = x\eta$ and canonical transformation $\chi(y,\eta) = (y,\eta)$.

One should mention that the use of almost diagonal estimates in proving continuity results goes back to the pioneering work [30], where the Calderón-Zygmund class of operators was studied via such a technique, and this was achieved by working with wavelets. Also, observe that simple cases of FIOs are the Fourier multipliers, where $\Phi(x,\eta) = x\eta + \tilde{\Phi}(\eta)$ and σ does not depend on x, and the usefulness of Gabor frames was first exhibited for these operators in [5].

The applications of Theorem 1.1 presented in the following are twofold. On the one hand, in Section 3 we shall apply Theorem 1.1 to the boundedness of the

FIO T, providing a review, and light improvement, of large part of the results in this connection, cf. [1, 3, 8, 50, 13, 14, 15, 16, 17, 19, 20]). The strategy is to translate first the definition of L^2 and modulation spaces in terms of Gabor coefficients, cf. Theorem 2.1, and then argue on the Gabor matrix of T, by using the estimates (10) and a Schur-type lemma.

On the other hand, in Section 4 we shall apply Theorem 1.1 to Numerical Analysis, namely we shall prove that Gabor frames provide optimally sparse representation of the FIO T, cf. [19].

We refer to [9, 10, 11, 39, 47], concerning the counterpart of this result for curvelets and wave propagators.

In Section 5 we shall give some numerical examples. In particular, as a novelty, we treat the case of the Harmonic Oscillator. Figures 1–5 show that the time-frequency content of the solution is strongly preserved in time, as evident from the explicit solution and its periodicity. As a consequence, the obtained coefficient decay is faster compared to [11] for curvelets and wave propagators, and this highlights that Gabor frames provide an efficient tool for the Numerical Analysis of the Schrödinger propagators.

Notation. We define $t^2 = t \cdot t$, for $t \in \mathbb{R}^d$, and $xy = x \cdot y$ is the scalar product on $\mathbb{R}^d$.

The Schwartz class is denoted by $\mathcal{S}(\mathbb{R}^d)$, the space of tempered distributions by $\mathcal{S}'(\mathbb{R}^d)$. We use the brackets $\langle f, g \rangle$ to denote the extension to $\mathcal{S}(\mathbb{R}^d) \times \mathcal{S}'(\mathbb{R}^d)$ of the inner product $\langle f, g \rangle = \int f(t)\overline{g(t)}dt$ on $L^2(\mathbb{R}^d)$. The Fourier transform is normalized to be $\hat{f}(\omega) = \mathcal{F}f(\omega) = \int f(t)e^{-2\pi i t \omega}dt$.

We denote by c_0 the space of sequences vanishing at infinity.

Throughout the paper, we shall use the notation $A \lesssim B$ to indicate $A \leq cB$ for a suitable constant $c > 0$, whereas $A \asymp B$ if $A \leq cB$ and $B \leq kA$, for suitable $c, k > 0$.

2. Time-frequency methods

2.1. Modulation spaces

We recall from the introduction that the short-time Fourier transform (STFT) of a distribution $f \in \mathcal{S}'(\mathbb{R}^d)$ with respect to a non-zero window $g \in \mathcal{S}(\mathbb{R}^d)$ is

$$V_g f(x, \omega) = \langle f, M_\omega T_x g \rangle = \int_{\mathbb{R}^d} f(t)\,\overline{g(t-x)}\,e^{-2\pi i \omega t}\,dt\,.$$

The STFT $V_g f$ is defined on many pairs of Banach spaces. For instance, it maps $L^2(\mathbb{R}^d) \times L^2(\mathbb{R}^d)$ into $L^2(\mathbb{R}^{2d})$ and $\mathcal{S}(\mathbb{R}^d) \times \mathcal{S}(\mathbb{R}^d)$ into $\mathcal{S}(\mathbb{R}^{2d})$. Furthermore, it can be extended to a map from $\mathcal{S}'(\mathbb{R}^d) \times \mathcal{S}(\mathbb{R}^d)$ into $\mathcal{S}'(\mathbb{R}^{2d})$.

For the quantitative description of decay properties, we use weight functions on the time-frequency plane. In the sequel v will always be a continuous, strictly positive, even, submultiplicative weight function (in short, a submultiplicative weight), hence, up to a multiplicative factor, $v(0) = 1$, $v(z) = v(-z)$, and $v(z_1 + z_2) \leq v(z_1)v(z_2)$, for all $z, z_1, z_2 \in \mathbb{R}^{2d}$. A positive weight function μ on $\mathbb{R}^{2d}$ belongs to $\mathcal{M}_v$, that is, is v-moderate if $\mu(z_1 + z_2) \leq Cv(z_1)\mu(z_2)$ for all $z_1, z_2 \in \mathbb{R}^{2d}$.

For our investigation of FIOs we assume $v \in \mathcal{S}'(\mathbb{R}^{2d})$ and we shall mostly use the polynomial weights defined by

$$v_s(z) = v_s(x, \omega) = \langle z \rangle^s = (1 + |x|^2 + |\omega|^2)^{s/2}, \quad z = (x, \omega) \in \mathbb{R}^{2d}.$$

Banach modulation spaces were introduced by H. Feichtinger in 1980; for their basic properties we refer, for instance, to [24], [32, Ch. 11–13] and the original literature quoted there. The quasi-Banach cases were first introduced in [31], then studied in [45, 48]. In particular, these spaces are examples of the general coorbit space theory for quasi-Banach spaces studied in [48]. There it is shown that also the quasi-Banach modulation spaces enjoy the same definition and the main properties of the Banach cases.

Given a non-zero window $g \in \mathcal{S}(\mathbb{R}^d)$, $\mu \in \mathcal{M}_v$, and $0 < p, q \leq \infty$, the *modulation space* $M_\mu^{p,q}(\mathbb{R}^d)$ consists of all tempered distributions $f \in \mathcal{S}'(\mathbb{R}^d)$ such that $V_g f \in L_\mu^{p,q}(\mathbb{R}^{2d})$ (weighted mixed-(quasi-)norm spaces). The (quasi-)norm on $M_\mu^{p,q}$ is

$$\|f\|_{M_\mu^{p,q}} = \|V_g f\|_{L_\mu^{p,q}} = \left(\int_{\mathbb{R}^d} \left(\int_{\mathbb{R}^d} |V_g f(x, \omega)|^p \mu(x, \omega)^p \, dx \right)^{q/p} d\omega \right)^{1/p} \tag{11}$$

(with obvious changes when $p = \infty$ or $q = \infty$). If $p = q$, we write M_μ^p instead of $M_\mu^{p,p}$, and if $\mu(z) \equiv 1$ on $\mathbb{R}^{2d}$, then we write $M^{p,q}$ and M^p for $M_\mu^{p,q}$ and $M_\mu^{p,p}$ respectively.

Then $M_\mu^{p,q}(\mathbb{R}^d)$ is a Banach space for $1 \leq p, q \leq \infty$ and a quasi-Banach space in the other cases, whose definition is always independent of the choice of the window g. Moreover, if $\mu \in \mathcal{M}_v$, $1 \leq p, q \leq \infty$, and $g \in M_v^1 \setminus \{0\}$, then $\|V_g f\|_{L_\mu^{p,q}}$ is an equivalent norm for $M_\mu^{p,q}(\mathbb{R}^d)$ (see [32, Thm. 11.3.7]):

$$\|f\|_{M_m^{p,q}} \asymp \|V_g f\|_{L_\mu^{p,q}}.$$

The class of modulation spaces contains the following well-known function spaces:

Weighted L^2-spaces: $M_{\langle x \rangle^s}^2(\mathbb{R}^d) = L_s^2(\mathbb{R}^d) = \{f : f(x)\langle x \rangle^s \in L^2(\mathbb{R}^d)\}, s \in \mathbb{R}$.

Sobolev spaces: $M_{\langle \omega \rangle^s}^2(\mathbb{R}^d) = H^s(\mathbb{R}^d) = \{f : \hat{f}(\omega)\langle \omega \rangle^s \in L^2(\mathbb{R}^d)\}, s \in \mathbb{R}$.

Shubin-Sobolev spaces [51, 7]: $M_{v_s(x,\omega)}^2(\mathbb{R}^d) = L_s^2(\mathbb{R}^d) \cap H^s(\mathbb{R}^d) = Q_s(\mathbb{R}^d), s \geq 0$.

Feichtinger's algebra: $M^1(\mathbb{R}^d) = S_0(\mathbb{R}^d)$.

The Schwartz class and the space of tempered distributions are characterized as $\mathcal{S}(\mathbb{R}^d) = \bigcap_{s \geq 0} M_{v_s}^1(\mathbb{R}^d)$ and $\mathcal{S}'(\mathbb{R}^d) \bigcup_{s \geq 0} M_{v_{-s}}^\infty(\mathbb{R}^d)$ respectively.

In the sequel we will also denote by $\mathcal{M}_\mu^{p,q}(\mathbb{R}^d)$ the closure of the Schwartz space $\mathcal{S}(\mathbb{R}^d)$ with respect to the quasi-norm of $M_\mu^{p,q}(\mathbb{R}^d)$. Hence $\mathcal{M}_\mu^{p,q}(\mathbb{R}^d) = M_\mu^{p,q}(\mathbb{R}^d)$ if $p, q < \infty$. Moreover we set $\mathcal{M}_\mu^p(\mathbb{R}^d) = \mathcal{M}_\mu^{p,p}(\mathbb{R}^d)$.

2.2. Gabor frames [26]

Fix a function $g \in L^2(\mathbb{R}^d)$ and a lattice $\Lambda = \alpha\mathbb{Z}^d \times \beta\mathbb{Z}^d$, for $\alpha, \beta > 0$. For $(m, n) \in \Lambda$, define $g_{m,n} := M_n T_m g$. The set of time-frequency shifts $\mathcal{G}(g, \alpha, \beta) = \{g_{m,n}, (m, n) \in \Lambda\}$ is called Gabor system. Associated to $\mathcal{G}(g, \alpha, \beta)$ we define the coefficient operator C_g, which maps functions to sequences as follows:

$$(C_g f)_{m,n} = (C_g^{\alpha,\beta} f)_{m,n} := \langle f, g_{m,n} \rangle, \quad (m, n) \in \Lambda, \tag{12}$$

the synthesis operator

$$D_g c = D_g^{\alpha,\beta} c = \sum_{(m,n) \in \Lambda} c_{m,n} T_m M_n g, \quad c = \{c_{m,n}\}_{(m,n) \in \Lambda}$$

and the Gabor frame operator

$$S_g f = S_g^{\alpha,\beta} f := D_g S_g f = \sum_{(m,n) \in \Lambda} \langle f, g_{m,n} \rangle g_{m,n}. \tag{13}$$

The set $\mathcal{G}(g, \alpha, \beta)$ is called a Gabor frame for the Hilbert space $L^2(\mathbb{R}^d)$ if S_g is a bounded and invertible operator on $L^2(\mathbb{R}^d)$. Equivalently, C_g is bounded from $L^2(\mathbb{R}^d)$ to $l^2(\alpha\mathbb{Z}^d \times \beta\mathbb{Z}^d)$ with closed range, i.e., $\|f\|_{L^2} \asymp \|C_g f\|_{l^2}$, cf. (4). If $\mathcal{G}(g, \alpha, \beta)$ is a Gabor frame for $L^2(\mathbb{R}^d)$, then the so-called *dual window* $\gamma = S_g^{-1} g$ is well defined and the set $\mathcal{G}(\gamma, \alpha, \beta)$ is a frame (the so-called canonical dual frame of $\mathcal{G}(g, \alpha, \beta)$). Every $f \in L^2(\mathbb{R}^d)$ possesses the frame expansion

$$f = \sum_{(m,n) \in \Lambda} \langle f, g_{m,n} \rangle \gamma_{m,n} = \sum_{(m,n) \in \Lambda} \langle f, \gamma_{m,n} \rangle g_{m,n} \tag{14}$$

with unconditional convergence in $L^2(\mathbb{R}^d)$, and norm equivalence:

$$\|f\|_{L^2} \asymp \|C_g f\|_{l^2} \asymp \|C_\gamma f\|_{l^2}.$$

This result is contained in [32, Proposition 5.2.1]. In particular, if $\gamma = g$ and $\|g\|_{L^2} = 1$ the frame is called *normalized tight* Gabor frame and the expansion (14) reduces to

$$f = \sum_{(m,n) \in \Lambda} \langle f, g_{m,n} \rangle g_{m,n}. \tag{15}$$

If we ask for more regularity on the window g, then the previous result can be extended to modulation (quasi-)Banach spaces, as shown below ([26, 36] and [48, Theorem 8.3]).

Theorem 2.1. *Let $\mu \in \mathcal{M}_v$, $\mathcal{G}(g, \alpha, \beta)$ be a frame for $L^2(\mathbb{R}^d)$, with lattice $\Lambda = \alpha\mathbb{Z}^d \times \beta\mathbb{Z}^d$, and $g \in \mathcal{S}$. Define $\tilde{\mu} = \mu_{|_\Lambda}$.*

(i) *For every $0 < p, q \leq \infty$, $C_g : M_\mu^{p,q} \to l_{\tilde{\mu}}^{p,q}$ and $D_g : l_{\tilde{\mu}}^{p,q} \to M_\mu^{p,q}$ continuously and, if $f \in M_\mu^{p,q}$, then the Gabor expansions (14) converge unconditionally in $M_\mu^{p,q}$ for $0 < p, q < \infty$, and weak*-M_μ^∞ unconditionally if $p = \infty$ or $q = \infty$.*

(ii) *The following (quasi-)norms are equivalent on $M_\mu^{p,q}$:*

$$\|f\|_{M_\mu^{p,q}} \asymp \|C_g f\|_{l_{\tilde{\mu}}^{p,q}}. \tag{16}$$

3. Fourier Integral Operators: almost diagonalization via Gabor frames and boundedness results

Now we study the FIO T defined in (6), having phase satisfying (7) and (8) and symbol enjoying (9).

If we set

$$\begin{cases} y = \nabla_\eta \Phi(x, \eta) \\ \xi = \nabla_x \Phi(x, \eta), \end{cases} \tag{17}$$

and solve with respect to (x, ξ), we obtain a mapping χ, defined by $(x, \xi) = \chi(y, \eta)$, which is a smooth bi-Lipschitz canonical transformation. This means that
- χ is a smooth diffeomorphism on $\mathbb{R}^{2d}$;
- both χ and χ^{-1} are uniformly Lipschitz continuous;
- χ preserves the symplectic form, i.e.,

$$dx \wedge d\xi = dy \wedge d\eta.$$

Indeed, under the above assumptions, the global inversion function theorem (see, e.g., [46]) allows us to solve the first equation in (17) with respect to x, and substituting in the second equation yields the smooth map χ. The bounds on the derivatives of χ, which give the Lipschitz continuity, follow from the expression for the derivatives of an inverse function combined with the bounds in (7) and (8). The symplectic nature of the map χ is classical, see, e.g., [12]. Similarly, solving the second equation in (17) with respect to η one obtains the map χ^{-1} with the desired properties.

In this section we present an almost diagonalization result for FIOs as above, with respect to Gabor frames. For simplicity, we consider a normalized tight frame $\mathcal{G}(g, \alpha, \beta)$, with $g \in \mathcal{S}(\mathbb{R}^d)$. We have the following result (see [18, Thm. 3.3]).

Theorem 3.1. *Consider a phase function* Φ *satisfying* (7) *and* (8) *and a symbol satisfying* (9). *Let* $g \in \mathcal{S}(\mathbb{R}^d)$. *There exists a constant* $C_N > 0$ *such that*

$$|\langle T g_{m,n}, g_{m',n'} \rangle| \leq C_N \langle \chi(m, n) - (m', n') \rangle^{-2N}, \tag{18}$$

where χ *is the canonical transformation generated by* Φ.

The proof uses the action of the Fourier transform over the time-frequency shifts: $(T_x f)^\wedge = M_{-x} \hat{f}$ and $(M_\omega f)^\wedge = T_\omega \hat{f}$, the smoothness of Φ (via a Taylor expansion of $\Phi(x, \eta)$ at (m', n)), and repeated integration by parts.

This result shows that the matrix representation of a FIO with respect a Gabor frame is well organized. More precisely, if $\sigma \in S_{0,0}^0$, namely if (9) is satisfied for every $N \in \mathbb{N}$, then the Gabor matrix of T is highly concentrated along the graph of χ.

The previous result, together with the following lemma, are the key to prove the continuity of these FIOs on the modulation spaces M_μ^p, with a weight function $\mu \in \mathcal{M}_{v_s}$, $s \geq 0$. Observe that the first part of the lemma is the classical Schur's test (see, e.g., [32, Lemma 6.2.1]), the second part is also classical (see, e.g., [19, Lemma 4.1]). The continuity on $c_0(\Lambda)$ is proved in [18, Lemma 4.1].

Lemma 3.1. *Consider an operator K defined on sequences by the matrix $K_{\lambda,\nu}$. Then,*

$$\|Kc\|_{l^p} \leq \begin{cases} \left(\sup_\nu \sum_\lambda |K_{\lambda,\nu}|^p\right)^{1/p} \|c\|_{l^p}, & 0 < p \leq 1, \\ \left(\sup_\lambda \sum_\nu |K_{\lambda,\nu}|\right)^{1/p'} \left(\sup_\nu \sum_\lambda |K_{\lambda,\nu}|\right)^{1/p} \|c\|_{l^p}, & 1 \leq p \leq \infty, \end{cases}$$

whenever the right-hand sides are finite. Moreover, the last condition implies that K maps the space $c_0(\Lambda)$ of sequences vanishing at infinity into itself.

We can now state the following boundedness result, proved in [18, Theorem 4.1] when $p \geq 1$ and in [19, Theorem 4.1] for $0 < p < 1$. A more general result for $p > 1$ was proved in [14, Theorem 2.1].

We sketch the proof to allow the reader to understand how Gabor frames are employed to attain the goal; see also [35, 56], where Gabor expansion were used to establish continuity properties of pseudodifferential operators.

Theorem 3.2. *Consider a phase function satisfying (7) and (8) and a symbol satisfying (9).*

(1) *Let $0 \leq s < 2N - 2d/p$, $0 < p < 1$, and $\mu \in \mathcal{M}_{v_s}$. Then T extends to a continuous operator from $M^p_{\mu \circ \chi}$ into M^p_μ.*

(2) *Let $0 \leq s < 2N - 2d$, and $\mu \in \mathcal{M}_{v_s}$. For every $1 \leq p \leq \infty$, T extends to a continuous operator from $M^p_{\mu \circ \chi}$ into M^p_μ.*

Recall that $\mathcal{M}^p_\mu$ is the closure of $\mathcal{S}(\mathbb{R}^d)$ in M^p_μ. Moreover, observe that $\mu \circ \chi \in \mathcal{M}_{v_s}$. Indeed, $v_s \circ \chi \asymp v_s$, due to the bi-Lipschitz property of χ.

Proof. We detail the case $1 \leq p \leq \infty$ and first prove the theorem in the case $p < \infty$.

For $T = C_g \circ T_{m',n',m,n} \circ D_g$, the following diagram is commutative:

$$\begin{array}{ccc} \mathcal{M}^p_{\mu \circ \chi} & \xrightarrow{\ T\ } & \mathcal{M}^p_\mu \\ {\scriptstyle C_g}\downarrow & & \uparrow{\scriptstyle D_g} \\ l^p_{\widetilde{\mu \circ \chi}} & \xrightarrow{\ T_{m',n',m,n}\ } & l^p_{\widetilde{\mu}} \end{array}$$

where T is viewed as an operator with dense domain $\mathcal{S}(\mathbb{R}^d)$. Whence, it is enough to prove the continuity of the infinite matrix $T_{m',n',m,n}$ from $l^p_{\widetilde{\mu \circ \chi}}$ into $l^p_{\widetilde{\mu}}$.

This follows from Schur's test (Lemma 3.1) if we prove that, upon setting

$$K_{m',n',m,n} = T_{m',n',m,n} \frac{\mu(m',n')}{\mu(\chi(m,n))},$$

we have

$$K_{m',n',m,n} \in l^\infty_{m,n} l^1_{m',n'}, \tag{19}$$

and

$$K_{m',n',m,n} \in l^\infty_{m',n'} l^1_{m,n}. \tag{20}$$

In view of (18) we have

$$|K_{m',n',m,n}| \lesssim \langle \chi(m,n) - (m',n') \rangle^{-2N+s} \frac{\mu(m',n')}{\langle \chi(m,n) - (m',n') \rangle^s \mu(\chi(m,n))}. \quad (21)$$

Now, the last quotient in (21) is bounded because μ is v_s-moderate, so we deduce (19).

Finally, since χ is a bi-Lipschitz function we have

$$|\chi(m,n) - (m',n')| \asymp |(m,n) - \chi^{-1}(m',n')| \quad (22)$$

so that (20) follows as well.

The case $p = \infty$ follows analogously by using the last part of the statement of Lemma 3.1. $\quad\square$

Remark 3.3. Theorem 3.2 with $v \equiv 1$ gives, in particular, continuity on the unweighted modulation spaces $\mathcal{M}^p$ and on $L^2 = \mathcal{M}^2$.

Also, Theorem 3.2 applies to $\mu = v_t$, with $|t| \le s$. In that case we obtain continuity on $\mathcal{M}^p_{v_t}$, because $v_t \circ \chi \asymp v_t$.

3.1. Boundedness on $\mathcal{M}^{p,q}(\mathbb{R}^d)$, for $p \ne q$

This part exhibits new conditions which guarantee the boundedness on $\mathcal{M}^{p,q}(\mathbb{R}^d)$, for $p \ne q$.

The first result is in fact a generalization of [3, Theorem 11] and [18, Theorem 5.2] to the case of rougher symbols and is contained in [17].

Theorem 3.4. *Consider a phase function Φ satisfying (7) and (8), and a symbol $\sigma \in M^{\infty,1}(\mathbb{R}^{2d})$. Suppose, in addition, that*

$$\sup_{x,x',\eta \in \mathbb{R}^d} |\nabla_x \Phi(x,\eta) - \nabla_x \Phi(x',\eta)| < \infty. \quad (23)$$

Then, the corresponding Fourier integral operator T extends to a bounded operator on $\mathcal{M}^{p,q}(\mathbb{R}^d)$, for every $1 \le p, q \le \infty$.

The condition (23) is seen to be essential for the conclusion to hold. In fact it was shown in [18, Proposition 7.1] that the pointwise multiplication operator by $e^{-\pi i |x|^2}$ (which has phase $\Phi(x,\eta) = x\eta - \frac{|x|^2}{2}$ and symbol $\sigma \equiv 1$) is not bounded on any $\mathcal{M}^{p,q}$, with $p \ne q$.

If we drop the condition (23), we need some further decay condition on the symbol, as explained by the next result.

For $s_1, s_2 \in \mathbb{R}$, we define the weight function $v_{s_1,s_2}(x,\eta) := \langle x \rangle^{s_1} \langle \eta \rangle^{s_2}$, $(x,\eta) \in \mathbb{R}^{2d}$.

Theorem 3.5. *Consider a phase Φ satisfying (7), (8), and a symbol $\sigma \in M^{\infty,1}_{v_{s_1,s_2} \otimes 1}(\mathbb{R}^{2d})$, $s_1, s_2 \in \mathbb{R}$.*

(i) *Let $1 \le p \le \infty$. If $s_1, s_2 \ge 0$, T extends to a bounded operator on $\mathcal{M}^p(\mathbb{R}^d)$.*

(ii) *Let $1 \le q < p \le \infty$. If $s_1 > d\left(\frac{1}{q} - \frac{1}{p}\right)$, $s_2 \ge 0$, T extends to a bounded operator on $\mathcal{M}^{p,q}(\mathbb{R}^d)$.*

(iii) Let $1 \leq p < q \leq \infty$. If $s_1 \geq 0$, $s_2 > d\left(\frac{1}{p} - \frac{1}{q}\right)$, T extends to a bounded operator on $\mathcal{M}^{p,q}(\mathbb{R}^d)$.

In all cases,

$$\|Tf\|_{\mathcal{M}^{p,q}} \lesssim \|\sigma\|_{M^{\infty,1}_{v_{s_1,s_2} \otimes 1}} \|f\|_{\mathcal{M}^{p,q}}. \tag{24}$$

Counterexamples for the thresholds arising in Theorem 4.2 are given in [17]: they show that the thresholds are in fact the expected ones.

For the proofs we refer to the above-mentioned paper. Let us just underline that boundedness results dealing with FIOs having symbols in weighted modulation spaces and acting on unweighted modulation spaces could be rephrased as boundedness results for FIOs with symbols in unweighted spaces and acting on weighted spaces, as explained below.

Proposition 3.6. *Let T be a FIO with symbol σ and $\tilde{T}$ a FIO with the same phase as T and symbol*

$$\tilde{\sigma}(x,\eta) := \langle x \rangle^{s_1} \sigma(x,\eta) \langle \eta \rangle^{s_2}, \quad x, \eta \in \mathbb{R}^d, \ s_i \in \mathbb{R}, \ i = 1, 2.$$

Then,

(i) *the operator T is bounded from $\mathcal{M}^{p,q}$ into $\mathcal{M}^{\tilde{p},\tilde{q}}$ if and only if the operator $\tilde{T}$ is bounded from $\mathcal{M}^{p,q}_{v_{0,s_2}}$ into $\mathcal{M}^{\tilde{p},\tilde{q}}_{v_{-s_1,0}}$.*

(ii) *It holds true*

$$\sigma \in M^{\infty,1}_{v_{s_1,s_2} \otimes 1}(\mathbb{R}^{2d}) \iff \tilde{\sigma} \in M^{\infty,1}(\mathbb{R}^{2d}). \tag{25}$$

Proof. The proof is an immediate consequence of [55, Theorem 2.2, Corollary 2.3]. Indeed, they guarantee that the vertical arrows of the following commutative diagram define isomorphisms:

$$
\begin{array}{ccc}
\mathcal{M}^{p,q} & \xrightarrow{\ T\ } & \mathcal{M}^{\tilde{p},\tilde{q}} \\
{\scriptstyle \langle D \rangle^{s_2}} \Big\uparrow & & \Big\downarrow {\scriptstyle \text{pointwise product by } \langle x \rangle^{s_1}} \\
\mathcal{M}^{p,q}_{v_{0,s_2}} & \xrightarrow{\ \tilde{T}\ } & \mathcal{M}^{\tilde{p},\tilde{q}}_{v_{-s_1,0}}
\end{array}
\tag{26}
$$

Moreover, the product by the weight function $\omega_0(x,\eta,\zeta_1,\zeta_2) := \langle x \rangle^{s_1} \langle \eta \rangle^{s_2}$, x, η, $\zeta_1, \zeta_2 \in \mathbb{R}^d$, defined on $\mathbb{R}^{4d}$, is an isomorphism from $M^{\infty,1}_{v_{s_1,s_2} \otimes 1}(\mathbb{R}^{2d})$ to $M^{\infty,1}(\mathbb{R}^{2d})$, that is (ii). $\qquad \square$

Thanks to the commutativity of the diagram (26), the results of Theorem 3.5 may be equivalently stated as the action of a FIO T on weighted modulation spaces:

Theorem 3.7. *Consider a phase Φ satisfying (7), (8), and a symbol $\sigma \in M^{\infty,1}(\mathbb{R}^{2d})$.*

(i) *If $1 \leq q < p \leq \infty$, $s_1 < -d\left(\frac{1}{q} - \frac{1}{p}\right)$, and $s_2 \geq 0$, then T extends to a bounded operator from $\mathcal{M}^{p,q}_{v_{0,s_2}}(\mathbb{R}^d)$ to $\mathcal{M}^{p,q}_{v_{s_1,0}}(\mathbb{R}^d)$.*

(ii) *If $1 \leq p < q \leq \infty$, $s_1 \leq 0$, and $s_2 > d \left(\frac{1}{p} - \frac{1}{q} \right)$, T extends to a bounded operator from $\mathcal{M}^{p,q}_{v_0, s_2}(\mathbb{R}^d)$ to $\mathcal{M}^{p,q}_{v_{s_1}, 0}(\mathbb{R}^d)$.*
In both cases,

$$\|Tf\|_{\mathcal{M}^{p,q}_{v_{s_1},0}} \lesssim \|\sigma\|_{M^{\infty,1}} \|f\|_{\mathcal{M}^{p,q}_{v_0,s_2}}.$$

3.2. The case of quadratic phases: metaplectic operators

We briefly discuss the particular case of quadratic phases, namely phases of the type

$$\Phi(x,\eta) = \frac{1}{2}Ax \cdot x + Bx \cdot \eta + \frac{1}{2}C\eta \cdot \eta + \eta_0 \cdot x - x_0 \cdot \eta, \tag{27}$$

where $x_0, \eta_0 \in \mathbb{R}^d$, A, C are real symmetric $d \times d$ matrices and B is a real $d \times d$ nondegenerate matrix.

It is easy to see that, if we take the symbol $\sigma \equiv 1$ and the phase (27), the corresponding FIO T is (up to a constant factor) a metaplectic operator. This can be seen by means of the easily verified factorization

$$T = M_{\eta_0} U_A D_B \mathcal{F}^{-1} U_C \mathcal{F} T_{x_0}, \tag{28}$$

where U_A and U_C are the multiplication operators by $e^{\pi i Ax \cdot x}$ and $e^{\pi i C\eta \cdot \eta}$ respectively, and D_B is the dilation operator $f \mapsto f(B \cdot)$. Each of the factors is (up to a constant factor) a metaplectic operator (see, e.g., the proof of [43, Theorem 18.5.9]), so T is.

For the benefit of the reader, some important special cases are detailed in the table below.

operator	phase $\Phi(x,\eta)$	canonical transformation
T_{x_0}	$(x - x_0) \cdot \eta$	$\chi(y,\eta) = (y + x_0, \eta)$
M_{η_0}	$(\eta + \eta_0) \cdot x$	$\chi(y,\eta) = (y, \eta + \eta_0)$
D_B	$Bx \cdot \eta$	$\chi(y,\eta) = (B^{-1}y, {}^t B\eta)$
U_A	$x \cdot \eta + \frac{1}{2}Ax \cdot x$	$\chi(y,\eta) = (y, \eta + Ax)$

We end up by observing that there are metaplectic operators, as the Fourier transform, which cannot be expressed as FIOs of this type.

We refer to [16] for the action of metaplectic operators on other spaces arising in Time-frequency Analysis.

4. Sparsity of the Gabor matrix and nonlinear approximation

Given a FIO T as above, and $f \in M^q$ we are going to present a formula to compute Tf by nonlinear approximation, namely a formula which expresses Tf as a finite sum of Gabor atoms, modulo an error whose size is estimated in terms of its time-frequency concentration.

This is the issue of [19, Section 5]. Let $\alpha, \beta > 0$, $g \in \mathcal{S}$, be such that the set

$$\mathcal{G}(g, \alpha, \beta) = \{g_{m,n} := T_m M_n g, \quad (m, n) \in \Lambda = \alpha\mathbb{Z}^d \times \beta\mathbb{Z}^d\},$$

is a frame of $L^2(\mathbb{R}^d)$. Assume the phase Φ satisfies $(7), (8)$ and the symbol σ fulfills (9) for every $N \in \mathbb{N}$.

Set $\lambda = (m', n') \in \Lambda$ and $\nu = (m, n) \in \Lambda$. Let moreover

$$T_{\lambda,\nu} = \langle Tg_\nu, g_\lambda \rangle.$$

Then, Theorem 3.1, i.e., [18, Theorem 3.3], shows that, for every $N' \in \mathbb{N}$,

$$|T_{\lambda,\nu}| \leq C_{N'} \langle \lambda - \chi(\nu) \rangle^{-N'}. \tag{29}$$

This gives at once that the Gabor matrix $T_{\lambda,\nu}$ is sparse, in the sense precised by the following proposition (cf. [10, 39]).

Proposition 4.1. *The Gabor matrix $T_{\lambda,\nu}$ is sparse. Namely, let a be any column or raw of the matrix, and let $|a|_n$ be the n-largest entry of the sequence a. Then, for each $M > 0$, $|a|_n$ satisfies*

$$|a|_n \leq C_M n^{-M}.$$

Indeed,

$$n^{1/p} \cdot |a|_n \leq \|a\|_{l^p},$$

for every $0 < p \leq \infty$. Hence it suffices to prove that every column or raw is in l^p for arbitrarily small p. This follows immediately from (29).

We now turn to the nonlinear approximation problem.

Given $N \in \mathbb{N}$ and $B \geq 1$, we associate to any function $f \in M_\mu^q$ two indices sets J_N and $I_{N,B}$ defined as follows. Let b_ν be the sequence of weighted Gabor coefficients of f, namely $b_\nu = \langle f, g_\nu \rangle \mu(\nu)$, and let $|b_{\nu_1}| \geq |b_{\nu_2}| \geq \cdots$ be a non-increasing rearrangement. We then define

$$J_N = \{\nu_1, \ldots, \nu_N\}$$

and

$$I_{N,B} = \{\lambda \in \Lambda : \exists \nu \in J_N, \ |\lambda - \chi(\nu)| \leq B\},$$

where χ is the canonical transformation generated by Φ. The approximation result can be formulated as follows.

Theorem 4.2. *Let $0 < p < q \leq \infty$, and $f \in M_{\mu \circ \chi}^p$, with $\mu \in \mathcal{M}_{v_s}$, $s \geq 0$. Let J_N, and $I_{N,B}$ be the index sets associated to f as above. Then, for every $N', N \in \mathbb{N}$, with $N' > d \max\{1, 1/q\} + s$, and $B \geq 1$, we have*

$$Tf = \sum_{\lambda \in I_{N,B}} \sum_{\nu \in I_N} T_{\lambda,\nu} \langle f, g_\nu \rangle \, g_\lambda + e_{N,B},$$

where the error $e_{N,B}$ satisfies the estimate

$$\|e_{N,B}\|_{M_\mu^q} \lesssim B^{d \max\{1, 1/q\} + s - N'} \|f\|_{M_{\mu \circ \chi}^q} + N^{1/q - 1/p} \|f\|_{M_{\mu \circ \chi}^p}.$$

Here the constant implicit in the notation $\lesssim$ is independent of f, B and N.

Hence, given f, the error term decays faster of any negative fixed power of N and B, as $N, B \to +\infty$. This is seen by fixing N' conveniently large and p small enough.

For the proof we refer to [19, Theorem 5.2].

5. Application to the solution of the Schrödinger equation

We now turn our attention to some important examples of the class of FIOs studied here, represented by solution operators to the Cauchy problem for Schrödinger equations.

5.1. The free particle

Consider the Cauchy problem for the Schrödinger equation

$$\begin{cases} i\partial_t u + \Delta u = 0 \\ u(0, x) = u_0(x), \end{cases} \tag{30}$$

with $x \in \mathbb{R}^d$, $d \geq 1$. The explicit formula for the solution is

$$u(t, x) = (K_t * u_0)(x), \tag{31}$$

where

$$K_t(x) = \frac{1}{(4\pi i t)^{d/2}} e^{i|x|^2/(4t)}. \tag{32}$$

The solution $u(t, x)$ can be as well expressed by means of a FIO:

$$u(t, x) = \int_{\mathbb{R}^d} e^{2\pi i (x\eta - 2\pi t |\eta|^2)} \widehat{u_0}(\eta) d\eta,$$

for $\mathcal{F}((ai)^{-d/2} e^{-\pi |x|^2/(ai)})(\eta) = e^{-\pi a i |\eta|^2}$, $a \in \mathbb{R} \setminus \{0\}$. Whence, $u(t, x) = T u_0(x)$, where T is the FIO with phase $\Phi(x, \eta) = x\eta - 2\pi t |\eta|^2$ and symbol $\sigma \equiv 1$.

The smooth bi-Lipschitz canonical transformation χ associated with the phase Φ is given by

$$\chi(y, \eta) = (y + 4\pi t \eta, \eta).$$

Hence, the matrix decay (29), for $\lambda = (m', n')$, $\nu = (m, n)$, is given by

$$|T_{\lambda, \nu}| \leq C_{N'} (1 + |(m' - (m + 4\pi tn), n' - n)|)^{-N'}.$$

In [19, Section 6] the matrix entries are computed explicitly when the initial datum is a Gabor atom with the Gaussian window function. Namely,

Proposition 5.1. *Let $u(t, x)$ be the solution of the Cauchy problem (32), with $u_0(x) = M_n T_m e^{-\pi |x|^2}$. Then,*

$$u(t, x) = (1 + 4\pi i t)^{-d/2} e^{-\frac{4\pi^2 nt(2m+in)}{1+4\pi it}} M_{\frac{n}{1+4\pi it}} T_m e^{-\frac{\pi}{1+4\pi it}|x|^2}. \tag{33}$$

Hence, we have a more explicit formula for the matrix entries

$$\{T_{m',n',m,n}\}_{m',n',m,n} = \langle T(M_n T_m g_0), M_{n'} T_{m'} g_0 \rangle.$$

Indeed,

$$T_{m',n',m,n} = (1+4\pi it)^{-d/2} e^{-\frac{4\pi^2 nt(2m+in)}{1+4\pi it}} \langle M_{\frac{n}{1+4\pi it}} T_m e^{-\frac{\pi}{1+4\pi it}|\cdot|^2}, M_{n'} T_{m'} g_0 \rangle. \quad (34)$$

We may go on computing the brackets, to obtain a more explicit formula for $T_{m',n',m,n}$. Details are left to the reader.

Numerical examples concerning (30) in dimension $d = 1$ and $d = 2$ are presented in [19]. The same type of decay in dimension $d = 2$ was obtained in [11, Figure 15] for the curvelet representation of the wave propagator.

5.2. The harmonic oscillator

This is the case of the Cauchy problem for the Schrödinger equation below:

$$\begin{cases} i\dfrac{\partial u}{\partial t} - \dfrac{1}{4\pi}\Delta u + \pi |x|^2 u = 0 \\ u(0,x) = u_0(x). \end{cases} \quad (35)$$

The solution is the metaplectic operator (hence FIO) given by [16, 29, 40]

$$u(t,x) = (\cos t)^{-d/2} \int_{\mathbb{R}^d} e^{2\pi i[\frac{1}{\cos t}x\eta + \frac{\tan t}{2}(x^2+\eta^2)]} \hat{u}_0(\eta)\, d\eta, \quad t \neq \frac{\pi}{2} + k\pi,\ k \in \mathbb{Z}, \quad (36)$$

see below for $t = \frac{\pi}{2} + k\pi,\ k \in \mathbb{Z}$. Here $u(t,x) = Tu_0$, where T is a FIO with symbol $\sigma = (\cos t)^{-d/2}$ and phase

$$\Phi(x,\eta) = \frac{1}{\cos t}x\eta + \frac{\tan t}{2}(x^2+\eta^2).$$

The canonical transformation χ is then obtained by solving (17). We get

$$\chi(y,\eta) = \begin{pmatrix} (\cos t)I & (-\sin t)I \\ (\sin t)I & (\cos t)I \end{pmatrix} \begin{pmatrix} y \\ \eta \end{pmatrix}.$$

Since

$$|\det \partial^2_{x,\eta}\Phi(x,\eta)| \frac{1}{|\cos t|^d} > 0, \quad t \neq \frac{\pi}{2} + k\pi,\ k \in \mathbb{Z},$$

the assumptions of Theorem 3.1 are fulfilled and we get the estimate

$$|\langle Tg_{m,n}, g_{m',n'} \rangle| \leq C_N (1 + |(\cos t)m + (\sin t)n - m', -(\sin t)m + (\cos t)n - n'|)^{-N}, \quad (37)$$

for every $N \in \mathbb{N}$, if $t \neq \frac{\pi}{2} + k\pi,\ k \in \mathbb{Z}$. The same formula continues to hold for $t = \frac{\pi}{2} + k\pi,\ k \in \mathbb{Z}$, as the more explicit result in Proposition 5.3 below shows. Indeed, the time singularity at those points is artificial and becomes from the representation of the propagator in the form (36), whose validity fails for $t = \frac{\pi}{2} + k\pi,\ k \in \mathbb{Z}$.

As for the free particle, in what follows we find the matrix coefficients $\{T_{m',n',m,n}\}_{m',n',m,n}$ for the time-frequency shifts of the Gaussian datum. Then we present numerical simulations by Maple software. For the Maple algorithms and more detailed computations we refer to [4]. First, let us recall the following formula.

Lemma 5.1. *Let a, b two complex numbers such that $\operatorname{Re} a < 0$. Then*

$$\int_{\mathbb{R}^d} e^{ax^2+bx}\, \mathrm{d}x = \left(-\frac{\pi}{a}\right)^{d/2} e^{-\frac{b^2}{4a}}. \tag{38}$$

We use this formula to calculate the solution to the Cauchy problem (35), with $u_0(x) := M_n T_m e^{-\pi|x|^2}$, for $x \in \mathbb{R}^d$ and $(m, n) \in \alpha \mathbb{Z}^d \times \beta \mathbb{Z}^d$.

Proposition 5.2. *Let $u_0 := M_n T_m e^{-\pi|x|^2}$ and $u(t, x)$ as in (35), (36). Then*

$$u(t, x) = e^{itd/2} e^{-\pi x^2} e^{-\pi(m+in)^2 \cos(t) e^{it}} e^{2\pi x(m+in)e^{it}} e^{2\pi m\cdot n - \pi n^2}. \tag{39}$$

Proof. Using $u_0(x) := M_n T_m e^{-\pi|x|^2}$ in Equation (36) we get

$$u(t, x) = (\cos t)^{-d/2} \int_{\mathbb{R}^d} e^{2\pi i\left[\frac{1}{\cos t} x\eta + \frac{\tan t}{2}(x^2+\eta^2)\right]} \mathcal{F}(M_n T_m e^{-\pi|x|^2})(\eta)\, d\eta$$

$$= (\cos t)^{-d/2} \int_{\mathbb{R}^d} e^{2\pi i\left[\frac{1}{\cos t} x\eta + \frac{\tan t}{2}(x^2+\eta^2)\right]} T_n M_{-m} \mathcal{F}(e^{-\pi x^2})(\eta)\, d\eta$$

$$= (\cos t)^{-d/2} \int_{\mathbb{R}^d} e^{2\pi i\left[\frac{1}{\cos t} x\eta + \frac{\tan t}{2}(x^2+\eta^2)\right]} T_n M_{-m}(e^{-\pi \eta^2})(\eta)\, d\eta$$

$$= (\cos t)^{-d/2} \int_{\mathbb{R}^d} e^{2\pi i\left[\frac{1}{\cos t} x\eta + \frac{\tan t}{2}(x^2+\eta^2)\right]} e^{-2\pi im\cdot(\eta-n)} e^{-\pi(\eta-n)^2}\, d\eta$$

$$= (\cos t)^{-d/2} \int_{\mathbb{R}^d} e^{2\pi i\left[\frac{1}{\cos t} x\eta + \frac{\tan t}{2}(x^2+\eta^2)\right]} e^{-2\pi im\cdot(\eta-n)} e^{-\pi(\eta^2-2\eta n+n^2)}\, d\eta$$

$$= (\cos t)^{-d/2} e^{\frac{2\pi i \tan t}{2} x^2 + 2\pi im\cdot n - \pi n^2} \int_{\mathbb{R}^d} e^{\pi(i\tan t - 1)\eta^2 + 2\pi i\left(\frac{x}{\cos t} - m - in\right)\eta}\, d\eta.$$

Now, using Formula (38),

$$u(t, x) = \left(\frac{\cos t \cdot e^{it}}{\cos t}\right)^{d/2} e^{\frac{2\pi i \tan t}{2} x^2 + 2\pi im\cdot n - \pi n^2} e^{-\pi e^{it} \cos t \cdot \left(\frac{x^2}{\cos^2 t} - \frac{2x(m+in)}{\cos t} + (m+in)^2\right)}$$

$$= e^{\frac{itd}{2}} e^{\frac{\pi i \sin t}{\cos t} x^2 + 2\pi im\cdot n - \pi n^2} e^{-\pi e^{it} \cos t \cdot \left(\frac{x^2}{\cos^2 t} - \frac{2x(m+in)}{\cos t} + (m+in)^2\right)}$$

$$= e^{\frac{itd}{2}} e^{-\pi x^2} e^{2\pi im\cdot n - \pi n^2} e^{2\pi x e^{it}(m+in)} e^{-\pi(m+in)^2 e^{it} \cos t},$$

as desired. □

Using similar computations as above (see [4]), one can compute the matrix $\{T_{m',n',m,n}\}_{m',n',m,n}$:

Proposition 5.3. *Let $u(t, x) = Tu_0$ be the solution of (35), with $u_0 = M_n T_m e^{-\pi x^2}$, then*

$$\{T_{m,n,m',n'}\}_{m,n,m',n'} = \langle T(u_0), M_{n'} T_{m'} e^{-\pi x^2}\rangle$$

$$= \left(\frac{e^{it}}{2}\right)^{d/2} e^{\varphi(m,n,m',n')} \cdot e^{i\pi\Psi(m,n,m',n')}, \tag{40}$$

where

$$\varphi(m,n,m',n') = -\frac{\pi}{2}\left[|(m,n)|^2 + |(m',n')|^2 - 2\chi(m,n)\cdot \left(\begin{smallmatrix} m' \\ n' \end{smallmatrix}\right)\right]$$

and

$$\Psi(m,n,m',n') = \cos(t)nm' - \cos(t)mn' + \sin(t)mm' + \sin(t)nn' + mn - m'n'.$$

Observe that

$$|T_{m,n,m',n'}| = 2^{-d/2}e^{\varphi(m,n,m',n')}. \tag{41}$$

This equation is used to develop the numerical calculations.

We choose the Gabor frame $\{M_nT_mg_0\}$, with $(m,n) \in \mathbb{Z}^d \times (1/2)\mathbb{Z}^d$ and $g_0(x) = e^{-\pi|x|^2}$ [32, Theorem 7.5.3].

First, we study the one-dimensional case $(d = 1)$, taking the initial datum $u_0(x) = M_{1/2}T_1g_0(x) = e^{\pi i x}e^{-\pi(x-1)^2}$.

Figure 1 represents the magnitude of the coefficients (sorted in decreasing order) of one column of the matrix $T_{m',n',m,n}$, namely, that obtained by fixing $m = 1, n = 1/2$. Note we need only about 2200 coefficients to reach the threshold 10^{-300}, and the coefficient decay is preserved in time, as expected from (41).

Figure 2(a) shows a similar analysis in dimension $d = 2$, for the column corresponding to $m = (1,1), n = (1/2, 1/2)$. Here we need about 10^6 coefficients to reach the threshold 10^{-300}: the coefficient decay is much faster that the corresponding one for the free particle [19] and that obtained by means of the curvelet frames [11, Figure 15]. Finally, Figure 2(b) shows the column decay for $m = (1,1,1)$ and $n = (1/2, 1/2, 1/2)$, in dimension $d = 3$. Notice that we need about 64000 coefficients to reach the threshold 10^{-70}.

Figures 3–5 show the magnitude of the STFT $|V_g u(t,\cdot)(x,\omega)|$ of the solution $u(t,x)$ of (35), with initial datum $u_0(x) = M_{1/2}T_1g_0(x) = e^{\pi i x}e^{-\pi(x-1)^2}$ and window $g(x) = e^{-\pi|x|^2}$. The dimension is $d = 1$.

Acknowledgement

We express our thanks to the anonymous referee, for some useful remarks and comments.

FIGURE 1. Coefficient magnitude of $T_{m',n',1,1}$ at different instant time t for $d = 1$.

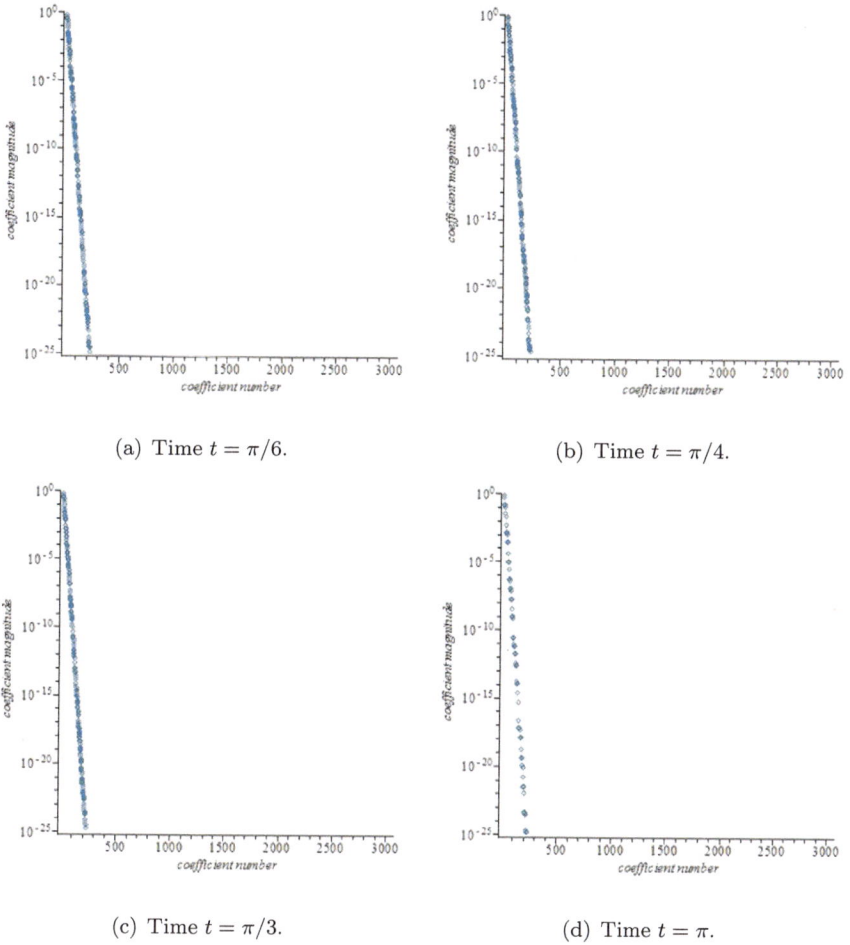

(a) Time $t = \pi/6$.

(b) Time $t = \pi/4$.

(c) Time $t = \pi/3$.

(d) Time $t = \pi$.

References

[1] K. Asada and D. Fujiwara. On some oscillatory transformation in $L^2(\mathbb{R}^n)$. *Japan J. Math.*, 4:299–361, 1978.

[2] A. Bényi, K. Gröchenig, C. Heil, and K. Okoudjou. Modulation spaces and a class of bounded multilinear pseudodifferential operators, *J. Operator Theory*, 54:389–401, 2005.

[3] A. Bényi, K. Gröchenig, K.A. Okoudjou and L.G. Rogers. Unimodular Fourier multipliers for modulation spaces. *J. Funct. Anal.*, 246(2):366–384, 2007.

FIGURE 2. Coefficient magnitude of $T_{m',n',0,0}$ at $t = \pi/3$ in dimensions $d = 2$ and $d = 3$.

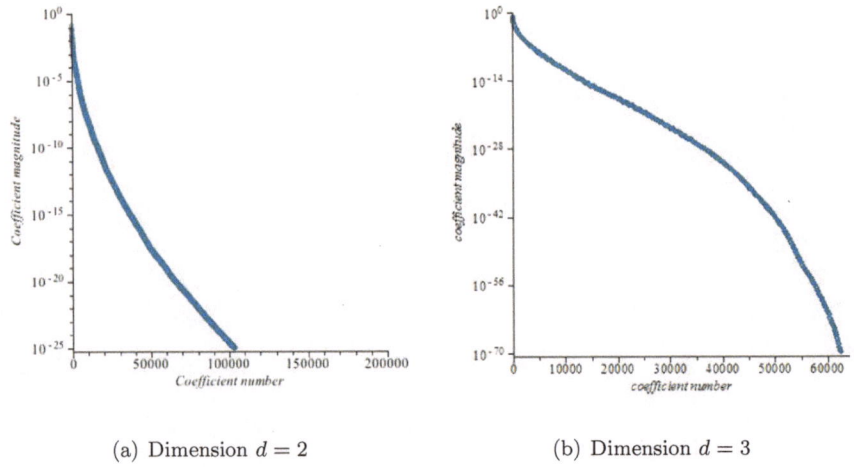

(a) Dimension $d = 2$ (b) Dimension $d = 3$

[4] M. Berra, Time-frequency analysis of Fourier Integral Operators and Applications. Master Thesis, July 2011.

[5] A. Bényi, L. Grafakos, K. Gröchenig and K. Okoudjou. A class of Fourier multipliers for modulation spaces, *Applied and Computational Harmonic Analysis*, 19:131–139, 2005.

[6] F.A. Berezin and M.A. Shubin. *The Schrödinger equation*. Mathematics and its Applications (Soviet Series), 66, Kluwer Academic Publishers Group, 1991.

[7] P. Boggiatto, E. Cordero, and K. Gröchenig. Generalized Anti-Wick operators with symbols in distributional Sobolev spaces. *Integral Equations and Operator Theory*, 48:427–442, 2004.

[8] A. Boulkhemair. Remarks on a Wiener type pseudodifferential algebra and Fourier integral operators. *Math. Res. Lett.* 4:53–67, 1997.

[9] E.J. Candés and L. Demanet. Curvelets and Fourier Integral Operators, *C. R. Math. Acad. Sci. Paris*, 336(5):395–398, 2003.

[10] E.J. Candés and L. Demanet. The curvelet representation of wave propagators is optimally sparse, *Comm. Pure Appl. Math.*, 58:1472–1528, 2005.

[11] E.J. Candés, L. Demanet and L. Ying. Fast computation of Fourier integral operators, *SIAM J. Sci. Comput.*, 29(6):2464–2493, 2007.

[12] C. Carathéodory. Variationsrechnung und partielle Differentialglichungen erster Ordnung. Teubner, Berlin, 1935, Leipzig 1956. English transl.: Holden-Day, San Francisco, 1965.

[13] F. Concetti and J. Toft. Schatten-von Neumann properties for Fourier integral operators with non-smooth symbols I, *Ark. Mat.*, 47(2):295–312, 2009.

[14] F. Concetti, G. Garello, J. Toft. Schatten–von Neumann properties for Fourier integral operators with non-smooth symbols II. *Osaka J. Math.*, 47(3):739–786, 2010.

FIGURE 3. Contourplots of the STFT magnitude of $u(t, x)$ at different instant time t.

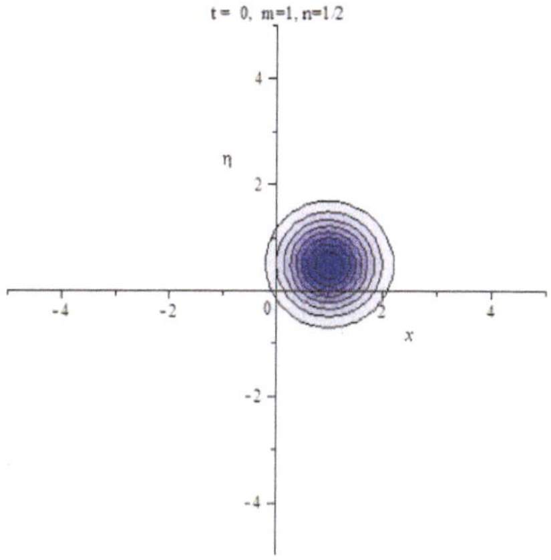

(a) Time $t = 0$.

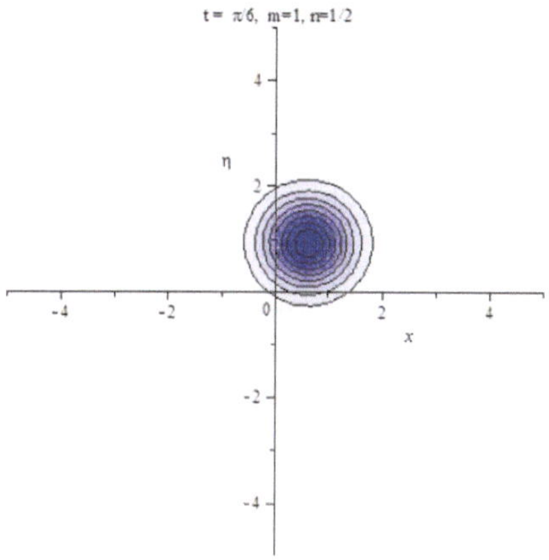

(b) Time $t = \pi/6$.

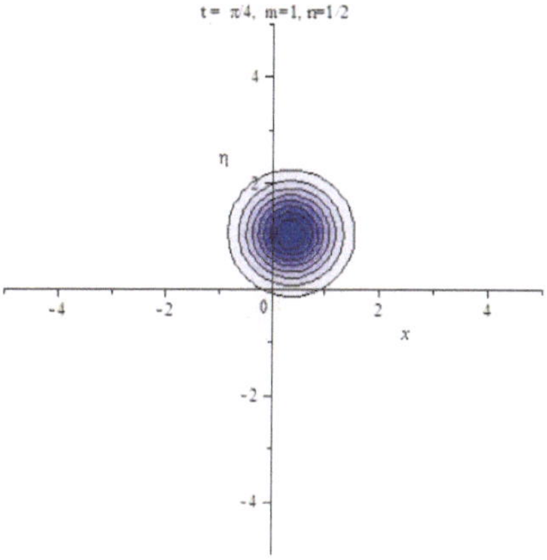

(c) Time $t = \pi/4$.

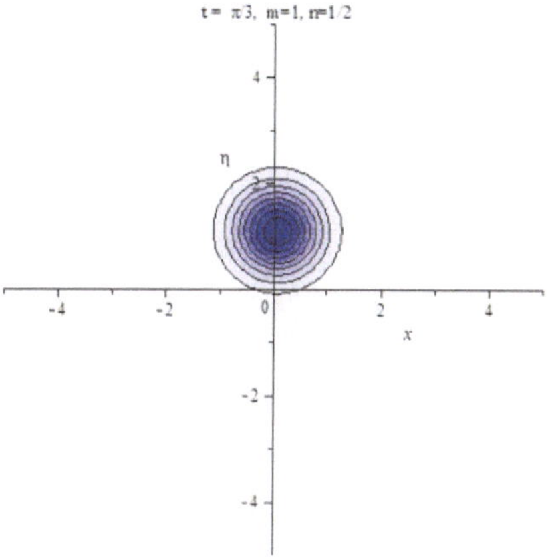

(d) Time $t = \pi/3$.

[15] F. Concetti, G. Garello, J. Toft. Trace ideals for Fourier integral operators with non-smooth symbols III. *Preprint*, available at arXiv:0802.2352.

[16] E. Cordero and F. Nicola. Metaplectic representation on Wiener amalgam spaces and applications to the Schrödinger equation. *J. Funct. Anal.*, 254:506–534, 2008.

[17] E. Cordero and F. Nicola. Boundedness of Schrödinger Type Propagators on Modulation Spaces. *J. Fourier Anal. Appl.*, 16:311–339, 2010.

[18] E. Cordero, F. Nicola and L. Rodino. Time-frequency Analysis of Fourier Integral Operators. *Comm. Pure Appl. Anal.*, 9(1):1–21, 2010.

FIGURE 4. Contourplots of the STFT magnitude of $u(t,x)$ at different instant time t.

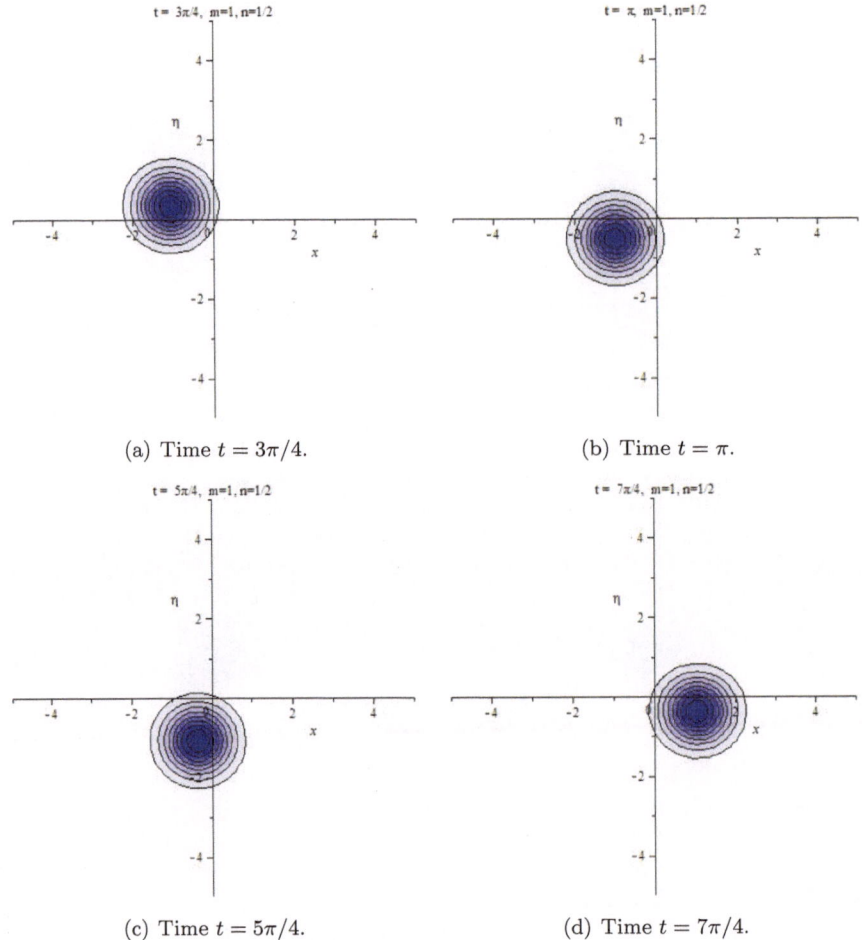

(a) Time $t = 3\pi/4$.   (b) Time $t = \pi$.

(c) Time $t = 5\pi/4$.   (d) Time $t = 7\pi/4$.

[19] E. Cordero, F. Nicola and L. Rodino, Sparsity of Gabor representation of Schrödinger propagators. *Appl. Comput. Harmon. Anal.*, 26(3):357–370, 2009.

[20] E. Cordero, F. Nicola and L. Rodino. Boundedness of Fourier Integral Operators on $\mathcal{F}L^p$ spaces. *Trans. Amer. Math. Soc.*, 361(11):6049–6071, 2009.

[21] A. Córdoba and C. Fefferman. Wave packets and Fourier integral operators. *Comm. Partial Differential Equations*, 3(11):979–1005, 1978.

[22] J.J. Duistemaat and V.W. Guillemin. The spectrum of positive elliptic operators and periodic bicharacteristics. *Invent. Math.*, 29:39–79, 1975.

FIGURE 5. Magnitude of the STFT of the solution $u(t,x)$ at different instant time t.

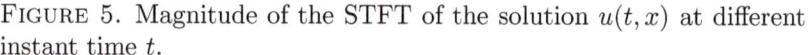

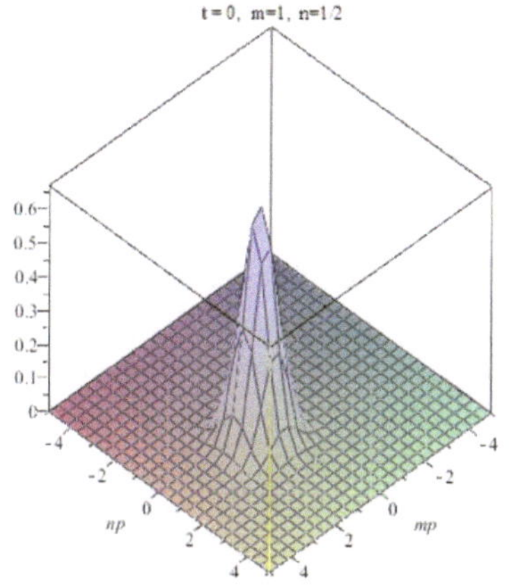

(a) Time $t = 0$. 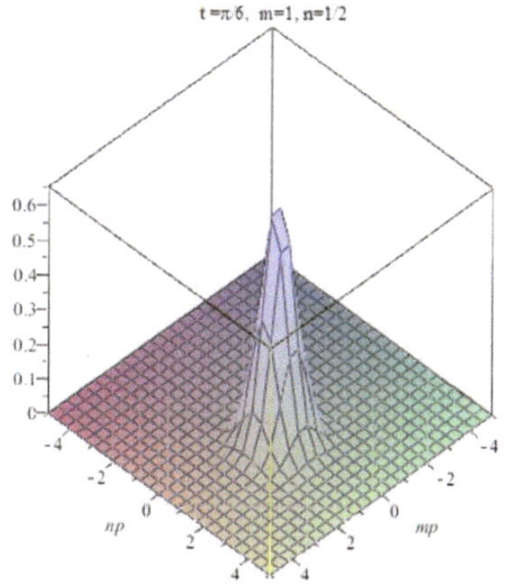   (b) Time $t = \pi/6$.

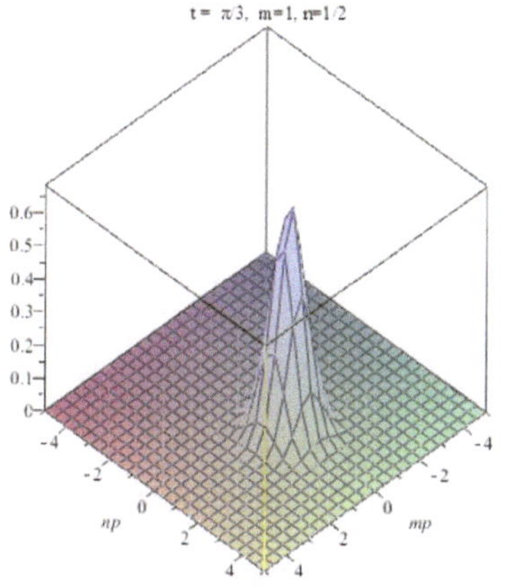

(c) Time $t = \pi/3$. 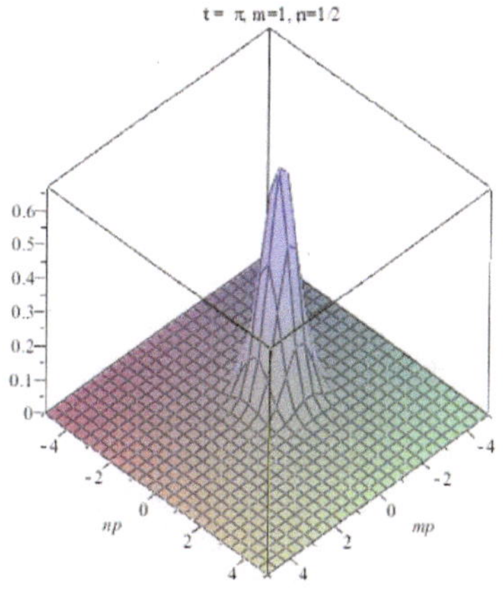   (d) Time $t = \pi$.

[23] J.J. Duistemaat and L. Hörmander. Fourier integral operators II. *Acta Math.*, 128:183–269, 1972.

[24] H.G. Feichtinger, Modulation spaces on locally compact abelian groups, *Technical Report, University Vienna, 1983*, and also in *Wavelets and Their Applications*, M. Krishna, R. Radha, S. Thangavelu, editors, Allied Publishers, 99–140, 2003.

[25] H.G. Feichtinger. On a new Segal algebra. *Monatsh. Math.*, 92(4):269–289, 1981.

[26] H.G. Feichtinger and K. Gröchenig. Gabor frames and time-frequency analysis of distributions. *J. Funct. Anal.*, 146(2):464–495, 1997.

[27] H.G. Feichtinger and K. Gröchenig. Banach spaces related to integrable group representations and their atomic decompositions I. *J. Funct. Anal.*, 86(2) 307–340, 1989.

[28] H.G. Feichtinger and K.H. Gröchenig. Banach spaces related to integrable group representations and their atomic decompositions II. *Monatsh. f. Math.*, 108:129–148, 1989.

[29] G.B. Folland. *Harmonic Analysis in Phase Space*. Princeton Univ. Press, Princeton, NJ, 1989.

[30] M. Frazier and B. Jawerth. A discrete transform and decomposition of distribution spaces. *J. Funct. Anal.*, 93:34–170, 1990.

[31] Y.V. Galperin and S. Samarah. Time-frequency analysis on modulation spaces $M_m^{p,q}$, $0 < p, q \leq \infty$. *Appl. Comp. Harm. Anal.*, 16:1–18, 2004.

[32] K. Gröchenig. *Foundations of Time-Frequency Analysis*. Birkhäuser, Boston, 2001.

[33] K. Gröchenig. An uncertainty principle related to the Poisson summation formula. *Studia Math.*, 121(1):87–104, 1996.

[34] K. Gröchenig. Time-Frequency Analysis of Sjöstrand's Class *Rev. Mat. Iberoamericana*, 22(2):703–724, 2006.

[35] K. Gröchenig and C. Heil. Modulation spaces and pseudodifferential operators. *Integral Equations Operator Theory*, 34(4):439–457, 1999.

[36] K. Gröchenig and M. Leinert. Wiener's lemma for twisted convolution and Gabor frames. *J. Amer. Math. Soc.*, 17:1–18, 2004.

[37] K. Gröchenig, Z. Rzeszotnik, and T. Strohmer. Quantitative estimates for the finite section method. *Preprint*, 2008.

[38] K. Gröchenig and S. Samarah. Nonlinear Approximation with Local Fourier Bases. *Constr. Approx.*, 16:317–331, 2000.

[39] K. Guo and D. Labate. Sparse shearlet representation of Fourier integral operators *Electron. Res. Announc. Math. Sci.*, 14:7–19, 2007.

[40] B. Helffer. *Théorie Spectrale pour des Opérateurs Globalement Elliptiques*. *Astérisque*, Société Mathématique de France, 1984.

[41] B. Helffer and D. Robert. Comportement Asymptotique Precise du Spectre d'Opérateurs Globalement Elliptiques dans $\mathbb{R}^d$. *Sem. Goulaouic-Meyer-Schwartz* 1980–81, École Polytechnique, Exposé II, 1980.

[42] L. Hörmander. Fourier integral operators I. *Acta Math.*, 127:79–183, 1971.

[43] L. Hörmander. *The Analysis of Linear Partial Differential Operators*, Vol. III, IV. Springer-Verlag, 1985.

[44] P. Lax. Asymptotic solutions of oscillatory initial value problems. *Duke Math. J.*, 24:627–646, 1957.

[45] M. Kobayashi. Modulation spaces $\mathcal{M}^{p,q}$, for $0 < p, q \leq \infty$. *J. Func. Spaces Appl.*, 4(2):329–341, 2006.

[46] S.G. Krantz and H.R. Parks. *The implicit function theorem*. Birkhäuser Boston Inc., Boston, 2002.

[47] G. Kutyniok and D. Labate. Resolution of the Wavefront Set using Continuous Shearlets. *Trans. Amer. Math. Soc.*, 361(5):2719–2754, 2009.

[48] H. Rauhut. Coorbit Space Theory for Quasi-Banach Spaces. *Studia Mathematica*, 180(3):237–253, 2007.

[49] R. Rochberg and K. Tachizawa. Pseudodifferential operators, Gabor frames, and local trigonometric bases. In *Gabor Analysis and Algorithms,* Appl. Numer. Harmon. Anal., Birkhäuser Boston, Boston, MA, 171–192, 1998.

[50] M. Ruzhansky and M. Sugimoto. Global L^2-boundedness theorems for a class of Fourier integral operators. *Comm. Partial Differential Equations,* 31(4-6):547–569, 2006.

[51] M.A. Shubin. *Pseudodifferential Operators and Spectral Theory.* Springer-Verlag, Berlin, second edition, 2001. Translated from the 1978 Russian original by Stig I. Andersson.

[52] G. Staffilani and D. Tataru. Strichartz estimates for a Schrödinger operator with nonsmooth coefficients. *Comm. Partial Differential Equations,* 27:1337–1372, 2002.

[53] E.M. Stein.*Harmonic Analysis.* Princeton University Press, Princeton, 1993.

[54] D. Tataru. Strichartz estimates for second-order hyperbolic operators with nonsmooth coefficients III. *J. Amer. Math. Soc.,* 15:419–442, 2002.

[55] J. Toft. Continuity properties for modulation spaces, with applications to pseudodifferential calculus. II. *Ann. Global Anal. Geom.,* 26(1):73–106, 2004.

[56] J. Toft. Continuity and Schatten properties for pseudo-differential operators on modulation spaces. In J. Toft, M.W. Wong, H. Zhu (eds.) *Modern Trends in Pseudo-Differential Operators,* Operator Theory: Advances and Applications, Birkhäuser Verlag Basel, 2007, 173–206.

[57] F. Treves. *Introduction to Pseudodifferential Operators and Fourier Integral Operators,* Vol. I, II., Plenum Publ. Corp., New York, 1980.

Elena Cordero and Luigi Rodino
Department of Mathematics
University of Torino
via Carlo Alberto 10
I-10123 Torino, Italy
e-mail: `elena.cordero@unito.it`
        `luigi.rodino@unito.it`

Fabio Nicola
Dipartimento di Matematica
Politecnico di Torino
corso Duca degli Abruzzi 24
I-10129 Torino, Italy
e-mail: `fabio.nicola@polito.it`

Progress in Mathematics, Vol. 301, 87–102

Geometric Regularization on Riemannian and Lorentzian Manifolds

Shantanu Dave, Günther Hörmann and Michael Kunzinger

Abstract. We investigate regularizations of distributional sections of vector bundles by means of nets of smooth sections that preserve the main regularity properties of the original distributions (singular support, wavefront set, Sobolev regularity). The underlying regularization mechanism is based on functional calculus of elliptic operators with finite speed of propagation with respect to a complete Riemannian metric. As an application we consider the interplay between the wave equation on a Lorentzian manifold and corresponding Riemannian regularizations, and under additional regularity assumptions we derive bounds on the rate of convergence of their commutator. We also show that the restriction to underlying space-like foliations behaves well with respect to these regularizations.

Mathematics Subject Classification. Primary 58J37; Secondary 46F30, 46T30, 35A27, 53C50.

Keywords. Regularization of distributions, complete Riemannian manifolds, globally hyperbolic Lorentz manifolds, algebras of generalized functions.

1. Introduction

We consider regularization processes to smooth out distributions on Riemannian and globally hyperbolic Lorentzian manifolds and investigate their compatibility with the wave-equation. The Riemannian setting has been addressed in [13] and we include an introduction to this approach. For a globally hyperbolic manifold we pick a splitting of the metric as obtained by [2] which provides us with a globally defined time function and a foliation by space-like hypersurfaces. The associated Riemannian metric naturally allows us to construct regularization processes on distributional sections of tensor-bundles and differential forms. We show that these regularizations interact nicely with the wave-equation on the Lorentzian manifold and with the foliation provided by the metric splitting.

By a regularization process we mean a net of smoothing operators that assigns a net of approximating smooth functions (or sections in a vector bundle) to any given distribution (or distributional section). We are interested in preserving a maximal set of regularity properties of the distribution in this process (support, singular support, wavefront set, Sobolev regularity). To assign such properties to approximating nets of smooth objects we employ the language of algebras of generalized functions in the sense of Colombeau ([9, 10, 31, 20]). In this approach any such regularization process provides an embedding of distributions into a space $\mathcal{G}(M)$ of nonlinear generalized functions (given by a quotient construction on spaces of approximating nets). This process preserves the regularity and singularity structure of the distributions as described in Section 2 below. The main interest in employing these regularizations is in studying non-smooth curved space-times.

Our approach is motivated by work on wave equations on non-smooth curved space-times. C.J.S. Clarke ([6, 7, 8]) suggested to study physical fields for understanding the singularity structure of the space-time itself, i.e., to consider wave equations in low regularity. His work gave rise to a number of further studies. In particular, in the framework of generalized functions this line of research was pursued in [34, 19, 33]. For a more detailed introduction we refer to [25].

Thus let us consider a model situation and analyze the generalized solutions to wave equations corresponding to a generalized Lorentz metric $\tilde{g}$ on the smooth manifold M (cf. [20, 27, 19]), i.e., a global Cauchy problem with initial data $a, b \in \mathcal{D}'(S)$ on a suitable initial value surface S in the form

$$\Box_{\tilde{g}} u = 0, \quad u|_S = \iota(a), \quad \nabla_n u|_S = \iota(b),$$

where ι is assumed to be an embedding of distributions into the algebra of generalized functions.

Suppose that the generalized Lorentzian structure induced by $\tilde{g}$ allows a splitting of the wave operator in the form $\Box_{\tilde{g}} = \gamma \cdot (\Box_g + Q)$, where g is a smooth "background" metric, γ is a positive measure or a strictly positive generalized function, and Q is a partial differential operator with distributions or generalized functions as coefficients, but whose coefficient singularities are concentrated in certain space-time regions. A simple example is a generalized Robertson-Walker space-time with $\tilde{g} = -dt^2 + (1 + \mu(t))h$ on $M = \mathbb{R} \times S$, where h is a smooth Riemannian metric on the smooth manifold S and μ is a nonnegative generalized function. In this case $g = -dt^2 + h$ can serve as a smooth background metric and the above splitting would involve $\gamma = 1/(1 + \mu)$ and $Q = Q(t, \partial_t)$, which acts only on the one-dimensional factor.

If we impose the splitting assumption as above, then we have for any generalized function $u \in \mathcal{G}(M)$ that the equation $\Box_{\tilde{g}} u = 0$ is equivalent to $\Box_g u + Qu = 0$. Let $u \in \mathcal{G}(M)$ be a solution to the Cauchy problem stated above. In attempting to extract its distributional aspects or assess its singularity structure we could adopt the following strategy. Let v be the distributional solution (assuming that it exists and is unique) to

$$\Box_g v = 0, \qquad v|_S = a, \quad \nabla_n v|_S = b,$$

and put $w := \iota(v) - u \in \mathcal{G}(M)$ as a comparison of the generalized function u with its "background distributional aspect". Now, *if the embedding ι commutes with restriction to S and with $\Box_g$*, then simple manipulations allow us to draw the conclusion that w satisfies the following Cauchy problem:

$$\Box_g w = Qu, \qquad w|_S = 0, \quad \nabla_n w|_S = 0.$$

Thus, w satisfies an inhomogeneous Cauchy problem corresponding to the smooth background metric g and with generalized functions occurring only on the right-hand side of the equation. In particular, w will be smooth in regions where null geodesics corresponding to g emanating from S do not intersect the $\mathcal{C}^\infty$-singular support of Qu.

In the more general situation of a globally hyperbolic Lorentz manifold provided with a foliation into spacelike hypersurfaces via a splitting of the metric we shall construct an embedding ι and show that the embedding almost commutes with $\Box_g$ and with restriction to space-like slices of the foliation. The extent of failure to commute is roughly measured by the divergence of the volume element on the slices along the time direction. Although we state our results in the scalar case, the extension to the case of wave-equations on differential forms is obvious.

2. Regularization of distributions on complete Riemannian manifolds

In this section we describe a geometric regularization procedure for distributions (or distributional sections of vector bundles) on complete Riemannian manifolds. This procedure encodes regularity and singularity features in terms of asymptotic behavior. Our approach is based on [13], to which we refer for further details.

We first fix some notations from the theory of distributions. Let M be an orientable complete Riemannian manifold of dimension n with Riemannian metric g. The space $\mathcal{D}'(M)$ of Schwartz distributions on M is the dual of the space $\Omega_c^n(M)$ of compactly supported n-forms on M. Further, $\mathcal{D}(M)$ is the space of smooth compactly supported functions on M. We identify $\mathcal{D}(M)$ with $\Omega_c^n(M)$ via $f \mapsto f \cdot dg$, with dg the Riemannian volume form induced by g. Thus $\mathcal{D}'(M)$ can be viewed as the dual space of $\mathcal{D}(M)$. We consider $L_{loc}^1(M)$ (hence in particular $\mathcal{C}^\infty(M)$) a subspace of M via $f \mapsto [\varphi \mapsto \int_M f\varphi dg]$. If E is a vector bundle over M then $\mathcal{D}'(M : E)$, the space of E-valued distributions on M is defined by $\mathcal{D}'(M : E) = \mathcal{D}'(M) \otimes_{\mathcal{C}^\infty(M)} \Gamma^\infty(M : E)$, with $\Gamma^\infty(M : E)$ the space of smooth sections of E. By $\mathcal{E}'(M : E)$ we denote the space of compactly supported distributional sections of E. We shall assume that E is endowed with a Hermitian inner product so that the distributional sections $\mathcal{D}'(M : E)$ of E can be identified with the dual $\Gamma_c^\infty(M : E)'$ of the space of compactly supported smooth sections. Regularity properties of distributions we are interested in are, in particular, encoded in the singular support singsupp(w), the wavefront set WF(w), and the Sobolev regularity of any given $w \in \mathcal{D}'(M)$.

In order to be able to track these regularity properties in terms of regularizations of a given distribution we need a conceptual framework that allows to assign geometrical and analytical properties to regularizations, i.e., to nets of smooth functions. Algebras of generalized functions in the sense of J.F. Colombeau ([9, 10, 31]) indeed provide a well-developed theory of this kind and below we shall briefly review some basic definitions, based mainly on [20, 16].

The basic idea of Colombeau's approach is to assign to any given locally convex space X a space $\mathcal{G}_X$ of generalized functions as follows. We define the space $\mathcal{M}_X$ of moderate nets to consist of those maps $\varepsilon \mapsto x_\varepsilon$ ($\varepsilon \in I := (0, 1]$) that are smooth and for any seminorm ρ of X satisfy $\rho(X_\varepsilon) = O(\varepsilon^N)$ for some integer N as $\varepsilon \to 0$. Similarly we call a net x_ε negligible if for all seminorms ρ and all integers N the asymptotic relation $\rho(X_\varepsilon) \sim O(\varepsilon^N)$ holds. The space $\mathcal{G}_X$ is defined as the quotient of the space of all moderate nets by the space $\mathcal{N}_X$ of all negligible nets. The class represented by a net x_ε shall be denoted by $[x_\varepsilon]$. In case M is a smooth manifold we call $\mathcal{G}(M) := \mathcal{G}_{\mathcal{C}^\infty(M)}$ the standard (special) Colombeau algebra of generalized functions on M ([9, 12, 20]). If $E \to M$ is a vector bundle then we set $\mathcal{G}(M : E) := \mathcal{G}_{\Gamma^\infty(M:E)}$. For $E = \mathbb{C}$ the space $\mathcal{G}_\mathbb{C}$ inherits a ring structure from $\mathbb{C}$. It is therefore called the space of generalized numbers, denoted by $\tilde{\mathbb{C}}$. Every space $\mathcal{G}_E$ is naturally a $\tilde{\mathbb{C}}$-module, and hence is called the $\tilde{\mathbb{C}}$-module associated with E ([16]).

Similarly we shall also consider the subspace $\mathcal{G}_X^\infty$ of regular elements of $\mathcal{G}_X$, defined as the space of all those elements of $\mathcal{G}_X$ that can be represented by a net x_ε for which one can find an integer N in the above relations independent of any seminorm ρ on X. Thus $\mathcal{G}_X^\infty$ is the space of uniformly controlled asymptotics. Again we set $\mathcal{G}^\infty(M) := \mathcal{G}_{\mathcal{C}^\infty(M)}^\infty$. In regularity theory, $\mathcal{G}^\infty(M)$ is the analogue of $\mathcal{C}^\infty(M)$ in the theory of distributions ([31, 24, 14, 17]). This is based on the fundamental result ([31]) that for open subsets Ω of $\mathbb{R}^n$, $\iota(\mathcal{D}'(\Omega)) \cap \mathcal{G}^\infty(\Omega) = \mathcal{C}^\infty(\Omega)$, where ι is the standard embedding of $\mathcal{D}'$ into $\mathcal{G}$ via convolution with a standard mollifier. We will introduce further notions of regularity theory based on $\mathcal{G}^\infty$ below.

The assignment $X \mapsto \mathcal{G}_X$ is obviously functorial, so that any continuous linear map $\phi : X \to Y$ naturally induces a map $\phi_* : \mathcal{G}_X \to \mathcal{G}_Y$. Thus a smooth map $f : M \to N$ gives rise to a pull-back $f^* : \mathcal{G}(M) \to \mathcal{G}(M)$. In particular $\mathcal{G}(M)$ defines a fine sheaf of algebras and similarly $\mathcal{G}(M : E)$ is a fine sheaf of $\mathcal{G}(M)$-modules.

We finally introduce the notion of association, which provides a concept of weak equality between elements of Colombeau spaces as well as a standardized way of assigning distributional limits to certain Colombeau generalized functions. For $u, v \in \mathcal{G}(M : E)$ we say that u is associated (or weakly equal) to v, $u \approx v$ if $u_\varepsilon - v_\varepsilon \to 0$ in $\mathcal{D}'(M : E)$ for (some, hence any) representatives of u, v. We say that u possesses $w \in \mathcal{D}'(M : E)$ as a distributional shadow if $u_\varepsilon \to w$ in $\mathcal{D}'(M : E)$.

To construct the desired regularization process on $\mathcal{D}'(M : E)$ we need two pieces of data:

1. A Schwartz function $F \in \mathscr{S}(\mathbb{R})$ such that $F \equiv 1$ near the origin.
2. An elliptic symmetric differential operator D on E such that the speed of propagation C_D, defined as

$$C_D = \sup\{\|\sigma_D(x,\xi)\| \mid x \in M, \|\xi\| = 1\} \tag{2.1}$$

is finite. Here σ_D denotes the principal symbol of D. Such an operator D will be referred to as *admissible*.

As a consequence of the finite speed of propagation, the symmetric operator D is essentially self-adjoint ([23, 10.2.11]). Therefore the equation

$$\frac{\partial}{\partial t}u = iDu \qquad u(\,.\,,0) = u_0, \tag{2.2}$$

has a unique solution for all times t for any initial datum $u_0 \in \Gamma_c^\infty(M : E)$. Indeed it follows from functional calculus that $e^{itD}u_0$ is a solution, and uniqueness can be established using energy estimates.

We note that functional calculus defines a map

$$\mathscr{S}(\mathbb{R}) \ni f \mapsto f(D) \in \mathcal{B}(L^2(M : E))$$

where the operator $f(D)$ can also be expressed by the Fourier inversion formula:

$$f(D) := \frac{1}{2\pi}\int_{\mathbb{R}} \hat{f}(s)e^{isD}ds \tag{2.3}$$

(in the sense of strong operator convergence). Since the operator D is elliptic, it follows that $f(D)$ has a smooth kernel by elliptic regularity.

Let $F \in \mathscr{S}(\mathbb{R})$ with $F \equiv 1$ near the origin and set $F_\varepsilon(x) := F(\varepsilon x)$. Then we shall obtain the desired regularizing procedure from $F_\varepsilon(D)$. To do this we first take a closer look at singularity properties available in generalized functions.

As has been mentioned above the space $\mathcal{G}(M : E)$ is a fine sheaf over M, hence provides a notion of support. Furthermore, the notion of wave-front set can be defined for generalized functions both locally and in invariant global terms analogous to distributions. For $\Omega \subseteq \mathbb{R}^n$ open we call $u \in \mathcal{G}(\Omega)$ $\mathcal{G}^\infty$-microlocally regular at $(x_0, \xi_0) \in T^*\Omega \setminus 0$ if there exists some test function $\varphi \in \mathcal{D}(\Omega)$ with $\varphi(x_0) = 1$ and a conic neighborhood $\Gamma \subseteq \mathbb{R}^n \setminus 0$ of ξ_0 such that the Fourier transform $\mathcal{F}(\varphi u)$ is rapidly decreasing in Γ, i.e., there exists N such that for all l,

$$\sup_{\xi \in \Gamma}(1 + |\xi|)^l |(\varphi u_\varepsilon)^\wedge(\xi)| = O(\varepsilon^{-N}) \qquad (\varepsilon \to 0).$$

The generalized wave front set of u, $\mathrm{WF}_g(u)$, is the complement of the set of points (x_0, ξ_0) where u is $\mathcal{G}^\infty$-microlocally regular. By [21], for any $u \in \mathcal{G}(M)$, $\mathrm{WF}_g(u)$ can naturally be viewed as a subset of $T^*M \setminus 0$. It is equivalently defined as (cf.[17]),

$$\mathrm{WF}_g(u) = \bigcap_{Pu \in \mathcal{G}^\infty(M)} \mathrm{char}(P) \qquad (P \in \Psi_{cl}^0(M)).$$

Here char$(P) \subseteq T^*M$ is the characteristic set of the order 0 classical pseudodifferential operator P. The singular support of u, singsupp$_g(u)$, is the complement of the maximal open set on which $u \in \mathcal{G}^\infty$. It then follows that singsupp$_g(u) =$ pr$_1$(WF$_g$).

We are now ready to define a class of regularization procedures for distributional sections of a vector bundle.

2.1. Regularizations

Definition 2.1. A parametrized family $(T_\varepsilon)_{\varepsilon \in I}$ of properly supported smoothing operators (in the sense of [4, Ch. 1.4]) is called an *optimal regularization process* if

1. The regularization of any compactly supported distributional section $s \in \mathcal{E}'(M : E)$ is of moderate growth: For any continuous seminorm ρ on $\Gamma^\infty(M : E)$, there exists some integer N such that

$$\rho(T_\varepsilon s) = O(\varepsilon^N) \qquad (\varepsilon \to 0).$$

2. The net (T_ε) is an approximate identity: for each $s \in \mathcal{E}'(M : E)$,

$$\lim_{\varepsilon \to 0} T_\varepsilon s = s \quad \text{in } \mathcal{D}'(M : E).$$

3. If $u \in \Gamma_c^\infty(M : E)$ is a smooth compactly supported section of E then for all continuous seminorms ρ and given any integer m,

$$\rho(T_\varepsilon u - u) = O(\varepsilon^m).$$

4. The induced map $\iota_T : \mathcal{E}'(M : E) \to \mathcal{G}(M : E)$ preserves support, and extends to a sheaf map $\mathcal{D}'(M : E) \to \mathcal{G}(M : E)$ that satisfies,

$$\iota_T(\mathcal{D}'(M : E)) \cap \mathcal{G}^\infty(M : E) = \Gamma^\infty(M : E). \qquad (2.4)$$

   This implies in particular that ι_T preserves singular support.

5. The map ι_T preserves wave-front sets in the sense that for any distribution $s \in \mathcal{D}'(M : E)$

$$\text{WF}(u) = \text{WF}_g(\iota_T(u)).$$

As mentioned already we shall use spectral properties of the elliptic differential operators to obtain regularizing processes satisfying the above conditions.

Our main result in this section is the following.

Theorem 2.2. *Let $F \in \mathscr{S}(\mathbb{R})$ be a Schwartz function such that $F \equiv 1$ near the origin. Let $F_\varepsilon(x) := F(\varepsilon x)$. Given an admissible differential operator D, the family of operators $(F_\varepsilon(D))_{\varepsilon \in I}$ provides an optimal regularization process in sense of Definition 2.1.*

The essential idea of the proof boils down to the following two steps.

First we prove Theorem 2.2 under the assumption that the underlying manifold M is compact. We then use finite propagation speed to extend the result to a general complete Riemannian manifold.

Thus let us for the time being assume that M is a closed manifold. Then the space of smoothing operators $\Psi^{-\infty}(M : E)$ is a Fréchet algebra. The functional calculus map defined by D induces a smooth map $\phi_D : \mathscr{S}(\mathbb{R}) \to \Psi^{-\infty}(M : E)$, hence the estimate 1 holds because F_ε is a moderate net and $F_\varepsilon(D) = (\phi_D)_*(F_\varepsilon)$, where $(\phi_D)_*$ is the induced map on the asymptotic spaces $\mathcal{G}_{\mathscr{S}(\mathbb{R})} \to \mathcal{G}_{\Psi^{-\infty}(M)}$. Condition 2 then follows as the ring map ϕ_D preserves approximate units.

When M is compact the operator D has discrete spectrum and the spectrum of D^2 satisfies Weyl's law, namely

$$N_{D^2}(\lambda) := \#\{\lambda_i \in \mathrm{sp}(D^2) | \ \lambda_i \leq \lambda\} \sim C\lambda^{\frac{\dim(M)}{2}}. \tag{2.5}$$

The Weyl estimates in conjunction with the fact that $F \equiv 1$ near the origin now provides the compatibility condition 3 (see [11]).

To describe the regularity of a distributional section $u \in \mathcal{D}'(M : E)$ we first note that any distribution provides a map between two Fréchet spaces, namely between the smoothing operators $\Psi^{-\infty}(M : E)$ and the smooth sections $\Gamma^\infty(M : E)$ by evaluation. More precisely, to any $u \in \mathcal{D}'(M : E)$ we associate the map

$$\Theta_u : \Psi^{-\infty}(M) \to \Gamma^\infty(M : E) \qquad \Theta_u(T) := T(u).$$

The mapping properties of the maps Θ_u imply that if $u \notin H^k(M : E)$ for every $k > t$ then given any $\delta > 0$, $\|F_\varepsilon(D)u\|^2_{L^2(M:E)}$ is not $O(\varepsilon^{\frac{\dim M}{2}+t+\delta})$ (see [11, Lemma 7.4]). This direct description of Sobolev regularity is in fact stronger than condition 4 and therefore,

$$\iota_{F_\varepsilon(D)}(\mathcal{D}'(M : E)) \cap \mathcal{G}^\infty(M : E) = \Gamma^\infty(M : E).$$

The sheaf property of the embedding $\iota_{F_\varepsilon(D)}$ as well as the proof of the result for the more general case of a not necessarily compact complete manifold rely on the finite speed of propagation of D, to which we turn next.

2.2. Finite speed of propagation

We shall reduce the computation of asymptotics on a complete Riemannian manifold to certain compact manifolds obtained as doubles of suitable compact submanifolds with boundaries. If X is a compact manifold with boundary, a double of X, denoted here by DX is a closed manifold obtained by gluing two copies of X along the boundary ∂X (see, e.g., [26]). If X is a compact manifold with boundary embedded in a Riemannian manifold M of the same dimension and if U is an open subset of M such that $\bar{U} \subset \mathrm{interior}(X)$, then there exists a Riemannian metric on DX such that the inclusion $j : U \hookrightarrow DX$ becomes an isometric embedding. Moreover, for any vector bundle $E \to M$ there exists a corresponding vector bundle $E_X \to DX$ such that $E_X|_U$ is canonically isomorphic to $E_{|U}$. Also, there exists a symmetric elliptic operator D_X on E_X that coincides with D on U.

Let $u \in \mathcal{E}'(M : E)$ and fix a constant $c > 0$. Then by the Hopf-Rinow theorem the open ball $U := B_{2c \cdot C_D}(\mathrm{supp}(u))$ (with C_D as defined in (2.1)) is relatively compact and therefore contained in a compact manifold with boundary $X \subseteq M$. By the above u can be identified with a distributional section of $E_X \to DX$.

Proposition 2.3. *With assumptions on u, c and F as above $F(D)u$ and $F(D_X)u$ are both supported in U and*

$$F(D)u = F(D_X)u.$$

Proof. The restrictions of D and D_X to the open set U coincide, hence uniqueness of solutions to (2.2) implies that $e^{isD}u$ and $e^{isD_X}u$ agree for $s \leq c$. Hence the claim follows from the Fourier Inversion Formula (2.3). □

From these observations the proof of Theorem 2.2 can be deduced along the following lines (cf. [13, Sec. 4]):

Let $\phi \in \mathcal{D}(-c, c)$ satisfy $\phi \equiv 1$ in a neighborhood of 0 and let $u \in \mathcal{E}'(M : E)$. Then

$$F_\varepsilon(D)u = \frac{1}{2\pi} \int_{-\infty}^{\infty} \phi(s)\hat{F}_\varepsilon(s)e^{isD}u \, ds = j^* \left(\frac{1}{2\pi} \int_{-\infty}^{\infty} \phi(s)\hat{F}_\varepsilon(s)e^{isD_X}u \, ds \right).$$

Based on this calculation we observe that the estimates required for Definition 2.1 in the general case of $F_\varepsilon(D)$ follow from that of $F_\varepsilon(D_X)$ which have already been established in case of the closed manifold DX. Furthermore the sheaf properties of $\iota_{F_\varepsilon(D)}$ can be obtained as a consequence of the calculations in Proposition 2.3. The support of u coincides with the generalized support of $[F_\varepsilon(D)u]$. This implies that the embedding extends to a sheaf morphism $i_{F_\varepsilon} : \mathcal{D}'(M : E) \to \mathcal{G}(M : E)$. Preservation of wavefront sets is more involved. We refer to [13, Th. 3.10] for a complete proof of this property.

Remark 2.4. In the scalar case, an alternative proof (not relying on the above doubling-technique) of Theorem 2.2 can be found in [13], Section 3.

2.3. Applications

As a first application consider $E = \bigwedge^* M$, the exterior bundle over a Riemannian manifold M and $D = d + d^*$ where d^* is the Hodge adjoint of the de Rham differential d. Then D is symmetric, elliptic and has propagation speed $C_D = 1$ since $\sigma_D(x, \xi)^2 = -\|\xi\|^2 \text{id}$, hence is essentially self-adjoint. Letting $\Delta := D^2$ be the Laplace-Beltrami operator on $\Omega^*(M)$, it follows from functional calculus that $\cos(sD) = \cos(s\sqrt{\Delta})$ on $L^2(M)$.

Now let F be an even Schwartz function which has germ 1 at the origin. Then

$$F(\sqrt{\Delta}) = \frac{1}{2\pi} \int_{-\infty}^{\infty} \hat{F}(s) \cos(s\sqrt{\Delta}) \, ds$$

as a Bochner integral in $\mathcal{B}(L^2(M))$.

Finally, let $c > 0$ and pick an even test function ϕ_c with support in $(-2c, 2c)$ and such that $\phi_c \equiv 1$ on $(-c, c)$. Now set

$$T_\varepsilon(\sqrt{\Delta}) := \frac{1}{2\pi} \int_{-\infty}^{\infty} \phi_c(s)(F_\varepsilon)^\wedge(s) \cos(s\sqrt{\Delta}) \, ds. \qquad (2.6)$$

Then each $T_\varepsilon(\sqrt{\Delta})$ is a properly supported smoothing operator and $(T_\varepsilon(\sqrt{\Delta}))_{\varepsilon \in I}$ is an optimal regularization process in the sense of Definition 2.1. From the construction of T in terms of the functional calculus of the Laplace operator we conclude the following invariance properties of the corresponding embedding ι_T:

Corollary 2.5.

(i) Let $f : M \to M$ be an isometry. Then for any $u \in \mathcal{D}'(M)$, $\iota_T(f^*u) = f^*\iota_T(u)$.

(ii) If Ψ is a pseudodifferential operator commuting with Δ, then Ψ commutes with ι_T.

For the special case $\mathbb{R}^n$ with the Euclidean metric the above construction gives

$$T_\varepsilon u = \mu_\varepsilon * u, \quad \text{with} \quad \widehat{\mu_\varepsilon}(\xi) = \frac{1}{2\pi}\mathcal{F}(\phi\frac{1}{\varepsilon}\widehat{F}(\frac{\cdot}{\varepsilon}))(|\xi|).$$

In particular, for the one-dimensional case $n = 1$ we obtain $\mu_\varepsilon = \phi_c \cdot \widehat{F}(./\varepsilon)/(2\pi\varepsilon)$. Note that $\widehat{F}(./\varepsilon)/(2\pi\varepsilon)$ is a standard mollifier, i.e., a Schwartz function with unit integral and all higher moments vanishing. Thus the regularization process reduces precisely to the usual Colombeau embedding via convolution ([10, 31, 20]).

A further consequence of the above construction is that it relates naturally to isomorphisms of vector bundles. Thus let $\phi : E_1 \to E_2$ be a vector-bundle isomorphism (covering the identity map on M) and let D_1 be an order one admissible operator on E_1. We choose a Hermitian structure on E_2 which makes ϕ an isometry. Then the push-forward operator $D_2 := \phi D_1 \phi^{-1}$ is isospectral to D_1. The naturality of functional calculus then gives that for any Schwartz function F with $F \equiv 1$ near the origin the embedding and the bundle map ϕ commute, that is

$$\phi \circ F_\varepsilon(D_1) = F_\varepsilon(D_2) \circ \phi.$$

We note that if $r_1 + s_1 = r_2 + s_2$ then any Riemannian metric provides an isomorphism of the tensor bundles $\mathfrak{T}^{r_1}_{s_1} \to \mathfrak{T}^{r_2}_{s_2}$ by 'raising or lowering of indices'. Thus if we pick the connection Laplace operators on the tensor bundles and an even Schwartz function F as above we obtain a regularization process that is well behaved with respect to raising and lowering of indices.

3. Regularization on globally hyperbolic space-times

We now return to the situation described at the end of Section 1. Thus, throughout this section, (M, g) will be a smooth space-time, i.e., a connected time-oriented Lorentz manifold. We first review the concept of metric splitting for globally hyperbolic space-times. Building on this we construct invariant regularizations of distributions on smooth globally hyperbolic space-times.

3.1. The metric splitting of globally hyperbolic space-times and associated Riemann metrics

The original definition, due to J. Leray of global hyperbolicity of a space-time M appeared in [28]. It requires that the set of causal curves connecting two points p, $q \in M$ be compact in the space of all rectifiable paths with respect to a suitable metric topology ([5, Ch. XII, Sec. 8, 9]). In what follows, we will use an equivalent definition, cf. [22, Sec. 6.6], and [5, Ch. XII, Th. 10.2].

Thus we call a space-time *globally hyperbolic* if it satisfies (a) strong causality and (b) for any $p, q \in M$ the intersection $J^+(p) \cap J^-(q)$ of the causal future $J^+(p)$ of p with the causal past $J^-(q)$ of q is a compact subset of M. Thanks to [3] condition (a) may be weakened to causality, i.e., non-existence of closed time-like curves.

As shown by Geroch (cf. [18]) global hyperbolicity is equivalent to the existence of a Cauchy hypersurface and in turn provides a foliation of M by Cauchy hypersurfaces. Further development of these constructions and techniques led to the following result due to Bernal-Sánchez (cf. [2]) on the so-called metric splitting of a globally hyperbolic space-time (M, g): There exists a Cauchy hypersurface S in M and an isometry of (M, g) with the Lorentz manifold $(\mathbb{R} \times S, \lambda)$ with Lorentz metric λ given by

$$\lambda = -\beta \, dt^2 + h_t, \tag{3.1}$$

where $\beta \in C^\infty(\mathbb{R} \times S)$ is positive, $(h_t)_{t \in \mathbb{R}}$ is a smoothly parametrized family of Riemannian metrics on S, and t denotes (slightly ambiguously) both the global time function $(t, x) \mapsto t$ and its values. In other words, in order to construct a regularization (or embedding) for distributions on the globally hyperbolic Lorentzian manifold (M, g) we may as well assume that $(M, g) = (\mathbb{R} \times S, \lambda)$.

The specific structure of the Lorentz metric λ in (3.1) suggests to associate with it the Riemann metric

$$\rho := \beta \, dt^2 + h_t \tag{3.2}$$

on $\mathbb{R} \times S$ and to simply use the regularization and embedding of distributions on $\mathbb{R} \times S$ based on this Riemann metric ρ. The regularization construction on Riemannian manifolds described above requires completeness of the Riemann metric ρ. It may happen that ρ is not complete, however we may then introduce an appropriate conformal factor to obtain a complete Riemannian metric (cf. [30]). Note that the latter would amount to introducing the exact same conformal factor for the Lorentz metric λ and would not change the class in the so-called causal hierarchy of space-times according to Minguzzi-Sánchez [29]. Therefore we assume henceforth that ρ is complete.

Denoting by Δ_ρ the Laplacian w.r.t. ρ, for any $s \in \mathbb{R}$ and any $u \in \mathcal{D}(M)$ we set

$$\|u\|_s := \|(1 + \Delta_\rho)^{s/2} u\|_{L^2(M)}.$$

The Sobolev space $H^s(M)$ of order s is the completion of $\mathcal{D}(M)$ with respect to this norm.

Example. (i) Consider $S = \mathbb{R}$ and $\lambda = -dt^2 + \frac{dx^2}{1+t^6x^6}$ as Lorentz metric on $\mathbb{R}^2$. It is elementary to check that $(\mathbb{R}^2, \lambda)$ is globally hyperbolic, e.g., by showing that $\{0\} \times \mathbb{R}$ is a Cauchy hypersurface. The associated Riemann metric $\rho = dt^2 + \frac{dx^2}{1+t^6x^6}$ is not complete, since the hypersurfaces $\{t\} \times \mathbb{R}$ with $t \neq 0$ are closed and bounded but not compact in $(\mathbb{R}^2, \rho)$. Multiplying ρ by the function $\alpha \in C^\infty(\mathbb{R}^2)$, $\alpha(t,x) = 1+t^6x^6$, yields a conformal metric which is complete (by the Hopf-Rinow theorem, since lengths of curves w.r.t. $\alpha \cdot \rho$ are greater than or equal to their Euclidean lengths).

(ii) Let (S, h_0) be a connected Riemannian manifold. For the Robertson-Walker space-time $\mathbb{R} \times S$ with Lorentz metric of the form $\lambda = -dt^2 + f(t)^2 h_0$, where $f \in C^\infty(\mathbb{R})$ is positive, we have (cf. [1, Lemma A.5.14]):

($\mathbb{R} \times S, \lambda$) is globally hyperbolic if and only if (S, h_0) is complete.

If this is the case, then the corresponding Riemannian metric $\rho = dt^2 + f(t)^2 h_0$ on $\mathbb{R} \times S$ is complete (see [32], Lemma 7.40).

3.2. Regularization and embedding via the associated Riemannian structure

Let $(T_\varepsilon)_{\varepsilon \in I}$ be the regularization (i.e., family of properly supported smoothing operators) and $\iota: \mathcal{D}'(\mathbb{R} \times S) \hookrightarrow \mathcal{G}(\mathbb{R} \times S)$ be the embedding associated with the complete Riemann metric ρ as in (2.6). We recall that the embedding thus respects the differential algebraic structure of $C^\infty(\mathbb{R} \times S)$ and the wave front sets of distributions in the strong sense of generalized functions, that is, with respect to equality on the level of $\mathcal{G}(\mathbb{R} \times S)$. Furthermore, ι is also invariant under isometries of the Riemannian structure and commutes with the action of the Laplace operator Δ_ρ corresponding to ρ on distributions and generalized functions, respectively.

Let $u \in \mathcal{D}'(\mathbb{R} \times S)$. Then according to (3.1) the difference between the wave operator $\Box_\lambda$ and the Laplace operator Δ_ρ acts on u as

$$\Box_\lambda u - \Delta_\rho u = \frac{-2}{\sqrt{\beta \det h_t}} \partial_t \left(\sqrt{\frac{\det h_t}{\beta}} \partial_t u \right) := -2\Theta u. \tag{3.3}$$

From this equation and the commutation property of the embedding ι with Δ_ρ we obtain for any $u \in \mathcal{D}'(\mathbb{R} \times S)$

$$\iota(\Box_\lambda u) - \Box_\lambda \iota(u) = \iota((\Delta_\rho - 2\Theta)u) - (\Delta_\rho - 2\Theta)\iota(u) = 2(\Theta\iota(u) - \iota(\Theta u)). \tag{3.4}$$

Thus, the precise invariance properties of the embedding with respect to the wave operator can be reduced to investigating the corresponding behavior upon interchanging ι with Θ.

As a first simple, but useful, observation we point out that ι commutes with $\Box_\lambda$ in the distributional sense (i.e., in the sense of association in $\mathcal{G}$), which follows from the fact that $\Box_\lambda$ is a differential operator with smooth symbol. This proves the following statement.

Proposition 3.1. *For any $u \in \mathcal{D}'(\mathbb{R} \times S)$ we have $\iota(\Box_\lambda u) \approx \Box_\lambda \iota(u)$.*

We will now show that under additional regularity assumptions on the distribution u a stronger asymptotic property holds.

Theorem 3.2. *If $u \in H^3(\mathbb{R} \times S)$ is compactly supported then*

$$\| [T_\varepsilon, \Box_\lambda] u \|_{L^2} = \mathcal{O}(\varepsilon^2) \qquad (\varepsilon \to 0).$$

Proof. As in (3.4) we have the basic relation

$$[T_\varepsilon, \Box_\lambda] u = 2[\Theta, T_\varepsilon] u.$$

Now recalling the Bochner-integral defining the action of T_ε on L^2-functions we have

$$\Theta T_\varepsilon u - T_\varepsilon \Theta u = \frac{1}{2\pi} \int_{\mathbb{R}} \phi_c(s) \frac{1}{\varepsilon} \widehat{F}(\frac{s}{\varepsilon}) \left(\Theta \cos(s\sqrt{-\Delta_\rho})u - \cos(s\sqrt{-\Delta_\rho})\Theta u \right) ds.$$

Hence everything boils down to deriving asymptotic estimates for the commutator of Θ with the operator $S(s) := \cos(s\sqrt{-\Delta_\rho})$ ($s \in \mathbb{R}$). For any $v \in L^2(\mathbb{R} \times S)$ the function $w \in \mathcal{C}^\infty(\mathbb{R}, L^2(\mathbb{R} \times S))$, $w(s) := S(s)v$ ($s \in \mathbb{R}$), is the mild solution to the following Cauchy problem on $\mathbb{R} \times \mathbb{R} \times S$:

$$\partial_s^2 w - \Delta_\rho w = 0, \quad w(0) = v, \quad \partial_s w(0) = 0. \qquad (*)$$

In the above integral formula the term $\cos(s\sqrt{-\Delta_\rho})\Theta u$ corresponds to the solution with $v = \Theta u$, whereas the term $\Theta \cos(s\sqrt{-\Delta_\rho})u$ is just the application of Θ to the solution $w(s)$ corresponding to $v = u$. Applying Θ to $(*)$ we obtain

$$\partial_s^2 \Theta w - \Delta_\rho \Theta w = [\Theta, \Delta_\rho] w =: f$$

and

$$\Theta w \mid_{s=0} = \Theta(w(0)) = \Theta u, \quad \partial_s \Theta w \mid_{s=0} = \Theta(\partial_s w(0)) = 0.$$

Therefore the Duhamel principle yields

$$\Theta \cos(s\sqrt{-\Delta_\rho})u = \Theta w(s)$$

$$= \cos(s\sqrt{-\Delta_\rho})\Theta u + \int_0^s (s-r)\, \mathrm{sinc}\big((s-r)\sqrt{-\Delta_\rho}\big) f(r)\, dr,$$

where $\mathrm{sinc} \colon \mathbb{R} \to \mathbb{R}$ is given by $\mathrm{sinc}(z) = \frac{\sin(z)}{z}$. In summary, we obtain

$$\Theta T_\varepsilon u - T_\varepsilon \Theta u = \frac{1}{2\pi} \int_{\mathbb{R}} \phi_c(s) \frac{1}{\varepsilon} \widehat{F}(\frac{s}{\varepsilon}) \int_0^s (s-r)\, \mathrm{sinc}\big((s-r)\sqrt{-\Delta_\rho}\big) f(r)\, dr\, ds,$$

where

$$f(r) = [\Theta, \Delta_\rho] w(r) = [\Theta, \Delta_\rho] \cos(r\sqrt{-\Delta_\rho})u.$$

Since $[\Theta, \Delta_\rho]$ is of third order and maps H^3_{comp} into L^2 and $\cos(r\sqrt{-\Delta_\rho})$ has operator norm 1 on every Sobolev space there exists a constant $C_1 > 0$ such that $\|f(r)\|_{L^2} \leq C_1 \|u\|_{H^3}$ for every $r \in \mathbb{R}$. Furthermore, the operator norm of $(s-r)\, \mathrm{sinc}\big((s-r)\sqrt{-\Delta_\rho}\big)$ is bounded by $\sup_{z\in\mathbb{R}} |(s-r)\,\mathrm{sinc}((s-r)z)| = |s-r|$. Combining these upper bounds we estimate

$$\|\Theta T_\varepsilon u - T_\varepsilon \Theta u\|_{L^2} \leq \frac{C_1}{2\pi} \int_{\mathbb{R}} |\phi_c(s)| \frac{1}{\varepsilon} |\widehat{F}(\frac{s}{\varepsilon})| \frac{s^2}{2}\, ds \cdot \|u\|_{H^3}$$

$$= \frac{C_1}{4\pi} \|u\|_{H^3} \int_{\mathbb{R}} |\phi_c(\varepsilon\sigma)| |\widehat{F}(\sigma)| \sigma^2\, d\sigma \cdot \varepsilon^2 \leq \frac{C_1}{4\pi} \|u\|_{H^3} \|\phi_c\|_{L^\infty} \|\widehat{F''}\|_{L^1} \cdot \varepsilon^2. \quad \Box$$

Remark 3.3. Applying the reasoning of the proof of Theorem 3.2 to the operators ∂_t or M_α (multiplication by α) instead of Θ gives the following additional asymptotic properties:

(i) If $u \in H^2(\mathbb{R} \times S)$ then $\| T_\varepsilon \partial_t u - \partial_t T_\varepsilon u \|_{L^2} = \mathcal{O}(\varepsilon^2)$ ($\varepsilon \to 0$).

(ii) Let $\alpha \in \mathcal{C}^\infty(\mathbb{R} \times S)$ and let $u \in H^1(\mathbb{R} \times S)$ have compact support. Then

$$\| T_\varepsilon(\alpha u) - \alpha T_\varepsilon u \|_{L^2} = \mathcal{O}(\varepsilon^2) \qquad (\varepsilon \to 0).$$

We have constructed the regularization operators $(T_\varepsilon)_{\varepsilon \in I}$ and the embedding ι of distributions on $\mathbb{R} \times S$ based on the Riemannian metric ρ given by (3.2). However, the construction itself does not directly reflect the metric splitting in (3.2) or (3.1) and so far we have not paid special attention to the foliation by the space-like Cauchy hypersurfaces $\{t\} \times S$ ($t \in \mathbb{R}$). However, this foliation becomes essential in case of distributions that allow restriction to these hypersurfaces. The latter is true in particular for distributional solutions to the wave equation $\Box_\lambda u = f$, where $f \in \mathcal{C}^\infty(\mathbb{R}, \mathcal{D}'(S))$. In these situations we automatically have $u \in \mathcal{C}^\infty(\mathbb{R}, \mathcal{D}'(S))$, since $\mathrm{Char}(\Box_\lambda)$ does not contain any elements of the form $(t, x; \pm 1, 0) \in T^*(\mathbb{R} \times S)$ (cf. [15], 23.65.5). Thus for every $t \in \mathbb{R}$ the value $u(t)$ is an element of $\mathcal{D}'(S)$ and the metric splitting (3.1) provides us with a Riemannian metric h_t on S, which can be used to regularize or embed $u(t)$ according to our general construction. For every $t \in \mathbb{R}$ let $(T_\varepsilon^{h_t})_{\varepsilon \in I}$ denote the regularization obtained from the Riemann metric h_t on S and let ι_{h_t} denote the corresponding embedding $\mathcal{D}'(S) \hookrightarrow \mathcal{G}(S)$. The following statements compare these with the global constructions on $\mathbb{R} \times S$.

Theorem 3.4.

(i) *The embedding* $\iota \colon \mathcal{D}'(\mathbb{R} \times S) \hookrightarrow \mathcal{G}(\mathbb{R} \times S)$ *respects the metric splitting* (3.1) *in the weak sense: If* $u \in \mathcal{C}^\infty(\mathbb{R}, \mathcal{D}'(S))$ *then we have*

$$\forall t \in \mathbb{R} : \ \iota(u)(t) \approx \iota_{h_t}(u(t)).$$

(ii) *If* $u \in \mathcal{C}^\infty(\mathbb{R}, H^2(S))$, *then we have for every compact subset* $Z \subseteq \mathbb{R}$

$$\sup_{t \in Z} \| (T_\varepsilon u)(t) - T_\varepsilon^{h_t} u(t) \|_{L^2(S)} = \mathcal{O}(\varepsilon^2) \quad (\varepsilon \to 0).$$

Proof. (ii) Pick $\chi \in \mathcal{D}(\mathbb{R})$ such that $\chi \equiv 1$ near Z and let $\tilde{u}(t) := \chi(t)u(t)$. It then follows from [13], Proposition 3.7 that $(T_\varepsilon \tilde{u} - T_\varepsilon u)_\varepsilon$ is negligible on $\mathrm{supp}(\tilde{u} - u)^c \supseteq Z$. Thus we may without loss of generality assume that the support of u is bounded in t, so $u \in L^2(\mathbb{R} \times S)$. This allows us to employ the integral formulae defining T_ε and $T_\varepsilon^{h_t}$ and obtain

$$(T_\varepsilon u)(t) - T_\varepsilon^{h_t} u(t)$$
$$= \frac{1}{2\pi} \int_\mathbb{R} \phi_c(s) \frac{1}{\varepsilon} \widehat{F}\left(\frac{s}{\varepsilon}\right) \Big(\underbrace{(\cos(s\sqrt{-\Delta_\rho})u)(t)}_{=:w(s,t)} - \underbrace{\cos(s\sqrt{-\Delta_{h_t}})u(t)}_{=:v(s,t)} \Big) \, ds,$$

where $v, w \in \mathcal{C}^\infty(\mathbb{R}^2, H^2(S))$ solve the following Cauchy problems, respectively:

$$(\partial_s^2 - \Delta_\rho)w = 0 \quad (\text{in } \mathcal{D}'(\mathbb{R} \times \mathbb{R} \times S)), \qquad w\mid_{s=0} = u, \qquad \partial_s w\mid_{s=0} = 0,$$
$$(\partial_s^2 - \Delta_{h_t})v(.,t) = 0 \quad (\text{in } \mathcal{D}'(\mathbb{R} \times S)), \qquad v(0,t) = u(t), \qquad \partial_s v(0,t) = 0.$$

Therefore we have at arbitrary, but fixed $t \in \mathbb{R}$

$$\partial_s^2 w(.,t) - \Delta_{h_t} w(.,t) = \Delta_\rho w(.,t) - \Delta_{h_t} w(.,t) =: f_t,$$

where $f_t \in \mathcal{C}^\infty(\mathbb{R}, L^2(S))$, since u (hence w) has values in $H^2(S)$. We observe that $\psi_t := w(.,t) - v(.,t)$ satisfies the Cauchy problem

$$(\partial_s^2 - \Delta_{h_t})\psi_t = f_t \quad (\text{in } \mathcal{D}'(\mathbb{R} \times S)), \quad \psi_t(0) = 0, \quad \partial_s \psi_t = 0,$$

which implies

$$\psi_t(s) = \int_0^s (s-r) \operatorname{sinc}((s-r)\sqrt{-\Delta_{h_t}}) f_t(r)\, dr.$$

Hence we obtain

$$\|\psi_t(s)\|_{L^2(S)} \leq \left| \int_0^s (s-r)\|f_t(r)\|_{L^2(S)}\, dr \right| \leq \frac{s^2}{2} \cdot \sup_{|\sigma| \leq |s|} \|f_t(\sigma)\|_{L^2(S)}$$

and therefore arrive at

$$\|(T_\varepsilon u)(t) - T_\varepsilon^{h_t} u(t)\|_{L^2(S)} \leq \frac{1}{2\pi} \int_{\mathbb{R}} |\phi_c(s)| \frac{1}{\varepsilon} |\widehat{F}(\tfrac{s}{\varepsilon})| \|\psi_t(s)\|_{L^2(S)}\, ds$$

$$\leq \frac{\varepsilon^2}{4\pi} \int_{-2c/\varepsilon}^{2c/\varepsilon} |\phi_c(\varepsilon\tau)| |\widehat{F}(\tau)| \tau^2 \sup_{|\sigma| \leq \varepsilon|\tau|} \|f_t(\sigma)\|_{L^2(S)}\, d\tau$$

$$\leq \frac{\varepsilon^2}{4\pi} \|\phi_c\|_{L^\infty} \sup_{|\sigma| \leq 2c} \|f_t(\sigma)\|_{L^2(S)} \|\widehat{F''}\|_{L^1}$$

$$= O(\varepsilon^2) \quad (\varepsilon \to 0)$$

uniformly when t varies in a compact set.

    (i) Since association is checked by action on test functions in $\mathcal{D}(S)$ we may reduce to the case that $\operatorname{supp}(u(t))$ is compact. Then $u(t) \in H^l(S)$ for some $l \in \mathbb{Z}$ and since $(1-\Delta_{h_t})^{\pm l/2}$ commutes weakly with ι and strongly with ι_{h_t} the assertion follows from (ii).   □

Acknowledgment

G.H. thanks the organizers for support by an EPSRC Pathway to Impact Award. We also acknowledge the support of FWF-projects Y237, P20525, and P23714. Finally, we would like to thank the referee for several comments that have led to improvements in the paper.

References

[1] Ch. Bär, N. Ginoux, and F. Pfäffle. *Wave equations on Lorentzian manifolds and quantization.* ESI Lectures in Mathematics and Physics. European Mathematical Society (EMS), Zürich, 2007.

[2] A.N. Bernal and M. Sánchez. Smoothness of time functions and the metric splitting of globally hyperbolic spacetimes. *Commun. Math. Phys.*, 257(1):43–50, 2005.

[3] A.N. Bernal and M. Sánchez. Globally hyperbolic spacetimes can be defined as 'causal' instead of 'strongly causal'. *Classical Quantum Gravity*, 24(3):745–749, 2007.

[4] J. Chazarain, A. Piriou. *Introduction to the theory of linear partial differential equations.* Studies in Mathematics and Its Applications, Vol. 14. North-Holland, 1982.

[5] Y. Choquet-Bruhat. *General relativity and the Einstein equations.* Oxford Mathematical Monographs. Oxford University Press, Oxford, 2009.

[6] C.J.S. Clarke. *The Analysis of Space-Time Singularities*, Cambridge University Press, Cambridge, 1993.

[7] C.J.S. Clarke. Singularities: boundaries or internal points?, in Singularities, Black Holes and Cosmic Censorship, Joshi, P.S. and Raychaudhuri, A.K., eds., IUCCA, Bombay, 1996, pp. 24–32.

[8] C.J.S. Clarke. Generalized hyperbolicity in singular spacetimes. *Class. Quantum Grav.* 15 (1998), pp. 975–984.

[9] J.F. Colombeau. *New generalized functions and multiplication of distributions.* North-Holland, Amsterdam, 1984.

[10] J.F. Colombeau. *Elementary introduction to new generalized functions.* North-Holland, Amsterdam, 1985.

[11] S. Dave. Geometrical embeddings of distributions into algebras of generalized functions, *Math. Nachr.*, 283 (2010), no. 11, 1575–1588.

[12] J.W. de Roever, M. Damsma. Colombeau algebras on a C^∞-manifold. *Indag. Math. (N.S.)* 2 (1991), no. 3, 341–358.

[13] S. Dave, G. Hörmann, and M. Kunzinger. Optimal regularization processes on complete Riemannian manifolds. 2010. arXiv:1003.3341 [math.FA].

[14] N. Dapić, S. Pilipović, D. Scarpalezos. Microlocal analysis of Colombeau's generalized functions: propagation of singularities. *J. Anal. Math.* 75 (1998), 51–66.

[15] J. Dieudonné. *Treatise on analysis. Vol. VIII*, volume 10 of *Pure and Applied Mathematics*. Academic Press Inc., Boston, MA, 1993.

[16] Garetto, C. Topological structures in Colombeau algebras: Topological $\widetilde{\mathbb{C}}$-modules and duality theory. *Acta Appl. Math.* 88, No. 1, 81–123 (2005).

[17] C. Garetto, G. Hörmann. Microlocal analysis of generalized functions: pseudodifferential techniques and propagation of singularities. *Proc. Edinb. Math. Soc.* (2) 48 (2005), no. 3, 603–629.

[18] R. Geroch. Domain of dependence. *Jour. Math. Phys.*, 11:437–449, 1970.

[19] J.D.E. Grant, E. Mayerhofer, R. Steinbauer. The wave equation on singular spacetimes. *Comm. Math. Phys.* 285 (2009), no. 2, 399–420.

[20] M. Grosser, M. Kunzinger, M. Oberguggenberger, R. Steinbauer. *Geometric theory of generalized functions*, Kluwer, Dordrecht, 2001.

[21] S. Haller. Microlocal analysis of generalized pullbacks of Colombeau functions. *Acta Appl. Math.* 105, no. 1, 83–109 (2009).

[22] S.W. Hawking, G.F.R. Ellis. *The large scale structure of space-time.* Cambridge Monographs on Mathematical Physics, No. 1. Cambridge University Press, 1973.

[23] N. Higson, J. Roe. *Analytic K-homology.* Oxford Mathematical Monographs, Oxford, 2000.

[24] G. Hörmann. Integration and microlocal analysis in Colombeau algebras of generalized functions. *J. Math. Anal. Appl.* 239 (1999), no. 2, 332–348.

[25] G. Hörmann, M. Kunzinger, R. Steinbauer. Wave equations on non-smooth spacetimes, this volume, pp. 163–186.

[26] A.A. Kosinski. *Differential manifolds.* Pure and Applied Mathematics 138, Academic Press (1993).

[27] M. Kunzinger, R. Steinbauer. Generalized pseudo-Riemannian geometry. *Trans. Amer. Math. Soc.* 354 (2002), no. 10, 4179–4199.

[28] J. Leray. *Hyperbolic Differential Equations*, Lecture Notes, IAS, Princeton, 1953.

[29] E. Minguzzi and M. Sánchez. The causal hierarchy of spacetimes. In *Recent developments in pseudo-Riemannian geometry*, ESI Lect. Math. Phys., pages 299–358. Eur. Math. Soc., Zürich, 2008.

[30] K. Nomizu and H. Ozeki. The existence of complete Riemannian metrics. *Proc. Am. Math. Soc.*, 12:889–891, 1961.

[31] M. Oberguggenberger. *Multiplication of distributions and applications to partial differential equations.* Pitman Research Notes in Mathematics 59. Longman, New York, 1992.

[32] B. O'Neill. *Semi-Riemannian geometry*, volume 103 of *Pure and Applied Mathematics*. Academic Press, New York, 1983.

[33] R. Steinbauer and J. Vickers, *The use of generalized functions and distributions in general relativity*, Class. Quantum Grav. 23 (2006), pp. R91–R114.

[34] J.A. Vickers and J.P. Wilson, *Generalized hyperbolicity in conical spacetimes*, Class. Quantum Grav. 17 (2000), pp. 1333–1260.

Shantanu Dave, Günther Hörmann and Michael Kunzinger
Faculty of Mathematics
University of Vienna
Nordbergstraße 15
A-1090 Wien, Austria

e-mail: shantanu.dave@univie.ac.at
       guenther.hoermann@univie.ac.at
       michael.kunzinger@univie.ac.at

Progress in Mathematics, Vol. 301, 103–114

A Remark on the Uniqueness for Backward Parabolic Operators with non-Lipschitz-continuous Coefficients

Daniele Del Santo

Abstract. Using Bony's paramultiplication we improve a result obtained in [6] for operators having coefficients non-Lipschitz-continuous with respect to t but $\mathcal{C}^2$ with respect to x, showing that the same result is valid when $\mathcal{C}^2$ is replaced by $\mathcal{C}^{1,\varepsilon}$, with $\varepsilon > 0$.

Mathematics Subject Classification. Primary 35K25, 35K35, 35A05.

Keywords. Backward parabolic operators, non-Lipschitz-continuous coefficients, paramultiplication.

1. Introduction

Let μ be a modulus of continuity and let a be a bounded function on $\mathbb{R}$ such that $a \in \mathcal{C}^\mu$, i.e.,

$$|a(t_1) - a(t_2)| \leq C\mu(|t_1 - t_2|),$$

for all t_1, $t_2 \in \mathbb{R}$ such that $|t_1 - t_2| \leq 1$. The main result in [6] states that if $1/\lambda_0 \geq a(t) \geq \lambda_0 > 0$ for all $t \in [0, T]$ and the modulus of continuity satisfies the so-called Osgood condition, then the backward parabolic operator

$$L = \partial_t + a(t)\partial_x^2 + \textit{lower-order terms}$$

has the property of uniqueness in the Cauchy problem for solutions in $H^1((0, T), L^2(\mathbb{R})) \cap L^2((0, T), H^2(\mathbb{R}))$.

The question of uniqueness in backward parabolic problems goes back to a classical example of Tychonoff [13] and a rather complete answer was given at the beginning of the 60's by Lions and Malgrange in [8], under the hypothesis of Lipschitz regularity of the coefficient a (see also [1] and [7] for other more refined and abstract results). Some years after Miller [11] showed that the uniqueness property depends essentially on the regularity of a and this point was enlightened by another non-uniqueness example in [9].

Actually in [8], [1], [7], [6], the backward uniqueness was proved in the case of the coefficient a depending also on the space variable, i.e., for the operator

$$\tilde{L} = \partial_t + \partial_x(a(t,x)\partial_x) + \textit{lower-order terms}.$$

In [6] a technical difficulty in the estimate of a commutator led to imposing on a the C^μ regularity with respect to t, uniformly with respect to x, together with the C^2 regularity with respect to x, uniformly with respect to t.

In the present note we improve this statement, showing that under the Osgood condition for μ, the C^μ regularity with respect to t, uniformly with respect to x together with the Hölder $C^{1,\varepsilon}$ regularity with respect to x, uniformly with respect to t, for the coefficient a, is sufficient for the same uniqueness result.

The proof follows the same path as that one in [6], the only difference being in the introduction of a paradifferential operator (actually a simple paramultiplication) at the place of the second-order part of the operator $\tilde{L}$.

The estimate of the commutator, in the present case, is made more effective by a theorem due to Coifman and Meyer [3, Th. 35] (see also, for a similar use of that theorem, [5, Prop. 3.7]).

2. Main result

Definition 2.1. A function μ is said to be a *modulus of continuity* if μ is continuous, concave and strictly increasing on $[0,1]$, with $\mu(0) = 0$ and $\mu(1) = 1$. Let $I \subseteq \mathbb{R}$ and let $\varphi : I \to \mathcal{B}$, where $\mathcal{B}$ is a Banach space. We say that $\varphi \in C^\mu(I,\mathcal{B})$ if $\varphi \in L^\infty(I,\mathcal{B})$ and

$$\sup_{\substack{0<|t-s|<1 \\ t,s \in I}} \frac{\|\varphi(t) - \varphi(s)\|_\mathcal{B}}{\mu(|t-s|)} < +\infty.$$

It is immediate to verify the following statements

- $\mu(s) \geq s$ for all $s \in [0,1]$;
- the function $s \mapsto \mu(s)/s$ is decreasing on $]0,1]$;
- there exists $\lim_{s\to 0^+} \mu(s)/s$;
- the function $\sigma \mapsto \sigma\mu(1/\sigma)$ is increasing on $[1,+\infty[$;
- the function $\sigma \mapsto 1/(\sigma^2\mu(1/\sigma))$ is decreasing on on $[1,+\infty[$.

Definition 2.2. A modulus of continuity is said to satisfy the *Osgood condition* if

$$\int_0^1 \frac{1}{\mu(s)} \, ds = +\infty. \tag{2.1}$$

Theorem 2.3. *Let μ be a modulus of continuity satisfying the Osgood condition and let $\varepsilon \in]0,1[$. Suppose $a \in C^\mu([0,T], C_b(\mathbb{R}_x)) \cap C([0,T], C^{1,\varepsilon}(\mathbb{R}_x))$, where $C_b(\mathbb{R}_x)$ is the space of bounded continuous functions and $C^{1,\varepsilon}(\mathbb{R}_x)$ is the usual Hölder space. Suppose moreover that $1/\lambda_0 \geq a(t,x) \geq \lambda_0 > 0$ for all $(t,x) \in [0,T] \times \mathbb{R}$. Let b and c bounded coefficients on $[0,T] \times \mathbb{R}$.*

Then the operator

$$\tilde{L} := \partial_t + \partial_x(a(t,x)\partial_x) + b(t,x)\partial_x + c(t,x) \tag{2.2}$$

has the $\mathcal{H}$-uniqueness property, with $\mathcal{H} := H^1((0,T), L^2(\mathbb{R})) \cap L^2((0,T), H^2(\mathbb{R}))$, i.e., if $u \in \mathcal{H}$, $\tilde{L}u = 0$ in $[0,T] \times \mathbb{R}$ and $u(0,x) = 0$ in $\mathbb{R}$, then $u = 0$ in $[0,T] \times \mathbb{R}$.

We remark that the statement of Theorem 2.3 is given for operators in only one space variable: this is done to avoiding cumbersome notations. There are no difficulties for treating the general case of n space variables (see [6]).

3. Proof

The proof of Theorem 2.3 is very similar to that of [6, Th. 1]. For the reader's convenience we repeat here the main points.

3.1. Weight function and Carleman estimate

Theorem 2.3 will follow from a Carleman estimate. The weight function in the Carleman estimate will be obtained from the modulus of continuity. This crucial idea goes back to the paper [12] in which a uniqueness result for elliptic operators with non-Lipschitz-continuous coefficients is proved. We define

$$\phi(t) = \int_{\frac{1}{t}}^{1} \frac{1}{\mu(s)} \, ds.$$

The function ϕ is a strictly increasing C^1 function. From (2.1) we have $\phi([1,+\infty[) = [0,+\infty[$; moreover $\phi'(t) = 1/(t^2\mu(1/t)) > 0$ for all $t \in [1,+\infty[$. We set

$$\Phi(\tau) = \int_0^\tau \phi^{-1}(s) \, ds.$$

We obtain $\Phi'(\tau) = \phi^{-1}(\tau)$ and consequently $\lim_{\tau \to +\infty} \Phi'(\tau) = +\infty$. Moreover

$$\Phi''(\tau) = (\Phi'(\tau))^2 \mu\left(\frac{1}{\Phi'(\tau)}\right) \tag{3.1}$$

for all $\tau \in [0,+\infty[$ and, as the function $\sigma \mapsto \sigma\mu(1/\sigma)$ is increasing on $[1,+\infty[$, we deduce that

$$\lim_{\tau \to +\infty} \Phi''(\tau) = \lim_{\tau \to +\infty} (\Phi'(\tau))^2 \mu\left(\frac{1}{\Phi'(\tau)}\right) = +\infty. \tag{3.2}$$

Now we can state the Carleman estimate.

Proposition 3.1. *There exist γ_0, $C > 0$ such that*

$$\int_0^{\frac{T}{2}} e^{\frac{2}{\gamma}\Phi(\gamma(T-t))} \|\partial_t u + \partial_x(a(t,x)\partial_x u)\|_{L^2}^2 \, dt$$

$$\geq C\gamma^{\frac{1}{2}} \int_0^{\frac{T}{2}} e^{\frac{2}{\gamma}\Phi(\gamma(T-t))} (\|\partial_x u\|_{L^2}^2 + \gamma^{\frac{1}{2}}\|u\|_{L^2}^2) \, dt \tag{3.3}$$

for all $\gamma > \gamma_0$ and for all $u \in C_0^\infty(\mathbb{R}^2)$ such that $\operatorname{supp} u \subseteq [0, T/2] \times \mathbb{R}$.

The way of obtaining the $\mathcal{H}$-uniqueness from the inequality (3.3) is a standard procedure and the details are in [6, Par. 3.4].

3.2. Littlewood-Paley decomposition, Bony's paramultiplication and commutators

We review some known results on Littlewood-Paley decomposition and related topics. More details can be found in [2], [10, Ch. 4 and Ch. 5] and [5, Par. 3].

Let $\chi \in C_0^\infty(\mathbb{R})$, $0 \le \chi \le 1$, even and such that $\chi(\xi) = 1$ for $|\xi| \le 11/10$ and $\chi(\xi) = 0$ for $|\xi| \ge 19/10$. For $k \in \mathbb{N}$, let us consider $\chi_k(\xi) = \chi(2^{-k}\xi)$, $\tilde{\chi}_k(x)$ its inverse Fourier transform, the operators

$$S_k u := \tilde{\chi}_k * u = \chi_k(D_x)u, \quad \text{and} \quad \Delta_0 u := S_0 u,$$

and, for $k \ge 1$,

$$\Delta_k u := S_k u - S_{k-1} u \quad \text{and} \quad S_{-k} u := 0.$$

In the following propositions we recall the characterization of Sobolev and Hölder spaces via Littlewood-Paley decomposition (see [10, Prop. 4.1.11 and Prop. 4.1.16] and [5, Prop. 3.1 and Prop. 3.2]).

Proposition 3.2. *Let $s \in \mathbb{R}$. A temperate distribution u is in H^s if and only if the following two conditions hold*

  i) *for all $k \ge 0$, $u_k := \Delta_k u \in L^2$;*
  ii) *the sequence $\delta_k := 2^{ks}\|u_k\|_{L^2}$ is in l^2.*

Moreover the norm of u in H^s is equivalent to the norm of $(\delta_k)_k$ in l^2.

Proposition 3.3. *Let $s \in \mathbb{R}$ and $R \ge 2$. Let $(u_k)_k$ a sequence of functions in L^2 such that*

  i) *the support of the Fourier transform of u_0 is contained in $\{|\xi| \le R\}$ and the support of the Fourier transform of u_k is contained in $\{\frac{1}{R} 2^k \le |\xi| \le R2^k\}$, for all $k \ge 1$;*
  ii) *the sequence $\delta_k := 2^{ks}\|u_k\|_{L^2}$ is in l^2.*

Then $u := \sum_k u_k$ belongs to H^s and the norm of u in H^s is equivalent to the norm of $(\delta_k)_k$ in l^2. When $s > 0$ it is sufficient to assume that the Fourier transform of u_k is contained in $\{|\xi| \le R2^k\}$, for all $k \ge 1$.

Proposition 3.4. *Let $\rho > 0$, $\rho \notin \mathbb{N}$. A temperate distribution a is in $C^{[\rho], \rho - [\rho]}$ (here $[\rho]$ denotes the integer part of ρ) if and only if the following two conditions hold*

  i) *for all $k \ge 0$, $a_k := \Delta_k a \in L^\infty$;*
  ii) *the sequence $\eta_k := 2^{k\rho}\|a_k\|_{L^\infty}$ is in l^∞.*

Moreover the norm of a in $C^{[\rho], \rho - [\rho]}$ is equivalent to the norm of $(\eta_k)_k$ in l^∞.

Let $a \in L^\infty$. The Bony's paraproduct of a and $u \in H^s$ (see [2, Par. 2]) is defined as

$$T_a u := \sum_{k=3}^{+\infty} S_{k-3} a \Delta_k u.$$

Useful properties of the paraproduct are contained in the following proposition (also see [10, Prop. 5.2.1] and [5, Prop. 3.4])

Proposition 3.5. *Let $a \in L^\infty$ and $s \in \mathbb{R}$. Then T_a maps H^s into H^s and*

$$\|T_a u\|_{H^s} \leq C\|a\|_{L^\infty}\|u\|_{H^s}.$$

Let $a \in C^{1,\varepsilon}$ with $\varepsilon \in {]}0,1[$. Then the operator $u \mapsto au - T_a u$ maps L^2 into H^1 and

$$\|au - T_a u\|_{H^1} \leq C_\varepsilon \|a\|_{C^{1,\varepsilon}}\|u\|_{L^2}.$$

Proof. We prove only the second part of the statement. We have

$$au - T_a u = \sum_{k=3}^{+\infty} \Delta_k a S_{k-3} u + \sum_{k=0}^{+\infty}\left(\sum_{\substack{j \geq 0 \\ |j-k| \leq 2}} \Delta_j a \Delta_k u\right).$$

We remark that the support of the Fourier transform of $\Delta_k a S_{k-3} u$ is contained in $\{2^{k-2} \leq |\xi| \leq 2^{k+2}\}$. Moreover, by Proposition 3.4, we have that

$$\|\Delta_k a S_{k-3} u\|_{L^2} \leq \|\Delta_k a\|_{L^\infty}\|S_{k-3} u\|_{L^2} \leq C 2^{-k}\delta_k \|a\|_{C^{1,\varepsilon}}\|u\|_{L^2}$$

where $\delta_k = 2^{-k\varepsilon}$. Since $(\delta_k)_k$ is in l^2 then, by Proposition 3.3, we have that $\sum_{k=3}^{+\infty} \Delta_k a S_{k-3} u \in H^1$ and

$$\left\|\sum_{k=3}^{+\infty} \Delta_k a S_{k-3} u\right\|_{H^1} \leq C_\varepsilon \|a\|_{C^{1,\varepsilon}}\|u\|_{L^2}. \tag{3.4}$$

Next, we see that

$$\sum_{\substack{j \geq 0 \\ |j-k| \leq 2}} \Delta_j a \Delta_k u = \Delta_{k-2} a \Delta_k u + \Delta_{k-1} a \Delta_k u + \Delta_k a \Delta_k u + \Delta_{k+1} a \Delta_k u + \Delta_{k+2} a \Delta_k u,$$

so that

$$\sum_{k=0}^{+\infty}\left(\sum_{\substack{j \geq 0 \\ |j-k| \leq 2}} \Delta_j a \Delta_k u\right) = \sum_{k=2}^{+\infty} \Delta_{k-2} a \Delta_k u + \cdots + \sum_{k=0}^{+\infty} \Delta_{k+2} a \Delta_k u.$$

We have that the support of the Fourier transform of $\Delta_{k-2} a \Delta_k u$ is contained in $\{|\xi| \leq 2^{k+3}\}$ and similarly for the other four terms. Moreover

$$\|\Delta_{k-2} a \Delta_k u\|_{L^2} \leq \|\Delta_{k-2} a\|_{L^\infty}\|\Delta_k u\|_{L^2} \leq C 2^{-k(1+\varepsilon)}\|a\|_{C^{1,\varepsilon}}\delta_k,$$

where $(\delta_k)_k = (\|\Delta_k u\|_{L^2})_k \in l^2$ and consequently $\|(\delta_k)_k\|_{l^2} \leq C\|u\|_{L^2}$. From Proposition 3.3 we have that $\sum_{k=2}^{+\infty} \Delta_{k-2} a \Delta_k u \in H^{1+\varepsilon}$ and

$$\left\|\sum_{k=2}^{+\infty} \Delta_{k-2} a \Delta_k u\right\|_{H^{1+\varepsilon}} \leq C\|a\|_{C^{1,\varepsilon}}\|u\|_{L^2},$$

Arguing similarly for the other terms and using the obvious inclusion in Sobolev spaces we have that $\sum\limits_{k=0}^{+\infty}\left(\sum\limits_{\substack{j\geq 0 \\ |j-k|\leq 2}}\Delta_j a\Delta_k u\right) \in H^1$ and

$$\left\|\sum_{k=0}^{+\infty}\left(\sum_{\substack{j\geq 0 \\ |j-k|\leq 2}}\Delta_j a\Delta_k u\right)\right\|_{H^1} \leq C\|a\|_{C^{1,\epsilon}}\|u\|_{L^2}. \tag{3.5}$$

The conclusion of the proof of the proposition is reached putting together (3.4) and (3.5). $\qquad\square$

We end this subsection with a property of commutation which will be crucial in the proof of the Carleman estimate (see [5, Prop.3.7]).

Proposition 3.6. *Let a be a bounded Lipschitz-continuous function and let $u \in H^1$. Then*

$$\Big(\sum_{\nu=0}^{+\infty}\|\partial_x([\Delta_\nu, T_a]\partial_x u)\|_{L^2}^2\Big)^{1/2} \leq C\|a\|_{Lip}\|u\|_{H^1} \tag{3.6}$$

(here $[A, B]$ denotes the commutator between the operators A and B, i.e., $[A, B]w := A(Bw) - B(Aw)$).

Proof. We start remarking that

$$[\Delta_\nu, T_a]w = \sum_{k=3}^{+\infty}[\Delta_\nu, S_{k-3}a]\Delta_k w,$$

and consequently

$$\partial_x([\Delta_\nu, T_a]\partial_x u) = \partial_x\Big(\sum_{k=3}^{+\infty}[\Delta_\nu, S_{k-3}a]\Delta_k(\partial_x u)\Big).$$

In fact Δ_ν and Δ_k commute so that

$$\Delta_\nu(S_{k-3}a\Delta_k w)) - S_{k-3}a\Delta_k(\Delta_\nu w) = \Delta_\nu(S_{k-3}a\Delta_k w)) - S_{k-3}a\Delta_\nu(\Delta_k w).$$

Let us consider

$$\partial_x\Big(\sum_{k=3}^{+\infty}[\Delta_\nu, S_{k-3}a]\Delta_k(\partial_x u)\Big) = \partial_x\Big(\sum_{k=3}^{+\infty}[\Delta_\nu, S_{k-3}a]\partial_x(\Delta_k u)\Big).$$

Looking at the support of the Fourier transform, it is possible to see that

$$[\Delta_\nu, S_{k-3}a]\partial_x(\Delta_k u)$$

is identically 0 if $|k - \nu| \geq 4$. Consequently the sum in k is reduced to at most 7 terms: $\partial_x([\Delta_\nu, S_{\nu-6}a]\partial_x(\Delta_{\nu-3}u)) + \cdots \partial_x([\Delta_\nu, S_\nu a]\partial_x(\Delta_{\nu+3}u))$, each of them having the support of the Fourier transform contained in $\{|\xi| \leq C2^\nu\}$. Let us consider one of these terms, e.g., $\partial_x([\Delta_\nu, S_{\nu-3}a]\partial_x(\Delta_\nu u))$, the computation for the other ones being similar. We have, from Bernstein's inequality

$$\|\partial_x([\Delta_\nu, S_{\nu-3}a]\partial_x(\Delta_\nu u))\|_{L^2} \leq C2^\nu\|[\Delta_\nu, S_{\nu-3}a]\partial_x(\Delta_\nu u)\|_{L^2}.$$

On the other hand, using [3, Th. 35], we have that

$$\|[\Delta_\nu, S_{\nu-3}a]\partial_x(\Delta_\nu u)\|_{L^2} \le C\|a\|_{Lip}\|\Delta_\nu u\|_{L^2},$$

where C does not depend on ν. Consequently

$$\|\partial_x([\Delta_\nu, S_{\nu-3}a]\partial_x(\Delta_\nu u))\|_{L^2} \le C2^\nu \|a\|_{Lip}\|\Delta_\nu u\|_{L^2}.$$

Since $u \in H^1$, by Proposition 3.2, we have that $(2^\nu \|\Delta_\nu u\|)_\nu$ is in l^2 with l^2-norm equivalent to $\|u\|_{H^1}$, so that

$$\sum_\nu \|\partial_x([\Delta_\nu, S_{\nu-3}a]\partial_x(\Delta_\nu u))\|_{L^2}^2 \le C\|a\|_{Lip}^2 \|u\|_{H^1}^2.$$

We obtain

$$\sum_{\nu=0}^{+\infty} \left\| \partial_x\left(\sum_{k=3}^{+\infty}[\Delta_\nu, S_{k-3}a]\Delta_k(\partial_x u)\right)\right\|^2 \le C\|a\|_{Lip}^2\|u\|_{H^1}^2,$$

and the proof is completed. $\qquad\qquad\qquad\qquad\qquad\qquad\qquad\qquad\square$

3.3. Carleman estimate and approximation

We set $v(t,x) = e^{\frac{1}{\gamma}\Phi(\gamma(T-t))}u(t,x)$. The inequality (3.3) becomes

$$\int_0^{\frac{T}{2}} \|\partial_t v + \partial_x(a(t,x)\partial_x v) + \Phi'(\gamma(T-t))v\|_{L^2}^2 \, dt \tag{3.7}$$

$$\ge C\gamma^{\frac{1}{2}}\int_0^{\frac{T}{2}} (\|\partial_x v\|_{L^2}^2 + \gamma^{\frac{1}{2}}\|v\|_{L^2}^2)\, dt.$$

Using the second part of Proposition 3.5, the inequality (3.7) will be deduced from the following

$$\int_0^{\frac{T}{2}} \|\partial_t v + \partial_x(T_a(\partial_x v)) + \Phi'(\gamma(T-t))v\|_{L^2}^2 \, dt \tag{3.8}$$

$$\ge C\gamma^{\frac{1}{2}}\int_0^{\frac{T}{2}} (\|\partial_x v\|_{L^2}^2 + \gamma^{\frac{1}{2}}\|v\|_{L^2}^2)\, dt,$$

as the quantity $\|\partial_x((a(t,x) - T_a)\partial_x u)\|_{L^2}$ can be absorbed by the right-hand side part of (3.7), possibly taking different C and γ_0.

Let us go back to the Littlewood-Paley decomposition; a consequence of Proposition 3.2 is that there exists $K > 0$ such that

$$\frac{1}{K}\sum_\nu \|u_\nu\|_{L^2}^2 \le \|u\|_{L^2}^2 \le K\sum_\nu \|u_\nu\|_{L^2}^2$$

for all $u \in L^2$. We have

$$\int_0^{\frac{T}{2}} \|\partial_t v + \partial_x(T_a \partial_x v) + \Phi'(\gamma(T-t))v\|_{L^2}^2 \, dt$$

$$\geq \frac{1}{K} \int_0^{\frac{T}{2}} \sum_\nu \|\Delta_\nu(\partial_t v + \partial_x(T_a \partial_x v) + \Phi'(\gamma(T-t))v)\|_{L^2}^2 \, dt$$

$$\geq \frac{1}{K} \int_0^{\frac{T}{2}} \sum_\nu \|\partial_t v_\nu + \partial_x(T_a \partial_x v_\nu) + \Phi'(\gamma(T-t))v_\nu + \partial_x([\Delta_\nu, T_a]\partial_x v)\|_{L^2}^2 \, dt$$

$$\geq \frac{1}{2K} \int_0^{\frac{T}{2}} \sum_\nu \|\partial_t v_\nu + \partial_x(T_a \partial_x v_\nu) + \Phi'(\gamma(T-t))v_\nu\|_{L^2}^2 \, dt$$

$$-\frac{1}{K} \int_0^{\frac{T}{2}} \sum_\nu \|\partial_x([\Delta_\nu, T_a]\partial_x v)\|_{L^2}^2 \, dt.$$

From the result of Proposition 3.6 is then immediate that (3.8) will be deduced from the same estimate from below for

$$\int_0^{\frac{T}{2}} \sum_\nu \|\partial_t v_\nu + \partial_x(T_a \partial_x v_\nu) + \Phi'(\gamma(T-t))v_\nu\|_{L^2}^2 \, dt,$$

and finally, again using the second part of Proposition 3.5, we have that (3.8) will be implied by the following: there exists C and γ_0 positive constants, such that

$$\int_0^{\frac{T}{2}} \sum_\nu \|\partial_t v_\nu + \partial_x(a(t,x)\partial_x v_\nu) + \Phi'(\gamma(T-t))v_\nu\|_{L^2}^2 \, dt \qquad (3.9)$$

$$\geq C\gamma^{\frac{1}{2}} \int_0^{\frac{T}{2}} (\|\partial_x v\|_{L^2}^2 + \gamma^{\frac{1}{2}}\|v\|_{L^2}^2) \, dt,$$

for all $\gamma > \gamma_0$ and for all $v \in C_0^\infty(\mathbb{R}^2)$ such that $\operatorname{supp} v \subseteq [0, T/2] \times \mathbb{R}$. In fact

$$\sum_\nu \|\partial_x((a(t,x) - T_a)\partial_x v_\nu)\|_{L^2}^2 \leq \sum_\nu C^2 \|a\|_{C^{1,\epsilon}}^2 \|v_\nu\|_{H^1}^2 \leq C^2 \|a\|_{C^{1,\epsilon}}^2 \|v\|_{H^1}^2$$

and consequently $\sum_\nu \|\partial_x((a(t,x) - T_a)\partial_x v_\nu)\|_{L^2}^2$ can be absorbed by the right-hand side part of (3.8). From this point on the proof is an exact repetition of that one of the main result in [6]. We have

$$\int_0^{\frac{T}{2}} \sum_\nu \|\partial_t v_\nu \partial_x(a(t,x)\partial_x v_\nu) + \Phi'(\gamma(T-t))v_\nu\|_{L^2}^2 \, dt$$

$$= \int_0^{\frac{T}{2}} \sum_\nu \Big(\|\partial_t v_\nu\|_{L^2}^2 + \|\partial_x(a\partial_x v_\nu) + \Phi'(\gamma(T-t))v_\nu\|_{L^2}^2 \qquad (3.10)$$

$$+ \gamma\, \Phi''(\gamma(T-t))\|v_\nu\|_{L^2}^2 + 2\operatorname{Re}\langle \partial_t v_\nu, \partial_x(a\partial_x v_\nu)\rangle_{L^2} \Big) \, dt.$$

Since a is not Lipschitz-continuous, it is not possible to use the integration by parts to estimate the last term in (3.10). We use the well-known approximation technique developed in [4]. Let $\rho \in C_0^\infty(\mathbb{R})$ with $\operatorname{supp}\rho \subseteq [-1/2,\, 1/2]$, $\int_\mathbb{R} \rho(s)\, ds = 1$ and $\rho(s) \geq 0$ for all $s \in \mathbb{R}$; we set

$$a_\varepsilon(t,x) = \int_\mathbb{R} a(s,x) \frac{1}{\varepsilon} \rho\left(\frac{t-s}{\varepsilon}\right)\, ds$$

for $\varepsilon \in \,]0, 1/2]$. We obtain that there exist $C > 0$ such that

$$|a_\varepsilon(t,x) - a(t,x)| \leq C\mu(\varepsilon) \qquad \text{and} \qquad \partial_t a_\varepsilon(t,x) \leq C\,\frac{\mu(\varepsilon)}{\varepsilon}.$$

We have

$$\int_0^{\frac{T}{2}} 2\operatorname{Re} \langle \partial_t v_\nu,\ \partial_x(a\partial_x v_\nu)\rangle_{L^2}\, dt = -2\operatorname{Re} \int_0^{\frac{T}{2}} \langle \partial_x\partial_t v_\nu,\ a\partial_x v_\nu\rangle_{L^2}\, dt$$

$$= -2\operatorname{Re} \int_0^{\frac{T}{2}} \langle \partial_x\partial_t v_\nu,\ (a - a_\varepsilon)\partial_x v_\nu\rangle_{L^2}\, dt - 2\operatorname{Re} \int_0^{\frac{T}{2}} \langle \partial_x\partial_t v_\nu,\ a_\varepsilon\partial_x v_\nu\rangle_{L^2}\, dt.$$

We remark that $\|\partial_x v_\nu\|_{L^2} \leq 2^{\nu+1}\|v_\nu\|_{L^2}$ and $\|\partial_x\partial_t v_\nu\|_{L^2} \leq 2^{\nu+1}\|\partial_t v_\nu\|_{L^2}$ for all $\nu \in \mathbb{N}$ so that

$$\left| 2\operatorname{Re} \int_0^{\frac{T}{2}} \langle \partial_x\partial_t v_\nu,\ (a - a_\varepsilon)\partial_x v_\nu\rangle_{L^2}\, dt \right| \leq 2C\mu(\varepsilon) \int_0^{\frac{T}{2}} \|\partial_x\partial_t v_\nu\|_{L^2}\, \|\partial_x v_\nu\|_{L^2}\, dt$$

$$\leq \int_0^{\frac{T}{2}} \|\partial_t v_\nu\|_{L^2}^2\, dt + C\, 2^{4(\nu+1)}\mu(\varepsilon) \int_0^{\frac{T}{2}} \|v_\nu\|_{L^2}^2\, dt$$

(remark that $\mu(\varepsilon^2) \leq \mu(\varepsilon)$), and similarly

$$\left| 2\operatorname{Re} \int_0^{\frac{T}{2}} \langle \partial_x\partial_t v_\nu,\ a_\varepsilon\partial_{x_k} v_\nu\rangle_{L^2}\, dt \right| = \left| \int_0^{\frac{T}{2}} \langle \partial_x v_\nu,\ \partial_t a_\varepsilon\partial_x v_\nu\rangle_{L^2}\, dt \right|$$

$$\leq C\, 2^{2(\nu+1)}\frac{\mu(\varepsilon)}{\varepsilon} \int_0^{\frac{T}{2}} \|v_\nu\|_{L^2}^2\, dt.$$

We deduce that, for all $\nu \in \mathbb{N}$,

$$\int_0^{\frac{T}{2}} \|\partial_t v_\nu + \partial_x(a\partial_{x_k} v_\nu) + \Phi'(\gamma(T-t))v_\nu\|_{L^2}^2\, dt$$

$$\geq \int_0^{\frac{T}{2}} \left(\|\partial_x(a\partial_x v_\nu) + \Phi'(\gamma(T-t))v_\nu\|_{L^2}^2 + \gamma\Phi''(\gamma(T-t))\|v_\nu\|_{L^2}^2 \right. \tag{3.11}$$

$$\left. -C\left(2^{4(\nu+1)}\mu(\varepsilon) + 2^{2(\nu+1)}\frac{\mu(\varepsilon)}{\varepsilon} \right) \|v_\nu\|_{L^2}^2 \right)\, dt.$$

Let $\nu = 0$. From (3.2) we can choose $\gamma_0 > 0$ such that $\Phi''(\gamma(T-t)) \geq 1$ for all $\gamma > \gamma_0$ and for all $t \in [0, T/2]$. Taking now $\varepsilon = 1/2$ we obtain from (3.11) that

$$\int_0^{\frac{T}{2}} \|\partial_t v_0 + \partial_x(a\partial_x v_0) + \Phi'(\gamma(T-t))v_0\|_{L^2}^2\, dt \geq \int_0^{\frac{T}{2}} \left(\gamma - 24C\mu\left(\frac{1}{2}\right) \right) \|v_0\|_{L^2}^2\, dt$$

for all $\gamma > \gamma_0$. Possibly choosing a larger γ_0 we have, again for all $\gamma > \gamma_0$,

$$\int_0^{\frac{T}{2}} \|\partial_t v_0 + \partial_x(a\partial_x v_0) + \Phi'(\gamma(T-t))v_0\|_{L^2}^2 \, dt \geq \frac{\gamma}{2} \int_0^{\frac{T}{2}} \|v_0\|_{L^2}^2 \, dt. \qquad (3.12)$$

Let now $\nu \geq 1$. We recall that in this case $\|\partial_x v_\nu\|_{L^2} \geq 2^{\nu-1}\|v_\nu\|_{L^2}$. We take $\varepsilon = 2^{-2\nu}$. We obtain from (3.11) that

$$\int_0^{\frac{T}{2}} \|\partial_t v_\nu + \partial_x(a\partial_x v_\nu) + \Phi'(\gamma(T-t))v_\nu\|_{L^2}^2 \, dt$$

$$\geq \int_0^{\frac{T}{2}} ((\|\partial_x(a\partial_x v_\nu)\|_{L^2} - \Phi'(\gamma(T-t))\|v_\nu\|_{L^2})^2$$
$$+\gamma\Phi''(\gamma(T-t))\|v_\nu\|_{L^2}^2 - C\,2^{4\nu}\,\mu(2^{-2\nu})\|v_\nu\|_{L^2}^2) \, dt.$$

On the other hand, since

$$\|\partial_x(a\partial_x v_\nu)\|_{L^2}\|v_\nu\| \geq |\langle \partial_x(a\partial_x v_\nu), v_\nu\rangle_{L^2}| \geq |\langle a\partial_x v_\nu, \partial_x v_\nu\rangle_{L^2}| \geq \lambda_0\|\partial_x v_\nu\|_{L^2}^2$$

we have

$$\|\partial_x(a\partial_x v_\nu)\|_{L^2} \geq \frac{\lambda_0}{4}\,2^{2\nu}\,\|v_\nu\|_{L^2}.$$

Suppose first that $\Phi'(\gamma(T-t)) \leq \frac{\lambda_0}{8}\,2^{2\nu}$. Then

$$\|\partial_x(a\partial_x v_\nu)\|_{L^2} - \Phi'(\gamma(T-t))\|v_\nu\|_{L^2} \geq \frac{\lambda_0}{8}\,2^{2\nu}\|v_\nu\|_{L^2}$$

and then, using also the fact that $\Phi''(\gamma(T-t)) \geq 1$, we obtain that

$$\int_0^{\frac{T}{2}} ((\|\partial_x(a\partial_x v_\nu)\|_{L^2} - \Phi'(\gamma(T-t))\|v_\nu\|_{L^2})^2$$
$$+\gamma\Phi''(\gamma(T-t))\|v_\nu\|_{L^2}^2 - C\,2^{4\nu}\,\mu(2^{-2\nu})\|v_\nu\|_{L^2}^2 \,)dt$$

$$\geq \int_0^{\frac{T}{2}} \left(\left(\frac{\lambda_0}{8}\,2^{2\nu}\right)^2 + \gamma - C\,2^{4\nu}\,\mu(2^{-2\nu})\|v_\nu\|_{L^2}^2 \right) dt$$

$$\geq \int_0^{\frac{T}{2}} \left(\left(\frac{1}{2}\left(\frac{\lambda_0}{8}\right)^2 - C\mu(2^{-2\nu})\right) 2^{4\nu} + \frac{\gamma}{3} \right) \|v_\nu\|_{L^2}^2 \, dt$$

$$+ \int_0^{\frac{T}{2}} \left(\frac{1}{2}\left(\frac{\lambda_0}{8}\right)^2 2^{4\nu} + \frac{2}{3}\gamma \right) \|v_\nu\|_{L^2}^2 \, dt.$$

Since $\lim_{\nu \to +\infty} \mu(2^{-2\nu}) = 0$, there exists $\gamma_0 > 0$ such that

$$\left(\frac{1}{2}\left(\frac{\lambda_0}{8}\right)^2 - C\mu(2^{-2\nu}) \right) 2^{4\nu} + \frac{\gamma}{3} \geq 0$$

for all $\gamma \geq \gamma_0$ and for all $\nu \geq 1$. Consequently there exist γ_0 and $c > 0$ not depending on ν such that

$$\int_0^{\frac{T}{2}} ((\|\partial_x(a\partial_x v_\nu)\|_{L^2} - \Phi'(\gamma(T-t))\|v_\nu\|_{L^2})^2$$

$$+ \gamma\Phi''(\gamma(T-t))\|v_\nu\|_{L^2}^2 - C\,2^{4\nu}\,\mu(2^{-2\nu})\|v_\nu\|_{L^2}^2\,)dt \tag{3.13}$$

$$\geq \int_0^{\frac{T}{2}} \left(\frac{1}{2}\left(\frac{\lambda_0}{8}\right)^2 2^{4\nu} + \frac{2}{3}\gamma\right)\|v_\nu\|_{L^2}^2\,dt \geq \int_0^{\frac{T}{2}} \left(\frac{\gamma}{2} + c\gamma^{\frac{1}{2}}\,2^{2\nu}\right)\|v_\nu\|_{L^2}^2\,dt$$

for all $\gamma \geq \gamma_0$.

If on the contrary $\Phi'(\gamma(T-t)) \geq \frac{\lambda}{8}\,2^{2\nu}$ then, using (3.1), the fact that $\lambda_0 \leq 1$ and the properties of μ (see Subsection 3.1),

$$\Phi''(\gamma(T-t)) = (\Phi'(\gamma(T-t))^2\mu\left(\frac{1}{\Phi'(\gamma(T-t))}\right)$$

$$\geq \left(\frac{\lambda_0}{8}\right)^2 2^{4\nu}\mu\left(\frac{8}{\lambda_0}\,2^{-2\nu}\right) \geq \left(\frac{\lambda_0}{8}\right)^2 2^{4\nu}\mu(2^{-2\nu}).$$

Hence also in this case there exist γ_0 and $c > 0$ such that

$$\int_0^{\frac{T}{2}} ((\|\partial_x(a\partial_x v_\nu)\|_{L^2} - \Phi'(\gamma(T-t))\|v_\nu\|_{L^2})^2$$

$$+ \gamma\Phi''(\gamma(T-t))\|v_\nu\|_{L^2}^2 - C\,2^{4\nu}\,\mu(2^{-2\nu})\|v_\nu\|_{L^2}^2)\,dt \tag{3.14}$$

$$\geq \int_0^{\frac{T}{2}} \left(\frac{\gamma}{2} + \left(\frac{\gamma}{2}\left(\frac{\lambda_0}{8}\right)^2 - C\right)2^{4\nu}\mu(2^{-2\nu})\right)\|v_\nu\|_{L^2}^2 dt \geq \int_0^{\frac{T}{2}} \left(\frac{\gamma}{2} + c\gamma\,2^{2\nu}\right)\|v_\nu\|_{L^2}^2 dt$$

for all $\gamma \geq \gamma_0$ and for all $\nu \geq 1$. Putting together (3.13) and (3.14), we have that there exist γ_0 and $c > 0$ such that

$$\int_0^{\frac{T}{2}} \|\partial_t v_\nu + \partial_x(a\partial_x v_\nu) + \Phi'(\gamma(T-t))v_\nu\|_{L^2}^2\,dt$$

$$\geq \int_0^{\frac{T}{2}} \left(\frac{\gamma}{2} + c\gamma^{\frac{1}{2}}\,2^{2\nu}\right)\|v_\nu\|_{L^2}^2\,dt \tag{3.15}$$

for all $\nu \geq 1$ and for all $\gamma \geq \gamma_0$.

Form (3.12) and (3.15) we get that there exist γ_0 and $\tilde{c} > 0$ such that

$$\int_0^{\frac{T}{2}} \sum_\nu \|\partial_t v_\nu + \partial_x(a\partial_x v_\nu) + \Phi'(\gamma(T-t))v_\nu\|_{L^2}^2\,dt$$

$$\geq \tilde{c}\gamma^{\frac{1}{2}} \int_0^{\frac{T}{2}} \sum_\nu (\gamma^{\frac{1}{2}}\|v_\nu\|_{L^2}^2 + \|\nabla v_\nu\|_{L^2}^2)\,dt$$

for all $\gamma \geq \gamma_0$. This inequality is equivalent to (3.9) and it concludes the proof of the Proposition 3.1.

Acknowledgment

The author would like to thank Prof. Marius Paicu for useful and interesting discussions on topics related to this paper.

References

[1] C. Bardos and L. Tartar, *Sur l'unicité rétrograde des équations paraboliques et quelques questions voisines,* Arch. Rational Mech. Anal. **50** (1973), 10–25.

[2] J.-M. Bony, *Calcul symbolique et propagations des singularités pour les équations aux dérivées partielles non linéaires,* Ann. Sci. École Norm. Supér. (4) **14** (1981), 209–246.

[3] R. Coifman and Y. Meyer, "Au delà des opérateurs pseudo-différentiels", Astérisque **57**, Société Mathématique de France, Paris, 1978.

[4] F. Colombini, E. De Giorgi, S. Spagnolo, *Sur les équations hyperboliques avec des coefficients qui ne dépendent que du temps,* Ann. Scuola Norm. Sup. Pisa Cl. Sci. **6** (1979), 511–559.

[5] F. Colombini and G. Métivier, *The Cauchy problem for wave equations with non Lipschitz coefficients; application to continuation of solutions of some nonlinear wave equations* Ann. Sci. École Norm. Supér. (4) **41** (2008), no. 2, 177–220.

[6] D. Del Santo and M. Prizzi, *Backward uniqueness for parabolic operators whose coefficients are non-Lipschitz continuous in time,* J. Math. Pures Appl. (9) **84** (2005), no. 4, 471–491.

[7] J.-M. Ghidaglia, *Some backward uniqueness results,* Nonlinear Anal. **10** (1986), no. 8, 777–790.

[8] J.-L. Lions, B. Malgrange, *Sur l'unicité rétrograde dans les problèmes mixtes paraboliques,* Math. Scand. **8** (1960), 277–286.

[9] N. Mandache, *On a counterexample concerning unique continuation for elliptic equations in divergence form,* Math. Phys. Anal. Geom. **1**(1998), 273–292.

[10] G. Métivier, "Para-differential calculus and applications to the Cauchy problem for nonlinear systems", Centro di Ricerca Matematica "Ennio De Giorgi" (CRM) Series, **5**, Edizioni della Normale, Pisa (2008).

[11] K. Miller, *Nonunique continuation for uniformly parabolic and elliptic equations in self–adjoint divergence form with Hölder continuous coefficients,* Arch. Rational Mech. Anal. **54** (1973), 105–117.

[12] S. Tarama, *Local uniqueness in the Cauchy problem for second-order elliptic equations with non-Lipschitzian coefficients,* Publ. Res. Inst. Math. Sci. **33** (1997), 167–188.

[13] A. Tychonoff, *Théorème d'unicité pour l'équation de la chaleur,* Math. Sbornik **42** (1935), 199–215.

Daniele Del Santo
Dipartimento di Matematica e Geoscienze
Università di Trieste
Via Valerio 12/1
I-34217 Trieste
e-mail: delsanto@units.it

Progress in Mathematics, Vol. 301, 115–143

Dispersive Properties of Schrödinger Operators in the Absence of a Resonance at Zero Energy in 3D

Vladimir Georgiev and Mirko Tarulli

To Bruna

Abstract. In this paper we study spectral properties associated to the Schrödinger operator $-\Delta - W$, with potential W that is an exponentially decaying C^1 function. As applications we prove local energy decay for solutions to the perturbed wave equation and lack of resonances for the NLS.

Mathematics Subject Classification. Primary 35B34; Secondary 35L05.

Keywords. Wave equation, Schrödinger equation, local energy decay, solitary solutions, resonances.

1. Introduction

In this paper we study the problem of resonances at zero energy for the operator

$$S(\psi)(x) = -\Delta\psi(x) - W(|x|)\psi(x), \qquad (1)$$

with $W(|x|)$ being a positive real-valued measurable radial function, decreasing sufficiently rapidly at infinity. There exists a vast literature concerning the theory of resonances, we cite here [2], [6], [29], [31] (and references therein). The resonances of an operator were introduced in physics and defined as the poles of its resolvent operator function taken in some generalized way. More precisely one can observe that, if we choose a radial function $u(|x|)$ in $\mathbf{R}^3$, we have the relation

$$\Delta\left(\frac{u(|x|)}{|x|}\right) = \frac{u''(|x|)}{|x|}.$$

Therefore, picking up $\psi(x) = u(x)/|x|$ and $S(\psi) = P(u)/|x|$, we can rewrite the operator (1) as

$$P(u)(r) = -u''(r) - W(r)u(r) \qquad (2)$$

on the semi-line $(0, \infty)$ together with the Dirichlet condition

$$u(0) = 0$$

at the origin, that is a selfadjoint unbounded operator in $L^2(0, \infty)$ with domain

$$\{u \in H^2(0, \infty); u(0) = 0\}.$$

A reasonable assumption on the potential that allows to apply the Friedrichs extension for P is

$$|W(r)| \leq \frac{C}{r^{2-\varepsilon}}, \quad r > 0. \tag{3}$$

For simplicity we shall suppose in this work that $W(r) \in C^1([0, \infty))$ that satisfies the assumption

$$|W'(r)| + |W(r)| \leq Ce^{-\varepsilon_0 r}, \quad r \geq 0 \tag{4}$$

for some $C > 0$ and $\varepsilon_0 > 0$.

It is well known that if the potential is of short range type, then the set of eigenvalues of P is finite, contained in $(-\infty, 0)$, with each eigenvalue of finite multiplicity.

Recall that the resolvent of P

$$R(\mu^2) = (P - \mu^2)^{-1}$$

(considered as an operator from C_0^∞ to C^∞) is a meromorphic operator in some subset of the complex plane. The poles of $R(\mu^2)$ are called resonances of P. The main goal of this work is to present an argument that gives a sufficient condition for the non-existence of resonances. This means the following,

Theorem 1.1. *Suppose that the potential $W(r) \in C^1([0, \infty))$ is a positive decreasing function satisfying (for some $C > 0$ and $\varepsilon_0 > 0$) the estimates (4).*
Then there exists a positive δ such that there are no resonances in

$$\{\mu \in \mathbf{C}; |\mu| \leq \delta\}.$$

Moreover, if μ with $\operatorname{Im} \mu > -\delta$ is a resonance, then μ is a real negative number and μ is an eigenvalue of P.

Remark 1.1. *The fact that all resonances in the domain of type*

$$\{\mu \in \mathbf{C}; 0 < |\mu| < R, |\operatorname{Im} \mu| < \delta\}$$

are eigenvalues is a well-known and follows from resolvent estimates leading to limiting absorption principle (see [1] for example). The fact that this domain can be extended taking $R \to \infty$ is also well known (see [28] for example). Therefore, the key information in this theorem is the lack of resonances at the origin $\mu = 0$.

The resonances can be considered in some manner like eigenvalues. The existence of a non-trivial solution of the equation $Pu = 0$, is a typical obstacle to find dispersive properties of the time evolution group associated with the Schrödinger operator $-\Delta - W(x)$. Therefore, as a first application of the above Theorem 1.1, we shall look for dispersive properties to the solution of the following (see [4], [5], [31], [33] for further details),

$$u_{tt} - \Delta u - Wu = 0, \qquad (t,x) \in \mathbf{R} \times \mathbf{R}^3$$
$$u(0,x) = u_0(x),\, u_t(0,x) = v_0(x), \tag{5}$$

where the potential W satisfies the assumptions (4). In our setting, we get local energy decay, see Theorem 5.1, for the above problem (5).

Consider now the solution u of the Schrödinger type equation

$$i\partial_t u + \Delta u + |u|^{p-1}u = 0. \tag{6}$$

The existence of solitary type solutions to the Schrödinger equation (6) is a well-studied problem. One can see for example [9] and [32] the existence results for $1 < p \le 1 + 4/3$.

The natural functional associated with this problem is

$$\mathcal{E}(\chi) = \frac{1}{2}\|\nabla\chi\|_{L^2}^2 - \frac{1}{p+1}\int_{\mathbf{R}^3} |\chi|^{p+1}dx. \tag{7}$$

The corresponding minimization problem is associated with the quantity

$$I_N = \inf\{\mathcal{E}(\chi); \chi \in H^1, \|\chi\|_{L^2}^2 = N\}. \tag{8}$$

We have the following result, see [9] and [10] for more details,

Lemma 1.1. 1. *For any* $\omega > 0$ *there exists a unique positive solution* $\chi(x) = \chi_\omega(x) \in H^1$ *of the equation* (53), *such that*

i) *the function* $\chi(x) = \chi(|x|)$ *is radially symmetric,*
ii) *the function* $\omega \in (0,\infty) \to \|\chi_\omega\|_{L^2}$ *is a strictly increasing one and belongs to* $C^1(0,\infty)$.

Remark 1.2. *To check* ii) *we use the following argument used in* [9]. *Any positive radial solution of*

$$-\Delta\chi + \omega\chi = \chi^p,\ \chi \in H^1(\mathbf{R}^3),\ \chi > 0$$

is given by $\chi(x) = \omega^{1/(p-1)}\chi_1(\sqrt{\omega}x)$ *where* χ_1 *is the unique radial solution of*

$$-\Delta\chi_1 + \chi_1 = \chi_1^p,\ \chi_1 \in H^1(\mathbf{R}^3),\ \chi_1 > 0$$

Thus

$$\|\chi\|_{L^2(\mathbf{R}^3)} = \omega^{1/(p-1)-3/4}\|\chi_1\|_{L^2(\mathbf{R}^3)}$$

so $N = $ *const* $\omega^{1/(p-1)-3/4}$ *where* $1/(p-1) - 3/4 > 0$. *This yields the characterization of* N *as a function of* ω.

One can see that the solutions χ_ω of Lemma 1.1 are radial and $\chi_\omega(x)$ is rapidly decreasing in x as $|x| \to \infty$ provided $N(\omega) \le 1$.

In particular property ii) of Lemma 1.1 guarantees that one can find a unique χ^* that is a radial positive function $\chi_* = \chi_*(|x|)$ that is a minimizer of I_1. Therefore, one can find a unique $\omega_* > 0$ so that

$$\Delta \chi_* + \omega_* \, \chi_* = \chi_*^p,$$

$$\int_{\mathbf{R}^3} |\chi_*|^2 = 1. \tag{9}$$

A standard linearization of the Nonlinear Schrödinger equation (6) around the solitary solution leads to the necessity to use some spectral properties of the following operator

$$-\Delta - \chi_*^p(x).$$

Note that the operator

$$L_-(\omega^*) = -\Delta - \chi_*^p(x) + \omega^*$$

introduced in [34] (with rescaled choice $\omega^* = 1$) has a nontrivial kernel and plays an important role in the study of modulational stability of ground states of nonlinear Schrödinger equations.

It is well known from the results in [13], [18], [7], [8], [24], [26] that asymptotic stability around solitary waves is closely connected with the existence of resonances at the origin. More precisely, the following assumption is frequently used in these articles:

(H1)   0 is not a resonance of $-\Delta - \chi_*^p(x)$.

The main goal of this work is to present an argument that proves the assumption (H1) in the general case and therefore the above-cited results can be established without this additional assumption and this is contained in Theorem 6.1.

The asymptotic expansions of the resolvent near zero energy are discussed in various works. For example in [19] this expansion is obtained and the possible resonance at zero energy is associated with the corresponding projection operator. For the general potential perturbation of the free Laplace operator $-\Delta - W(|x|)$ the resonances can be connected (see Definition 4 below) with nontrivial solutions $u \in C(\mathcal{R}^n)$, to the equation $-\Delta u - W(|x|)u = 0$ satisfying the appropriate weak decay at infinity. If the space dimension is $n \geq 5$ and the potential decays sufficiently rapidly at infinity, then a remark in [20] clarifies that $u \in L^2$ so resonance at 0 means that 0 is an eigenvalue. Under a suitable sign condition assumption one can establish directly that 0 is not a resonance even in the 3-dimensional case (see [16] for example. Under a suitable compact support assumption (which one can see in [3]) it is possible to extend in a continuous way the resolvent even for the perturbation of the principal symbol of the Laplacian, the smallness assumption on the positive part of W can guarantee also that 0 is not a resonance (see [4], [5] for the potential case and [15] for the perturbation with magnetic fields).

The scheme of the paper is the following: in Section 2 we distinguish between strong and weak resonances, defining the last one. We also give an asymptotic expansions of the corresponding solution. In Section 3 we give the main Theorem 3.1 on lack of strong resonances at the spectral point zero, in the radial case. In Section 4 we extend the result to the non-radial case. In Sections 5 and 6 we obtain some application, local energy decay for a wave equation perturbed by a potential (Theorem 5.1), and lack of resonance (and eigenvalue) for the linearized operator of NLS around its ground states (Theorem 6.1). Finally in Section 7 we furthermore focalize on the weak resonances.

Recall the definition of the norm in the Sobolev spaces H^s

$$\|u\|_{H^s} = \|\langle \Delta \rangle^s u\|_{L^2(\mathbf{R}^3)},$$

where $s \in \mathbf{R}$ and $\langle \Delta \rangle = (1 + |\Delta|^2)^{1/2}$ Moreover, given any two positive real numbers a, b, we write $a \lesssim b$ to indicate $a \leq Cb$, with $C > 0$.

2. Resonance at the spectral origin

To study the poles of the perturbed resolvent $R(\mu^2) = (P - \mu^2)^{-1}$ we start with an explicit representation of the free resolvent $R_0(\mu^2) = (P_0 - \mu^2)^{-1}$, where

$$P_0(u)(r) = -u''(r)$$

with $r \in (0, \infty)$ together with the Dirichlet condition

$$u(0) = 0$$

at the origin. Then

$$R_0^{\pm}(\mu^2)(f)(r) = \int_0^r e^{\pm i\mu r} \frac{\sin(\mu s)}{\mu} f(s)ds + \int_r^\infty e^{\pm i\mu s} \frac{\sin(\mu r)}{\mu} f(s)ds, \qquad (10)$$

are well defined for $f \in C_0^\infty$. A direct calculation shows that

$$u_\pm(r) = R_0^\pm(\mu^2)(f)(r)$$

satisfies the equation

$$P_0(u) - \mu^2 u = f$$

as well as the Dirichlet condition $u(0) = 0$. Choosing the sign $+$ in the above representations we see that

$$R_0(\mu^2)(f)(r) = \int_0^r e^{i\mu r} \frac{\sin(\mu s)}{\mu} f(s)ds + \int_r^\infty e^{i\mu s} \frac{\sin(\mu r)}{\mu} f(s)ds$$

can be extended as an operator in L^2 provided $\operatorname{Im} \mu > 0$. Moreover, this is a holomorphic operator-valued function for $\operatorname{Im} \mu > 0$. The representation formula guarantees also that the operator

$$\varphi(r)(P_0 - \mu^2)^{-1}\varphi(r), \quad \varphi(r) = e^{-\delta r}$$

is a holomorphic operator-valued (in L^2) function in the larger domain $\operatorname{Im} \mu > -\delta$.

The perturbed resolvent $R(\mu^2) = (P - \mu^2)^{-1}$ satisfies on the upper half-plane of $\mathbb{C}$ the relation

$$(P - \mu^2)^{-1} = \left(I - (P_0 - \mu^2)^{-1}W\right)^{-1}(P_0 - \mu^2)^{-1}, \qquad (11)$$

provided the operator $(I - (P_0 - \mu^2)^{-1}W)$ is invertible. This relation implies

$$\varphi(P - \mu^2)^{-1}\varphi = \left(I - \varphi(P_0 - \mu^2)^{-1}W\frac{1}{\varphi}\right)^{-1}\varphi(P_0 - \mu^2)^{-1}\varphi. \qquad (12)$$

Applying the Fredholm alternative, we see that

$$\varphi(P - \mu^2)^{-1}\varphi$$

is holomorphic except the points μ such that the equation

$$f = \varphi(P_0 - \mu^2)^{-1}W\frac{f}{\varphi}$$

has a nontrivial solution $f \in L^2(0, +\infty)$. These complex numbers μ are the resonances and we give a more detailed description of the resonances in the following.

Lemma 2.1. *Assume $W(r) \leq Ce^{-\varepsilon_0 r}$ and $\varphi(r) = e^{-\delta r}$ with $0 < \delta < \varepsilon_0/2$. For any complex μ with $0 \geq \mathrm{Im}\,\mu > -\delta$ the following conditions are equivalent:*

i) *μ is a resonance, i.e., there is $f \in L^2(0, \infty)$ so that f is not identically zero and*

$$f = \varphi(P_0 - \mu^2)^{-1}W\frac{f}{\varphi};$$

ii) *there is $u \in C[0, \infty)$ not identically zero, $e^{\mathrm{Im}\,\mu\, r}u$ is bounded and*

$$u = (P_0 - \mu^2)^{-1}Wu;$$

iii) *there is $u \in C[0, \infty) \cap C^2(0, \infty)$ so that u is not identically zero, $e^{\mathrm{Im}\,\mu\, r}u$ is bounded and u solves*

$$Pu = \mu^2 u, \quad u(0) = 0.$$

Proof. i) $\Leftrightarrow$ ii). The proof follows from the standard substitution $u = f\varphi^{-1}$ and classical estimates. We get $u = (P_0 - \mu^2)^{-1}Wf\varphi^{-1}$. Now by the assumption on the complex number μ, we can write

$$\|e^{\mathrm{Im}\,\mu\, r}u\|_{L^\infty} \lesssim \|u\|_{L^\infty} \lesssim \|(P_0 - \mu^2)^{-1}Wf\varphi^{-1}\|_{L^\infty}.$$

The resolvent estimates $\|(P_0 - \mu^2)^{-1}g\|_{L^\infty} \lesssim \|\langle x\rangle^{1/2+\varepsilon}g\|_{L^2}$, (see [16] and reference therein), and the bound $\langle x\rangle \lesssim \varphi^{-1}$, yield the implication i)$\Rightarrow$ii). The other implication follows from

$$|f(x)| = |\varphi(x)u(x)| \leq Ce^{-(\delta+\mathrm{Im}\,\mu)|x|} \in L^2$$

with

$$C = \sup_{r>0} e^{\mathrm{Im}\,\mu\, r}|u(r)|.$$

ii) $\Rightarrow$ iii). Applying the operator P on both sides of the identity $u = (P_0-\mu^2)^{-1}Wu$, we get the result. Moreover the representation (10) says that $u(0) = 0$.

iii) $\Rightarrow$ ii). The proof follows by an application of the Limiting Absorption Principle (see [1]) and by the bound $\varphi \lesssim \langle x\rangle^{-1}$. $\qquad\square$

The above lemma reduces the study of resonance to the study of the solutions to the problem

$$Pu = \mu^2 u, \quad u(0) = 0,$$

satisfying the bound

$$|u(r)| \le C e^{-\operatorname{Im}\mu \, r}, \quad 0 \ge \operatorname{Im}\mu > -\delta.$$

A more general question to solve is the existence of nontrivial solutions satisfying the weaker bound

$$|u(r)| \le C e^{\delta r}, \quad 0 \ge \operatorname{Im}\mu > -\delta.$$

We distinguish these two cases: we call the resonances from Lemma 2.1 strong resonances and define the weak ones as follows:

Definition 1. *A complex number μ with $0 \ge \operatorname{Im}\mu > -\delta$ is called a weak resonance of*

$$P = -\left(\frac{d}{dr}\right)^2 - W(r)$$

if there exists a $u \in C([0, \infty))$, such that $u(0) = 0$, $u(r)$ is not identically zero, $P(u) = \mu^2 u$ in the distribution sense in $(0, \infty)$ and the solution u satisfies the inequality

$$|u(r)| \le C e^{\delta|r|}. \tag{13}$$

We recall that, as underlined for example in [11] and [25], that the resonances are defined as functions not in L^2 but in larger L^2 weighted spaces. Moreover these functions are L^∞-bounded and behave asymptotically as $\langle x \rangle^{-1}$, in the spatial dimension three. These resonances will be called strong resonances (see Definition 2 for precise notion). But, for completeness, we study also the weak ones.

Moreover, as it was mentioned in the introduction, we have to study only the existence of resonance at the origin.

3. Strong resonance at the spectral origin

Definition 2. *A real number λ is called a strong resonance of*

$$P = \left(\frac{d}{dr}\right)^2 + W(r)$$

if there exists $u \in C([0, \infty))$, such that $u(0) = 0$, $u(s)$ is not identically zero, $P(u) = \lambda u$ in the distribution sense in $(0, \infty)$ and the solution u satisfies the inequality

$$|u(r)| \le C (1 + r)^a. \tag{14}$$

with some $0 \le a < 1$.

We shall need the asymptotic expansions of Lemma 7.1. Without loss of generality we can assume that u is real-valued. Multiplying the equation

$$P(u) \equiv u''(r) + W(r)u = 0$$

by u' and integrating over (r, ∞), we find

$$-\frac{|u'(r)|^2}{2} - \frac{W(r)|u(r)|^2}{2} - \int_r^\infty \frac{W'(\tau)|u(\tau)|^2}{2}\, d\tau = 0. \tag{15}$$

Take any function $g(r) \in C([0, \infty))$, such that $g(r)$ tends to 0 at infinity. We multiply further the last relation by $-g(r)$ and integrate over $(0, \infty)$

$$\int_0^\infty \frac{g(r)|u'(r)|^2}{2}\, dr + \int_0^\infty \frac{g(r)W(r)|u(r)|^2}{2}\, dr$$
$$+ \int_0^\infty \frac{G(r)W'(r)|u(r)|^2}{2}\, dr = 0, \tag{16}$$

where $G(r) = \int_0^r g(\tau)d\tau$.

Take a C^1 function $h(r)$ on $(0, \infty)$ such that $h''(r)$ exists, is continuous on $(0, r_0) \cup (r_0, +\infty)$ and has a finite jump

$$-\infty < \lim_{r \to r_0^+} h''(r) - \lim_{r \to r_0^-} h''(r) < \infty.$$

We shall require further that $h'(r)$ tends to zero at infinity and $h''(r)$ is decaying sufficiently rapidly (for example exponentially) at infinity. We multiply the equation $Pu = 0$ by hu and integrate over $(0, \infty)$, so we get

$$\int_0^\infty \frac{h''(r)|u(r)|^2}{2}\, dr - \int_0^\infty h(r)|u'(r)|^2\, dr$$
$$+ \int_0^\infty h(r)W(r)|u(r)|^2\, dr = 0. \tag{17}$$

Choosing $h = g/2$ and summing the above two relations, we obtain

$$\int_0^\infty \Phi(r)\, |u(r)|^2\, dr = 0, \tag{18}$$

where

$$\Phi(r) = 2g(r)W(r) + G(r)W'(r) + \frac{g''(r)}{2}.$$

The starting definition of the function $g''(r)$ is the following one

$$g''(r) = \begin{cases} r^M, & \text{if } 0 \le r < r_0; \\ e^{-\delta r}, & \text{if } r > r_0. \end{cases} \tag{19}$$

We take $r_0 = M^2$ and shall define the positive parameters M, δ later on.

For $0 \le r < r_0$ we have

$$g(r) = \frac{r^{M+2}}{(M+2)(M+1)}, \quad G(r) = \frac{r^{M+3}}{(M+3)(M+2)(M+1)}.$$

From these relations we deduce the following.

Lemma 3.1. *Suppose the potential $W(r)$ is a positive decreasing function in $C^1([0,\infty))$, such that the assumption (4) is satisfied for some positive constants C, ε_0. Then one can find a positive constant M^* depending on C, ε_0, such that for any $r \in (0, M^2)$ and $M > M^*$ we have*

$$\Phi(r) \geq 2gW.$$

Proof. We have the estimates

$$G(r)W'(r) + \frac{g''(r)}{2} \geq -\frac{C^* r^{M+3} e^{-\varepsilon_0 r}}{(M+3)(M+2)(M+1)} + \frac{r^M}{2}$$

$$= \frac{r^M}{2}\left(1 - \frac{2C^* r^3 e^{-\varepsilon_0 r}}{(M+3)(M+2)(M+1)}\right).$$

Now we are in the position to choose $M^* = M^*(C^*, \varepsilon_0)$ so large that

$$1 - \frac{2C^* r^3 e^{-\varepsilon_0 r}}{(M+3)(M+2)(M+1)} \geq 0$$

for $r \leq M^2$ and $M \geq M^*$. Hence

$$\Phi(r) = 2g(r)W(r) + G(r)W_2'(r) + \frac{g''(r)}{2} \geq 2g(r)W(r).$$

This completes the proof. $\qquad\qquad\square$

For $r > r_0 = M^2$ we shall use the relations

$$g(r) = \int_0^r (r-\tau)g''(\tau)d\tau, \quad G(r) = \int_0^r \frac{(r-\tau)^2}{2} g''(\tau)d\tau.$$

Here and below we used the property $g'(0) = 0$.

Lemma 3.2. *Suppose that the potential $W(r)$ is a positive decreasing function in $C^1([0,\infty))$, such that there exist positive constants C, ε_0 so that (4) is fulfilled. Then one can find a positive constant M^* depending on C, ε_0, such that for any $M \geq M^*$ and for any $r \in (M^2, \infty)$ we have*

$$|\Phi(r)| \leq 2e^{-\varepsilon_0 r/2}.$$

Proof. Take $r > M^2$. First we evaluate

$$g(r) = \int_0^{M^2} (r-\sigma)\sigma^M d\sigma + \int_{M^2}^r (r-\sigma)e^{-\delta\sigma}d\sigma$$

$$\leq r\int_0^{M^2} \sigma^M d\sigma + r\int_0^\infty e^{-\delta\sigma}d\sigma \leq r\left(M^{2M+2} + \frac{1}{\delta}\right) \leq \frac{2r^2 r^{\sqrt{r}}}{\delta}$$

for $\delta < 1$ and $M > 1$. In a similar way we get

$$2G(r) = \int_0^{M^2} (r-\sigma)^2 \sigma^M d\sigma + \int_{M^2}^r (r-\sigma)^2 e^{-\delta\sigma}d\sigma$$

$$\leq r^2\int_0^{M^2} \sigma^M d\sigma + r^2\int_0^\infty e^{-\delta\sigma}d\sigma \leq r^2\left(M^{2M+2} + \frac{1}{\delta}\right) \leq \frac{2r^3 r^{\sqrt{r}}}{\delta}.$$

From these estimates and the assumptions on the decay of W we derive

$$|\Phi(r)| = \left|2g(r)W(r) + G(r)W_2'(r) + \frac{g''(r)}{2}\right| \leq \frac{4C^* e^{-\varepsilon_0 r} r^2 r \sqrt{r}}{\delta} + e^{-\delta r}.$$

Taking $\delta = \varepsilon_0/2$ and using the fact that

$$\lim_{r \to \infty} e^{-\varepsilon_0 r/2} r^2 r \sqrt{r} = 0$$

we find $M^* = M^*(C^*, \varepsilon_0)$ so that

$$|\Phi(r)| \leq 2e^{-\varepsilon_0 r/2}$$

for $r > M^2$ and $M > M^*$. This completes the proof of the lemma. $\square$

Now we can state in a precise form the main Theorem 1.1:

Theorem 3.1. *Suppose that the potential $W(r)$ is a positive decreasing function in $C^1([0, \infty))$, such that the assumption (4) is satisfied for some positive constants C, ε_0. The zero is not a strong resonance for $Pu = u''(r) + W(r)u(r)$.*

Proof. Suppose that $u(r)$ is a solution to $u''(r) + W(r)u(r) = 0$, such that $u(r)$ is not identically zero, i.e.,

$$\int_1^2 |u(r)|^2 dr \neq 0.$$

We choose M^* so that the conclusions of Lemma 3.1 and Lemma 3.2 are fulfilled. Then the identity (18) implies that

$$\int_0^{M^2} \Phi(r)\,|u(r)|^2\,dr = -\int_{M^2}^{\infty} \Phi(r)\,|u(r)|^2\,dr \tag{20}$$

for any $M \geq M^*$. We can apply Lemma 7.1 and conclude that

$$u(s) = u(1) + \int_1^s (\tau - 1)W(\tau)u(\tau)d\tau + \int_s^{\infty} (s - 1)W(\tau)u(\tau)d\tau.$$

This relation and the exponential decay of the potential $W(\tau)$ implies

$$S = \sup |u(s)| \leq |u(1)| + \int_0^{\infty} (\tau + 1)W(\tau)|u(\tau)|d\tau < \infty.$$

Applying Lemma 3.1, we see that

$$\int_0^{M^2} \Phi(r)\,|u(r)|^2\,dr \geq \int_0^{M^2} g(r)W(r)\,|u(r)|^2\,dr$$

$$\geq \int_1^2 |u(r)|^2 dr\, g(1)W(1) = \frac{1}{(M+2)(M+1)}\,W(1)\int_1^2 |u(r)|^2 dr$$

The right-hand side of the identity (20) can be evaluated by the aid of Lemma 3.2 and we get

$$-\int_{M^2}^{\infty} \Phi(r)\,|u(r)|^2\,dr \leq 2\int_{M^2}^{\infty} e^{-\varepsilon_0 r/2}|u(r)|^2 dr \leq 2S^2\,\frac{e^{-\varepsilon_0 M^2/2}}{\varepsilon_0}$$

so we arrive at the inequality

$$W(1) \int_1^2 |u(r)|^2 dr \leq 2S^2 \frac{(M+1)(M+2)e^{-\varepsilon_0 M^2/2}}{\varepsilon_0}$$

that obviously leads to a contradiction with

$$\int_1^2 |u(r)|^2 dr \neq 0$$

if $M > M^*$ is sufficiently large. The contradiction implies $u = 0$. This completes the proof.   □

4. The non-radial case: zero is resonance implies zero is an eigenvalue

Along the previous sections we treated the non-existence of radial resonances. Our next step is to treat the general case, that is the resonances in the non-radial case. One can use standard projections on spherical harmonics and reduce the analysis to the proof that zero is not a resonance for the following operator (see [11] and [34]),

$$P(u)(r) = -u''(r) - W(r)u(r), \quad r \in (0, +\infty). \tag{21}$$

Here and below we shall assume that $W(r) = W_1(r) + W_2(r)$, where

$$W_1(r) = -\frac{\alpha(\alpha+1)}{r^2}, \quad \alpha \geq 0, \tag{22}$$

while $W_2(r)$ is a $C^1(0, \infty)$ positive strictly decreasing function such that for some positive constants C, ε_0 it satisfies the estimate (4) for some positive numbers C, ε_0.

It is clear that we need for the applications only the case when $\alpha(\alpha + 1)$ is an eigenvalue of the Laplace-Beltrami operator on the sphere $\mathbf{S}^2$. The arguments from this section are valid for any $\alpha > 0$.

One can have

Definition 3. *A real number λ is called eigenvalue of P if there exists $u \in H^1(0, \infty)$ such that $u(0) = 0$, $u(r)$ is not identically zero and $P(u) = \lambda u$ in the distribution sense in $(0, +\infty)$.*

Suppose that there exists a real-valued function $u(r) \in H^1(0, \infty)$ so that $P(u) = 0$. Our goal will be to show that u is identically zero.

The Sobolev embedding on $(0, \infty)$ implies that $u(r) \in C([0, \infty))$. Then the equation $Pu = 0$ guarantees that $u \in H^2(R, \infty) \subset C^1([R, \infty))$ for any $R > 0$. To analyze the behavior of the solution at infinity, we integrate the equation $Pu = 0$ in the interval (R, R_1) and find

$$|u'(R) - u'(R_1)| \leq \left(\int_R^{R_1} W(r)^2 \, dr \right)^{1/2} \|u\|_{L^2} \leq CR^{-3/2},$$

since $W(r)$ is bounded from above by constant times r^{-2}. The assumption $u \in H^1(0, \infty)$ easily yields

$$|u'(R)| \leq CR^{-3/2}, \qquad |u(R)| \leq CR^{-1/2}. \tag{23}$$

From the asymptotic expansion obtained in Lemma 7.3 we have also

$$\left| u'(R) + \frac{\alpha}{R} u(R) \right| \leq C e^{-\varepsilon_0 R/2}. \tag{24}$$

Once we have proved the lack of a zero-energy eigenvalue, we shall prove the following:

Lemma 4.1. *Suppose that $\alpha > 1/2$. If $\lambda = 0$ is a strong resonance (in the sense of Definition 2) of P, then $\lambda = 0$ is an eigenvalue of P.*

Proof. We can use the asymptotic formulae (68) at infinity, combined with the estimates (79), (80) near $r = 0$. If the operator P has in the spectral point $\lambda = 0$ a resonance, then these asymptotic estimates imply that $u(r)$ is a function in $H^1(0, \infty)$, so 0 is an eigenvalue of P. This completes the proof of the lemma. $\square$

Finally, we may study the resonances of the operator

$$-\Delta - W(|x|), \quad x \in \mathbf{R}^3.$$

Definition 4. *A real number λ is called a strong resonance of $-\Delta - W(|x|)$ if there exists a $u \in C(\mathcal{R}^3)$, such that $u(x)$ is not identically zero, $-\Delta - W(|x|)u = \lambda u$ in the distribution sense in $\mathcal{R}^3$ and the solution u satisfies the inequality*

$$|u(x)| \leq C (1 + |x|)^{-\varepsilon} \tag{25}$$

with some $\varepsilon > 0$.
If $u(x) = u(|x|)$ is a radial function, we shall call λ a radial resonance.

Theorem 4.1. *Suppose that the potential $W(r)$ is a positive decreasing function, such that there exist positive constants C, ε so that (4) is fulfilled. Then zero is not a radial resonance for $-\Delta - W(|x|)$. If $\lambda = 0$ is a (non-radial) resonance, then $\lambda = 0$ is an eigenvalue.*

5. Resolvent estimates and local energy decay for wave equations with potential

Along this section we will prove the main resolvent estimates concerning the perturbed operator (1). Let us indicate by

$$R_0(\mu) = (-\Delta - \mu^2)^{-1},$$

the resolvent of the operator $-\Delta$, and set $R_0^+(\mu) = R_0(\mu)$ if $\text{Im}\,\mu > 0$ and respectively $R_0^-(\mu) = R_0(\mu)$ for $\text{Im}\,\mu < 0$. Take into account now the initial value problem

$$\begin{cases} \partial_t^2 u - \Delta u = 0 \quad t \in \mathbf{R}, \quad x \in \mathbf{R}^3 \\ (u(0, x), \partial_t u(0, x)) = (0, g(x)), \end{cases} \tag{26}$$

Once we pick $\operatorname{Im}\mu > C > 0$, we have, by an application of the Laplace transform,

$$R_0(\mu) = \int_0^\infty e^{i\mu t}\mathcal{U}_0(t)dt \tag{27}$$

where $\mathcal{U}_0(t) = \frac{\sin t\sqrt{-\Delta}}{\sqrt{-\Delta}}$ is the evolution operator associated to (26). This means that the resolvent associated to (26) is a well-defined operator in $\mathcal{B}(H^{m-1}, H^m)$, and depends analytically on μ, once one notices that (see [21] and [31] for more details)

$$\|\mathcal{U}_0(t)f\|_{H^m} \lesssim \|f\|_{H^{m-1}}. \tag{28}$$

for any Schwartz function f. This aims to the following inequality, after an integration by parts of (27) and by (28), to the bound,

$$\|R_0^+(\mu)f\|_{H^m} \lesssim \frac{1}{\langle\mu\rangle}\|f\|_{H^m}, \tag{29}$$

for any Schwartz f. Moreover one could get from the identity,

$$R_0^+(\mu) = -\frac{1}{\mu^2} - \frac{1}{\mu^2}\Delta R_0^+(\mu),$$

the estimate,

$$\|R_0^+(\mu)f\|_{L^2} \lesssim \frac{1}{\langle\mu\rangle^2}\|f\|_{H^1}. \tag{30}$$

Other relevant estimates obtained easily from (27) are,

$$\left\|\frac{d}{d\mu}R_0^+(\mu)f\right\|_{L^2} \lesssim \frac{1}{\langle\mu\rangle^2}\|f\|_{L^2}, \tag{31}$$

$$\left\|\frac{d^2}{d\mu^2}R_0^+(\mu)f\right\|_{L^2} \lesssim \frac{1}{\langle\mu\rangle^3}\|f\|_{L^2}, \tag{32}$$

and

$$\left\|\frac{d^2}{d\mu^2}R_0^+(\mu)f\right\|_{L^2} \lesssim \frac{1}{\langle\mu\rangle^2}\|f\|_{H^{-1}}. \tag{33}$$

Recall the classical resolvent identities (11), (12) and set

$$R(\mu) = (-\Delta - W - \mu^2)^{-1}, \quad A(\mu) = \varphi(-\Delta - \mu^2)^{-1}W\varphi^{-1}.$$

One has the following compactness result in the spaces

$$L_{-\delta}^2 = \{f \in L_{loc}^2, (1 + |x|)^{-\delta}f \in L^2\}.$$

Lemma 5.1. *The operators* $A(\mu)$ *are compact in the space* $B(L_{-\delta}^2, L_{-\delta'}^2)$, *for* $\operatorname{Im}\mu > -\delta$ *and* $\delta, \delta' > 0$. *Moreover the following estimate is satisfied:*

$$\|A(\mu)\|_{B(L_{-\delta}^2, L_{-\delta'}^2)} \to 0,$$

as $|\mu| \to \infty$.

This lemma is a well-known standard resultm so we skip the proof. It suffices to notice that $\varphi(-\Delta - \mu^2)^{-1}W$ is analytic on the zone $\operatorname{Im}\mu > -\delta$, and that the potential W is of the short range type (see [1] and references therein). The continuity of the multiplication operator φ^{-1} in $B(L^2_{-\delta}, L^2_{-\delta'})$, with the estimate (29), give the result.

Lemma 5.2. *Let us assume that the potential W satisfies (4). The cutoff resolvent operator $\varphi R^+(\mu)\varphi$ has a meromorphic extension from $\operatorname{Im}\mu > 0$ to $\operatorname{Im}\mu > -\delta$. Moreover for each $\delta > 0$ there exists a real constant $C > 0$ such that the following estimates are true:*

$$\|\varphi P_{ac}R^+(\mu)\varphi f\|_{L^2} \leq \frac{C}{\langle\mu\rangle}\|f\|_{L^2}, \tag{34}$$

$$\|\varphi P_{ac}R^+(\mu)\varphi f\|_{L^2} \leq \frac{C}{\langle\mu\rangle^2}\|f\|_{H^1}, \tag{35}$$

$$\|\varphi\frac{d}{d\mu}P_{ac}R^+(\mu)\varphi f\|_{L^2} \leq C\frac{C}{\langle\mu\rangle^2}\|f\|_{L^2}, \tag{36}$$

$$\|\varphi\frac{d^2}{d\mu^2}P_{ac}R^+(\mu)\varphi f\|_{L^2} \leq C\frac{C}{\langle\mu\rangle^2}\|f\|_{H^{-1}}, \tag{37}$$

for any Schwartz function f.

Proof. We start to prove the first claim. Denote by P_{ac} the projection on the absolutely continuous part of the operator

$$-\Delta - W.$$

The perturbed resolvent $R(\mu^2) = (P - \mu^2)^{-1}$ satisfies in $\operatorname{Im}\mu > 0$ the relation

$$P_{ac}(-\Delta - W - \mu^2)^{-1} = P_{ac}\left(I - (\Delta - \mu^2)^{-1}W\right)^{-1}(-\Delta - \mu^2)^{-1} \tag{38}$$

provided the operator $(I - (P_0 - \mu^2)^{-1}W)$ is invertible. This relation implies

$$\varphi P_{ac}(-\Delta - W - \mu^2)^{-1}\varphi = \varphi P_{ac}\varphi^{-1}\left(I - \varphi(P_0 - \mu^2)^{-1}W\frac{1}{\varphi}\right)^{-1}\varphi(P_0 - \mu^2)^{-1}\varphi. \tag{39}$$

We can assume that $\mu_1 < \cdots < \mu_N \leq 0$ are the eigenvalues of the operator $-\Delta - W$ with corresponding eigenvectors (normalized in L^2)

$$f_1(x), \ldots, f_N(x).$$

They decay exponentially and this fact implies

$$(I - P_{ac})f = \sum_{j=1}^{N} f_j\langle f, f_j\rangle_{L^2}.$$

Hence taking $\delta > 0$ sufficiently small in $\varphi(x) = e^{-\delta|x|}$ we get

$$\|(I - P_{ac})\varphi^{-1}f\|_{L^2} \leq C\|f\|_{L^2}.$$

Hence the operator $\varphi P_{\mathrm{ac}} \varphi^{-1}$ is bounded in L^2 and the relation (39) can be used in combination with the Analytic Fredholm Theory and Theorem 3.1 concerning the lack of strong resonances in $\mu = 0$. We are able to say that the operator

$$\left(I - \varphi(-\Delta - \mu^2)^{-1} W \frac{1}{\varphi} \right)^{-1},$$

is analytic in $\operatorname{Im} \mu > -\delta$, excluded a discrete subset where there are the eigenvalues of (1). Moreover, the application of Theorem 4.1 and the remark after the theorem guarantees that

$$\varphi P_{\mathrm{ac}} \varphi^{-1} \left(I - \varphi(-\Delta - \mu^2)^{-1} W \frac{1}{\varphi} \right)^{-1}$$

is analytic in $\operatorname{Im} \mu > -\delta$.

In this way we obtain the inequality

$$\|\varphi P_{\mathrm{ac}} R^+(\mu) \varphi f\|_{L^2}$$
$$\lesssim \left\| \varphi P_{\mathrm{ac}} \varphi^{-1} \left(I - \varphi(-\Delta - \mu^2)^{-1} W \frac{1}{\varphi} \right)^{-1} \right\|_{B(L^2_{-\delta}, L^2_{-\delta'})} \|\varphi R_0^+(\mu) \varphi f\|_{L^2}.$$

Considering $\varphi \in L^1(\mathbf{R})$, the right-hand side of the previous estimate could be bounded in several different ways: by (29), it does not exceed $C\langle\mu\rangle^{-1}\|f\|_{L^2}$, while, from (30), we get that it is less than $C\langle\mu\rangle^{-2}\|f\|_{L^2}$. In that way the resolvent estimates (34) and (35) are obtained. After integrations by parts and following the same lines of the proof for the above estimates, by using (31), (32) and (33) we finally get (36) and (37). We notice that the meromorphic extension of the cutoff resolvent $\varphi(P - \mu^2)^{-1}\varphi$ guarantees that the estimates for $\operatorname{Im} \mu > 0$ remain valid also in the domain $\operatorname{Im} \mu > -\delta$. □

Remark 5.1. *Since the operator* $-\Delta - W(|x|)$ *has no resonances, the operator* $R^+(\mu)$ *has the form,*

$$R^+(\mu) = \sum_{j>0} \frac{B_j}{(\mu - \mu_j)} + Q(\mu) \tag{40}$$

where the B_j *are projection operators on the eigenspaces associated to the eigenvalues* $\mu = \mu_j$, *while* $Q(\mu)$ *is analytic in* $\operatorname{Im} \mu > -\delta$. *The operator*

$$P_{\mathrm{ac}} R^+(\mu) = P_{\mathrm{ac}} Q(\mu),$$

is also analytic in $\operatorname{Im} \mu > -\delta$.

We now finally give

Theorem 5.1. *Suppose that the potential* W *satisfies the properties* (4), u_0 *and* v_0 *are smooth compactly supported functions and* $\varphi(x) = e^{-\delta|x|}$ *for some* $\delta > 0, \delta \in (0, \varepsilon_0)$. *Then there exists an* $a > 0$, *such that the solution to* (5) *satisfies the following estimates*

$$\|\varphi P_{\mathrm{ac}} u(t, x)\|_{L^2} \lesssim e^{-at}(\|\varphi^{-1} u_0\|_{L^2} + \|\varphi^{-1} v_0\|_{H^{-1}}), \tag{41}$$

and

$$\|\varphi \partial_t P_{\mathrm{ac}} u(t, x)\|_{L^2} \lesssim e^{-at}(\|\varphi^{-1} u_0\|_{H^1} + \|\varphi^{-1} v_0\|_{L^2}). \tag{42}$$

Proof. The proof of the main theorem follows the one of Vainberg in [33, 31].

Proof of (41). We lose no generality if we assume $u(0,x) = 0$ in (5) and u, v_0 are real valued functions. Let $\mathcal{H}$ be a general Hilbert space and denote by $L_\nu(\mathbf{R}, \mathcal{H}) = e^{-\nu r} L(\mathbf{R}, \mathcal{H})$. We have, by (27), the inversion formula

$$u(t,x) = \frac{1}{2\pi} \operatorname{Re} \int_{-\infty+i\nu}^{\infty+i\nu} e^{-i\mu t} R^+(\mu) v_0 d\mu, \tag{43}$$

and the above integral converges in $L_\nu(\mathbf{R}, L^2)$. Let $\overline{R}^+(\mu)$ denote the meromorphic extension of the cutoff resolvent $\varphi R^+(\mu)\varphi$. Set

$$\widetilde{R}^+(\mu) = \overline{R}^+(\mu) - \frac{\varphi^2}{\mu - i(\nu-1)}. \tag{44}$$

From the fact that

$$\operatorname{Re} \int_{-\infty+i\nu}^{\infty+i\nu} \frac{e^{-i\mu t}}{\mu - i(\nu-1)} d\mu = 0, \tag{45}$$

we can rewrite (43) as

$$\varphi u(t,x) = \frac{1}{2\pi} \operatorname{Re} \int_{-\infty+i\nu}^{\infty+i\nu} e^{-i\mu t} \widetilde{R}^+(\mu)\varphi^{-1} v_0 d\mu. \tag{46}$$

Consider now, in the complex plane, the path[1] $\Gamma \cup \Gamma_1 \cup \Gamma_2 \cup \Gamma_3$, where

$$\Gamma := \{z \in \mathbf{C} \,|\, z = s + i\nu, -R \le s \le R\},$$
$$\Gamma_1 := \{z \in \mathbf{C} \,|\, z = R + is, -a \le s \le \nu\},$$
$$\Gamma_2 := \{z \in \mathbf{C} \,|\, z = s - ia, -R \le s \le R\}.$$
$$\Gamma_3 := \{z \in \mathbf{C} \,|\, z = -R + is, -a \le s \le \nu\}.$$

By the Cauchy theorem and integrating along the above path we yield

$$\frac{1}{2\pi} \int_{-\infty+i\nu}^{\infty+i\nu} e^{-i\mu t} \widetilde{R}^+(\mu)\varphi^{-1} v_0 d\mu = \sum_{j=1}^{3} \frac{1}{2\pi} \int_{\Gamma_j} e^{-i\mu t} \widetilde{R}^+(\mu)\varphi^{-1} v_0 d\mu = \sum_{j=1}^{3} \Lambda_j. \tag{47}$$

First we notice that from inequality (35) we achieve

$$\|\Lambda_1\|_{L^2} \lesssim \int_{R-ia}^{R+i\nu} \|e^{-ist} \widetilde{R}^+(s)\varphi^{-1} v_0\|_{L^2} ds \lesssim c(t,\nu) R^{-2} \|\varphi^{-1} v_0\|_{H^1}. \tag{48}$$

where $c(t,\nu)$ is a positive measurable function depending (exponentially) on t and ν. A similar estimate is valid also for the term Γ_3. Taking $R \to \infty$, we easily see that $\|\Lambda_l\|_{L^2}$ approaches to 0 for both $l = 1, 3$. What concerns the remaining integral Λ_2, we obtain by two integrations by parts,

$$\Lambda_2 = N(R)\varphi^{-1} v_0 + \frac{1}{2t^2\pi} \int_{-R-ia}^{R-ia} e^{-i\mu t} \frac{d^2}{ds^2} \widetilde{R}^+(\mu)\varphi^{-1} v_0 d\mu, \tag{49}$$

---

[1] As pointed out in Remark 5.1, we do not include such eigenvalues in our proof because we project the solution of (5) along the absolutely continuous spectrum of the Schrödinger operator $-\Delta - W$. This does not affect the generality of the results.

where $\mu = s - ia$ and $N(R)$ denotes the expression involving all boundary integral terms. If $\overline{\Lambda}_2$ denotes the second term on the right-hand side of the previous identity, we may write

$$\|\Lambda_2\|_{L^2} \lesssim \|N(R)\varphi^{-1}v_0\|_{L^2} + \|\overline{\Lambda}_2\|_{L^2}. \qquad (50)$$

By estimate (34) and (35), it is easy to see that the first term on the right-hand side of the above identity can be bounded by $C(R)\|\varphi^{-1}v_0\|_{H^1}$, where $C(R)$ is a constant depending on R approaching to 0 as $R \to 0$. The remaining term can be handled, for t large enough, as

$$\|\overline{\Lambda}_2\|_{L^2} \lesssim \int_{-R-ia}^{R-ia} e^{-at}\|\tilde{R}^+(s)\varphi^{-1}v_0\|_{L^2}ds$$

$$\lesssim e^{-at}\|\varphi^{-1}v_0\|_{H^{-1}} \int_{-R}^{R} \langle s \rangle^{-2}ds \lesssim e^{-at}\|\varphi^{-1}v_0\|_{H^{-1}}, \qquad (51)$$

and this completes the proof of the first part of the theorem.

Proof of (42). It is enough to see that the function $\partial_t u(t, x)$ satisfies equation (5), with initial data v_0 and Δv_0. Taking into account the estimate (41), the remaining part of the proof is fulfilled.   □

6. The nonlinear Schrödinger equation

Consider the nonlinear Schrödinger equation (NLS),

$$iu_t(t, x) = -\Delta u(t, x) - |u|^{p-1}u(t, x) = 0, \quad (t, x) \in \mathbf{R} \times \mathbf{R}^3,$$

$$u(0, x) = u_0(x), \qquad (52)$$

where $1 < p < 1 + 4/3$, this means in the domain where the problem is globally well posed. In fact for $p > 1 + 4/3$, there exist solutions with H^1 norms blowing up in a finite time interval. Solitary waves associated with the NLS type equation have the form

$$\psi_s(t, x) = \chi_\omega(x)e^{-i\omega t}, \quad t \in \mathbf{R}, x \in \mathbf{R}^3,,$$

with $\omega \subset \mathcal{O}$, for some open interval $\mathcal{O} \in \mathbf{R}$, where χ_ω satisfies the equation

$$-\Delta\chi_\omega - \omega\chi_\omega + |\chi_\omega|^{p-1}\chi_\omega = 0, \quad x \in \mathbf{R}^3, \qquad (53)$$

$$\int_{\mathbf{R}^3} \chi^2 = 1. \qquad (54)$$

We recall some well-known facts about the linearization at a ground state. Let us write the ansatz

$$u(t, x) = e^{it}(\chi_\omega(x) + r(t, x)), \qquad (55)$$

Inserting it into equation (55) we get

$$i\partial_t r = \mathcal{A}r + (\text{nonlinear terms}), \qquad (56)$$

and

$$\mathcal{A}(\omega)r = (-\Delta + \omega - \tfrac{p+1}{2}\chi_\omega^p)r - \tfrac{p-1}{2}\chi_\omega^p \overline{r}.$$

Because of the presence of the variable $\bar{r}$, we write the above as a system. This yields

$$i\partial_t R = H(\omega)R \tag{57}$$

where

$$H(\omega) = \begin{pmatrix} 0 & L_+(\omega) \\ -L_-(\omega) & 0, \end{pmatrix} \tag{58}$$

and with

$$L_+(\omega) = -\Delta + \omega - p\chi_\omega^p, \qquad L_-(\omega) = -\Delta + \omega - \chi_\omega^p, \tag{59}$$

having in mind also that the essential spectrum of $H(1)$ consists of $(-\infty, -\omega] \cup [\omega, +\infty)$, and that 0 is its isolated eigenvalue. Furthermore it is easy to see that $L(\omega)_\pm$ are self-adjoint operators with a continuous spectrum in $[\omega, \infty)$, that $L(\omega)_-$ is nonnegative, while $L(\omega)_+$ has exactly one negative eigenvalue (see the paper [11] and [34] for more details). Let us mention some recent works [14], [22], [12] where computer assisted proofs on the complete information of the spectrum of linearized operators for specific nonlinear terms are presented. We make the following application of the results of the previous sections.

Theorem 6.1. *The operators $L_\pm(\omega)$ have no radial resonance (see Definition 4) at the spectral point ω. If ω is a (non-radial) resonance of $L_\pm(\omega)$, then ω is an eigenvalue of $L_\pm(\omega)$.*

Before starting the proof of the above theorem we need to give some preliminary lemmas. By a rescaling argument we can pick $\omega = 1$, and focalize our attention on the operator

$$L_- = -\Delta\chi_1 + \chi_1 - \chi_1^p, \ \chi_1 \in H^2(R^3), \ \chi_1 > 0$$

because all results can be proved in the same manner for L_+. It is well known that positive radial solutions exist and they are exponentially decaying (see, for instance, [17]). Here we briefly sketch the proof for completeness and make a better asymptotic expansion. First we note that the Sobolev embedding implies

$$|\chi_1(x)| \leq C. \tag{60}$$

A better decay estimate follows from an argument of Strauss (see page 155, Section 2 of [32]). The classical Strauss lemma (Radial Lemma 1 in [32]) gives

$$r\chi_1(r) \leq \|\chi_1\|_{H^1} = C,$$

so

$$\chi_1(r) - \chi_1^p(r) > (1 - \delta)\chi_1(r)$$

for any positive δ and for $r > 0$ large enough. Setting $u = \chi_1(r)r$, we see that $u''(r) \geq (1 - \delta)u(r)$ so

$$\left(\frac{u^2}{2}\right)'' \geq (u')^2 + (1 - \delta)u^2 \geq (1 - \delta)u^2.$$

This differential inequality shows that the quantity

$$e^{-\sqrt{2(1-\delta)}r}(w' + \sqrt{2(1-\delta)}w), \quad w(r) - u^2(r)$$

is non-decreasing and has to be non-negative, since w and w' are integrable on $(0, \infty)$. This implies the decay estimate

$$u^2(r) \leq Ce^{-\sqrt{2(1-\delta)}r}, \quad r \to \infty.$$

Using the argument of Remark 1.2, we arrive at the following.

Lemma 6.1. *For any $\omega > 0$ there exists a unique positive solution $\chi(x) = \chi_\omega(x) \in H^1$ of the equation (53) and a positive $\delta_0 = \delta_0(\omega)$, such that*

$$|\chi_\omega(r)| \leq Ce^{-\delta_0 r}.$$

To obtain a more precise estimate we rewrite

$$-\Delta\chi_1 + \chi_1 = F(r) \equiv \chi_1^p, \ \chi_1 \in H^1(R^3), \ \chi_1 > 0$$

as follows

$$\chi_1(|x|) = c \int_{\mathbb{R}^3} \frac{e^{-|x-y|}}{|x-y|} F(|y|) dy.$$

Introducing polar coordinates, we find

$$\chi_1(|x|) = c \int_0^\infty \int_{\mathbb{S}^2} \frac{e^{-r}}{r} F(|x+r\omega|) d\omega r^2 dr.$$

Now we can use the following identity

$$\int_{\mathbb{S}^2} F(|x+r\omega|) d\omega = \frac{c}{|x|r} \int_{||x|-r|}^{|x|+r} F(\lambda) \lambda d\lambda$$

so

$$\chi_1(|x|) = \frac{c}{|x|} \int_0^\infty e^{-r} \int_{||x|-r|}^{|x|+r} F(\lambda) \lambda d\lambda dr.$$

One can see that

$$\chi_1(|x|) = \frac{c}{|x|} \left(K_1(F)(|x|) + K_2(F)(|x|) \right) \tag{61}$$

with

$$K_1(F)(|x|) = \int_0^{|x|} e^{-|x|} \sinh(\lambda) F(\lambda) \lambda d\lambda.$$

$$K_2(F)(|x|) = \int_{|x|}^\infty \sinh(|x|) e^{-\lambda} F(\lambda) \lambda d\lambda.$$

If one substitutes $F(\lambda)$ with $\chi_1^p(\lambda)$ and notes that $F(y) = \chi_1^p(y) \leq C$ for $|y|$ bounded due to (60) and moreover the estimate of Lemma 6.1 implies $F(y) \leq Ce^{-\delta_0 p|y|}$ for $|y| \geq 1$, we deduce

$$|K_1(F)(|x|)| \leq Ce^{-B|x|}, \ B = \min(1, \delta_0 p)$$

$$|K_2(F)(|x|)| \leq Ce^{-B|x|}, \ B = \min(1, \delta_0 p),$$

and we find

$$|x||\chi_1(|x|)| \leq Ce^{-B|x|}, \ B = \min(1, \delta_0 p).$$

If $\delta_0 p < 1$ we derive $F(|x|) \leq C e^{-pB|x|}$ and making further iterations we get

Lemma 6.2. *For any $\omega > 0$ there exists a unique positive solution $\chi(x) = \chi_\omega(x) \in H^1$ of equation (53) and a positive $C = C(\omega)$, such that*

$$r|\chi_\omega(r)| \leq C e^{-\sqrt{\omega} r}$$

for $r > 0$.

To get an asymptotic expansion, we use (61). If one substitutes $F(\lambda)$ with $\chi_1^p(\lambda)$ and applies the estimate of Lemma 6.2 (with $\omega = 1$) one can obtain the asymptotic expansions

$$K_1(\chi_1^p)(|x|) = C_0 e^{-|x|} + O\left(e^{-(p-\delta)|x|}\right), \qquad |x| > 1,$$

$$K_1'(\chi_1^p)(|x|) = -C_0 e^{-|x|} + O\left(e^{-(p-\delta)|x|}\right), \qquad |x| > 1,$$

$$K_2(\chi_1^p)(|x|) = O\left(e^{-(p-\delta)|x|}\right), \qquad\qquad |x| > 1,$$

$$K_2'(\chi_1^p)(|x|) = O\left(e^{-(p-\delta)|x|}\right), \qquad\qquad |x| > 1,$$

where δ is any positive number. After rescaling the argument we get

Lemma 6.3. *For any $\omega > 0$ there exists a unique positive solution $\chi(|x|) = \chi_\omega(x) \in H^2$ of equation (53) and a positive $C_0 = C(\omega)$, such that*

$$r\chi_\omega(r) = C_0 e^{-\sqrt{\omega} r} + O\left(e^{-(p-\delta)\sqrt{\omega} r}\right), \qquad r > 1,$$

$$(r\chi_\omega(r))' = -\sqrt{\omega} C_0 e^{-\sqrt{\omega} r} + O\left(e^{-(p-\delta)\sqrt{\omega} r}\right), \quad r > 1.$$

Theorem 6.2. *For any $\omega > 0$ there exists a unique positive solution $\chi(|x|) = \chi_\omega(x) \in H^1$ of equation (53) so that*

$$0 > \frac{\partial_{|x|} \chi_\omega(x)}{\chi_\omega(x)} > -C_1$$

for some positive constant C_1. Moreover there exists a positive $C_0 = C(\omega)$ constant, such that

$$r\chi_\omega(r) = C_0 e^{-\sqrt{\omega} r} + O\left(e^{-(p-\delta)\sqrt{\omega} r}\right), \qquad r > 1,$$

$$(r\chi_\omega(r))' = -\sqrt{\omega} C_0 e^{-\sqrt{\omega} r} + O\left(e^{-(p-\delta)\sqrt{\omega} r}\right), \quad r > 1.$$

Proof of Theorem 6.1. The proof is easy, so we reduce it to few lines; it is a consequence of the results stated in the previous sections. By Theorem 6.2, we obtain that the operator $L_- - 1$ has the form of $-\Delta - W$, where W satisfies the assumption of Theorem 4.1. This assures that we have no radial resonance (or eigenvalue) at zero energy; the same is valid for $L_+ - 1$. Moreover, if $\lambda = 0$ is a (non-radial) resonance, then $\lambda = 0$ is an eigenvalue. $\qquad\square$

7. Appendix 1: Asymptotics for solutions to some ODE

Lemma 7.1. *Assume that $W(s) > 0, \forall s > 0$ and*

$$(1+s)^2 W(s) \in L^1(0, \infty).$$

If $u \in C^2(\mathbb{R})$ is a solution of $u'' + Wu = 0$ and there exists $s_0 \geq 0$ so that $u(s) > 0$ for $s > s_0$, then there exists a non-negative number C_1 so that we have the relations

$$u'(s) = C_1 + \int_s^\infty W(\tau)u(\tau)d\tau, \tag{62}$$

$$u(s) = u(s_0) + \int_{s_0}^\infty (\tau - s_0)W(\tau)u(\tau)d\tau + C_1(s - s_0) - \int_s^\infty (\tau - s)W(\tau)u(\tau)d\tau, \tag{63}$$

as well as the asymptotic expansions (valid for $s \to \infty$)

$$u(s) = C_0 + C_1 s + O(g(s)), \quad u'(s) = C_1 + O(h(s)),$$

where

$$h(s) = \int_s^\infty (1+s)W(s)ds, \quad g(s) = \int_s^\infty (1+s)^2 W(s)ds.$$

Proof. We have

$$u''(s) = -W(s)u(s) < 0, \quad s \geq s_0.$$

The Taylor expansion gives

$$0 < u(s) = D_0 + D_1 s + \frac{u''(\xi)(s - s_0)^2}{2} < D_0 + D_1 s,$$

where

$$s > s_0, \xi \in (s_0, s), \quad D_1 = u'(s_0), \quad D_0 = u(s_0) - s_0 u'(s_0).$$

The inequality

$$0 < u(s) < C(1 + s)$$

and the assumption

$$(1+s)^2 W(s) \in L^1(0, \infty)$$

show that

$$u'(s) - u'(t) = -\int_t^s W(\tau)u(\tau)d\tau \tag{64}$$

is small when $s_0 < t < s$ and s, t are large enough. This argument shows the existence of the limit

$$\lim_{s \to \infty} u'(s) = C_1$$

as well as the asymptotic expansion

$$u'(s) = C_1 + O(h(s)).$$

Integrating this relation and using the fact

$$\int_s^\infty h(\tau)d\tau = O(g(s)),$$

we obtain the desired expansion

$$u(s) = C_0 + C_1 s + O(g(s)).$$

The fact that $C_1 \geq 0$ follows from the positivity of $u(s)$ for $s > s_0$. Finally, to prove (63) we use (64) and integrating (63) we find (62).

This completes the proof. □

A slight modification is the following

Lemma 7.2. *Assume that $W(s) > 0, \forall s > 0$ and*

$$(1+s)W(s) \in L^1(0, \infty).$$

If $u \in C^2(\mathbb{R})$ is a solution of $u'' + Wu = 0$ and there exists $s_0 > 0$ so that $u(s) > 0$ for $s > s_0$, then

$$u(s) \leq C(1+s)$$

and the limit

$$\lim_{s \to \infty} u'(s)$$

exists.

Proof. As in the proof of the previous lemma we have

$$u''(s) = -W(s)u(s) < 0.$$

The Taylor expansion gives

$$0 < u(s) = D_0 + D_1 s + \frac{u''(\xi)(s - s_0)^2}{2} < D_0 + D_1 s,$$

where

$$s > s_0, \xi \in (s_0, s), \quad D_1 = u'(s_0), \quad D_0 = u(s_0) - s_0 u'(s_0).$$

The inequality

$$0 < u(s) < C(1+s)$$

and the assumption

$$(1+s)W(s) \in L^1(0, \infty)$$

show that

$$u'(s) - u'(t) = -\int_t^s W(\tau)u(\tau)d\tau$$

is small when $s_0 < t < s$ and s, t are large enough. This argument shows the existence of the limit

$$\lim_{s \to \infty} u'(s).$$

This completes the proof. □

Our next step is to consider equation (21) with the potential $W(r) = W_1(r) + W_2(r)$, where

$$W_1(r) = -\frac{\alpha(\alpha+1)}{r^2}, \quad \alpha \geq 0, \tag{65}$$

while $W_2(r)$ is a $C^1(0, \infty)$ positive strictly decreasing function such that for some positive constants C, ε_0 it satisfies the estimate (4).

The first step is to obtain asymptotic expansions of the solution and for this aim, by Definitions 3 and 2, we give the following lemma.

Lemma 7.3. *Suppose* (4) *is true.*

a) *If* 0 *is an eigenvalue of* P *and* $Pu = 0$, *in the sense of Definition 3, then one can find a real number* C *so that*

$$u(r) = \frac{C}{r^\alpha} + O\left(e^{-\varepsilon_0 r/2}\right) \tag{66}$$

and

$$u'(r) = -\frac{C\alpha}{r^{\alpha+1}} + O\left(e^{-\varepsilon_0 r/2}\right) \tag{67}$$

as $r \to \infty$.

b) *If* 0 *is a strong resonance of* P *and* $Pu = 0$ *in the sense of Definition 2, then there exists a real number* $C > 0$ *so that*

$$u(r) = \frac{C}{r^\alpha} + O\left(e^{-\varepsilon_0 r/2}\right) \tag{68}$$

and

$$u'(r) = -\frac{C\alpha}{r^{\alpha+1}} + O\left(e^{-\varepsilon_0 r/2}\right) \tag{69}$$

as $r \to \infty$.

Proof. First we prove a). One can rewrite the equation $Pu = 0$ as

$$\left[r^{-\alpha}\left(u'(r) + \frac{\alpha}{r}u(r)\right)\right]' + r^{-\alpha}W_2(r)u(r) = 0 \tag{70}$$

or as

$$\left[r^{-2\alpha}\left(r^\alpha u(r)\right)'\right]' + r^{-\alpha}W_2(r)u(r) = 0. \tag{71}$$

Note that the assumption $W_2(r) \in C^1(0,\infty)$ combined with the equation $Pu = 0$ imply that $u \in C^2(R_1, R_2)$ for any $0 < R_1 < R_2$. Integrating (71) in the interval (R_1, R_2), we find

$$\left|R_2^{-2\alpha}(R_2^\alpha u(R_2))' - R_1^{-2\alpha}(R_1^\alpha u(R_1))'\right| = \left|\int_{R_1}^{R_2} \tau^{-\alpha}W_2(\tau)u(\tau)\,d\tau\right|, \tag{72}$$

so using assumption (4) together with the fact that u is bounded (since it belongs to $H^1(1,+\infty) \cap C^1(1,+\infty)$), and taking $R_1 > 1$, we find

$$\left|\int_{R_1}^{R_2} \tau^{-\alpha}W_2(\tau)u(\tau)\,d\tau\right| \leq Ce^{-\varepsilon_0 R_1/2}. \tag{73}$$

In this way we conclude that the limit

$$\lim_{r\to\infty} r^{-2\alpha}(r^\alpha u(r))', \tag{74}$$

exists and is equal to a real constant C_0. By this, we achieve the expansion

$$(r^\alpha u(r))' = C_0 r^{2\alpha} + O\left(e^{-\varepsilon_0 r/2}\right). \tag{75}$$

Consider now the function

$$g(r) = r^\alpha u(r) - C_0 \frac{r^{2\alpha+1}}{2\alpha+1},$$

then (75) implies that $g'(r) \in L^1(1, \infty)$. Moreover we can see that $g(r)$ has a limit (say C) as r goes to ∞ and

$$g(r) = C - \int_r^\infty g'(\tau)\,d\tau = C + O(e^{-\varepsilon_0 r/2}).$$

Thus we obtain

$$u(r) = C_0 \frac{r^{\alpha+1}}{2\alpha+1} + \frac{C}{r^\alpha} + O\big(e^{-\varepsilon_0 r/2}\big) \tag{76}$$

and

$$u'(r) = C_0 \frac{(\alpha+1)r^\alpha}{2\alpha+1} - \frac{C\alpha}{r^{\alpha+1}} + O\big(e^{-\varepsilon_0 r/2}\big) \tag{77}$$

as r moves to infinity. Comparing these asymptotic expansions with the fact that u is bounded, we see that $C_0 = 0$ and this completes the first part of the lemma.

The proof of b) can be obtained similarly to the above using assumption (14), so we skip it. $\qquad\square$

To study the asymptotic behavior of the solution to (21) we can use the integral equation

$$u(r) = cr^{-\alpha} \int_0^r \tau^{\alpha+1} W_2(\tau)u(\tau)d\tau + cr^{\alpha+1} \int_r^\infty \tau^{-\alpha} W_2(\tau)u(\tau)d\tau. \tag{78}$$

If 0 is a strong resonance of P and $Pu = 0$ in sense of Definition 2, then

$$u(r) = O(r^{\alpha+1}) \tag{79}$$

and

$$u'(r) = O(r^\alpha) \tag{80}$$

as $r \to 0$.

The above arguments suffice to get

Lemma 7.4. *If 0 is a weak resonance of P and $Pu = 0$, then one can find real numbers C_0, C_1 so that*

$$u(r) = C_0 \frac{r^{\alpha+1}}{2\alpha+1} + \frac{C_1}{r^\alpha} + O\big(e^{-\varepsilon_0 r/2}\big) \tag{81}$$

and

$$u'(r) = C_0 \frac{(\alpha+1)r^\alpha}{2\alpha+1} - \frac{C_1\alpha}{r^{\alpha+1}} + O\big(e^{-\varepsilon_0 r/2}\big) \tag{82}$$

as $r > 1$ tends to infinity.

8. Appendix 2

In this section we complete the discussion concerning the weak resonances and its connection with different types of potentials. It seems that the weak resonances cannot be never avoid, more precisely they are an intrinsic character of the structure of the differential equation involved in the description of such phenomena. In order to do that we look at large potentials and small potentials.

8.1. Large potentials do not generate weak resonance at the spectral origin

As in the previous section we shall assume $W(s) > 0, \forall s > 0$ and

$$(1+s)^2 W(s) \in L^1(0, \infty).$$

To show that all solutions $u \in C^2(\mathbf{R})$ to $u'' + Wu = 0$ having linear growth at infinity are identically zero, we can apply Lemma 7.1 so without loss of generality one can assume

$$u(0) = 0, u(s) > 0, \ \forall s > 0.$$

The key assumption that will guarantee that such solutions do not exist is

$$\frac{1}{M} \int_0^M s^2 W(s) ds \geq 1 \tag{83}$$

for some real $M > 0$.

Turning back to the integral equation of Lemma 7.1 we have the relations

$$u'(s) = C_1 + \int_s^\infty W(\tau) u(\tau) d\tau, \tag{84}$$

$$u(s) = \int_0^\infty \tau W(\tau) u(\tau) d\tau + C_1 s - \int_s^\infty (\tau - s) W(\tau) u(\tau) d\tau, \tag{85}$$

so we can introduce the operator

$$K(u)(s) = \int_0^\infty \tau W(\tau) u(\tau) d\tau - \int_s^\infty (\tau - s) W(\tau) u(\tau) d\tau$$
$$= \int_0^s \tau W(\tau) u(\tau) d\tau + s \int_s^\infty W(\tau) u(\tau) d\tau.$$

It is clear that $u \geq 0, \forall s \geq 0$ implies $K(u)(s) \geq 0 \ \forall s \geq 0$. Relation (84) shows that $u'(s) > 0$ for the interval $[0, M]$ so that $u'(s) \geq C^* = u'(M) > 0$ for $0 \leq s \leq M$. Then we have the estimate

$$u(s) \geq C^* s, \ \ \forall s \in [0, M].$$

For any $s \in [0, M]$ we have

$$K(u)(s) > K_M(u)(s) \equiv \int_0^s \tau W(\tau) u(\tau) d\tau + s \int_s^M W(\tau) u(\tau) d\tau$$

so

$$\frac{K(u)(s)}{s} \geq C_* \left(\frac{1}{s} \int_0^s \tau^2 W(\tau) d\tau + \int_s^\delta \tau W(\tau) d\tau \right).$$

The function

$$\frac{1}{s}\int_0^s \tau^2 W(\tau)d\tau + \int_s^M \tau W(\tau)d\tau$$

is decreasing so

$$\frac{K(u)(s)}{s} \geq C_* \left(\frac{1}{M}\int_0^M \tau^2 W(\tau)d\tau\right).$$

Assumption (83) implies

$$D = \frac{1}{M}\int_0^M \tau^2 W(\tau)d\tau \geq 1.$$

So if u solves the equation $u(s) = C_1 s + K(u)(s)$ we obtain

$$\frac{u(s)}{s} \geq C^* \ \forall s \in [0, M] \implies \frac{u(s)}{s} = C_1 + C^* D > C^* \ \forall s \in [0, M],$$

and we find

$$\lim_{s\to 0+} \frac{u(s)}{s} = u'(0) \geq C^* + NC_1,$$

but this is a contradiction since N is arbitrary.

8.2. Small potentials generate weak resonance at the spectral origin

To construct nontrivial solutions $u \in C^2(\mathbf{R}_+)$ to $u'' + Wu = 0$ having linear growth at infinity, we assume

$$\sum_{M>0} \frac{1}{M}\int_0^M s^2 W(s)ds = D < 1. \tag{86}$$

Remark 8.1. *One sufficient condition for (86) is*

$$\int_0^\infty sW(s)ds < 1,$$

since we have the estimate

$$\frac{1}{M}\int_0^M s^2 W(s)ds \leq \int_0^M sW(s)ds.$$

Consider the integral equation

$$u(s) = C_1 s + K(u)(s),$$

where

$$K(u)(s) = \int_0^\infty \tau W(\tau)u(\tau)d\tau - \int_s^\infty (\tau - s)W(\tau)u(\tau)d\tau$$

$$= \int_0^s \tau W(\tau)u(\tau)d\tau + s\int_s^\infty W(\tau)u(\tau)d\tau.$$

We plan to show that this equation has a solution in the Banach space obtained as a closure of the linear space L formed by the functions $u(s) \in C([0, \infty))$, such that $u(0) = 0$ and

$$|||u||| = \sup_{s>0} \frac{|u(s)|}{s} < \infty.$$

To be more precise, B is the closure of L with respect to the norm $|||u|||$.

To show this fact it is sufficient to notice that

$$|||K(u)||| \le D|||u|||$$

so the assumption $D < 1$ enables one to apply a contraction argument for the equation

$$u(s) = C_1 s + K(u)(s).$$

Acknowledgement

The first author was supported by the Italian National Council of Scientific Research (project PRIN No. 2008BLM8BB) entitled: "Analisi nello spazio delle fasi per E.D.P."

The second author is supported by an INdAM grant. Currently he is a Academic Visitor at Department of Mathematics of the Imperial College London.

References

[1] S. Agmon, *Spectral properties of Schrödinger operators and scattering theory*, Ann. Scuola Norm. Sup. Pisa Cl. Sci. **2/2** (1975), 151–218.

[2] S. Agmon *A perturbation theory of resonances*, Comm. Pure Appl. Math. **51** (1998), no. 11-12, 1255–1309.

[3] M. Ben-Artzi *Spectral theory for divergence-form operators*, in "Spectral and Scattering Theory and Related Topics", Ed. H. Ito, RIMS Kokyuroku 1607 (2008), 77–84.

[4] N. Burq, F. Planchon, J.G. Stalker, A.S. Tahvildar-Zadeh, *Strichartz estimates for the wave and Schrödinger equations with the inverse-square potential*, J. Funct. Anal., **203**, (2003), no. 2, 519–549.

[5] N. Burq, F. Planchon, J.G. Stalker, A.S. Tahvildar-Zadeh, *Strichartz estimates for the wave and Schrödinger equations with potentials of critical decay*. Indiana Univ. Math. J., **53** (2004), no. 6, 1665–1680.

[6] N. Burq, M. Zworski, *Resonance expansions in semi-classical propagation*. Commun. Math. Phys., **223 (1)**, (2001), 1–12.

[7] V.S. Buslaev, C. Sulem, *On asymptotic stability of solitary waves for nonlinear Schrödinger equations*, Annales de l'Institut Henri Poincaré (C) Non Linear Analysis, **20/3** (2003), 419–475.

[8] V.S. Buslaev, G.S. Perelman, *On the stability of solitary waves for nonlinear Schrödinger equations*, Amer. Math. Soc. Trans., **164/2** (1995), 75–98.

[9] T. Cazenave, P.L. Lions, *Orbital stability of standing waves for some nonlinear Schrödinger equations*, Comm. Math. Physics, **85/4** (1982), 549–561.

[10] T. Cazenave, *Semilinear Schrödinger equations*, Courant Lecture Notes in Mathematics, 10. New York University, Courant Institute of Mathematical Sciences, New York; American Mathematical Society, Providence, RI, 2003. xiv+323 pp.

[11] S. Chang, S. Gustafson, K. Nakanishi, T.P. Tsai, *Asymptotic Stability and Completeness in the Energy Space for Nonlinear Schrödinger Equations with Small Solitary Waves*, Siam J. Math. Anal. **39** (2007), 1070–1111.

[12] O. Costin, H. Min and W. Schlag, *On the spectral properties of $L_\pm$ in three dimensions*, preprint (2011) arXiv:1107.0323v1.

[13] S. Cuccagna, T. Mizumachi, *On asymptotic stability in energy space of ground states for Nonlinear Schrödinger equations*, Comm. Math. Phys., **284** (2008), 51–87.

[14] L. Demanet, W. Schlag, *Numerical verification of a gap condition for a linearized nonlinear Schrödinger equation*, Nonlinearity **19** 2006, no. 4, 829–852.

[15] L. Fanelli, *Non-trapping magnetic fields and Morey–Campanato estimates for Schrödinger operators*, Journal of Mathematical Analysis and Applications, 357(1) 2009, 1–14.

[16] V. Georgiev, N. Visciglia, *Decay estimates for the wave equation with potential*, Comm. Partial Differential Equations, *28* (7-8), (2003), 1325–1369.

[17] B. Gidas, W.M. Ni, L. Nirenberg, *Symmetry of positive solutions to nonlinear elliptic equations in $\mathbf{R}^n$*, Mathematical Analysis and Applications, Part A, Advances in Mathematics, Supplementary Studies, vol 7A (1981) 369–402.

[18] S. Gustafson, K. Nakanishi, T.P. Tsai, *Asymptotic Stability and Completeness in the Energy Space for Nonlinear Schrödinger Equations with Small Solitary Waves*, Int. Math. Res. Notices, **66** (2004), 3559–3584.

[19] A. Jensen, T. Kato, *Spectral properties of Schrödinger operators and time-decay of the wave functions*, Duke Math. Journal, **46(3)** (1979), 583–611.

[20] A. Jensen, *Spectral properties of Schrödinger operators and time-decay of the wave functions in $L^2(\mathcal{R}^m), m \geq 5$*, Duke Math. Journal, **47(1)** (1980), 57–80.

[21] M. Keel and T. Tao, *Endpoint Strichartz estimates*, Amer. J. Math., **120(5)**, 955–980, 1998.

[22] J.L. Marzuola and G. Simpson, *Spectral analysis for matrix Hamiltonian operators* Nonlinearity **24** (2011) 389–429.

[23] J. Metcalfe and C.D. Sogge, *Hyperbolic trapped rays and global existence of quasilinear wave equations*, Invent. Math., **159(1)** (2005), 75–117.

[24] R. Pego, M. Weinstein, *Asymptotic Stability of Solitary Waves*, Commun. Math. Phys., **164** (1994), 305–349.

[25] G. Perelman, *On the formation of singularities in solutions of the critical nonlinear Schrödinger equation*, Ann. Henri Poincaré **2**, (2001), no. 4, 605–673.

[26] G. Perelman, *Asymptotic Stability of multi-soliton solutions for nonlinear Schrödinger equation*, Communications in Partial Differential Equations, **29/7,8** (2004), 1051–1095.

[27] B. Perthame and L. Vega, *Morrey Campanato Estimates for Helmholtz Equations*, Journal of Functional Analysis, **164** (1999), 340–355.

[28] A.G. Ramm, *Domain where the resonances are absent in the three-dimensional scattering problem*, Sov. Phys.-Doklady, **166** (1966), 1319–1322.

[29] A. Sà Barreto, M. Zworski, *Distribution of resonances for spherical black holes*. Math. Res. Lett. **4(1)**, (1997), 103–121.

[30] W. Schlag, *Stable manifolds for an orbitally unstable nonlinear Schrödinger equation*, Ann. of Math. **2** (2009), no. 1, 139–227.

[31] J. Sjöstrand, *Lectures on resonances*, homepage of J. Sjöstrand at `www.math.polytechnique.fr/~sjoestrand`.

[32] W. Strauss, *Existence of Solitary Waves in Higher Dimension*, Comm. Math. Physics **55**, (1977), 149–162.

[33] B.R. Vainberg, *Asymptotic methods in equations of mathematical physics*. Translated from the Russian by E. Primrose. Gordon & Breach Science Publishers, New York, 1989. viii+498 pp.

[34] M. Weinstein, *Modulational Stability of Ground States of Nonlinear Schrodinger Equations*, SIAM J. Math. Anal. **16** (1985), no. 3, 472–491.

Vladimir Georgiev and Mirko Tarulli
Department of Mathematics
Faculty of Sciences
Pisa University
Largo Pontecorvo 5
I-56127 Pisa, Italy
e-mail: `georgiev@dm.unipi.it`
     `tarulli@mail.dm.unipi.it`

Progress in Mathematics, Vol. 301, 145–162
© 2012 Springer Basel

Decay Estimates for the Supercritical 3-D Schrödinger Equation with Rapidly Decreasing Potential

Vladimir Georgiev and Bozhidar Velichkov

Abstract. We establish an almost optimal decay estimate for the 3-D Schrödinger equation with non-negative potential decaying exponentially and nonlinearity of power $p > 1 + 2/3 = 5/3$. The key point is the introduction of an appropriate analogue of the generators of the pseudoconformal group for the free Schrödinger equation.

Mathematics Subject Classification. Primary 35Q55, 35P10; Secondary 42B37, 42B35.

Keywords. Semilinear, Schrödinger equation, decay estimates.

1. Introduction

We consider the following Schrödinger type equation

$$(i\partial_t + \Delta_x - W(x))u + |u|^{p-1}u = 0. \tag{1.1}$$

where $t \geq 1$ and $x \in \mathbb{R}^3$ and $W(x)$ is a non-negative potential.

Potential type perturbation for the classical Schrödinger equation

$$(i\partial_t + \Delta_x)U + |U|^{p-1}U = 0$$

appears in a natural way, after linearization around solitary type solutions, i.e., a solution of the form

$$U = e^{i\omega t}\chi(x),$$

where $\chi \in H^1(\mathbb{R}^3)$ is a critical point of the functional

$$E(\chi) = \frac{1}{2}\|\nabla\chi\|_{L^2}^2 - \frac{1}{p+1}\int_{\mathbb{R}^3}|\chi(x)|^{p+1}dx$$

The first author was supported by the Italian National Council of Scientific Research (project PRIN No. 2008BLM8BB) entitled: "Analisi nello spazio delle fasi per E.D.P.".

subject to the constraint

$$\|\chi\|_{L^2}^2 = 1.$$

Restricting the attention to the minimum of $E(\chi)$ subject to the same constraint one can work with so-called ground states (see [6] and the references therein) and see the existence and the fact that ground states are positive, radial and exponentially decaying functions. The linearization of type

$$U = (u + e^{i\omega t}\chi)$$

leads to an equation of type (1.1), where W is a bounded (possibly non-selfadjoint) operator in L^2 expressed in terms of $\chi(x)$.

For simplicity in this work we consider the case when $W(x)$ is a real-valued non-negative function having the same decay properties as the ground state $\chi(x)$. More precisely, we assume the following hypotheses on W:

(H1) W is a non-negative Schwartz function decaying exponentially at infinity and such that there exist positive constants $c_0 > c_1 > 0$ so that for any $x \in \mathbb{R}^3$

$$0 \le c_0 W(x) \le -\partial_r W(x),$$

and

$$W(x) + |\partial_r W(x)| \le Ce^{-c_1|x|}.$$

Let Σ^s be the Hilbert space defined as the closure of $C_0^\infty(\mathbb{R}^3)$ functions with respect to the norm

$$\|u\|_{\Sigma^s}^2 = \|u\|_{H^s(\mathbb{R}^3)}^2 + \||x|^s u\|_{L^2(\mathbb{R}^3)}^2. \tag{1.2}$$

It is well known that $p = 5/3$ is a critical value for the existence of asymptotic profiles and dispersive estimates for small data solutions in case of the potential $W = 0$ (see [11] for example). In this work we study the supercritical case $p > 5/3$ and obtain the following decay estimate.

Theorem 1.1. *Assume* (H1) *and the parameters* s, p *satisfy* $p > 5/3, s > 3/2$. *Then there is a constant* $\epsilon_0 > 0$, *so that for any* $\delta > 0$ *one can find a constant* $C_0 = C_0(\varepsilon_0, \delta) > 0$ *such that for any* $\epsilon \in (0, \epsilon_0)$ *the solution to* (1.1) *satisfies the inequality*

$$\|u(t)\|_{L^\infty(\mathbb{R}^3)} \le \frac{C_0}{t^{3/2-\delta}}\epsilon, \tag{1.3}$$

provided

$$\|u(1)\|_{\Sigma^s} \le \epsilon.$$

There is a long list of results concerning the Strichartz type estimates, $L^p - L^q$ estimates and similar dispersive estimates for potential (or magnetic) type perturbations of the linear Schrödinger or wave equation (see [16], [18], [7], [2], [14] for some of these results). However these types of estimates seem to have a non-obvious application if one tries to get the almost optimal decay rate for the nonlinear supercritical Schrödinger equation. The classical approach developed in [11] is based on the use of the generators of the pseudoconformal transform that enables us to get the optimal decay for the case of the potential $W = 0$. Since no reasonable definition of these generators is available for the case of potential we are forced to use a different approach.

Our approach is based on a direct application of the pseudoconformal transform. After this transform the global Cauchy problem with initial data at $t = 1$ becomes a local Cauchy problem for the Schrödinger equation with time-dependent potential. More precisely, we shall need an estimate of the solution to the problem

$$i\partial_T v + \Delta v - T^{-2} W\left(\frac{X}{T}\right) v = G, \quad T \in (0,1).$$

Taking $s \in [3/2, 2]$ for any $\delta > 0$ we shall be able to find positive a constant $C = C(s, \delta)$ so that

$$\|v(T, \cdot)\|_{H^s_X} \le C|T|^{3/2-s-\delta} \|v(1, \cdot)\|_{H^s_X} + + C|T|^{3/2-s-\delta} \|G\|_{L^1((T,1);H^s_X)}.$$

This is our key estimate to derive the local existence result for the nonlinear Schrödinger equation after the pseudoconformal transform.

2. Pseudo-conformal transform

The pseudo-conformal transform is defined as follows

$$(t, x, u) \Longrightarrow (T, X, v),$$

where $t = 1/T, x = X/T$ and

$$v(T, X) = \frac{1}{T^{n/2}} \overline{u}(\frac{1}{T}, \frac{X}{T}) e^{i\frac{X^2}{4T}}. \tag{2.1}$$

Then we have the relation

$$i\partial_T v(T, X) + \Delta_X v(T, X) = T^{-2-\frac{n}{2}} e^{i\frac{X^2}{4T}} \left(-i\partial_t \overline{u}(t, x) + \Delta_x \overline{u}(t, x)\right)|_{t=1/T, x=X/t}.$$

Since $\overline{u}$ satisfies the equation

$$(-i\partial_t + \Delta_x - W(x))\overline{u} + |\overline{u}|^{p-1}\overline{u} = 0,$$

we get

$$i\partial_T v(T, X) + \Delta_X v(T, X) = T^{-2-\frac{n}{2}} \left(W(x)\overline{u}(t, x) - |u(t, x)|^{p-1}\overline{u}(t, x)\right) e^{i\frac{X^2}{4T}}$$

$$= T^{-2} W\left(\frac{X}{T}\right) v(T, X) - T^{\frac{np-n-4}{2}} |v(T, X)|^{p-1} v(T, X).$$

Hence

$$i\partial_T v + \Delta_X v - T^{-2} W\left(\frac{X}{T}\right) v + T^{\frac{np-n-4}{2}} |v|^{p-1} v = 0 \tag{2.2}$$

for $0 < T \le 1$ and $X \in \mathbb{R}^n$. Note that the L^2 norm of u is constant, since u satisfied the nonlinear Schrödinger equation (1.1). On the other hand, the transform (2.1) preserves the L^2 norm, so we have

$$\frac{d}{dT} \int_{\mathbb{R}^n} (|v|^2) \, dX = 0.$$

In this way we obtain the following lemma.

Lemma 2.1. *If u satisfies the linear Schrödinger equation*

$$(i\partial_t + \Delta_x - W(x))u = F \text{ for } t \geq 1 \text{ and } x \in \mathbb{R}^n, n \geq 3, \tag{2.3}$$

then $v(T, X)$ defined according to (2.1) and

$$G(T, X) = \frac{1}{T^{2+n/2}} \overline{F}(\frac{1}{T}, \frac{X}{T}) e^{i\frac{X^2}{4T}}. \tag{2.4}$$

satisfy

$$i\partial_T v + \Delta_X v - W_T(X)v = G(T, X), \tag{2.5}$$

where

$$W_T(X) = T^{-2}W\left(\frac{X}{T}\right). \tag{2.6}$$

3. H^2 estimate

Setting

$$v_2(T, X) = \left(-\Delta + T^{-2}W\left(\frac{X}{T}\right)\right) v(T, X),$$

we have the following equation satisfied by v_2

$$i\partial_T v_2 + \Delta v_2 - T^{-2}W\left(\frac{X}{T}\right) v_2 = iW_1 v + G_2,$$

where

$$W_1(T, X) = \left[\partial_T \left(T^{-2}W\left(\frac{X}{T}\right)\right)\right]$$

and

$$G_2 = \left(-\Delta + T^{-2}W\left(\frac{X}{T}\right)\right) G.$$

Let us take $n = 3$. Applying the Strichartz estimates of Theorem 6.8, we get

$$\|v_2(T, \cdot)\|_{L_X^2} \leq C\left(\|v_2(1, \cdot)\|_{L_X^2} + \|W_1 v\|_{L^2((T,1);L_X^{6/5})} + \|G_2\|_{L^1((T,1);L_X^2)}\right)$$

The assumption (H1) guarantees that

$$W_1(T, X) \leq C\frac{W_0(T, X)}{T}, \tag{3.1}$$

where

$$W_0(T, X) = \frac{e^{-c_2|X|/T}}{T^2}.$$

We have a generalization of this estimate, given in the lemma below, where typically we shall assume that $b(s) - 1 + |\log s|$ or $b(s) = s^{-\delta}$ with $\delta > 0$ small.

Lemma 3.1. *Suppose $b(s)$ is a positive continuous function in $(0, \infty)$, such that b is decreasing in $(0, 1)$, satisfies the estimate $b(s^2) \leq Cb(s), s \in (0, 1)$ and*

$$\lim_{s \searrow 0} s^\varepsilon b(s) = 0$$

for any $\varepsilon > 0$ such that $\varepsilon < \min(c_2, a/2))$ and satisfies one of the following assumptions

a) *b is decreasing in $(1, \infty)$, and*

$$\lim_{s \nearrow \infty} s^\varepsilon b(s) = \infty$$

for some $\varepsilon > 0$ such that $\varepsilon < \min(c_2, a/2))$ or

b) *b is increasing in $(1, \infty)$, satisfies the estimate $b(s^{-1}) \leq Cb(s), s \in (0, 1)$ and*

$$\lim_{s \nearrow \infty} s^\varepsilon b(s) = \infty$$

for any $\varepsilon > 0$ such that $\varepsilon < \min(c_2, a/2))$.
Then for $T \in (0, 1)$ we have the estimate

$$b(|X|)W_1(T, X) \leq C\frac{b(T)\widetilde{W}_0(T, X)}{T^{1-a}|X|^a}, \quad a \in (0, 1), \tag{3.2}$$

where

$$\widetilde{W}_0(T, X) = \frac{e^{-(c_2 - \varepsilon)|X|/T}}{T^2}.$$

Proof. It is sufficient to verify the inequality

$$\frac{|X|^a b(|X|)}{b(T)T^a} \leq Ce^{\varepsilon|X|/T}.$$

If $T^2 < |X| < T$, then

$$\frac{|X|^a b(|X|)}{T^a b(T)} \leq \frac{|X|^a b(T^2)}{T^a b(T)} \leq C.$$

If $|X| \leq T^2$, then the condition

$$\lim_{s \searrow 0} s^\varepsilon b(s) = 0$$

implies $s^\varepsilon b(s) \leq C$ so

$$\frac{|X|^a b(|X|)}{T^a b(T)} \leq \frac{|X|^{a-\varepsilon}}{T^a b(T)} \leq \frac{T^{a-2\varepsilon}}{b(T)} \leq C.$$

For $T < |X| < 1$ we have

$$\frac{|X|^a b(|X|)}{T^a b(T)} \leq \frac{|X|^a}{T^a} \leq Ce^{\varepsilon|X|/T}$$

and the same argument works if $|X| > 1 > T$ and b is decreasing everywhere, i.e., a) holds. For b) we can separate the cases $1 < |X| < 1/T$ and $|X| > 1/T$. If $1 < |X| < 1/T$ then $b(|X|) \leq b(1/T) \leq Cb(T)$ and we have

$$\frac{|X|^a b(|X|)}{T^a b(T)} \leq \frac{C|X|^a}{T^a} \leq C_1 e^{\varepsilon|X|/T}.$$

If $|X| \geq 1/T$, then $b(T) \geq C^{-1}b(1/T) \geq C^{-1}b(1) = C_1$ and hence

$$\frac{|X|^a b(|X|)}{T^a b(T)} \leq C|X|^{2a} b(|X|) \leq C e^{\varepsilon|X|} \leq C e^{\varepsilon|X|/T}. \qquad \square$$

Note that

$$\|\widetilde{W}_0(T, \cdot)\|_{L^{n/2}} = O(1). \tag{3.3}$$

Applying the estimate of the previous lemma (with $n = 3, b(T) = |\log T|$) we get

$$\|W_1 v\|_{L^2((T,1);L_X^{6/5})} \leq C \left\| \frac{b(\tau)\widetilde{W}_0(\tau, X)}{\tau^{1/2} b(|X|)|X|^{1/2}} v \right\|_{L^2((T,1);L_X^{6/5})}.$$

Using the fact that

$$\|b(\tau)\tau^{-1/2}\|^2_{L^2((T,1))} = \int_T^1 \log^2(\tau) \frac{d\tau}{\tau} \sim |\log T|^3 = b(T)^2 |\log T|,$$

combined with the Hölder inequality

$$\left\| \widetilde{W}_0(\tau, X) g(x) \right\|_{L_X^{6/5}} \leq \|\widetilde{W}_0(T, \cdot)\|_{L_X^{3/2}} \|g(x)\|_{L_X^6} = C \|g(x)\|_{L_X^{6/5}},$$

we get

$$\|W_1 v\|_{L^2((T,1);L_X^{6/5})} \leq Cb(T)|\log T|^{1/2} \left\| \frac{v}{b(|X|)|X|^{1/2}} \right\|_{L^\infty((T,1);L_X^6)}$$

$$\leq Cb(T)|\log T|^{1/2}\|v\|_{L^\infty((T,1);L_X^\infty)}$$

so

$$\|W_1 v\|_{L^2((T,1);L_X^{6/5})} \leq C|\log T|^2 \|v\|_{L^\infty((T,1);L_X^\infty)}.$$

In this way we arrive at

$$\|(-\Delta + W_T)v(T, \cdot)\|_{L_X^2} \tag{3.4}$$
$$\leq C\|v(1,\cdot)\|_{H_X^2} + C|\log T|^2 \|v(1,\cdot)\|_{L_X^2} + C\|(-\Delta + W_T)G\|_{L^1((T,1);L_X^2)},$$

where W_t is defined according to (2.6).

4. Interpolation between $s = 2$ and $s = 0$

First we take any $\delta \in (0,1)$ and $v(1, x) = 0$ and consider the operator

$$M : G \in L^1((\delta, 1); L^2(\mathbb{R}^3)) \longrightarrow v \in L^\infty((\delta, 1); L^2(\mathbb{R}^3))$$

such that v solves the equation

$$i\partial_T v + \Delta v - T^{-2} W\left(\frac{X}{T}\right) v = G,$$

with zero data at $T = 1$. We take another small parameter $\delta_1 \in (0,1)$ and define for any complex z with $\mathrm{Re}\, z \in [0,1]$ the following operator

$$\mathbb{U}(z) = \mathbb{U}_{\delta_1}(z) = (-\Delta + W_T)^z T(\delta_1 - \Delta + W_T)^{-z}$$

$$\mathbb{U}(z) : L^1((\delta, 1); L^2(\mathbb{R}^3)) \longrightarrow v \in L^\infty((\delta, 1); L^2(\mathbb{R}^3)).$$

This is an analytic operator-valued operator. Using the fact that

$$(-\Delta + W_T)^{is} : L^2(\mathbb{R}^3) \longrightarrow L^2(\mathbb{R}^3)$$

is a bounded operator for a real s due to the spectral theorem and using the charge conservation law for the Schrödinger equation with real-valued potential, we see that

$$\|\mathbb{U}_{\delta_1}(z) F\|_{L^\infty((\delta,1);L^2(\mathbb{R}^3))} \le C\|F\|_{L^1((\delta,1);L^2(\mathbb{R}^3))} \tag{4.1}$$

provided $\mathrm{Re}\, z = 0$. Note that the constant $C > 0$ is independent of G, δ, δ_1.

Using the estimate (3.4), we see that

$$\|\mathbb{U}_{\delta_1}(z) F\|_{L^\infty((\delta,1);L^2(\mathbb{R}^3))} \le C\|F\|_{L^1((\delta,1);L^2(\mathbb{R}^3))} \tag{4.2}$$

provided $\mathrm{Re}\, z = 1$ and again the constant $C > 0$ is independent of G, δ, δ_1.

Applying the Stein interpolation theorem (in this simple case the three lines lemma) we see that for any $s \in (0,2)$ we have

$$\|\mathbb{U}_{\delta_1}(s/2) F\|_{L^\infty((\delta,1);L^2(\mathbb{R}^3))} \le C\|F\|_{L^1((\delta,1);L^2(\mathbb{R}^3))} \tag{4.3}$$

and using the definition of $\mathbb{U}_{\delta_1}(z)$ we set

$$F = (\delta_1 - \Delta + W_T)^{s/2} G, \quad G \in L^1((\delta, 1); H^s(\mathbb{R}^3))$$

and get

$$\|(-\Delta + W_T)^{s/2} v(T, \cdot)\|_{L^\infty(\delta,1);L^2_X)} \le C\|(\delta_1 - \Delta + W_T)^{s/2} G\|_{L^1((\delta,1);L^2_X)}. \tag{4.4}$$

With a constant $C > 0$ independent of G, δ, δ_1, letting δ and δ_1 tend to zero we find

$$\|(-\Delta + W_T)^{s/2} v(T, \cdot)\|_{L^\infty(0,1);L^2_X)} \le C\|(-\Delta + W_T)^{s/2} G\|_{L^1((0,1);L^2_X)}. \tag{4.5}$$

In a similar way one can consider the map

$$M_0 : f \in L^2(\mathbb{R}^3) \longrightarrow v \in L^\infty((\delta, 1); L^2(\mathbb{R}^3))$$

such that v solves the equation

$$i\partial_T v + \Delta v - T^{-2} W\left(\frac{X}{T}\right) v = 0,$$

with data $v(1, X) = f(X)$ at $T = 1$. The estimate (3.4) shows that we have

$$\|(-\Delta + W_T)v(T, \cdot)\|_{L_X^2} \leq C\|f\|_{H_X^2} + C|\log T|^2\|f\|_{L_X^2}. \tag{4.6}$$

Applying the interpolation argument as above, we find

$$\|(-\Delta + W_T)^{s/2}v(T, \cdot)\|_{L_X^2} \leq C(1 + |\log T|)^s\|f\|_{H_X^s} \tag{4.7}$$

for any $s \in [0, 2]$.

In this way we obtain the estimate.

Theorem 4.1. *Assume $s \in [0, 2]$. Then the solution to the equation*

$$i\partial_T v + \Delta v - T^{-2}W\left(\frac{X}{T}\right)v = G$$

satisfies the inequality

$$\begin{aligned} &\|(-\Delta + W_T)^{s/2}v(T, \cdot)\|_{L_X^2} \\ &\leq C\left(1 + |\log T|\right)^s \|v(1, \cdot)\|_{H_X^s} + C\|(-\Delta + W_T)^{s/2}G\|_{L^1((T,1);L_X^2)}. \end{aligned} \tag{4.8}$$

One can show that

$$\|(-\Delta + W_T)^{s/2}f\|_{L_X^2} \sim \|(-\Delta)^{s/2}f\|_{L_X^2},$$

for $0 \leq s < 3/2$. This fact is established in [9], [8] (see also Section 7 below where this is verified for completeness).

Then we arrive at

Theorem 4.2. *Assume $s \in [0, 3/2)$. Then the solution to the equation*

$$i\partial_T v + \Delta v - T^{-2}W\left(\frac{X}{T}\right)v = G$$

satisfies the inequality

$$\begin{aligned} &\|(-\Delta)^{s/2}v(T, \cdot)\|_{L_X^2} \\ &\leq C\left(1 + |\log T|\right)^s \|v(1, \cdot)\|_{H_X^s} + C\|(-\Delta)^{s/2}G\|_{L^1((T,1);L_X^2)}. \end{aligned} \tag{4.9}$$

For $s = 2$ we can use the maximum principle for $-\Delta + W$ and see that

$$\|(-\Delta + W)^{-1}f\|_{L^2} \leq \|(-\Delta)^{-1}f\|_{L^2}$$

so

$$\|(-\Delta + W)^{-1}\Delta f\|_{L^2} \leq C\|f\|_{L^2} \tag{4.10}$$

and by duality

$$\|\Delta(-\Delta + W)^{-1}f\|_{L^2} \leq C\|f\|_{L^2}.$$

From this estimate we find

$$\|\Delta f\|_{L^2} = \|\Delta(-\Delta + W)^{-1}(-\Delta + W)f\|_{L^2} \leq C\|(-\Delta + W)f\|_{L^2}. \tag{4.11}$$

Thus we can obtain the following

Theorem 4.3. *Assume $s \in [3/2, 2]$. Then the solution to the equation*

$$i\partial_T v + \Delta v - T^{-2} W\left(\frac{X}{T}\right) v = G$$

satisfies the inequality

$$
\begin{aligned}
&\|(-\Delta)^{s/2} v(T, \cdot)\|_{L_X^2} \\
&\leq C|T|^{3/2-s-\delta} \|v(1, \cdot)\|_{H_X^s} + C|T|^{3/2-s-\delta} \|(1-\Delta)^{s/2} G\|_{L^1((T,1);L_X^2)}.
\end{aligned}
\tag{4.12}
$$

Proof. It is sufficient to verify the estimate for $s = 2$ and then to apply the interpolation argument between $s = 2$ and $s < 3/2$ (established in the previous theorem). For $s = 2$ we use (3.4) as well as (4.11) and see that

$$
\begin{aligned}
\|(-\Delta) v(T, \cdot)\|_{L_X^2} &\leq C\|v(1, \cdot)\|_{H_X^2} + C|\log T|^2 \|v(1, \cdot)\|_{L_X^2} \\
&\quad + C\|(-\Delta) G\|_{L^1((T,1);L_X^2)} + C\|WG\|_{L^1((T,1);L_X^2)}.
\end{aligned}
\tag{4.13}
$$

Now the estimate

$$\|Wf\|_{L^2} \leq \frac{C}{T^{1/2+\delta}} \|(-\Delta)^{3/2+\delta} f\|_{L^2}.$$

Hence the desired estimate with $s = 2$ is fulfilled. This completes the proof. □

Corollary 4.4. *Assume $s \in [3/2, 2]$. Then for any $\delta > 0$ one can find a positive constant $C = C(s, \delta)$ so that the solution to the equation*

$$i\partial_T v + \Delta v - T^{-2} W\left(\frac{X}{T}\right) v = G, \quad T \in (0, 1)$$

satisfies the inequality

$$\|v(T, \cdot)\|_{H_X^s} \leq C|T|^{3/2-s-\delta} \|v(1, \cdot)\|_{H_X^s} + C|T|^{3/2-s-\delta} \|G\|_{L^1((T,1);H_X^s)}. \tag{4.14}$$

5. Proof of Theorem 1.1

Our goal is to solve the nonlinear problem (2.2)

$$i\partial_T v + \Delta_X v - T^{-2} W\left(\frac{X}{T}\right) v + T^{\frac{3p-7}{2}} |v|^{p-1} v = 0$$

with initial data

$$v(1, X) = \varphi(X) \in H^s, s > 3/2.$$

We shall assume that we deal with small initial data, i.e.,

$$\|\varphi\|_{H_X^s} \leq \varepsilon.$$

We shall apply the contraction mapping principle for the Banach space suggested by the estimate (4.14). Indeed, taking $s = 3/2 + \delta$, with $\delta > 0$, consider the norm

$$\||v\||_\delta = \sup_{0 \leq T \leq 1} T^{2\delta} \|v(T, \cdot)\|_{H_X^s} \tag{5.1}$$

and the corresponding Banach space B_δ. The estimates

$$\||v(T,\cdot)|^p\|_{H^s_X} \leq C\|v(T,\cdot)\|_{H^s_X}^p \tag{5.2}$$

$$\||v(T,\cdot) - w(T,\cdot)|^p\|_{H^s_X} \leq C\|v(T,\cdot)\|_{H^s_X}^{p-1}\|v(T,\cdot) - w(T,\cdot)\|_{H^s_X}$$
$$+ C\|w(T,\cdot)\|_{H^s_X}^{p-1}\|w(T,\cdot) - w(T,\cdot)\|_{H^s_X} \tag{5.3}$$

are fulfilled for any $s > 3/2$. A possible reference for these estimates is Theorem 1, Section 5.4.3 in [17]. Using the estimate of Corollary 4.4, one can define the sequence $v_k \in B_\delta$ so that v_0 is a solution to the linear Cauchy problem

$$i\partial_T v_0 + \Delta_X v_0 - T^{-2}W\left(\frac{X}{T}\right)v_0 = 0$$

with initial data

$$v_0 = \varphi(X).$$

Then given any $v_k \in B_\delta$ we define v_{k+1} as the unique solution to

$$i\partial_T v_{k+1} + \Delta_X v_{k+1} - T^{-2}W\left(\frac{X}{T}\right)v_{k+1} + T^{\frac{3p-7}{2}}|v_k|^{p-1}v_k = 0$$

with initial data

$$v(1,X) = \varphi(X) \in H^s.$$

Applying the estimate of Corollary 4.4 as well as (5.2), we find

$$\||\,v_{k+1}\,\||_{2\delta} \leq C\varepsilon + C \||\,v_k\,\||_{2\delta} \int_T^1 \tau^{3p-7-2p\delta}\,d\tau.$$

The assumption $p > 5/3$ guarantees that (taking $\delta > 0$ small enough)

$$\int_T^1 \tau^{3p-7-2p\delta}\,d\tau \leq C < \infty,$$

so

$$\||\,v_{k+1}\,\||_{2\delta} \leq C\varepsilon + C \||\,v_k\,\||_{2\delta}^p.$$

From this estimate we easily get

$$\||\,v_k\,\||_{2\delta} \leq C_1\varepsilon. \tag{5.4}$$

In a similar way, we can use (5.3) and derive

$$\||\,v_{k+1} - v_k\,\||_{2\delta} \leq C \||\,v_k - v_{k-1}\,\||_{2\delta}\left(\||\,v_k\,\||_{2\delta}^{p-1} + \||\,v_{k-1}\,\||_{2\delta}^{p-1}\right)$$
$$\leq C\varepsilon^{p-1} \||\,v_k - v_{k-1}\,\||_{2\delta},$$

so taking $\varepsilon > 0$ small enough, we can apply the contraction mapping principle and we find a solution

$$v \in B_\delta \subset L^\infty([0,1];H^s).$$

Turning back to the pseudoconformal transform (2.1), we see that for $T = 1$ we have

$$v(1,x) = \bar{u}(1,x)e^{i\frac{x^2}{4}}. \tag{5.5}$$

It is easy to see that the map

$$\psi(x) \Longrightarrow \varphi(x) = \overline{\psi(x)}e^{i\frac{x^2}{4}}$$

maps Σ^s in H^s and

$$\|\varphi\|_{H^s} \le C\|\psi\|_{\Sigma^s}.$$

This completes the proof of Theorem 1.1.

6. Resolvent and Strichartz type estimates

In this section we discuss briefly the dispersive and Strichartz type estimates using resolvent estimates. This link is possible in view of the following result due to Kato.

Theorem 6.1 (Kato [12]). *Let H be a self-adjoint operator on the Hilbert space $\mathcal{X}$, and for $\mu \in \mathbb{R}, \Im \mu \ne 0$, let*

$$(H - \mu)^{-1},$$

denote the resolvent. Suppose that A is a closed, densely defined operator, possibly unbounded, from $\mathcal{X}$ into a Hilbert space $\mathcal{Y}$. Suppose that

$$\Gamma := \sup\{\|A((H - \mu)^{-1})A^*f\|_{\mathcal{Y}}; \Im \mu \ne 0, f \in D(A^*), \|f\|_{\mathcal{X}} = 1\} < \infty.$$

Then A is H-smooth and

$$\|A\|_H^2 := \sup\left\{\frac{1}{2\pi}\int_{-\infty}^{\infty} \|Ae^{-itH}f\|_{\mathcal{Y}}^2 dt; f \in \mathcal{X}, \|f\|_{\mathcal{X}} = 1\right\} \le \frac{\Gamma^2}{\pi^2}.$$

A typical application for the Schrödinger equation is the choice $\mathcal{X} = L^2$ and A is the multiplication operator

$$\langle x \rangle^{-s} : f(x) \in L^2 \longrightarrow \langle x \rangle^{-s} f(x) \in L_s^2,$$

where here and below for any real s

$$L_s^2 = \{f \in L_{\text{loc}}^2, \langle x \rangle^s f \in L^2\}.$$

The study of the resolvent estimates is closely connected with the resonances of the operator

$$-\Delta + W(|x|), x \in \mathbb{R}^3.$$

Definition 6.2. A real number λ is called a strong resonance of $-\Delta + W(|x|)$ if there exists $u \in L_{-a}^2(\mathbb{R}^3)$ with $a > 1/2$, so that $u(x)$ is not identically zero and $-\Delta u + W(|x|)u = \lambda u$ in the distribution sense in $\mathbb{R}^3$.

Theorem 6.3 (see Theorem IX.2 in [9]). *Suppose that the potential $W(r)$ is a positive decreasing function, such that there exist positive constants C^*, ε so that (H_1) is fulfilled. Then zero is not a strong resonance for $-\Delta + W(|x|)$.*

Remark 6.4. Since W is exponentially decaying and real valued, the above result implies that $-\Delta + W(|x|)$ has no resonances.

In order to verify the resolvent estimate of the perturbed operator $-\Delta + W(|x|)$, denote

$$R_0(\mu) = (-\Delta - \mu^2)^{-1},$$

the resolvent of the operator $-\Delta$, and set $R_0^+(\mu) = R_0(\mu)$ if $\Im\mu > 0$ and respectively $R_0^-(\mu) = R_0(\mu)$ for $\Im\mu < 0$. The classical resolvent estimate (limiting absorbtion principle) is the following one

$$\lim_{\Im\mu \searrow 0} \|\langle x\rangle^{-s}(-\Delta - \mu^2)^{-1}\langle x\rangle^{-s}f\|_{L^2(\mathbb{R}^n)} \le C\|f\|_{L^2(\mathbb{R}^n)}, \qquad (6.1)$$

where $s > 1$, $\Re\mu \ge 0$, and the constant C is independent of μ. We have also the estimate

$$\|\langle x\rangle^{-s}\nabla(-\Delta - \mu^2)^{-1}\langle x\rangle^{-s}f\|_{L^2(\mathbb{R}^n)} \le C\|f\|_{L^2(\mathbb{R}^n)},$$

where

$$s > \frac{1}{2}, \quad \Re\mu \ge 0,$$

so we can claim that the operators

$$\langle x\rangle^{-s}(-\Delta - \mu^2)^{-1}\langle x\rangle^{-s}$$

are compact ones in L^2 provided $s > 1$ and $\Im\mu \ge 0$.

Hence

$$\|\langle x\rangle^{-s} R_0^+(\mu)\langle x\rangle^{-s}f\|_{L^2(\mathbb{R}^n)} \le C\|f\|_{L^2(\mathbb{R}^n)}, \quad s > 1, \Re\mu \ge 0, \Im\mu = 0. \qquad (6.2)$$

Set

$$R(\mu) = (-\Delta + W - \mu^2)^{-1}, \quad A(\mu) = \langle x\rangle^{-s}(-\Delta - \mu^2)^{-1}W\langle x\rangle^{s}.$$

One has the following compactness result:

Lemma 6.5. *The operators $A(\mu)$ are compact in the space $B(L^2, L^2)$, for*

$$\Im\mu \ge 0, \quad s > 1.$$

Moreover the following estimate is satisfied:

$$\|A(\mu)\|_{B(L^2,L^2)} \to 0,$$

as $\Im\mu \ge 0, \Re\mu \to \infty$.

This lemma is a well-known standard result so we give only the idea of the proof. It suffices to notice that $\langle x\rangle^{-s}(-\Delta - \mu^2)^{-1}\langle x\rangle^{-s}$ is continuous and compact as an operator in $B(L^2, L^2)$, in the zone $\Im\mu \ge 0$. Since the potential W is such that

$$\langle x\rangle^{s} W \langle x\rangle^{s}$$

is bounded in L^2, we have the desired result.

Lemma 6.6. *Let us assume that the potential W satisfies (H_1). For any $s > 1$ the weighted resolvent operator $\langle x\rangle^{-s} R^+(\mu)\langle x\rangle^{-s}$ has a continuous extension from $\Im\mu > 0$ to $\Im\mu \ge 0$. Moreover there exists a real constant $C > 0$ such that the following estimate is true:*

$$\|\langle x\rangle^{-s} R^+(\mu)\langle x\rangle^{-s}f\|_{L^2} \le C\|f\|_{L^2}. \qquad (6.3)$$

for any Schwartz function f.

Proof. The result is well known, so we briefly sketch the idea. The perturbed resolvent $R(\mu^2) = (-\Delta + W - \mu^2)^{-1}$ satisfies in $\Im\mu > 0$ the relation

$$(-\Delta + W - \mu^2)^{-1} = \left(I + (-\Delta - \mu^2)^{-1}W\right)^{-1}(-\Delta - \mu^2)^{-1} \tag{6.4}$$

provided the operator $(I - (P_0 - \mu^2)^{-1}W)$ is invertible. This relation implies

$$\begin{aligned}
&\langle x \rangle^{-s}(-\Delta + W - \mu^2)^{-1}\langle x \rangle^{-s} \\
&= \left(I + \langle x \rangle^{-s}(-\Delta - \mu^2)^{-1}W\langle x \rangle^{s}\right)^{-1}\langle x \rangle^{-s}(-\Delta - \mu^2)^{-1}\langle x \rangle^{-s}.
\end{aligned} \tag{6.5}$$

We can apply the Fredholm Theory and Theorem 6.3 that shows 0 is not a resonance, so we are able to say that the operator

$$\left(I + \langle x \rangle^{-s}(-\Delta - \mu^2)^{-1}W\langle x \rangle^{+s}\right)^{-1},$$

is continuous in $\Im\mu \geq 0$. $\qquad\square$

Once the resolvent estimate is established, one can use the approach from [5] and derive the Strichartz type estimate for the corresponding inhomogeneous Cauchy problem

$$i\partial_t u - \Delta u = F \ , \quad u(0) = f. \tag{6.6}$$

We shall call the pair $(\frac{1}{p}, \frac{1}{q})$ sharp admissible (see [13] for this notion and the properties of sharp admissible pairs), if it satisfies the condition:

$$\frac{n}{4} = \frac{1}{p} + \frac{n}{2q} \ , \quad 2 \leq p \leq \infty, (p, q, n) \neq (2, \infty, 2). \tag{6.7}$$

If $n = 3$, then we can choose the end point

$$p^* = 2, q^* = 6$$

as an admissible couple. Moreover, (6.7) becomes

$$\frac{3}{4} = \frac{1}{p} + \frac{3}{2q} \ , \quad 2 \leq p \leq \infty. \tag{6.8}$$

Then we have

Theorem 6.7. *If (p, q) and $(\tilde{p}, \tilde{q})$ satisfy (6.8), then the solution to the Cauchy problem*

$$i\partial_t u + \Delta u - Wu = F \ , \quad (t, x) \in (1, \infty) \times \mathbb{R}^3_x, \tag{6.9}$$
$$u(1, x) = f(x),$$

satisfies the estimate

$$\|u\|_{L^p((1,\infty);L^q_x)} + \|u\|_{C([1,\infty);L^2)} \leq C\left(\|F\|_{L^{\tilde{p}'}((1,\infty);L^{\tilde{q}'}_x)} + \|f\|_{L^2}\right). \tag{6.10}$$

Using the pseudoconformal transform, we make the substitution

$$v(T, X) = \frac{1}{T^{3/2}} \overline{u}(\frac{1}{T}, \frac{X}{T}) e^{i \frac{X^2}{4T}},$$

$$H(T, X) = \frac{1}{T^{2+3/2}} \overline{F}(\frac{1}{T}, \frac{X}{T}) e^{i \frac{X^2}{4T}} \tag{6.11}$$

$$h(X) = f(X) e^{i \frac{X^2}{4}}.$$

and see that v is a solution to the Cauchy problem

$$i\partial_T v + \Delta_X v - T^{-2} W \left(\frac{X}{T} \right) v = H , \quad (T, X) \in (0,1)) \times \mathbb{R}^3_x, \quad u(1, x) = h(x). \tag{6.12}$$

A simple computation shows that

$$\|u\|_{L^p((1,\infty);L^q_x)} = \|v\|_{L^p((0,1);L^q_X)},$$

and

$$\|F\|_{L^{\tilde{p}'}((1,\infty);L^{\tilde{q}'}_x)} = \|H\|_{L^{\tilde{p}'}((0,1);L^{\tilde{q}'}_X)},$$

provided the couples (p, q) and $(\tilde{p}, \tilde{q})$ are admissible ones. Since

$$\|u\|_{L^\infty((1,\infty);L^2)} = \|v\|_{L^\infty((0,1);L^2)},$$

we can take

$$p^* = 2, q^* = 6$$

as an admissible couple and we arrive at the following.

Theorem 6.8. *If $n = 3$, then the solution to the Cauchy problem (6.12) satisfies the estimates*

$$\|v\|_{L^2((0,1);L^6_X)} + \|v\|_{L^\infty((0,1);L^2_X)} \leq C \left(\|H\|_{L^2((0,1);L^{6/5}_X)} + \|h\|_{L^2_X} \right) \tag{6.13}$$

and

$$\|v\|_{L^2((0,1);L^6_X)} + \|v\|_{L^\infty((0,1);L^2_X)} \leq C \left(\|H\|_{L^1((0,1);L^2_X)} + \|h\|_{L^2_X} \right). \tag{6.14}$$

7. Equivalence of $\dot{H}^s_W$ and $\dot{H}^s$

Here we follow the argument of Section 5 in [8]. To show that $\dot{H}^s_{W_T} = \dot{H}^s$ for $s < \frac{3}{2}$ we will first prove the following

Lemma 7.1. $\dot{H}^1_{W_T} = \dot{H}^1$.

Proof. The positivity of W_T implies

$$\left\| (-\Delta + W_T)^{\frac{1}{2}} f \right\|^2_{L^2} = \langle (-\Delta + W_T)f, f \rangle_{L^2} \geq \langle (-\Delta)f, f \rangle_{L^2} = \left\| (-\Delta)^{\frac{1}{2}} f \right\|^2_{L^2}.$$

The assumption (H1) implies

$$W_T(X) = T^{-2} W \left(\frac{X}{T} \right) \leq \frac{C}{|X|^2}.$$

The Hardy inequality yields

$$\langle (W_T f, f)_{L^2} \leq C \left\| (-\Delta)^{\frac{1}{2}} f \right\|_{L^2}^2$$

so

$$\left\| (-\Delta + W_T)^{\frac{1}{2}} f \right\|_{L^2}^2 = \langle (-\Delta + W_T) f, f \rangle_{L^2} \leq C \left\| (-\Delta)^{\frac{1}{2}} f \right\|_{L^2}^2.$$

This completes the proof. $\qquad\square$

Lemma 7.2. *For $0 \leq s < 3/2$ we have*

$$\left\| (-\Delta + W_T)^{\frac{s}{2}} f \right\|_{L^2}^2 \sim \left\| (-\Delta)^{\frac{s}{2}} f \right\|_{L^2}^2.$$

Proof. Take $1 < s < \frac{3}{2}$. We shall use the identity

$$\left\| (-\Delta + W)^{\frac{s}{2}} f \right\|_{L^2}^2 = \left((-\Delta + W)^{s-1} f, (-\Delta + W) f \right)_{L^2}$$
$$= \left((-\Delta + W)^{s-1} f, (-\Delta) f \right)_{L^2} + \left((-\Delta + W)^{s-1} f, W f \right)_{L^2} \qquad (7.1)$$
$$= \left((-\Delta)^{1-\frac{s}{2}} (-\Delta + W)^{s-1} f, (-\Delta)^{\frac{s}{2}} f \right)_{L^2} + \left(|W|^{1-\frac{s}{2}} (-\Delta + W)^{s-1} f, W^{\frac{s}{2}} f \right)_{L^2}.$$

Let us set

$$I_1 = \left((-\Delta)^{1-\frac{s}{2}} (-\Delta + W)^{s-1} f, (-\Delta)^{\frac{s}{2}} f \right)_{L^2},$$
$$I_2 = \left(W^{1-\frac{s}{2}} (-\Delta + W)^{s-1} f, W^{\frac{s}{2}} f \right)_{L^2}.$$

Now we can apply Lemma 7.1 and using the fact that $\frac{1}{2} < 2 - s < 1$, we get

$$\left\| (-\Delta)^{\frac{2-s}{2}} g \right\|_{L^2} \leq C \left\| (-\Delta + V)^{\frac{2-s}{2}} g \right\|_{L^2}.$$

Taking now $g = (-\Delta + V)^{s-1} f$, we get

$$\left\| (-\Delta)^{\frac{2-s}{2}} (-\Delta + W)^{s-1} f \right\|_{L^2} \leq C \left\| (-\Delta + W)^{\frac{s}{2}} f \right\|_{L^2}. \qquad (7.2)$$

Hence

$$|I_1| \leq \| f \|_{\dot{H}_W^s} \| f \|_{\dot{H}^s}. \qquad (7.3)$$

Further we need the following

Lemma 7.3. *We have the estimate*

$$\left\| W^{\frac{s}{2}} f \right\|_{L^2} \leq C \| f \|_{\dot{H}^s},$$

where $W = W_T = T^{-2} W(X/T)$ and $0 \leq s < \frac{3}{2}$.

Proof. Applying the Hölder inequality for Lorentz spaces and using the fact that $\left\| |W|^{\frac{s}{2}} \right\|_{L^{(\frac{3}{s}, \infty)}} \leq C \| W \|_{L^{(\frac{3}{2}, \infty)}}^{\frac{s}{2}} \leq C_0^{\frac{s}{2}}$, we get

$$\left\| W^{\frac{s}{2}} f \right\|_{L^2} \leq C \left\| W^{\frac{s}{2}} \right\|_{L^{(\frac{3}{s}, \infty)}} \| f \|_{L^{(q,2)}}, \qquad (7.4)$$

$$\frac{1}{2} = \frac{s}{3} + \frac{1}{q}, \quad q = 6 \in (2, \infty). \qquad (7.5)$$

Now we can apply Sobolev's embedding (see [3]) $\dot{H}^s \subset L^{(q,2)}$ for $\frac{1}{2} = \frac{s}{3} + \frac{1}{q}$ and we get $\left\| |W|^{\frac{s}{2}} f \right\|_{L^2} \leq C_1 \|f\|_{\dot{H}^s}$.   □

Now we are ready to estimate the term I_2. We have

$$|I_2| \leq \left\| W^{\frac{2-s}{2}} (-\Delta + W)^{s-1} f \right\|_{L^2} \left\| (W)^{\frac{s}{2}} f \right\|_{L^2}. \qquad (7.6)$$

Since $2 - s \in (0, \frac{3}{2})$, we can apply Lemma 5.2 and get

$$\left\| |W|^{\frac{2-s}{2}} (-\Delta + W)^{s-1} f \right\|_{L^2} \leq \left\| (-\Delta)^{\frac{2-s}{2}} (-\Delta + W)^{s-1} f \right\|_{L^2} \qquad (7.7)$$

and $\left\| |W|^{\frac{s}{2}} f \right\|_{L^2} \leq C \|f\|_{\dot{H}^s}$. We estimate the right-hand side of (7.7) using (7.2) and find

$$\left\| W^{\frac{2-s}{2}} (-\Delta + W)^{s-1} f \right\|_{L^2} \leq C \left\| (-\Delta + W)^{\frac{s}{2}} f \right\|_{L^2}. \qquad (7.8)$$

From (7.2) (7.7) and (7.8) we obtain

$$|I_2| \leq C \|f\|_{\dot{H}^s_W} \|f\|_{\dot{H}^s}. \qquad (7.9)$$

This estimate, (7.2) and (7.1) lead to

$$\|f\|^2_{\dot{H}^s_W} \leq C \|f\|_{\dot{H}^s_W} \|f\|_{\dot{H}^s}.$$

Hence

$$\|f\|_{\dot{H}^s_W} \leq C \|f\|_{\dot{H}^s}, \qquad (7.10)$$

for $0 \leq s < \frac{3}{2}$.

To show the opposite inequality, we use the fact that W_T is a non-negative potential decaying faster than $|x|^{-2}$ at infinity, so one can apply Theorem 1.1 of the work [19] and get the following estimate of the heat kernel $K_W(t, x, y)$ of the heat operator

$$\partial_t - \Delta + W$$

$$|K_W(t, x, y)| \leq C K_0(t, x, y), \qquad (7.11)$$

where

$$K_0(t, x, y) = ct^{-3/2} e^{-c|x-y|^2/(4t)}$$

is the heat kernel of the free heat operator $\partial_t - \frac{1}{c}\Delta$.

It is important to notice that the potential $W = W_T$ depends on the parameter $T \in (0, 1)$, but the constant C in (7.11) is independent of this parameter, since we have the inequality

$$|W_T(x)| = |T^{-2} W(x/T)| \leq \frac{C}{|x|^3}$$

with some constant C independent of $T \in (0, 1)$. Given any sectorial operator A with spectrum $\sigma(A)$ satisfying

$$z \in \sigma(A) \implies \Re z \geq 0,$$

we can define the negative powers of A as follows (see for example Section 1.4 in [10])

$$A^{-k} = \frac{1}{\Gamma(k)} \int_0^\infty t^{k-1} e^{-At} dt. \tag{7.12}$$

Choosing $k = 1$,

$$A = -\Delta + W, \quad A_0 = -\frac{1}{c} \Delta$$

and comparing the kernels of

$$e^{-At}, \quad e^{-A_0 t}$$

by the aid of the estimate (7.12), we see

$$\|(-\Delta + W)^{-1} f\|_{L^2} \le C \|(-\Delta)^{-1} f\|_{L^2}.$$

This estimate shows that the operator

$$(-\Delta + W)^{-1}(-\Delta)$$

is L^2 bounded, so its dual

$$(-\Delta)(-\Delta + W)^{-1}$$

is also L^2 bounded and we see that

$$\|(-\Delta) f\|_{L^2} \le C \|(-\Delta + W) f\|_{L^2}.$$

Hence, by interpolation

$$\|f\|_{\dot{H}^s_W} \le C \|f\|_{\dot{H}^s}, \tag{7.13}$$

for $0 \le s \le 2$.

This completes the proof. $\square$

Acknowledgment

The first author is grateful to Michael Ruzhansky for the hospitality during the meeting "Asymptotic Properties of Solutions to Hyperbolic Equations" held at Imperial College London in March 2011.

References

[1] S. Agmon, *Spectral properties of Schrödinger operators and scattering theory*, Ann. Scuola Norm. Sup. Pisa Cl. Sci. **2/2** (1975), 151–218.

[2] M. Beceanu, *A Centre-Stable Manifold for the Focussing Cubic NLS in R^{1+3}*, Commun. Math. Phys. **280**, (2008), 145–205.

[3] J. Bergh and J. Löfström, *Interpolation spaces*, Springer-Verlag, 1976.

[4] N. Burq, F. Planchon, J.G. Stalker, A.S. Tahvildar-Zadeh, *Strichartz estimates for the wave and Schrödinger equations with the inverse-square potential*, J. Funct. Anal., **203 (2)**, (2003), 519–549.

[5] N. Burq, F. Planchon, J.G. Stalker, A.S. Tahvildar-Zadeh, *Strichartz estimates for the wave and Schrödinger equations with potentials of critical decay.* Indiana Univ. Math. J., **53 (6)** (2004), 1665–1680.

[6] Th. Cazenave, *Semilinear Schrödinger Equations*, Courant Lecture Notes, vol. 10, 2003.

[7] B. Erdogan, W. Schlag, *Dispersive estimates for Schrödinger operators in the presence of a resonance and/or an eigenvalue at zero energy in dimension three, II.* J. Anal. Math. **99** (2006), 199–248.

[8] V. Georgiev, A. Ivanov, *Existence and mapping properties of wave operator for the Schrödinger equation with singular potential*, Proc. Amer. Math. Soc. **133** (2005), 1993–2003.

[9] V. Georgiev, N. Visciglia, *Decay estimates for the wave equation with potential*, Comm. Part. Diff. Eq. **28 (7,8)** (2003), 1325–1369.

[10] D. Henry, *Geometric theory of semilinear parabolic equations*, Springer, 1981.

[11] N. Hayashi, P. Naumkin, *Asymptotics for large time of solutions to the nonlinear Schrödinger and Hartree equations* Amer. J. Math. **120 (2)** (1998), 369–389.

[12] T. Kato, *Wave operators and similarity for some non-selfadjoint operators.* Math. Ann., **162** (1966), 258–279.

[13] M. Keel, T. Tao *Endpoint Strichartz estimates*, Amer. J. Math. **120 (5)** (1998), 955–980.

[14] T. Matsuyama, *Dispersion for 3D wave equation with a potential in an exterior domain,* preprint (2010).

[15] K. Mochizuki, *Resolvent estimates for magnetic schrodinger operators and their applications to related evolution equations*, Rendiconti Univ. Trieste **42 Suppl.** (2010), 143–164.

[16] I. Rodnianski and W. Schlag, *Time decay for solutions of Schrödinger equations with rough and time-dependent potentials.* Invent. Math. **155, No. 3** (2004), 451-513.

[17] T. Runst and W. Sinckel, *Sobolev Spaces of Fractional Order, Nemitskij Operators and Nonlinear Partial Differential Equations*, Walter de Gruyter, Berlin (1996).

[18] K. Yajima, *Dispersive estimate for Schrödinger equations with threshold resonance and eigenvalue.* Commun. Math. Phys. **259(2)**, 475–509 (2005)

[19] Qi S. Zhang, *Large Time Behavior of Schrödinger Heat Kernels and Applications* Commun. Math. Phys. **210**, 371–398 (2000)

Vladimir Georgiev and Bozhidar Velichkov
Department of Mathematics
University of Pisa
Largo Bruno Pontecorvo 5
I-56127 Pisa, Italy
e-mail: georgiev@dm.unipi.it
        b.velichkov@sns.it

Progress in Mathematics, Vol. 301, 163–186

Wave Equations on Non-smooth Space-times

Günther Hörmann, Michael Kunzinger and Roland Steinbauer

Abstract. We consider wave equations on Lorentzian manifolds in case of low regularity. We first extend the classical solution theory to prove global unique solvability of the Cauchy problem for distributional data and right-hand side on smooth globally hyperbolic space-times. Then we turn to the case where the metric is non-smooth and present a local as well as a global existence and uniqueness result for a large class of Lorentzian manifolds with a weakly singular, locally bounded metric in Colombeau's algebra of generalized functions.

Mathematics Subject Classification. Primary: 58J45; Secondary: 35L05, 35L15, 35D99, 46F30.

Keywords. Wave equation, Cauchy problem, global hyperbolicity, distributional solutions, generalized solutions.

1. Introduction

In this note we are concerned with wave equations on Lorentzian manifolds, with a particular interest in the Cauchy problem in non-smooth situations. To achieve a self-contained presentation we first give a brief account on the classical solution theory in the smooth case, in particular discussing local well-posedness as well as global well-posedness on globally hyperbolic manifolds. We first extend this theory to the case of distributional data and right-hand sides. Then we turn to the case where the metric is non-smooth, that is we deal with normally hyperbolic operators with coefficients of low regularity. Actually the regularity class of the metric which we have in mind is below $C^{1,1}$ (i.e., the first derivative locally Lipschitz) – the largest class where standard differential geometric results such as existence and uniqueness of geodesics remain valid, and also below the Geroch-Traschen class of metrics – the largest class that allows for a consistent distributional treatment ([15, 26]). It is evident that no consistent distributional solution concept is available for these equations. Therefore we consider a large class of weakly singular, locally bounded metrics in the setting of generalized global analysis and nonlinear distributional geometry ([18, Sec. 3.2]) based on Colombeau algebras of generalized functions ([9, 10]). In particular, we present a local existence and uniqueness result in the spirit of [16] and extend it to a global result on space-times with a generalized metric that allows for a suitable globally hyperbolic metric splitting.

This line of research has drawn some motivation from general relativity: In [7] Chris Clarke suggested to replace the standard geometric definition of singularities by viewing them as obstructions to the well-posedness of the initial value problem for a scalar field. However, for many relevant examples as, e.g., impulsive gravitational waves, cosmic strings, and shell crossing singularities which have a metric of low regularity, the Cauchy problem cannot consistently be formulated in distribution theory and one has to use a more sophisticated solution concept. In case of shell crossing singularities Clarke himself used a cleverly designed weak solution concept to argue for local solvability of the wave equation ([8]). On the other hand Vickers and Wilson ([33]) used Colombeau generalized functions to show local well-posedness of the wave equation in conical space-times modelling a cosmic string. This result has been generalized to a class of locally bounded space-times in [16], also see [27] for the static case and [19] for an extension to non-scalar equations.

This work is organized in the following way. In Section 2 we recall the classical theory of normally hyperbolic operators on smooth Lorentzian manifolds, thereby essentially following a recent book by Bär et al. ([1]). We extend their global existence and uniqueness result (Th. 2.7) to the case of distributional data and right-hand side in Section 3. In Section 4 we recall the necessary background from global analysis based on Colombeau generalized functions ([18]). We devote Section 5 to presenting a variant of the local existence and uniqueness theorem for the wave equation in weakly singular space-times of [16]. Finally, in Section 6 we extend this result to a global theorem for weakly singular space-times which allow for a suitable globally hyperbolic splitting of the metric.

2. A review of wave equations on smooth Lorentzian manifolds

In this section we present the solution theory for wave equations on smooth Lorentzian manifolds with smooth right-hand side and data (cf. (2.3) below), mainly based on [1, 34] (see also [13]). We will at once formulate the theory for normally hyperbolic operators on sections of a vector bundle E over some Lorentzian manifold M.

We first fix some notations. Throughout we will assume (M, g) to be a smooth, connected, time-oriented Lorentz manifold and $E \to M$ a smooth vector bundle over M. By $\Gamma(M, E)$ or $\Gamma(E)$ we denote the smooth sections of $E \to M$, $\mathcal{D}(M, E)$ is the space of compactly supported sections, and $\Gamma^k(E)$ denotes the spaces of sections of finite differentiability. Given a finite-dimensional vector space

W, the space of W-valued distributions in E, $\mathcal{D}'(M, E, W)$ comprises the continuous linear maps $\mathcal{D}(M, E^*) \to W$. For example, given $x \in M$, the delta-distribution δ_x is the E_x^*-valued distribution in E given by $\delta_x : \mathcal{D}(M, E^*) \to E_x^*$, $\varphi \mapsto \varphi(x)$.

A differential operator $P : \Gamma(E) \to \Gamma(E)$ of second order is called *normally hyperbolic* if its principal symbol satisfies $\sigma_P(\xi)|_x = -\langle \xi, \xi \rangle \cdot \mathrm{id}_{E_x}$ ($\xi \in T_x^*M$). If E is equipped with a linear connection ∇ then an important example of a normally hyperbolic operator is given by the connection d'Alembertian

$$\Box^\nabla : \Gamma(E) \xrightarrow{\nabla} \Gamma(T^*M \otimes E) \xrightarrow{\nabla} \Gamma(T^*M \otimes T^*M \otimes E) \xrightarrow{-\mathrm{tr} \otimes \mathrm{id}_E} \Gamma(E),$$

where $\mathrm{tr} : T^*M \otimes T^*M \to \mathbb{R}$, $\mathrm{tr}(\xi \otimes \eta) = \langle \xi, \eta \rangle$ is the metric trace. By the Weitzenböck formula, for any normally hyperbolic operator on $\Gamma(E)$ there exists a unique connection ∇ on E and a unique $B \in \Gamma(\mathrm{Hom}(E, E))$ such that $P = \Box^\nabla + B$. For any differential operator $D : \Gamma(E) \to \Gamma(F)$, its formal adjoint $D^* : \Gamma(F^*) \to \Gamma(E^*)$ is uniquely characterized by

$$\int \psi(D\varphi) \, dV = \int (D^*\psi)(\varphi) \, dV$$

for all $\varphi \in \mathcal{D}(M, E)$ and all $\psi \in \mathcal{D}(M, F^*)$. Here dV is the Lorentzian volume density on M.

For notions from causality in Lorentz manifolds our main references are [28, 2]. In particular, for p, $q \in M$, by $p < q$ (resp. $p \ll q$) we mean that there exists a causal (resp. timelike) future directed curve from p to q. For $A \subseteq \Omega \subseteq M$ we write

$$I_+^\Omega(A) := \{ q \in \Omega \mid \exists p \in A \text{ s.t. } q \gg p \text{ in } \Omega \}$$

for the relative chronological future of A in Ω, and analogous for $I_-^\Omega(A)$. Also we write

$$J_+^\Omega(A) := \{ q \in \Omega \mid \exists p \in A \text{ s.t. } q \geq p \text{ in } \Omega \},$$

for the relative causal future and analogous for $J_-^\Omega(A)$. Here $q \geq p$ means either $p < q$ or $p = q$. A domain Ω is called causal if its closure $\overline{\Omega}$ is contained in a geodesically convex domain Ω' and if for all $p, q \in \overline{\Omega}$ the causal diamond $J_+^{\Omega'}(p) \cap J_-^{\Omega'}(q)$ is a compact subset of Ω'. The manifold M is said to satisfy the *causality condition* (CC) if there are no closed causal curves in M. It satisfies the *strong causality condition* (SCC) if for each $p \in M$ and each neighborhood U of p there exists a neighborhood $V \subseteq U$ such that no causal curve that starts and ends in V can leave U (i.e., there are no 'almost closed' causal curves in M).

A *Cauchy hypersurface* is a subset S of M that is intersected by each inextendible timelike curve exactly once. M is called *globally hyperbolic* if it satisfies SCC and for all $p, q \in M$ the causal diamond $J_+^M(p) \cap J_-^M(q)$ is compact. The following is a very useful characterization of global hyperbolicity, due in its final form to [3].

Theorem 2.1. *For any time-oriented Lorentzian manifold M, the following are equivalent:*

(i) *M is globally hyperbolic.*

(ii) *There exists a Cauchy hypersurface in M.*

(iii) *M is isometric to $(\mathbb{R} \times S, -\beta dt^2 + g_t)$, where $\beta = \beta(t, x)$ is smooth and strictly positive, $t \mapsto g_t$ is a smooth family of Riemannian metrics on S, and each $\{t\} \times S$ is a (smooth) spacelike Cauchy-hypersurface in M.*

We remark that by a recent result ([4]) global hyperbolicity is also characterized by the condition that results from replacing SCC by CC in the above definition.

Turning now to the problem of solving the initial value problem for a normally hyperbolic differential operator on $E \to M$, we first consider the case where $M = (V, \langle \, , \, \rangle)$ is a Lorentz vector space of dimension n (later on, the role of V will be played by a tangent space $T_x M$ to M). We denote by $\gamma : V \to \mathbb{R}$, $\gamma(X) := -\langle X, X \rangle$ the quadratic form associated with $\langle \, , \, \rangle$. The analytic centerpiece of the entire construction that is to follow is provided by the so-called *Riesz distributions*:

Definition 2.2. For $\alpha \in \mathbb{C}$ with $\mathrm{Re}(\alpha) > n$, let

$$R_\pm(\alpha)(X) := \begin{cases} C(\alpha, n)\gamma(X)^{\frac{\alpha-n}{2}} & \text{if } X \in J_\pm(0) \\ 0 & \text{else} \end{cases}$$

where $C(\alpha, n) := \frac{2^{1-\alpha} \pi^{\frac{2-n}{2}}}{\Gamma(\frac{\alpha}{2})\Gamma(\frac{\alpha-n}{2}+1)}$. $R_+(\alpha)$ $(R_-(\alpha))$ is called advanced (retarded) Riesz distribution on V.

$R_\pm(\alpha)$ is continuous on V, and using the fact that $\square R_\pm(\alpha + 2) = R_\pm(\alpha)$ for $\mathrm{Re}(\alpha) > n + 2$, $\alpha \mapsto R_\pm(\alpha)$ uniquely extends to a holomorphic family of distributions on all of $\mathbb{C}$. For all α, $\mathrm{supp}(R_\pm(\alpha)) \subseteq J_\pm(0)$ and $\mathrm{singsupp}(R_\pm(\alpha))$ is contained in the boundary $C_\pm(0)$ of $J_\pm(0)$. Moreover, $R_\pm(0) = \delta_0$.

The next step in the construction is to transport the Riesz distributions onto the Lorentz manifold M. To this end, let Ω be a normal neighborhood of $x \in M$ and define the smooth function $\mu_x : \Omega \to \mathbb{R}$ by $dV = \mu_x \cdot (\exp_x)_*(dz)$, where dz is the standard volume density on $T_x M$, and $\exp_x$ is the exponential map at x. In normal coordinates around x, $\mu_x = \sqrt{|\det g_{ij}|}$. Now we set $R_\pm^\Omega(\alpha, x) := \mu_x \cdot (\exp_x)_* R_\pm(\alpha)$, i.e.,

$$\forall \varphi \in \mathcal{D}(\Omega, \mathbb{C}) : \quad \langle R_\pm^\Omega(\alpha, x), \varphi \rangle = \langle R_\pm(\alpha), (\mu_x \cdot \varphi) \circ \exp_x \rangle.$$

$R_\pm^\Omega(\alpha, x)$ are called *advanced (retarded) Riesz distributions* on Ω. The analytical properties of $R_\pm(\alpha)$ carry over to the manifold setting, albeit in slightly more involved form, e.g., (setting $\Gamma_x := \gamma \circ \exp_x^{-1}$),

$$\square R_\pm^\Omega(\alpha + 2, x) = \left(\frac{1}{2\alpha} (\square \Gamma_x - 2n) + 1 \right) R_\pm^\Omega(\alpha, x) \quad \text{for } \alpha \neq 0. \qquad (2.1)$$

Moreover, $R_\pm^\Omega(0, x) = \delta_x$, $\mathrm{supp}(R_\pm^\Omega(\alpha, x)) \subseteq J_\pm^\Omega(x)$ and $\mathrm{singsupp}(R_\pm^\Omega(\alpha, x)) \subseteq C_\pm^\Omega(x)$.

For any normally hyperbolic operator P on $\Gamma(E)$, our aim is to construct a fundamental solution in the following sense:

Definition 2.3. Let $P : \Gamma(E) \to \Gamma(E)$ be normally hyperbolic. A distribution $F \in \mathcal{D}'(M, E, E_x^*)$ such that $PF = \delta_x$ (i.e., $\langle F, P^*\varphi \rangle = \varphi(x)$ for all $\varphi \in \mathcal{D}(M, E^*)$) is called a *fundamental solution* of P at $x \in M$. F is called *advanced (retarded)* if $\mathrm{supp}(F) \subseteq J_+^M(x)$ $(\subseteq J_-^M(x))$.

For example, in a Lorentz vector space V as above, $R_{\pm}(2)$ is an advanced (retarded) fundamental solution of $\Box$ at 0 since $\Box R_{\pm}(2) = R_{\pm}(0) = \delta_0$. In the manifold setting matters are more complicated, as already indicated in (2.1). We first make the following formal ansatz for a fundamental solution on a normal neighborhood Ω of x:

$$\mathcal{R}_{\pm}(x) := \sum_{k=0}^{\infty} V_x^k R_{\pm}^{\Omega}(2 + 2k, x), \tag{2.2}$$

where $V_x^k \in \Gamma(\Omega, E \otimes E_x^*)$, the so-called *Hadamard coefficients*, are to be determined. By formal termwise differentiation one finds that in order for $\mathcal{R}_{\pm}(x)$ to be a fundamental solution, the V_x^k have to satisfy the following transport equations:

$$\nabla_{\mathrm{grad}\,\Gamma_x} V_x^k - (\frac{1}{2}\Box\Gamma_x - n + 2k)V_x^k = 2kPV_x^{k-1} \quad (k \geq 0)$$

with $V_x^0(x) = \mathrm{id}_{E_x}$. The Hadamard coefficients are therefore uniquely determined as the solution to this problem.

Next, introducing convergence-generating factors into (2.2) we obtain an approximate fundamental solution $\tilde{\mathcal{R}}(x)$ in the following sense:

$$\exists K_{\pm} \in \Gamma(\bar{\Omega} \times \bar{\Omega}, E^* \boxtimes E) \text{ s.t. } P_{(2)}\tilde{\mathcal{R}}(x) = \delta_x + K_{\pm}(x, \,.\,).$$

Here, $E^* \boxtimes E$ denotes the exterior tensor product and $P_{(2)}$ indicates that P acts on the second variable. We now use the $K_{\pm}$ as integral kernels to define for any continuous section u of E^* over $\bar{\Omega}$:

$$(\mathcal{K}_{\pm} u)(x) := \int_{\bar{\Omega}} K_{\pm}(x, y) u(y)\, dV(y)\,.$$

For Ω a sufficiently small causal domain and any k, $\mathrm{id} + \mathcal{K}_{\pm} : \Gamma^k(\bar{\Omega}, E^*) \to \Gamma^k(\bar{\Omega}, E^*)$ is an isomorphism with bounded inverse given by the Neumann series

$$(\mathrm{id} + \mathcal{K}_{\pm})^{-1} = \sum_{j=0}^{\infty} (-\mathcal{K}_{\pm})^j\,.$$

Finally, for each $\varphi \in \mathcal{D}(\Omega, E^*)$, we set $F_{\pm}^{\Omega}(\,.\,)[\varphi] := (\mathrm{id} + \mathcal{K}_{\pm})^{-1}(\tilde{\mathcal{R}}_{\pm}(\,.\,)[\varphi]) \in \Gamma(E^*)$. Then

$$\Gamma(E^*) \to E_x^*$$
$$\varphi \mapsto F_{\pm}^{\Omega}(x)[\varphi]$$

is an advanced (retarded) fundamental solution for P at x. Indeed,

$$P_{(2)}F_{\pm}^{\Omega}(\,.\,)[\varphi] = F_{\pm}^{\Omega}(\,.\,)[P^*\varphi] = (\mathrm{id} + \mathcal{K}_{\pm})^{-1}(\tilde{\mathcal{R}}_{\pm}(\,.\,)[P^*\varphi])$$
$$= (\mathrm{id} + \mathcal{K}_{\pm})^{-1}(P_{(2)}\tilde{\mathcal{R}}_{\pm}(\,.\,)[\varphi]) = (\mathrm{id} + \mathcal{K}_{\pm})^{-1}(\varphi + \mathcal{K}_{\pm}\varphi) = \varphi$$

In addition, for each $\varphi \in \mathcal{D}(\Omega, E^*)$, the map $x' \mapsto F_\pm^\Omega(x')[\varphi]$ is in $\Gamma(\Omega, E^*)$. It can be shown that the approximate fundamental solution $\tilde{\mathcal{R}}_\pm(x)$ is in fact an asymptotic expansion of the true fundamental solution $F_\pm^\Omega(\,.\,)(x)$ in a suitable sense. Altogether, we obtain the following result on the solution of the inhomogeneous problem on small domains:

Theorem 2.4. *For any $x \in M$ there exists a relatively compact causal neighborhood Ω such that the following holds: given $v \in \mathcal{D}(\Omega, E)$ and defining $u_\pm$ by*

$$\langle u_\pm, \varphi \rangle := \int_\Omega F_\pm^\Omega(x)[\varphi] \cdot v(x)\, dV(x) \quad (\varphi \in \mathcal{D}(\Omega, E))$$

we have: $u_\pm \in \Gamma(\Omega, E)$, $Pu_\pm = v$, and $\mathrm{supp}(u_\pm) \subseteq J_\pm^\Omega(\mathrm{supp}(v))$.

Turning now to the global theory, the first step is to assure uniqueness of fundamental solutions. For this, we need certain restrictions on the causal structure of M:

Theorem 2.5. *Suppose that M satisfies the causality condition, that the relation $\leq$ is closed on M, and that the time separation function $\tau : M \times M \to \bar{\mathbb{R}}$ (see [28]) is finite and continuous. Let $u \in \mathcal{D}'(M, E)$ be a solution of $Pu = 0$ with future or past compact support $(\mathrm{supp}(v) \cap J_\pm^M(p)$ compact for all $p)$. Then $u = 0$.*

As an immediate corollary we obtain that under the above assumptions for each $x \in M$ there is at most one fundamental solution for P at x with past (future) compact support. The causality conditions in Theorem 2.5 are satisfied if M is globally hyperbolic. Under this assumption, by Theorem 2.1 there exists a spacelike Cauchy hypersurface S in M. We denote by $\hat{\xi}$ the future directed timelike unit normal vector field on S and consider the following Cauchy problem:

$$
\begin{aligned}
Pu &= f \quad \text{on } M \\
u &= u_0 \quad \text{on } S \\
\nabla_{\hat{\xi}} u &= u_1 \quad \text{on } S
\end{aligned}
\tag{2.3}
$$

where $f \in \mathcal{D}(M, E)$, and $u_0, u_1 \in \mathcal{D}(S, E)$. One first notes that for vanishing f, u_0, and u_1, this problem only has the trivial solution. The analytical core of this result is the observation that for each $\psi \in \Gamma(E^*)$ and each $v \in \Gamma(E)$, one has

$$\psi \cdot (Pv) - (P^*\psi) \cdot v = \mathrm{div}\,(W) \tag{2.4}$$

where the vector field W is uniquely characterized by

$$\langle W, X \rangle = (\nabla_X \psi) \cdot v - \psi \cdot (\nabla_X v) \quad (X \in \mathfrak{X}(M)).$$

This allows to control the solution of the homogeneous equation $Pu = 0$ by the Cauchy data on any Cauchy hypersurface. To prove existence of solutions one uses the above local theory to obtain solutions for f, u_0 and u_1 supported in sufficiently small causal domains. More precisely, we call a causal domain Ω an RCCSV-domain (for *relatively compact causal with small volume*) if it is relatively compact and so small that $\mathrm{vol}(\overline{\Omega}) \cdot \|K_\pm\|_{\mathcal{C}^0(\overline{\Omega} \times \overline{\Omega})} < 1$. Then we have:

Proposition 2.6. *Let Ω be an RCCSV-domain and suppose that f, u_0 and u_1 are compactly supported in Ω (resp. $\Omega \cap S$). Then the corresponding Cauchy problem in Ω is uniquely solvable.*

Combined with rather subtle causality arguments, this local result finally leads to the following main theorem on existence and uniqueness of solutions to (2.3):

Theorem 2.7. *Let E be a vector bundle over a globally hyperbolic Lorentz manifold M with a spacelike Cauchy hypersurface S, and let P be normally hyperbolic on $\Gamma(E)$. Then for each $f \in \mathcal{D}(M, E)$ and each $u_0, u_1 \in \mathcal{D}(S, E)$ there exists a unique solution $u \in \Gamma(E)$ satisfying (2.3). Moreover, $\mathrm{supp}(u) \subseteq J^M(\mathrm{supp}(f) \cup \mathrm{supp}(u_0) \cup \mathrm{supp}(u_1))$ and u depends continuously on (f, u_0, u_1).*

This result can immediately be utilized to show existence and uniqueness of fundamental solutions:

Theorem 2.8. *Under the assumptions of Theorem 2.7, for each $x \in M$ there exists a unique fundamental solution $F_+(x)$ ($F_-(x)$) for P at x with past (future) compact support. These fundamental solutions satisfy*

(i) $\mathrm{supp}(F_\pm(x)) \subseteq J_\pm^M(x)$.
(ii) $\forall \varphi \in \mathcal{D}(M, E^*)$, $x \mapsto F_\pm(x)[\varphi] \in \Gamma(E^*)$ *and* $P^*(F_\pm(\,.\,)[\varphi]) = \varphi$.

The corresponding fundamental kernels are called Green operators:

Theorem 2.9. *Under the assumptions of Theorem 2.7, there exist unique Green operators $G_\pm : \mathcal{D}(M, E) \to \Gamma(M, E)$ satisfying*

(i) $P \circ G_\pm = \mathrm{id}_{\mathcal{D}(M,E)}$
(ii) $G_\pm \circ P|_{\mathcal{D}(M,E)} = \mathrm{id}_{\mathcal{D}(M,E)}$
(iii) $\forall \varphi \in \mathcal{D}(M, E) : \mathrm{supp}(G_\pm \varphi) \subseteq J_\pm^M(\mathrm{supp}(\varphi))$

In fact, $(G_\pm \varphi)(x) = F_\mp(x)[\varphi]$. Moreover, denoting by $G_\pm^*$ the Green operators for P^*,

$$\int_M (G_\pm^* \varphi) \cdot \psi \, dV = \int_M \varphi \cdot (G_\mp \psi) \, dV \qquad (\varphi \in \mathcal{D}(M, E^*),\ \psi \in \mathcal{D}(M, E)).$$

3. Wave equations on smooth Lorentzian manifolds: the case of distributional data

In this section we will be concerned with the global Cauchy problem in the case where the metric is still smooth but the data and right-hand side are distributional. To keep the presentation simple we will restrict our attention to the wave operator.

Based on Theorem 2.1, we will assume that $M = \mathbb{R} \times S$ and the Lorentz metric on M is of the form $\lambda = -\beta dt^2 + g_t$. We denote by $\Box$ the d'Alembertian w.r.t. λ. The Cauchy problem we are considering can then be written as

$$\Box u = f \quad \text{on } M$$
$$u(0, \cdot) = u_0 \qquad\qquad (3.1)$$
$$\nabla_{\hat\xi} u(0, \cdot) = u_1$$

In the present setting, $\hat\xi = \frac{1}{\sqrt{\beta}}\partial_t$. To make this initial value problem meaningful in the distributional setting we suppose that f is smooth in the t-variable. More precisely, we assume that $f \in \mathcal{C}^\infty(\mathbb{R}, \mathcal{E}'(S)) \cap \mathcal{E}'(\mathbb{R} \times S)$. It then follows from non-characteristic regularity ([22, Th. 8.3.1]) that any $u \in \mathcal{D}'(M)$ with $\Box u = f$ has $\pm(1,0)$ not in the wave front set $\mathrm{WF}(u)|_{(t,x)}$ for all (t,x). Thus by [12, 23.65.5], $u \in \mathcal{C}^\infty(\mathbb{R}, \mathcal{D}'(S))$, so the initial value problem (3.1) indeed makes sense for u_0, $u_1 \in \mathcal{E}'(S)$.

To our knowledge, (3.1) has not been treated in full generality in the literature so far for f, u_0, u_1 as specified above. We therefore supply the necessary arguments.

Lemma 3.1. (*Uniqueness*) *There is at most one solution $u \in \mathcal{C}^\infty(\mathbb{R}, \mathcal{D}'(S))$ of* (3.1).

Proof. We first note that $\Box$ is strictly hyperbolic with respect to the level sets of the map $T : \mathbb{R} \times S \to \mathbb{R}$, $(t,x) \mapsto t$ (in the sense of [21, Def. 23.2.3]). In fact, for the principal symbol σ of $\Box$ we have

$$\sigma|_{(t,x)}(\tau, \xi) = \frac{1}{\beta(t,x)}\tau^2 - h_t^{-1}(x)(\xi, \xi)\,.$$

Thus fixing $(t,x) \in M$ and $\xi \in T_x^* S$, $\xi \neq 0$, the polynomial $p(z) := \sigma|_{(t,x)}(z, \xi)$ has the distinct real zeros $\pm\sqrt{\beta(t,x)h_t^{-1}(\xi, \xi)}$.

Suppose now that u and $\tilde{u}$ are solutions of (3.1) and set $w := u - \tilde{u}$. By strict hyperbolicity it follows that there exists some neighborhood V of S such that $w|_V = 0$ (see [12, 23.72.8]). Furthermore, by [21, Th. 23.2.9], $\mathrm{WF}(w) \subseteq \mathrm{Char}(\Box)$ and is invariant under the Hamiltonian flow of σ. Now $\mathrm{Char}(\Box)$ consists entirely of lightlike directions. Thus the projection of any (maximal) bicharacteristic onto M is an inextendible null-geodesic, hence intersects the Cauchy surface S. Since the wavefront set of w is empty in V and is transported along the bicharacteristics it therefore must be empty everywhere on M. Hence $w \in \mathcal{C}^\infty(M)$, and $w = 0$ follows from the uniqueness part of Theorem 2.7. $\qquad\square$

Turning now to the problem of existence, it clearly suffices to treat the following special cases of (3.1): on the one hand, the homogeneous problem ($f = 0$), which we denote by (CP1), and on the other hand the inhomogeneous problem with vanishing initial data, called (CP2).

Turning first to (CP2), from Theorem 2.9 it is straightforward to conclude that the Green operators $G_\pm : \mathcal{D}(M) \to \mathcal{C}^\infty(M)$ continuously extend to operators from $\mathcal{E}'(M) \to \mathcal{D}'(M)$ as transposed operators of $G_\mp^*$ (see [34, Th. 4.3.10]). Thus given $f \in \mathcal{C}^\infty(\mathbb{R}, \mathcal{E}'(S)) \cap \mathcal{E}'(\mathbb{R} \times S)$ we may set $w := G_+ f$ to obtain $\Box w = f$.

Our task is thereby reduced to proving solvability of (CP1) since adding the solution of (CP1) with $u_0 = -w(0, .)$ and $u_1 = -\nabla_{\hat{\xi}} w(0, .)$ we obtain the desired solution of (CP2) (note that since $\text{supp}(G_+ f) \subseteq J_+^M(\text{supp } f)$ by [34, (4.3.20)] both u_0 and u_1 are in $\mathcal{E}'(S)$ by [1, Cor. A.5.4]).

To obtain a solution of (CP1) we first observe the following consequence of (2.4), (cf. [34, Th. 4.3.20]): Denote by $G^* := G_+^* - G_-^*$ the *propagator* of the transposed operator $\Box^*$. Then any smooth solution u of the homogeneous equation $\Box u = 0$ satisfies for all $\varphi \in \mathcal{D}(M)$

$$\int_M \varphi \cdot u \, dV = \int_S (\nabla_{\hat{\xi}} G^*(\varphi)) \cdot u_0 - G^*(\varphi) \cdot u_1 \, dA \qquad (3.2)$$

where $u_0 = u|_S$, $u_1 = \nabla_{\hat{\xi}} u|_S$, and dA is the Riemannian surface element of the spacelike surface S (i.e., the Riemannian density w.r.t. g_0 in our case). For the distributional Cauchy problem (CP1), we take (3.2) as a starting point and for given $u_0, u_1 \in \mathcal{E}'(S)$ define $L(u_0, u_1) \in \mathcal{D}'(M)$ by

$$\langle L(u_0, u_1), \varphi \rangle := \langle u_0, (\nabla_{\hat{\xi}} G^*(\varphi))|_S \rangle - \langle u_1, G^*(\varphi)|_S \rangle .$$

In case u_0 and u_1 are the given Cauchy data $u := L(u_0, u_1)$ will be the desired solution to (CP1). We start establishing this fact by first deriving an explicit formula for $L(u_0, u_1)$ well suited to the (t, x)-splitting. This naturally has to involve the Green operators. First observe that by Theorem 2.9 (ii) we have $\Box L(u_0, u_1) = 0$, so the argument preceding Lemma 3.1 shows that in fact $L(u_0, u_1) \in \mathcal{C}^\infty(\mathbb{R}, \mathcal{D}'(S))$. Hence for $\varphi_0 \in \mathcal{D}(\mathbb{R})$, $\varphi_1 \in \mathcal{D}(S)$ we may write

$$\langle L(u_0, u_1), \varphi_0 \otimes \varphi_1 \rangle = \int_{\mathbb{R}} \varphi_0(t) \langle L(u_0, u_1)(t), \varphi_1 \rangle \, dt. \qquad (3.3)$$

Denote by $F_\pm^*(s, x)$ the fundamental solutions of $\Box^*$ at (s, x) according to Theorem 2.8. Setting $F_{s,x}^* := F_-^*(s, x) - F_+^*(s, x)$, we have $(G^* \varphi)(s, x) = \langle F_{s,x}^*, \varphi \rangle$ for all $(s, x) \in M$ and all $\varphi \in \mathcal{D}(M)$. Furthermore, $\Box F_{(s,x)} = \delta_{s,x} - \delta_{s,x} = 0$, so $F_{s,x}^* \in \mathcal{C}^\infty(\mathbb{R}, \mathcal{D}'(S))$ for each $(s, x) \in M$. Thus for each $t \in \mathbb{R}$ and each $\psi \in \mathcal{D}(S)$ we obtain

$$\langle L(u_0, u_1)(t), \psi \rangle = \langle u_0, \nabla_{\hat{\xi}} \langle F_{s,x}^*(t), \psi \rangle|_{s=0} \rangle - \langle u_1, \langle F_{0,x}^*(t), \psi \rangle \rangle . \qquad (3.4)$$

We are now in the position to show that u indeed attains the Cauchy data. To this end choose sequences $u_0^{(j)}$, $u_1^{(j)}$ in $\mathcal{D}(S)$ that converge to u_0 resp. u_1 in $\mathcal{E}'(S)$. By Theorem 2.7, for each j there exists a unique $u^{(j)} \in \mathcal{C}^\infty(M)$ such that

$$\Box u^{(j)} = 0, \ u^{(j)}(0) = u_0^{(j)}, \ \nabla_{\hat{\xi}} u^{(j)}(0) = u_1^{(j)}.$$

In addition, again by [34, Th. 4.3.20] we have $u^{(j)} = L(u_0^{(j)}, u_1^{(j)})$. Moreover, by (3.4) we obtain for all $\psi \in \mathcal{D}(S)$ and all $t \in \mathbb{R}$

$$\langle u^{(j)}(t), \psi \rangle = \langle L(u_0^{(j)}, u_1^{(j)})(t), \psi \rangle \to \langle L(u_0, u_1)(t), \psi \rangle = \langle u(t), \psi \rangle , \qquad (3.5)$$

i.e., for all t, $u(t) = \lim u^{(j)}(t)$ in $\mathcal{D}'(S)$. In particular, $u(0) = \lim u^{(j)}(0) = \lim u_0^{(j)} = u_0$, thereby verifying the first initial condition for u. To show that $\nabla_{\hat{\xi}} u(0) = u_1$ we first observe that by (3.4) we have for all $\psi \in \mathcal{D}(S)$ and all $t \in \mathbb{R}$

$$\langle \partial_t u^{(j)}(t), \psi \rangle = \langle L(u_0^{(j)}, u_1^{(j)})'(t), \psi \rangle$$
$$\to \langle L(u_0, u_1)'(t), \psi \rangle = \langle \partial_t u(t), \psi \rangle.$$

Now since for all $t \in \mathbb{R}$ we have $\nabla_{\hat{\xi}} u(t) = 1/\sqrt{\beta(t,.)}\, \partial_t u(t)$ we obtain

$$\nabla_{\hat{\xi}} u(0) = \frac{1}{\sqrt{\beta(0,.)}}\, \partial_t u(0) = \frac{1}{\sqrt{\beta(0,.)}} \lim \partial_t u^{(j)}(0)$$
$$= \lim \nabla_{\hat{\xi}} u^{(j)}(0) = \lim u_1^{(j)} = u_1,$$

thereby also verifying the second initial condition.

Finally, we demonstrate that the support of the unique solution of (3.1) satisfies the same inclusion relation as in the smooth case (Th. 2.7). To see this, we first note that for any $u \in \mathcal{C}^\infty(M)$ such that $\text{supp}(\Box u)$ is compact we have the following generalization of (3.2):

$$\int_M \varphi \cdot u \, dV = \int_M \Box u (G_+^* + G_-^*)(\varphi) \, dV + \int_S (\nabla_{\hat{\xi}} G^*(\varphi)) \cdot u_0 - G^*(\varphi) \cdot u_1 \, dA \quad (3.6)$$

(this follows by adapting the proof of [1, Lemma 3.2.2]). Now suppose that u is the unique solution of (3.1) and pick a sequence $u_m \in \mathcal{D}(M)$ such that $u_m(t) \to u(t)$ for each $t \in \mathbb{R}$. For any $\varphi \in \mathcal{D}(M)$ and each $t \in \mathbb{R}$, $\text{supp}((G_+^* + G_-^*)\varphi(t, .))$ is compact, so

$$\langle \Box u_m, (G_+^* + G_-^*)\varphi \rangle = \int_\mathbb{R} \langle \Box u_m(t), (G_+^* + G_-^*)\varphi(t, .) \rangle \, dt$$
$$\to \int_\mathbb{R} \langle \Box u(t), (G_+^* + G_-^*)\varphi(t, .) \rangle \, dt = \langle \Box u, (G_+^* + G_-^*)\varphi \rangle$$

Thus applying (3.6) to each u_m and letting $m \to \infty$ we obtain

$$\langle u, \varphi \rangle = \langle f, (G_+^* + G_-^*)\varphi \rangle + \langle u_0, (\nabla_{\hat{\xi}} G^*(\varphi))|_S \rangle - \langle u_1, G^*(\varphi)|_S \rangle.$$

From this and Theorem 2.9 (iii), the claimed support properties of u follow. Summing up, we have proved:

Theorem 3.2. *Given $f \in \mathcal{C}^\infty(\mathbb{R}, \mathcal{E}'(S)) \cap \mathcal{E}'(\mathbb{R} \times S)$ and $u_0, u_1 \in \mathcal{E}'(S)$ there exists a unique solution $u \in \mathcal{C}^\infty(\mathbb{R}, \mathcal{D}'(S))$ of the Cauchy problem (3.1). Moreover, $\text{supp}(u) \subseteq J^M(\text{supp}(f) \cup \text{supp}(u_0) \cup \text{supp}(u_1))$.*

4. Generalized global analysis

Colombeau algebras of generalized functions ([9, 10]) are differential algebras which contain the vector space of distributions and display maximal consistency with classical analysis in the light of the Schwartz impossibility result ([30]). Here we review global analysis based on the special Colombeau algebra $\mathcal{G}(M)$, for further details see [11, 24, 25] and [18, Sec. 3.2]. The basic idea of its construction is regularization of distributions by nets of smooth functions and the use of asymptotic estimates in terms of a regularization parameter.

Let M be a smooth, second countable Hausdorff manifold. Set $I = (0,1]$ and denote by $\mathcal{E}(M)$ the subset of $C^\infty(M)^I$ consisting of all nets depending smoothly on the parameter $\varepsilon \in I$. The algebra of generalized functions on M ([11]) is defined as the quotient $\mathcal{G}(M) := \mathcal{E}_M(M)/\mathcal{N}(M)$ of moderate modulo negligible elements of $\mathcal{E}(M)$, where the respective notions are defined by the following asymptotic estimates. Here $\mathcal{P}$ denotes the space of all linear differential operators on M.

$$\mathcal{E}_M(M) := \{(u_\varepsilon)_\varepsilon \in \mathcal{E}(M) : \forall K \subset\subset M \; \forall P \in \mathcal{P} \; \exists N : \sup_{p \in K} |Pu_\varepsilon(p)| = O(\varepsilon^{-N})\}$$

$$\mathcal{N}(M) := \{(u_\varepsilon)_\varepsilon \in \mathcal{E}(M) : \forall K \subset\subset M \; \forall P \in \mathcal{P} \; \forall m : \sup_{p \in K} |Pu_\varepsilon(p)| = O(\varepsilon^m)\}.$$

Elements of $\mathcal{G}(M)$ are denoted by $u = [(u_\varepsilon)_\varepsilon] = (u_\varepsilon)_\varepsilon + \mathcal{N}(M)$. With componentwise operations, $\mathcal{G}(M)$ is a fine sheaf of differential algebras where the derivations are Lie derivatives with respect to smooth vector fields defined by $L_X u := [(L_X u_\varepsilon)_\varepsilon]$, also denoted by $X(u)$.

There exist embeddings ι of $\mathcal{D}'(M)$ into $\mathcal{G}(M)$ that are sheaf homomorphisms and render $C^\infty(M)$ a subalgebra of $\mathcal{G}(M)$. Another, more coarse way of relating generalized functions in $\mathcal{G}(M)$ to distributions is based on the notion of association: $u \in \mathcal{G}(M)$ is called associated with $v \in \mathcal{G}(M)$, $u \approx v$, if $u_\varepsilon - v_\varepsilon \to 0$ in $\mathcal{D}'(M)$. A distribution $w \in \mathcal{D}'(M)$ is called the distributional shadow of u if $u \approx \iota(w)$.

The ring of constants in $\mathcal{G}(M)$ is the space $\tilde{\mathbb{C}}$ of generalized numbers, which form the natural space of point values of Colombeau generalized functions. These, in turn, are uniquely characterized by their values on so-called (compactly supported) generalized points (equivalence classes of bounded nets $(p_\varepsilon)_\varepsilon$ of points, where $(p_\varepsilon)_\varepsilon \sim (q_\varepsilon)_\varepsilon$ if $d(p_\varepsilon, q_\varepsilon) = O(\varepsilon^m)$ for each m).

An element $u \in \mathcal{G}(M)$ is called globally bounded, if there exists a representative $(u_\varepsilon)_\varepsilon$ and a $C > 0$ such that $|u_\varepsilon(x)| \leq C$ for all $x \in M$ and all $\varepsilon \in I$.

A similar construction is in fact possible for any locally convex space F: an analogous quotient construction in terms of the seminorms on F allows to assign a $\tilde{\mathbb{C}}$-module $\mathcal{G}_F$ to F in a natural way ([14]). A particularly important special case is obtained for $F = \Gamma(M, E)$, the space of smooth sections of a vector bundle $E \to M$ (again with representatives that are supposed to depend smoothly on ε). The resulting space $\Gamma_{\mathcal{G}}(M, E) := \mathcal{G}_F$ then is a $\mathcal{G}(M)$-module, called the space of generalized sections of the vector bundle E. A convenient algebraic description is as follows:

$$\Gamma_{\mathcal{G}}(M, E) = \mathcal{G}(M) \otimes_{C^\infty(M)} \Gamma(M, E) = L_{C^\infty(M)}(\Gamma(M, E^*), \mathcal{G}(M)). \qquad (4.1)$$

$\Gamma_{\mathcal{G}}$ is a fine sheaf of finitely generated and projective $\mathcal{G}$-modules. For the special case of generalized tensor fields of rank r, s we use the notation $\mathcal{G}_s^r(M)$. We have:

$$\mathcal{G}_s^r(M) \cong L_{\mathcal{G}(M)}(\mathcal{G}_0^1(M)^s, \mathcal{G}_1^0(M)^r; \mathcal{G}(M)).$$

Observe that this allows the insertion of generalized vector fields and one-forms into generalized tensors, a point of view which is essential when dealing with generalized metrics in the following sense.

Definition 4.1. An element g of $\mathcal{G}_2^0(M)$ is called a generalized pseudo-Riemannian metric if it is symmetric ($g(\xi, \eta) = g(\eta, \xi) \;\forall \xi, \eta \in \mathfrak{X}(M)$), its determinant $\det g$ is invertible in $\mathcal{G}$ (equivalently, for each compact subset K of a chart in M there exists some m such that $|\det(g_\varepsilon)_{ij}(p)| > \varepsilon^m$ for ε small and all $p \in K$), and it possesses a well-defined index ν (the index of g_ε equals ν for ε small).

Based on this definition, many notions from (pseudo-)Riemannian geometry can be extended to the generalized setting (cf. [25]). In particular, any generalized metric induces an isomorphism between generalized vector fields and one-forms, and there is a unique Levi-Civita connection corresponding to g. This provides a convenient framework for non-smooth pseudo-Riemannian geometry and the analysis of highly singular space-times in general relativity (see, e.g., [18, Ch. 5], [31]).

For the purposes of the present work we will also need the notion of a time-dependent generalized metric. To this end, let S be an n-dimensional smooth manifold and let $\mathrm{pr}_2 : \mathbb{R} \times S \to S$ denote the projection onto S. Also, let $\mathrm{pr}_2^*(T_2^0 S)$ be the corresponding pullback-bundle.

Definition 4.2. An element $h \in \Gamma_{\mathcal{G}}(\mathrm{pr}_2^*(T_2^0 S))$ is called t-dependent generalized pseudo-Riemannian metric if h_t possesses a well-defined index and if

(i) (Symmetry) $h_t(\xi, \eta) = h_t(\eta, \xi)$ in $\mathcal{G}(\mathbb{R} \times S)$ for all $\xi, \eta \in \mathfrak{X}(M)$.
(ii) (Non-degeneracy) $\det(h_t)$ is *strictly nonzero* in the following sense: for any $K \subset\subset \mathbb{R}$, and any L compact in some chart neighborhood on S there exists some m such that $|\det((h_\varepsilon)_t)_{ij}(x)| > \varepsilon^m$ for $(t, x) \in K \times L$ and ε small.

We conclude this section by the following globalization lemma.

Lemma 4.3. *Let $u : I \times M \to N$ be a smooth map and let (P) be a property attributable to values $u(\varepsilon, p)$ that is stable with respect to decreasing K and ε in the following sense: if $u(\varepsilon, p)$ satisfies (P) for all $p \in K \subset\subset M$ and all ε less than some $\varepsilon_K > 0$ then for any compact set $K' \subseteq K$ and any $\varepsilon_{K'} \leq \varepsilon_K$, u satisfies (P) on K' for all $\varepsilon \leq \varepsilon_{K'}$. Then there exists a smooth map $\tilde{u} : I \times M \to N$ such that (P) holds for all $\tilde{u}(\varepsilon, p)$ ($\varepsilon \in I$, $p \in M$) and for each $K \subset\subset M$ there exists some $\varepsilon_K \in I$ such that $\tilde{u}(\varepsilon, p) = u(\varepsilon, p)$ for all $(\varepsilon, p) \in (0, \varepsilon_K] \times K$.*

Proof. Let $(K_l)_l$ be a compact exhaustion of M with $K_l \subseteq K_{l+1}^\circ$ for all l and choose a smooth function $\eta : M \to \mathbb{R}$ with $0 < \eta(x) \leq \varepsilon_{K_l}$ for all $x \in K_l \setminus K_{l-1}^\circ$ ($K_0 := \emptyset$) (cf., e.g., [18, Lem. 2.7.3]). Moreover, let $\nu : \mathbb{R}_0^+ \to [0, 1]$ be a smooth function satisfying $\nu(t) \leq t$ for all t and

$$\nu(t) = \begin{cases} t & 0 \leq t \leq \frac{1}{2} \\ 1 & t \geq \frac{3}{2} \end{cases}$$

For $(\varepsilon, x) \in I \times M$ let $\mu(\varepsilon, x) := \eta(x)\nu\left(\frac{\varepsilon}{\eta(x)}\right)$ and set $\tilde{u}(\varepsilon, x) := u(\mu(\varepsilon, x), x)$. Then $\tilde{u}$ has all the required properties. $\qquad \square$

Remark 4.4. Lemma 4.3 allows to globalize properties of Colombeau-type general-ized functions provided that representatives depend smoothly on ε, as is assumed throughout this work. Indeed, if in the above situation it is additionally assumed that u is a representative of a Colombeau generalized function then by the very nature of the defining asymptotic estimates, $\tilde{u}$ is itself moderate and in fact con-stitutes a representative of the same generalized function possessing the required property (P) globally on M. In particular, any (time-dependent) generalized metric g possesses a representative $(g_\varepsilon)_\varepsilon$ such that each g_ε is a smooth (time-dependent) metric globally on M, a fact that will repeatedly be used in what is to follow.

5. Wave equations of non-smooth metrics I: The local theory

In this section we present a local existence and uniqueness result (closely related to the one in [16]) for the homogeneous wave equation of a class of generalized Lorentz metrics g, which will be extended to a global result in Section 6. We start by introducing this class of generalized weakly singular metrics.

5.1. Weakly singular Lorentzian metrics

Let g be a generalized Lorentzian metric on M. From now on we call the pair (M, g) a generalized space-time. To formulate asymptotic conditions on representatives $(g_\varepsilon)_\varepsilon$ of g let m be a background Riemannian metric on M and denote by $\| \ \|_m$ the norm induced on the fibers of the respective tensor bundle. To begin with we impose the following condition:

(A) For all compact sets K, for all orders of derivative $k \in \mathbb{N}_0$ and all k-tuples of smooth vector fields $\eta_1, \ldots, \eta_k$ and for any representative $(g_\varepsilon)_\varepsilon$ we have:

$$\sup_K \|\mathscr{L}_{\eta_1} \cdots \mathscr{L}_{\eta_k} g_\varepsilon\|_m = O(\varepsilon^{-k}) \quad (\varepsilon \to 0)$$

$$\sup_K \|\mathscr{L}_{\eta_1} \cdots \mathscr{L}_{\eta_k} g_\varepsilon^{-1}\|_m = O(\varepsilon^{-k}) \quad (\varepsilon \to 0). \tag{5.1}$$

A generalized metric with property (A) will be called a *weakly singular metric*. Note that here we use a somewhat different terminology as compared to [16].

Now with a view to formulating the local Cauchy problem of the wave op-erator for such a metric we consider a local foliation of M given by the level sets of some non-singular function $t \in C^\infty(U)$, where $U \subseteq M$ is open and relatively compact.

To exclude trivial cases we require the level sets $\Sigma_\tau = \{q \in U : t(q) = \tau\}$ to be space-like with respect to all g_ε. In fact, we suppose a uniform variant of this condition which can also be viewed as a suitable generalization of the classical notion of time-orientability. Moreover, we will need a uniform bound on the covariant derivative of the normal form which, in particular, contains a condition on the second fundamental form of the level sets. More precisely, we demand:

(B) Each $p \in M$ possesses a neighborhood U on which there exists a *local time-function*, that is a smooth function t with uniformly timelike differential $dt =: \sigma$, i.e.,

$$g_\varepsilon^{-1}(\sigma, \sigma) \le -C < 0 \quad \text{for some positive constant } C \tag{5.2}$$

and one (hence any) representative g_ε and all small ε. In addition, we have for all $K \subset\subset U$ that

$$\sup_K \|\nabla^\varepsilon \sigma\|_m = O(1) \quad (\varepsilon \to 0), \tag{5.3}$$

where ∇ is the covariant derivative of g.

Let us denote the normal vector field to Σ_τ by ξ (it is a *generalized* vector field defined via its representative $\xi_\varepsilon \in \mathfrak{X}(U)$, given by $\sigma = g_\varepsilon(\xi_\varepsilon, \cdot)$, i.e., $\xi_\varepsilon = \operatorname{grad}_\varepsilon t$). We observe that by pulling up the index, given (A), condition (B) is equivalent to $\sup_K \|\nabla^\varepsilon \xi_\varepsilon\|_m = O(1)$. Also for vector fields X, Y tangent to Σ_τ, we obtain $\nabla^\varepsilon \sigma(X, Y) = Y(\sigma(X)) - \sigma(\nabla_Y^\varepsilon X) = 0 - \sigma(\operatorname{nor}\nabla_Y^\varepsilon X)$, hence we obtain for the second fundamental form II_ε of the hypersurfaces Σ_τ

$$\|II_\varepsilon\|_m = O(1) \quad (\varepsilon \to 0) \quad \text{uniformly on compact sets.}$$

Remark 5.1. Conditions (A) and (B) are given in terms of the ε-asymptotics of the generalized metric. There is, however, the following close connection to the classical situation. Assume that we are given a space-time metric that is locally bounded but not necessarily $C^{1,1}$ or of Geroch-Traschen class (i.e., the largest class that allows a consistent distributional treatment, see [15, 26] and [32] for the relation with the present setting). We may then embed this metric into the space of generalized metrics essentially by convolution with a standard mollifier (for details again see [32]). From the explicit form of the embedding it is then clear that condition (A) holds.

Condition (B), in adapted coordinates, demands somewhat better asymptotics of the time-derivatives of the spatial part of the metric as well as the spatial derivative of the $(0,0)$-component. This condition, in fact, is satisfied by several relevant examples as well. In particular, we have:

- Conical space-times fall into our class since estimates (6) and (7) in [33] for the embedded metric imply our condition (A), while (B) is immediate from the staticity of the metric.
- Impulsive pp-waves (in "Rosen form")

$$ds^2 = -dudv + (1 + u_+)^2 \, dx^2 + (1 - u_+)^2 \, dy^2$$

as well as expanding impulsive waves with line element

$$ds^2 = 2dudv + 2v^2 \left| dz + \frac{u_+}{2v} \overline{H} d\bar{z} \right|^2$$

satisfy conditions (A) and (B). Here u_+ denotes the kink function and $H(z)$ is the Schwarzian derivative $H = h'''/(2h') - (3h''^2)/(4h'^2)$ of some analytic function $h(z)$ (which may be chosen arbitrarily). For details see [17, Ch. 20].

In both cases the metric is continuous and it will obey conditions (A) and (B) when embedded with a standard mollifier, or – more generally – if we use any regularization that converges locally uniformly to the original metric.

5.2. Local existence and uniqueness

We start by formulating the local Cauchy problem for the wave operator on weakly singular space-times. Let $p \in M$, and choose U to be an open and relatively compact neighborhood of p as in condition (B). Denote the corresponding foliation by $\Sigma_\tau = \{q \in U : t(q) = \tau\}$ ($\tau \in [-\gamma, \gamma]$) and suppose $p \in \Sigma_0 =: \Sigma$. In addition to the normal vector field ξ and the normal covector field σ whose asymptotics have already been discussed above we will need their corresponding normalized versions $\widehat{\xi} = [(\widehat{\xi}_\varepsilon)_\varepsilon] = [(\xi_\varepsilon/V_\varepsilon)_\varepsilon]$ and $\widehat{\sigma} = [(\widehat{\sigma}_\varepsilon)_\varepsilon] = g(\widehat{\xi}, \cdot)$, where we have set $V_\varepsilon^2 = -g_\varepsilon(\xi_\varepsilon, \xi_\varepsilon)$.

We are interested in the initial value problem

$$\begin{aligned} \Box u &= 0 &&\text{on } U \\ u &= u_0 &&\text{on } \Sigma \\ \nabla_{\widehat{\xi}} u &= u_1 &&\text{on } \Sigma, \end{aligned} \qquad (5.4)$$

where the initial data u_0, u_1 are supposed to be in $\mathcal{G}(\Sigma)$. Note that this, in particular, includes the case of distributional initial data. We are interested in finding a local solution $u \in \mathcal{G}$ on U or an open subset thereof.

A general strategy to solve PDEs in generalized functions is the following. First, solve the equation for fixed ε in the smooth setting and form the net $(u_\varepsilon)_\varepsilon$ of smooth solutions. This will be a candidate for a solution in $\mathcal{G}$, but particular care has to be taken to guarantee that the u_ε share a common domain of definition and depend smoothly on ε. In fact, as has recently been shown in [23] it suffices to verify continuous dependence on ε. In the second step, one shows that the solution candidate $(u_\varepsilon)_\varepsilon$ is a moderate net, hence obtaining existence of a solution $[(u_\varepsilon)_\varepsilon]$ in $\mathcal{G}$. Finally, to obtain uniqueness of solutions, one has to prove that changing representatives of the data leads to a solution that is still in the class $[(u_\varepsilon)_\varepsilon]$. Note that this amounts to an additional stability of the equation with respect to negligible perturbations of the initial data.

So, in the present situation we need a condition which provides us with the existence of a solution candidate:

(C) For each $p \in \Sigma$ there exists a neighborhood $V \subseteq U$ and a representative $(g_\varepsilon)_\varepsilon$ of the metric g on V such that V is, for each ε, an RCCSV-neighborhood in (M, g_ε) with $\Sigma \cap V$ a spacelike Cauchy hypersurface for V.

Indeed Proposition 2.6 now provides us with a solution candidate defined on V, that is a net (u_ε) with $\Box_\varepsilon u_\varepsilon = f_\varepsilon$ on V for some negligible net (f_ε) and moreover $u_\varepsilon|_{\Sigma \cap V} = u_{0\varepsilon}$, $\nabla_{\widehat{\xi}} u|_{\Sigma \cap V} = u_{1\varepsilon}$ for some representatives $(u_{0\varepsilon})$ $(u_{1\varepsilon})$ of the data. We note that continuous dependence of u on ε follows readily from the construction steps detailed in Section 2. Observe that the only part that exceeds the classical condition for existence and uniqueness of solutions is a certain uniformity in ε.

Heuristically this means that the light-cones of the metric g_ε neither vary to wildly with ε nor collapse as $\varepsilon \to 0$. In terms of regularizations of classical metrics which are locally bounded but not necessarily of $C^{1,1}$ or of Geroch-Traschen class, this condition will always be satisfied due to non-degeneracy of the classical metric.

Now we may state the main result of this section.

Theorem 5.2 (Local existence and uniqueness of generalized solutions). *Let (M, g) be a generalized space-time with a weakly singular metric and assume that conditions* (B) *and* (C) *hold. Then, for each $p \in \Sigma$ there exists an open neighborhood Ω such that for all compactly supported $u_0, u_1 \in \mathcal{G}(\Sigma \cap \Omega)$, the initial value problem* (5.4) *has a unique solution u in $\mathcal{G}(\Omega)$.*

The core of the proof of Theorem 5.2 consists of higher-order energy estimates for the solution candidate whose existence is secured by condition (C). The energy estimates, which rely on conditions (A) and (B), will be carried out in a geometric setting using ε-dependent energy momentum tensors and ε-dependent Sobolev norms. These notions have been introduced in [33] and a suitable generalization of this method will be presented in the next section.

5.3. Higher-order energy estimates

Let $U \supseteq V$ be the neighborhoods of p given by conditions (B) and (C). The solution candidate $(u_\varepsilon)_\varepsilon$ is defined on V and we are going to estimate u_ε on some suitable neighborhood of p.

We start by introducing some notation. Let Ω be an open neighborhood of p with the property that $\overline{\Omega} \subseteq V$, and such that the boundary of the region $\Omega \cap \{q \in U : t(q) \geq 0\}$ and the boundary of the region $\Omega \cap \{q \in U : t(q) \leq 0\}$ is space-like w.r.t. all g_ε. Indeed, such a neighborhood exists by condition (A). We now concentrate on the forward-in-time part of Ω (i.e., the part where $t \geq 0$) since the backward-in-time part can be dealt with analogously. To this end we set $S_\tau := \Sigma_\tau \cap \Omega$ and denote by Ω_τ the open part of Ω between Σ and Σ_τ. We denote the part of the boundary of Ω_τ with $0 < t < \tau$ by $S_{\Omega,\tau}$, so that $\partial \Omega_\tau = S_0 \cup S_\tau \cup S_{\Omega,\tau}$ (see Figure 1).

From now on we will adopt abstract index notation for (generalized) tensorial objects (see, e.g., [29]). In particular, representatives of the metric g_ε and its inverse will be denoted by g^ε_{ab} and g_ε^{ab}, respectively. In addition, to simplify the notation for tensors we are going to use capital letters to abbreviate tuples of indices, i.e., we will write T^I_J for $T^{p_1 \cdots p_r}_{q_1 \cdots q_s}$ with $|I| = r$, $|J| = s$.

We now define a generalized Riemannian metric by

$$e := g + 2\hat{\sigma} \otimes \hat{\sigma},$$

and use it in combination with the covariant derivative ∇ of g to define "ε-dependent" Sobolev norms and energies on U. Observe that by conditions (A) and (B) we have $\|e_\varepsilon\|_m = O(1)$ and $\|\nabla_\varepsilon e_\varepsilon\|_m = O(1)$ on compact sets. We will also frequently need tensor products of e and use the notation $e_{IJ} = e_{p_1 q_1} \cdots e_{p_r q_r}$ with $|I| = r = |J|$.

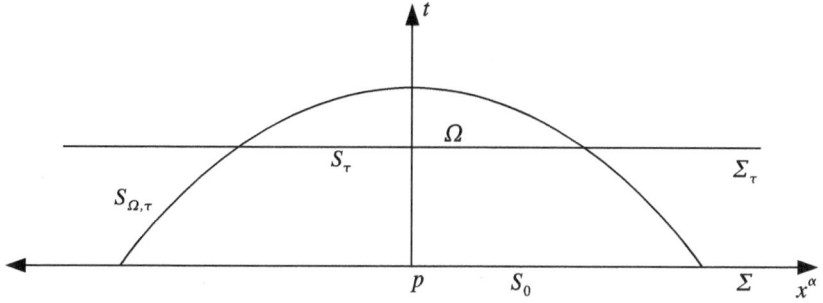

FIGURE 1. Local foliation of space-time

Definition 5.3 (Sobolev norms and energies). Let T_J^I be a smooth tensor field and u a smooth function on U, $0 \leq \tau \leq \gamma$, and $k, j \in \mathbb{N}_0$.

(i) We define the pointwise norm of T_J^I by $\|T_J^I\|_{e_\varepsilon}^2 := e_{KL}^\varepsilon e_\varepsilon^{IJ} T_I^K T_J^L$

   and the higher-order pointwise norm of u by $|\nabla_\varepsilon^{(j)} u|^2 := \|\nabla_{p_1}^\varepsilon \ldots \nabla_{p_j}^\varepsilon u\|_{e_\varepsilon}^2$.

(ii) On Ω_τ and S_τ we define the Sobolev norms

$$^\nabla \|u\|_{\Omega_\tau, \varepsilon}^k := \left(\sum_{j=0}^k \int_{\Omega_\tau} |\nabla_\varepsilon^{(j)}(u)|^2 \mu^\varepsilon \right)^{\frac{1}{2}},$$

and

$$^\nabla \|u\|_{S_\tau, \varepsilon}^k := \left(\sum_{j=0}^k \int_{S_\tau} |\nabla_\varepsilon^{(j)}(u)|^2 \mu_\tau^\varepsilon \right)^{\frac{1}{2}}$$

respectively. Here $\mu = [(\mu^\varepsilon)_\varepsilon]$ and $\mu_\tau = [(\mu_\tau^\varepsilon)_\varepsilon]$ denote the respective volume forms on Ω_τ and S_τ derived from g. Note that although in the second norm the integration is performed over the three-dimensional manifold S_τ only, derivatives are not confined to directions tangential to S_τ.

(iii) On Ω we define the energy momentum tensors by ($k > 0$)

$$T_\varepsilon^{ab,0}(u) := -\frac{1}{2} g_\varepsilon^{ab} u^2,$$

$$T_\varepsilon^{ab,k}(u) := \left(g_\varepsilon^{ac} g_\varepsilon^{bd} - \frac{1}{2} g_\varepsilon^{ab} g_\varepsilon^{cd} \right) e_\varepsilon^{p_1 q_1} \ldots e_\varepsilon^{p_{k-1} q_{k-1}}$$
$$\times (\nabla_c^\varepsilon \nabla_{p_1}^\varepsilon \ldots \nabla_{p_{k-1}}^\varepsilon u)(\nabla_d^\varepsilon \nabla_{q_1}^\varepsilon \ldots \nabla_{q_{k-1}}^\varepsilon u),$$

(iv) Finally, the energy integrals are defined by

$$E_{\tau, \varepsilon}^k(u) := \sum_{j=0}^k \int_{S_\tau} T_\varepsilon^{ab,j}(u) \xi_a \xi_b V_\varepsilon^{-1} \mu_\tau^\varepsilon. \tag{5.5}$$

A straightforward calculation shows that the tensor fields $T_\varepsilon^{ab,k}(u)$ satisfy the dominant energy condition hence an application of Stokes' theorem yields the basic energy estimate (see, e.g., [20, Sec. 4.3])

$$E_{\tau,\varepsilon}^k(u) \le E_{\tau=0,\varepsilon}^k(u) + \sum_{j=0}^k \int_{\Omega_\tau} \left(\xi_b \nabla_a^\varepsilon T_\varepsilon^{ab,j}(u) + T_\varepsilon^{ab,j}(u)\nabla_a^\varepsilon \xi_b\right)\mu_\varepsilon. \tag{5.6}$$

One key estimate in our approach is the equivalence of Sobolev norms and energies. Indeed using condition (A) and the estimate (5.2) one may derive ([16, Lemma 4.1]):

Lemma 5.4 (Energy integrals and Sobolev norms). *There exist constants C, C' such that for each $k \ge 0$ and all ε small*

$$C(^\nabla\|u\|_{S_\tau,\varepsilon}^k)^2 \le E_{\tau,\varepsilon}^k(u) \le C'(^\nabla\|u\|_{S_\tau,\varepsilon}^k)^2. \tag{5.7}$$

With this tool at hand we may derive the core estimate allowing to prove existence and uniqueness of solutions.

Proposition 5.5. *Let (u_ε) be a solution candidate on V. Then, for each $k \ge 1$, there exist positive constants C_k', C_k'', C_k''' such that for all $0 \le \tau \le \gamma$ we have*

$$E_{\tau,\varepsilon}^k(u_\varepsilon) \le E_{0,\varepsilon}^k(u_\varepsilon) + C_k'(^\nabla\|f_\varepsilon\|_{\Omega_\tau,\varepsilon}^{k-1})^2 + C_k'' \sum_{j=1}^{k-1} \frac{1}{\varepsilon^{2(1+k-j)}} \int_0^\tau E_{\zeta,\varepsilon}^j(u_\varepsilon)d\zeta$$

$$+ C_k''' \int_0^\tau E_{\zeta,\varepsilon}^k(u_\varepsilon)d\zeta. \tag{5.8}$$

Before sketching the proof of this statement, we draw the essential conclusions from it. Observe that the constant in front of the highest-order term on the r.h.s. does not depend on ε, hence we obtain, by an application of Gronwall's lemma:

Corollary 5.6. *Let (u_ε) be a solution candidate on V. Then, for each $k \ge 1$, there exist positive constants C_k', C_k'', C_k''' such that for all $0 \le \tau \le \gamma$,*

$$E_{\tau,\varepsilon}^k(u_\varepsilon) \le \left(E_{0,\varepsilon}^k(u_\varepsilon) + C_k'(^\nabla\|f_\varepsilon\|_{\Omega_\tau,\varepsilon}^{k-1})^2 + C_k'' \sum_{j=1}^{k-1} \frac{1}{\varepsilon^{2(1+k-j)}} \int_{\zeta=0}^\tau E_{\zeta,\varepsilon}^j(u_\varepsilon)d\zeta\right) e^{C_k'''\tau} \tag{5.9}$$

Consequently, if the initial energy $(E_{0,\varepsilon}^k(u_\varepsilon))_\varepsilon$ is a moderate (resp. negligible) net of real numbers, and $(f_\varepsilon)_\varepsilon$ is negligible then

$$\sup_{0\le\tau\le\gamma}(E_{\tau,\varepsilon}^k(u_\varepsilon))_\varepsilon$$

is moderate (resp. negligible).

Now we *sketch the proof of Proposition* 5.5: We have to estimate the right-hand side of the basic energy estimate (5.6). Starting with the second term under the integral we use condition (B) to obtain

$$\left|T_\varepsilon^{ab,j}(u_\varepsilon)\nabla_a^\varepsilon \xi_b\right| \le \|T_\varepsilon^{ab,j}(u_\varepsilon)\|_{e_\varepsilon} \|\nabla_{(a}^\varepsilon \xi_{b)}\|_{e_\varepsilon} \le C\|T_\varepsilon^{ab,j}(u_\varepsilon)\|_{e_\varepsilon}. \tag{5.10}$$

Observe that by condition (A), the Riemannian metric e is $O(1)$ hence $\| \ \|_m$ and $\| \ \|_{e_\varepsilon}$ are equivalent norms. We now may estimate $\|T_\varepsilon^{ab,j}(u_\varepsilon)\|_{e_\varepsilon}$ by the higher-order pointwise norm of u. After integration this gives

$$\left| \sum_{j=0}^k \int_{\Omega_\tau} T_\varepsilon^{ab,j}(u_\varepsilon) \nabla_a^\varepsilon \xi_b \mu_\varepsilon \right| \leq C \left({}^\nabla\|u_\varepsilon\|_{\Omega_\tau,\varepsilon}^k \right)^2. \tag{5.11}$$

Now turning to the divergence term on the right-hand side of (5.6) we start with orders $k = 0, 1$. Using the wave equation we find

$$\nabla_a^\varepsilon T_\varepsilon^{ab,0}(u_\varepsilon) = -u_\varepsilon \nabla_\varepsilon^b u_\varepsilon \quad \text{and} \quad \nabla_a^\varepsilon T_\varepsilon^{ab,1}(u_\varepsilon) = f_\varepsilon \nabla_\varepsilon^b u_\varepsilon,$$

which after integration clearly can be estimated by the squares of ${}^\nabla\|u_\varepsilon\|_{\Omega_\tau,\varepsilon}^1$ and ${}^\nabla\|f_\varepsilon\|_{\Omega_\tau,\varepsilon}^0$. Inserting this and (5.11) for $k = 1$ into (5.6) we obtain

$$E_{\tau,\varepsilon}^1(u_\varepsilon) \leq E_{0,\varepsilon}^1(u_\varepsilon) + C \left({}^\nabla\|f_\varepsilon\|_{\Omega_\tau,\varepsilon}^0 \right)^2 + C \left({}^\nabla\|u_\varepsilon\|_{\Omega_\tau,\varepsilon}^1 \right)^2.$$

Next we use Lemma 5.4 to estimate the Sobolev norm of u_ε in terms of its energy, i.e.,

$$\left({}^\nabla\|u_\varepsilon\|_{\Omega_\tau,\varepsilon}^1 \right)^2 = \int_0^\tau \left({}^\nabla\|u_\varepsilon\|_{S_\zeta,\varepsilon}^1 \right)^2 d\zeta \leq C \int_0^\tau E_{\zeta,\varepsilon}^1(u_\varepsilon) d\zeta,$$

which gives the claim for $k = 1$ (with $C_1'' = 0$).

Finally, one has to estimate the divergence of the higher (i.e., $k > 1$) order energy momentum tensors. The general strategy is, of course, to rewrite terms containing the $(k+1)^{\text{st}}$ order derivative of u_ε using the wave equation. This necessitates interchanging the order of covariant derivatives, which introduces additional curvature terms. These can be estimated using condition (A). Observe, however, that there also appear terms where the covariant derivatives falls on e. These terms of the form $\nabla_a^\varepsilon e_\varepsilon^{IJ}$ can be estimated thanks to condition (B). $\qquad\square$

We finally *sketch the proof of Theorem 5.2:* We have already noted the existence of a solution candidate (u_ε) on V. To prove that u_ε is moderate on Ω_γ we start from moderateness of the data u_0, u_1. Inductively using the wave equation this translates into moderateness of the initial energies $(E_{0,\varepsilon}^k(u_\varepsilon))_\varepsilon$. Now by Corollary 5.6 we obtain moderateness of the energies $(E_{\tau,\varepsilon}^k(u_\varepsilon))_\varepsilon$ for all $0 \leq \tau \leq \gamma$. Finally, we use the Sobolev embedding theorem (together with the fact the volume is $O(1)$ due to condition (A)) to estimate the sup-norm of u_ε in terms of the Sobolev norms, which in turn can be bounded by the energies $(E_{\tau,\varepsilon}^k(u_\varepsilon))_\varepsilon$ due to Lemma 5.4 (for details see [16, Lemma 6.2]). So we see that moderateness of the energies implies moderateness of (u_ε) and we have proved existence of solutions.

Uniqueness follows along the same lines replacing moderateness by negligibility. $\qquad\square$

6. Wave equations of non-smooth metrics II: The global theory

We extend the results of Theorem 5.2 to establish existence and uniqueness of global generalized solutions. As in the classical situation we have to impose additional global conditions on the generalized Lorentzian metric to control causality properties of space-time in the large. Thus, we begin by transferring the notion of global hyperbolicity to the setting of generalized space-times by appealing to a variant in terms of the metric splitting property stated in Theorem 2.1, (iii).

Definition 6.1. Let g be a generalized Lorentz metric on the smooth $(n + 1)$-dimensional manifold M. We say that (M, g) allows a *globally hyperbolic metric splitting* if there exists a $\mathcal{C}^\infty$-diffeomorphism $\psi \colon M \to \mathbb{R} \times S$, where S is an n-dimensional smooth manifold such that the following holds for the pushed forward generalized Lorentz metric $\lambda := \psi_* g$ on $\mathbb{R} \times S$:

(a) There is a representative $(\lambda_\varepsilon)_{\varepsilon \in I}$ of λ such that every λ_ε is a Lorentz metric and each slice $\{t_0\} \times S$ with arbitrary $t_0 \in \mathbb{R}$ is a (smooth, spacelike) Cauchy hypersurface for every λ_ε ($\varepsilon \in I$).
(b) We have the metric splitting of λ in the form

$$\lambda = -\beta dt^2 + h,$$

where $h \in \Gamma_{\mathcal{G}}(\mathrm{pr}_2^*(T_2^0 S))$ is a t-dependent generalized Riemannian metric (in the sense of Definition 4.2) and $\beta \in \mathcal{G}(\mathbb{R} \times S)$ is globally bounded and *locally uniformly positive*, i.e., for some (hence any) representative (β_ε) of β and for every $K \subset\subset \mathbb{R} \times S$ we can find a constant $C > 0$ such that $\beta_\varepsilon(x) \geq C$ holds for small $\varepsilon > 0$ and $x \in K$.
(c) For every $T > 0$ there exists a representative (h_ε) of h and a smooth complete Riemannian metric ρ on S which uniformly bounds h from below in the following sense: for all $t \in [-T, T]$, $x \in S$, $v \in T_x S$, and $\varepsilon \in I$

$$(h_\varepsilon)_t(v, v) \geq \rho(v, v).$$

Remark 6.2. Observe that the basic splitting structure and the requirements on lower bounds for β and h in the above definition display common features with the notion of regularly sliced space-times ([5, Ch. XII, Subsec. 11.4]) which provide sufficient conditions for global hyperbolicity in the smooth case ([6]).

Example. To obtain simple non-trivial examples of generalized space-times satisfying the conditions of Definition 6.1 we consider Robertson-Walker space-times. First, we briefly recall the classical situation: Let (S, h_0) be a connected Riemannian manifold, $f \colon \mathbb{R} \to \,]0, \infty[$ be smooth, and put $\lambda_{(t,x)} = -dt^2 + f(t)^2 (h_0)_x$ for every $(t, x) \in \mathbb{R} \times S$. Then the Lorentzian metric λ on $\mathbb{R} \times S$ is globally hyperbolic if and only if (S, h_0) is complete (cf. [1, Lemma A.5.14]). Moreover, if this is the case, then every slice $\{t_0\} \times S$ is a smooth, spacelike Cauchy hypersurface.

We generalize the Robertson-Walker space-time by allowing as warping function any $f \in \mathcal{G}(\mathbb{R})$ that is globally bounded and locally uniformly positive and replace h_0 by a generalized Riemannian metric on S which is bounded below by some smooth complete Riemannian metric (to guarantee condition (c)). By Lemma 4.3 on globalization techniques we may pick representatives (f_ε) of f and $(h_{0\varepsilon})$ of h_0 such that the smooth function f_ε is everywhere positive and $h_{0\varepsilon}$ is a Riemannian metric on S for every $\varepsilon \in I$. In addition, we assume that each $h_{0\varepsilon}$ is complete. The generalized Lorentz metric $\lambda := -dt^2 + f^2 h_0$ on $\mathbb{R} \times S$ then trivially satisfies condition (b). Putting $\lambda_\varepsilon := -dt^2 + f_\varepsilon^2 h_{0\varepsilon}$ ($\varepsilon \in I$) we obtain a representative of λ. By completeness of the Riemann metric $h_{0\varepsilon}$, the Lorentz metric λ_ε is globally hyperbolic and every slice $\{t_0\} \times S$ is a Cauchy hypersurface for every $\varepsilon \in I$. Thus condition (a) in Definition 6.1 is also satisfied.

From now on we consider only generalized space-times (M, g) which possess a globally hyperbolic metric splitting. To simplify notation we will henceforth suppress the diffeomorphism providing the splitting and assume that $M = \mathbb{R} \times S$ and $g = \lambda$ with S and λ as in the statement of Definition 6.1. Thus, the generalized space-time is represented by a family of globally hyperbolic space-times (M, g_ε) such that $S \cong \{0\} \times S$ is a Cauchy hypersurface for every g_ε ($\varepsilon \in I$).

Therefore we are provided with a suitable Cauchy hypersurface for the initial value problem for the wave operator $\Box$ corresponding to the generalized space-time metric g on M, i.e., the Cauchy problem

$$\Box u = 0 \quad \text{on } M$$
$$u = u_0 \quad \text{on } S \tag{6.1}$$
$$\nabla_{\widehat{\xi}} u = u_1 \quad \text{on } S.$$

Here the unit normal vector field of S is given by $\widehat{\xi} = \frac{1}{\sqrt{\beta}}\partial_t$ and the initial data u_0, u_1 are assumed to belong to $\mathcal{G}(S)$ and to have compact supports, e.g., arising by embedding distributional data from $\mathcal{E}'(S)$.

Now the key strategy to establish a global version of Theorem 5.2 on existence and uniqueness of solutions to the Cauchy problem (6.1) is as follows: From the classical existence and uniqueness result in Theorem 2.7 for every $\varepsilon \in I$ we obtain a solution candidate defined on all of M, which again depends continuously on ε. In this sense the global hyperbolic metric splitting of (M, g) replaces condition (C) used in the proof of Theorem 5.2 to produce a solution candidate. Then we aim at showing moderateness, thus existence of a generalized solution, as well as uniqueness by employing energy estimates as in the local constructions of Section 5. Therefore it is appropriate to suppose also condition (A), i.e., g to be weakly singular. In the present situation we may translate (5.1) into corresponding asymptotic conditions on β and h_t. As for condition (B), we see that the existence of a suitable (in this case even global) foliation is a consequence of the globally hyperbolic metric splitting. Indeed we globally have $g_\varepsilon^{-1}(dt, dt) = -1/\beta_\varepsilon \leq -C < 0$ for some positive C, which implies (5.2). The second asymptotic boundedness condition (5.3) in (B), i.e., $\sup_K \|\nabla^\varepsilon dt\|_m = O(1)$, now simply reads

$$\|d\beta_\varepsilon\|_m = O(1) \qquad (\varepsilon \to 0) \tag{6.2}$$

184 G. Hörmann, M. Kunzinger and R. Steinbauer

uniformly on compact sets. As in the local setting this implies for the extrinsic curvature of the hypersurfaces $\{t\} \times S$

$$\|II_\varepsilon\|_m = O(1) \qquad (\varepsilon \to 0) \quad \text{uniformly on compact sets.}$$

We may now state the main result of this section.

Theorem 6.3 (Global existence and uniqueness of generalized solutions). *Let (M, g) be a generalized space-time with a weakly singular metric admitting a globally hyperbolic metric splitting and assume that condition (5.3) holds. Then the Cauchy problem (6.1) has a unique solution $u \in \mathcal{G}(M)$ for all compactly supported $u_0, u_1 \in \mathcal{G}(S)$.*

Sketch of proof: Let $\Box_\varepsilon$ denote the wave operator corresponding to g_ε. Then Theorem 2.7 provides us with a global solution u_ε to $\Box_\varepsilon = 0$ with Cauchy data $u_{0\varepsilon}$ and $u_{1\varepsilon}$, thereby defining a solution candidate.

To prove existence we have to establish moderateness of the net $(u_\varepsilon)_\varepsilon$. Choose an exhaustive sequence of compact sets K_j ($j \in \mathbb{N}$) in S. Then it suffices to show moderateness of $(u_\varepsilon)_\varepsilon$ on $L_j := [-j, j] \times K_j$ for each $j \in \mathbb{N}$.

Fix $j \in \mathbb{N}$ and choose ρ as in Definition 6.1, (c) with $T = j$. We cover L_j by finitely many lens-shaped regions as in Figure 1 that are, in turn, contained in coordinate neighborhoods. By the explicit construction given in [27, Sec. 3.3.4] (based on condition (A)) the heights of these lenses are uniformly bounded below on L_j by some $\delta > 0$ due to the properties of ρ and β.

Employing the energy estimates from Section 4.3 we thereby derive moderateness estimates of u_ε on the strip $[0, \delta/2] \times K_j$ from the moderateness of $u_{0\varepsilon}$ and $u_{1\varepsilon}$. We may iterate this procedure to cover $[0, j] \times K_j$ in finitely many steps, and analogously for $[-j, 0] \times K_j$.

By the same token negligibility of $(u_{0\varepsilon})$ and $(u_{1\varepsilon})$ implies the corresponding property for (u_ε), which proves uniqueness. $\qquad\Box$

Finally, we remark that our methods also allow to treat the inhomogeneous equation as well as the inclusion of lower-order terms, which have to satisfy certain asymptotic bounds, see [19].

Also, condition (A) is actually a little stronger than what is needed to prove moderateness of the solution candidate resp. negligibility in case of negligible data. Indeed, again by a result of [19], it suffices to suppose (A) for $k = 0$. On the other hand we could use (A) to explicitly calculate ε-power bounds of (derivatives) of the solution, which encode additional regularity information of our generalized solutions.

Acknowledgment

G.H. thanks the organizers for support by an EPSRC Pathway to Impact Award. We also acknowledge the support of FWF-projects Y237, P20525, and P23714.

References

[1] C. Bär, N. Ginoux, and F. Pfäffle. *Wave equations on Lorentzian manifolds and quantization*. ESI Lectures in Mathematics and Physics. European Mathematical Society (EMS), Zürich, 2007.

[2] J.K. Beem, P.E. Ehrlich, and K.L. Easley. *Global Lorentzian geometry*. Marcel Dekker Inc., New York, 1996.

[3] A.N. Bernal and M. Sánchez. Smoothness of time functions and the metric splitting of globally hyperbolic spacetimes. *Commun. Math. Phys.*, 257(1):43–50, 2005.

[4] A.N. Bernal and M. Sánchez. Globally hyperbolic spacetimes can be defined as 'causal' instead of 'strongly causal'. *Classical Quantum Gravity*, 24(3):745–749, 2007.

[5] Y. Choquet-Bruhat. *General relativity and the Einstein equations*. Oxford Mathematical Monographs. Oxford University Press, Oxford, 2009.

[6] Y. Choquet-Bruhat and S. Cotsakis. Global hyperbolicity and completeness. *J. Geom. Phys.*, 43(4):345–350, 2002.

[7] C.J.S. Clarke. Singularities: boundaries or internal points? In P.S. Joshi and A.K. Raychaudhuri, editors, *Singularities, Black Holes and Cosmic Censorship*, pages 24–32. IUCCA, Bombay, 1996.

[8] C.J.S. Clarke. Generalized hyperbolicity in singular spacetimes. *Classical Quantum Gravity*, 15(4):975–984, 1998.

[9] J.F. Colombeau. *New generalized functions and multiplication of distributions*. North-Holland, Amsterdam, 1984.

[10] J.F. Colombeau. *Elementary introduction to new generalized functions*. North-Holland, 1985.

[11] J.W. de Roever and M. Damsma. Colombeau algebras on a C^∞-manifold. *Indag. Math. (N.S.)*, 2(3):341–358, 1991.

[12] J. Dieudonné. *Treatise on analysis. Vol. VIII*, volume 10 of *Pure and Applied Mathematics*. Academic Press Inc., Boston, MA, 1993.

[13] F.G. Friedlander. *The wave equation on a curved space-time*. Cambridge Monographs on Mathematical Physics, Cambridge University Press, 1975.

[14] C. Garetto. Topological structures in Colombeau algebras: topological $\tilde{\mathbb{C}}$-modules and duality theory. *Acta Appl. Math.*, 88(1):81–123, 2005.

[15] R. Geroch and J. Traschen. Strings and other distributional sources in general relativity. *Phys. Rev. D (3)*, 36(4):1017–1031, 1987.

[16] J. Grant, E. Mayerhofer, and R. Steinbauer. The wave equation on singular spacetimes. *Commun. Math. Phys.*, 285(2):399–420, 2009.

[17] J.B. Griffiths and J. Podolský. *Exact space-times in Einstein's general relativity*. Cambridge Monographs on Mathematical Physics. Cambridge University Press, Cambridge, 2009.

[18] M. Grosser, M. Kunzinger, M. Oberguggenberger, and R. Steinbauer. *Geometric theory of generalized functions*. Kluwer, Dordrecht, 2001.

[19] C. Hanel. Wave-type equations of low regularity. *Appl. Anal.*, to appear, 2011.

[20] S.W. Hawking and G.F.R. Ellis. *The large scale structure of space-time*. Cambridge Monographs of Mathematical Physics. Vol. I. London: Cambridge University Press. XI, 391 p., 1973.

[21] L. Hörmander. *The analysis of linear partial differential operators*, volume III. Springer-Verlag, 1985. Second printing 1994.

[22] L. Hörmander. *The analysis of linear partial differential operators*, volume I. Springer-Verlag, second edition, 1990.

[23] A. Burtscher, M. Kunzinger, Algebras of generalized functions with smooth parameter dependence. *Proc. Edinburgh Math. Soc.*, to appear.

[24] M. Kunzinger and R. Steinbauer. Foundations of a nonlinear distributional geometry. *Acta Appl. Math.*, 71:179–206, 2002.

[25] M. Kunzinger and R. Steinbauer. Generalized pseudo-Riemannian geometry. *Trans. Amer. Math. Soc.*, 354(10):4179–4199, 2002.

[26] Philippe G. Le Floch and Cristinel Mardare. Definition and stability of Lorentzian manifolds with distributional curvature. *Port. Math. (N.S.)*, 64(4):535–573, 2007.

[27] E. Mayerhofer. *The wave equation on static singular space-times*. PhD thesis, University of Vienna, 2006. http://arxiv.org/abs/0802.1616.

[28] B. O'Neill. *Semi-Riemannian geometry*, volume 103 of *Pure and Applied Mathematics*. Academic Press, New York, 1983.

[29] R. Penrose and W. Rindler. *Spinors and space-time. Vol. 1*. Cambridge Monographs on Mathematical Physics. Cambridge University Press, Cambridge, 1987. Two-spinor calculus and relativistic fields.

[30] L. Schwartz. Sur l'impossibilité de la multiplication des distributions. *C. R. Acad. Sci. Paris*, 239:847–848, 1954.

[31] R. Steinbauer and J.A. Vickers. The use of generalized functions and distributions in general relativity. *Classical Quantum Gravity*, 23(10):R91–R114, 2006.

[32] R. Steinbauer and J.A. Vickers. On the Geroch-Traschen class of metrics. *Classical Quantum Gravity*, 26(6):065001, 19, 2009.

[33] J.A. Vickers and J.P. Wilson. Generalized hyperbolicity in conical spacetimes. *Class. Quantum. Grav.*, **17**:1333–1360, 2000.

[34] S. Waldmann. Geometric wave equations. Lecture Notes, University of Freiburg. http://omnibus.uni-freiburg.de/~sw12/Lectures/Wellen0809/main.pdf, 2009.

Günther Hörmann, Michael Kunzinger and Roland Steinbauer
Faculty of Mathematics
University of Vienna
Nordbergstraße 15
A-1090 Wien, Austria
e-mail: guenther.hoermann@univie.ac.at
        michael.kunzinger@univie.ac.at
        roland.steinbauer@univie.ac.at

Progress in Mathematics, Vol. 301, 187–212

Lower Bounds for the Lifespan of Solutions to Nonlinear Wave Equations in Elasticity

Hideo Kubo

Abstract. In this paper we study the lifespan of solutions to nonlinear wave equations in elasticity with small initial data. Main step of our argument is to construct a good approximate solution. A natural choice of the approximation seems to be the leading term of solutions to the free elastic wave equation. However, it does not satisfy the nonlinear elastic wave equation in a suitable sense. For this reason, we modify the approximation by adding a higher-order term. Then, we are able to obtain a lower bound of the lifespan which is expressed in terms of initial data and a coefficient in the nonlinearity.

Mathematics Subject Classification. Primary 35L70; Secondary 35B40.

Keywords. Nonlinear wave equations in elasticity, lifespan, radiation field for elastic waves.

1. Introduction

In this paper we consider the Cauchy problem for homogeneous, isotropic, hyperelastic wave equations:

$$(\partial_t^2 - L)u(t, x) = F(\nabla u, \nabla^2 u), \quad (t, x) \in (0, T) \times \mathbf{R}^3, \tag{1.1}$$

$$u(0, x) = \varepsilon f(x), \quad (\partial_t u)(0, x) = \varepsilon g(x), \quad x \in \mathbf{R}^3, \tag{1.2}$$

where $u(t, x) = {}^t(u_1(t, x), u_2(t, x), u_3(t, x))$ is the displacement vector from the configuration, $\nabla u = (\partial_1 u, \partial_2 u, \partial_3 u)$, $\partial_j = \partial/\partial x_j$ $(j = 1, 2, 3)$, and

$$L = c_2^2 \Delta + (c_1^2 - c_2^2) \operatorname{grad} \operatorname{div}, \quad \Delta = \operatorname{div} \operatorname{grad}$$

with material constants c_1, c_2 satisfying $0 < c_2 < c_1$. Here grad and div stand for the spatial gradient and divergence, respectively. Besides, f, g are smooth

This research is partially supported by Grant-in-Aid for Challenging Exploratory Research (No. 22654017), JSPS.

functions with compact support and ε is a positive parameter. The nonlinearity F is assumed to be of the form

$$F(\nabla u, \nabla^2 u) = A_1 \operatorname{grad}(\operatorname{div} u)^2 + A_2 \operatorname{grad} |\operatorname{rot} u|^2 \qquad (1.3)$$
$$+ A_3 \operatorname{rot}((\operatorname{div} u)(\operatorname{rot} u)) + N(u).$$

Here, A_l $(l = 1, 2, 3)$ are real constants, and the ith component of N is expressed as

$$N_i(u) = \sum_{j,k,l=1}^{3} \sum_{a,b=1}^{3} B_{jkl}^{iab} (\partial_j u_a)(\partial_k \partial_l u_b)$$

with real constants B_{jkl}^{iab} satisfying $\sum_{j,k,l=1}^{3} B_{jkl}^{iab} \theta_j \theta_k \theta_l = 0$ for any $i, a, b \in \{1, 2, 3\}$ and $(\theta_1, \theta_2, \theta_3) \in S^2$. In addition, we assume that if we rewrite the ith component of F as

$$F_i(\nabla u, \nabla^2 u) = \sum_{j,k,l=1}^{3} \sum_{a,b=1}^{3} C_{jkl}^{iab} (\partial_j u_a)(\partial_k \partial_l u_b),$$

then $C_{jkl}^{iab} = C_{jlk}^{iab} = C_{jlk}^{bai}$ holds for any $i, j, k, l, a, b \in \{1, 2, 3\}$. This is actually an additional restriction only on $N(u)$ (for details, we refer to [1], [15]).

We denote the lifespan of the problem (1.1)–(1.2) by T_ε which is the supremum of all $T > 0$ such that the problem admits a unique smooth solution in $[0, T) \times \mathbf{R}^3$. In John [10] the lower bound for the lifespan $T_\varepsilon \geq e^{C/\varepsilon}$ with a positive number C was obtained for sufficiently small ε (see also [13]). Moreover, if $A_1 = 0$, then the global solvability of the problem for sufficiently small initial data was proved by Agemi [1] and Sideris [15], independently.

On the one hand, concerning the Cauchy problem for scalar wave equations:

$$(\partial_t^2 - \Delta)v(t, x) = \sum_{j,k,l=0}^{3} g_{jkl}(\partial_j v)(\partial_k \partial_l v), \quad (t, x) \in (0, T) \times \mathbf{R}^3, \qquad (1.4)$$

$$v(0, x) = \varepsilon\phi(x), \ (\partial_t v)(0, x) = \varepsilon\psi(x), \quad x \in \mathbf{R}^3, \qquad (1.5)$$

not only the estimate of the lifespan $\widetilde{T}_\varepsilon$ of this problem from below but also much precise information of $\widetilde{T}_\varepsilon$ are known (here, g_{jkl} are real constants satisfying $g_{jkl} = g_{jlk}$ for $j, k, l \in \{0, 1, 2, 3\}$, and $\phi, \psi \in C_0^\infty(\mathbf{R}^3)$). More explicitly, it was independently shown by Hörmander [5] and John [9] that

$$\liminf_{\varepsilon \to +0} \varepsilon \log \widetilde{T}_\varepsilon \geq \left(\max\{-2^{-1}G(\theta) \partial_s^2 \widetilde{\mathcal{R}}[\phi, \psi](s, \theta) \, ; \, s \in \mathbf{R}, \ \theta \in S^2\} \right)^{-1}, \quad (1.6)$$

provided the right-hand side is a finite number. Here, the functions G and $\widetilde{\mathcal{R}}[\phi, \psi]$ are defined by

$$G(\theta) = \sum_{j,k,l=0}^{3} g_{jkl} \theta_j \theta_k \theta_l \quad \text{with } \theta_0 = -1, \ (\theta_1, \theta_2, \theta_3) \in S^2,$$

$$\widetilde{\mathcal{R}}[\phi, \psi](s, \theta) = \frac{1}{4\pi}(\mathcal{R}[\psi](s, \theta) - \partial_s \mathcal{R}[\phi](s, \theta)), \quad (s, \theta) \in \mathbf{R} \times S^2,$$

where $\mathcal{R}[\phi]$ is the Radon transform of ϕ, that is,

$$\mathcal{R}[\phi](s,\theta) = \int_{\theta \cdot y = s} \phi(y)\, dS_y, \quad (s,\theta) \in \mathbf{R} \times S^2. \tag{1.7}$$

The counter part of the estimate (1.6) has been studied by Alinhac [2]. We remark that $G \equiv 0$ on S^2 is equivalent to the null condition introduced by Klainerman [12], and the condition implies $\widetilde{T}_\varepsilon = +\infty$ (see also [3]). On the other hand, $\widetilde{\mathcal{R}}[\phi, \psi] \equiv 0$ on $\mathbf{R} \times S^2$ is equivalent to $\phi \equiv \psi \equiv 0$ on $\mathbf{R}^3$.

Therefore, a natural question is if it is possible to derive an analogous estimate to (1.6) for the lifespan T_ε of the problem (1.1)–(1.2) or not. The difficulty for dealing with the elastic wave equation (1.1) comes from the fact that the equation has two distinct propagation speeds. For this, the hyperbolic boosts $x_j \partial_t + t \partial_j$ do not work well, and construction of a nonlinear approximate solution is not straightforward as in the case of the wave equation. Nevertheless, by using a higher-order approximation (see (6.36) below) together with careful treatments of the decay factor $(1 + |c_i t - |x||)^{-1}$, we are able to overcome the difficulty.

In order to state our result, we define

$$\widetilde{\mathcal{R}}_i[f,g](s,\theta) = \frac{1}{4\pi}\left(c_i^{-1}\mathcal{R}[g](s,\theta) - \partial_s\mathcal{R}[f](s,\theta)\right) \quad (i = 1,2) \tag{1.8}$$

for $(s,\theta) \in \mathbf{R} \times S^2$, where the Radon transform $\mathcal{R}[f]$ of $f = {}^t(f_1, f_2, f_3) \in \left(C_0^\infty(\mathbf{R}^3)\right)^3$ is given by $\mathcal{R}[f] = {}^t(\mathcal{R}[f_1], \mathcal{R}[f_2], \mathcal{R}[f_3])$. We note that $\widetilde{\mathcal{R}}_i[f,g]$ is bounded on $\mathbf{R} \times S^2$ and compactly supported in s for $f,\, g \in \left(C_0^\infty(\mathbf{R}^3)\right)^3$. In particular, if

$$p_0(s,\theta) := \theta \cdot \widetilde{\mathcal{R}}_1[f,g](s,\theta) \tag{1.9}$$

is not identically zero on $\mathbf{R} \times S^2$, then $\partial_s^2 p_0(s,\theta)$ takes both positive and negative values. Therefore, one can define a positive number

$$\tau_* = \left(\max\{-c_1^{-2}A_1\, \partial_s^2 p_0(s,\theta)\,;\, s \in \mathbf{R},\ \theta \in S^2\,\}\right)^{-1}, \tag{1.10}$$

provided $A_1 \neq 0$ and $p_0 \not\equiv 0$ on $\mathbf{R} \times S^2$.

Then, our main result is the following:

Theorem 1.1. *Let* $f,\, g \in \left(C_0^\infty(\mathbf{R}^3)\right)^3$. *If* $A_1 \neq 0$ *and* $p_0 \not\equiv 0$ *on* $\mathbf{R} \times S^2$, *then we have*

$$\liminf_{\varepsilon \to +0} \varepsilon \log T_\varepsilon \geq \tau_*. \tag{1.11}$$

Remark 1.2. (i) Unfortunately, we do not have the estimate in the opposite direction to (1.11), that is to say

$$\limsup_{\varepsilon \to +0} \varepsilon \log T_\varepsilon \leq \tau_* \tag{1.12}$$

in general. But, when the initial data take the following form:

$$f(x) = \phi(r)x, \quad g(x) = \psi(r)x, \quad x \in \mathbf{R}^3,$$

(1.12) was shown by John [8], provided $A_1 \neq 0$ and the corresponding p_0 does not identically vanish on $\mathbf{R} \times S^2$. Hence, the lower bound (1.11) seems to be optimal.

(ii) The number τ_* is related to the lifespan of the following Cauchy problem for $p = p(s, \theta, \tau)$:

$$2c_1^2 \, \partial_\tau p + A_1 (\partial_s p)^2 = 0 \quad \text{in } \mathbf{R} \times S^2 \times [0, \tau_*), \tag{1.13}$$

$$p(s, \theta, 0) = p_0(s, \theta) \quad \text{for } (s, \theta) \in \mathbf{R} \times S^2. \tag{1.14}$$

Indeed, it is known that the solution to the above problem uniquely exists in $\mathbf{R} \times S^2 \times [0, \tau_*)$ (for the proof, see Lemma 6.5.4 with $G(\omega) \equiv 2A_1/c_1^2$ in [6]).

This paper is organized as follows. In the next section we gather notation. In Section 3 we give some preliminaries. An approximate solution is constructed in Section 6, and useful estimates for the approximation are established in Proposition 6.5. Outline of the proof of Theorem 1.1 is given in Section 7. The detail of the proof will be published elsewhere.

2. Notation

In this section, we introduce notation which will be used throughout this paper. We denote $r = |x|$ and $\omega = x/r$. We set $\partial_r = \sum_{j=1}^3 (x_j/r)\partial_j$ and $O = {}^t(O_1, O_2, O_3) = x \wedge {}^t(\partial_1, \partial_2, \partial_3)$, where $\wedge$ stands for the outer product in $\mathbf{R}^3$. Then we have

$${}^t(\partial_1, \partial_2, \partial_3) = \omega \partial_r - r^{-1}\omega \wedge O. \tag{2.1}$$

We denote $Z = \{Z_0, Z_1, \ldots, Z_6\} = \{\partial_t, \partial_1, \partial_2, \partial_3, O_1, O_2, O_3\}$. We write Z^α for $Z_0^{\alpha_0} \cdots Z_6^{\alpha_6}$ with a multi-index $\alpha = (\alpha_0, \ldots, \alpha_6)$. Note that we have $[Z_a, \partial_t^2 - \Delta] = 0$ $(a = 0, \ldots, 6)$, where we have set $[A, B] = AB - BA$.

We also use $\widetilde{Z} = \{\widetilde{Z}_0, \widetilde{Z}_1, \ldots, \widetilde{Z}_6\} = \{\partial_t I, \partial_1 I, \partial_2 I, \partial_3 I, \widetilde{O}_1, \widetilde{O}_2, \widetilde{O}_3\}$ for $\mathbf{R}^3$-valued functions, where I is the 3×3 identity matrix and

$$\widetilde{O}_j = O_j I + U_j \quad (j = 1, 2, 3) \tag{2.2}$$

with

$$U_1 = \begin{pmatrix} 0 & 0 & 0 \\ 0 & 0 & 1 \\ 0 & -1 & 0 \end{pmatrix}, \quad U_2 = \begin{pmatrix} 0 & 0 & -1 \\ 0 & 0 & 0 \\ 1 & 0 & 0 \end{pmatrix}, \quad U_3 = \begin{pmatrix} 0 & 1 & 0 \\ -1 & 0 & 0 \\ 0 & 0 & 0 \end{pmatrix}.$$

The vector fields $\widetilde{O}_j$ is closely related to the fact that if $u(t, x)$ solves (1.1), then so does $A^{-1}u(t, Ax)$ for any orthogonal matrix A. This observation leads to the good algebraic relations $[\widetilde{Z}_a, L] = 0$ for $a = 0, \ldots, 6$. We write $\widetilde{Z}^\alpha$ for $\widetilde{Z}_0^{\alpha_0} \cdots \widetilde{Z}_6^{\alpha_6}$ with a multi-index $\alpha = (\alpha_0, \ldots, \alpha_6)$.

For functions of $(s, \theta, \tau) \in \mathbf{R} \times S^2 \times [0, \infty)$, we denote the differentiation with respect to s, θ and τ by

$$\Lambda_0 = \partial_s, \quad \Lambda_1 - o_1, \quad \Lambda_2 = o_2, \quad \Lambda_3 = o_3, \quad \Lambda_4 = \partial_\tau, \tag{2.3}$$

where differential operators o_i on S^2 are (formally) defined by ${}^t(o_1, o_2, o_3) = \theta \wedge {}^t(\partial_{\theta_1}, \partial_{\theta_2}, \partial_{\theta_3})$. We write Λ^β for $\Lambda_0^{\beta_0} \cdots \Lambda_4^{\beta_4}$ and $\Lambda_*^\gamma = \Lambda_0^{\gamma_0} \cdots \Lambda_3^{\gamma_3}$ with multi-indices $\beta = (\beta_0, \ldots, \beta_4)$ and $\gamma = (\gamma_0, \ldots, \gamma_3)$.

For a nonnegative integer k, and a real-valued smooth function $\varphi(t, x)$, we define

$$|\varphi(t,x)|_k = \sum_{|\alpha| \leq k} |(Z^\alpha \varphi)(t,x)|, \quad |\partial\varphi(t,x)|_k = \sum_{|\alpha| \leq k} \sum_{a=0}^{3} |(Z^\alpha \partial_a \varphi)(t,x)|.$$

For an $\mathbf{R}^3$-valued function $u(t, x)$ we use the same notation $|u(t,x)|_k$ and $|\partial u(t,x)|_k$ with Z replaced by $\widetilde{Z}$.

For $\nu \geq 0$, a nonnegative integer k, and $\phi \in \mathcal{S}(\mathbf{R}^3)$, we define

$$\|\phi\|_{k,\nu} = \left(\sup_{x \in \mathbf{R}^3} \sum_{|\alpha| \leq k} (1 + |x|^2)^\nu |\partial_x^\alpha \phi(x)|^2 \right)^{1/2}.$$

Here, $\mathcal{S}(\mathbf{R}^3)$ is the Schwartz class, the set of rapidly decreasing real-valued functions. Besides, for $f, g \in \left(\mathcal{S}(\mathbf{R}^3)\right)^3$, we set

$$A_{k,\nu}[f,g] = \sum_{j=1}^{3} \left(\|f_j\|_{k+1,\nu} + \|g_j\|_{k,\nu} \right). \tag{2.4}$$

As usual, various positive constants which may change line by line are denoted just by the same letter C throughout this paper.

3. Preliminaries

First we recall basic properties of the Radon transform discussed in the section 4 of [11] for the case of $n = 3$ and $\chi \equiv 1$ (note that when $\chi \equiv 1$, $\mathcal{S}_\chi(\mathbf{R}^3)$ and $\|\varphi\|_{\chi,k,\nu}$ in [11] become to $\mathcal{S}(\mathbf{R}^3)$ and $\|\varphi\|_{k,\nu}$, respectively). It holds that

$$\partial_s \mathcal{R}[\varphi](s, \theta) = \mathcal{R}[(\theta \cdot \mathrm{grad})\varphi](s, \theta), \tag{3.1}$$

$$o_i \mathcal{R}[\varphi](s, \theta) = \mathcal{R}[O_i\varphi](s, \theta), \quad i = 1, 2, 3, \tag{3.2}$$

$$\mathcal{R}[\partial_i\varphi](s, \theta) = \theta_i\, \partial_s \mathcal{R}[\varphi](s, \theta), \quad i = 1, 2, 3 \tag{3.3}$$

for a real-valued function $\varphi \in \mathcal{S}(\mathbf{R}^3)$. Moreover, for $\nu \geq 0$, a nonnegative integer k, and a multi-index α, we have

$$\left| \partial_s^k o^\alpha \mathcal{R}[\varphi](s, \theta) \right| \leq C\|\varphi\|_{k+|\alpha|, \nu+3+|\alpha|} (1 + s^2)^{-\frac{\nu}{2}} \tag{3.4}$$

for $(s, \theta) \in \mathbf{R} \times S^2$. Here $C = C(k, \nu, \alpha)$ is a positive constant.

Next we define

$$Q_\gamma[\varphi](t, x) = \frac{1}{4\pi} \int_{\theta \in S^2} \theta^\gamma \varphi(x + t\theta) dS_\theta', \quad (t, x) \in (0, \infty) \times \mathbf{R}^3 \tag{3.5}$$

for a multi-index $\gamma = (\gamma_1, \gamma_2, \gamma_3)$, a real-valued function $\varphi \in \mathcal{S}(\mathbf{R}^3)$. Here, dS'_θ is the area element on S^2. Note that $Q_0[\varphi]$ is the *spherical mean* of φ. We shall derive decay property of $Q_\gamma[\varphi]$.

Proposition 3.1. *Let k be a nonnegative integer, $\nu > 0$, and γ be a multi-index. Then there exists a positive constant C such that we have*

$$|\partial_t^k Q_\gamma[\varphi](t,x)| \le C\|\varphi\|_{k,\nu+2}(1+t+r)^{-2}(1+|r-t|)^{-\nu} \tag{3.6}$$

for $(t,x) \in (0,\infty) \times \mathbf{R}^3$ with $r = |x|$, provided that $\varphi \in \mathcal{S}(\mathbf{R}^3)$.

Proof. It follows that

$$\partial_t^k Q_\gamma[\varphi](t,x) = \sum_{|\alpha|=k} \frac{1}{4\pi} \int_{\theta \in S^2} c_\alpha \theta^{\gamma+\alpha} (\partial_x^\alpha \varphi)(x+t\theta) dS'_\theta \tag{3.7}$$

with some appropriate constants c_α. Therefore, we get

$$|\partial_t^k Q_\gamma[\varphi](t,x)| \le C\|\varphi\|_{k,\nu+2} \int_{\theta \in S^2} (1+|x+t\theta|)^{-\nu-2} dS'_\theta$$

$$= C\|\varphi\|_{k,\nu+2} \times \frac{2\pi}{tr} \int_{|t-r|}^{t+r} \lambda(1+\lambda)^{-\nu-2} d\lambda$$

(see, e.g., [14, Lemma 2.1] for the proof of the last equality). Hence, the desired estimate follows from

$$\frac{1}{tr} \int_{|t-r|}^{t+r} \lambda(1+\lambda)^{-\nu-2} d\lambda \le C(1+t+r)^{-2}(1+|r-t|)^{-\nu} \tag{3.8}$$

for $t, r > 0$. By symmetry, it suffices to show (3.8) for $0 < r \le t$.

First suppose $0 < r \le t < 1$. Then the desired estimate follows from

$$\frac{1}{tr} \int_{|t-r|}^{t+r} \lambda(1+\lambda)^{-\nu-2} d\lambda \le \frac{1}{tr} \int_{|t-r|}^{t+r} \lambda d\lambda = 2.$$

Next suppose $t \ge 1$ and $0 < r \le t$. Since $t \ge (t+r+1)/3$, we get

$$\frac{1}{tr} \int_{|t-r|}^{t+r} \lambda(1+\lambda)^{-\nu-2} d\lambda \le \frac{3}{(1+t+r)r} \int_{|t-r|}^{t+r} (1+\lambda)^{-\nu-1} d\lambda.$$

Observing that $t - r \ge (t+r)/3$ for $t \ge 2r$ and that $r \ge (t+r)/3$ for $t \le 2r$, we obtain (3.8). This completes the proof. $\square$

The following proposition shows that the leading term of $Q_\gamma[\varphi]$ is described by the Radon transform. Since the proof of the proposition is similar to that of Lemma 4.3 in [11], we omit it.

Proposition 3.2. *Let k be a nonnegative integer, $\nu \ge 0$, γ be a multi-index, and $c_* \ge 1$. Then there exist a positive constant C and an integer $N_0 (\ge \nu + 4)$ such that we have*

$$|t\partial_t^k Q_\gamma[\varphi](t,x) - (4\pi r)^{-1}(-\omega)^\gamma ((-\partial_s)^k \mathcal{R}[\varphi])(r-t,\omega)| \tag{3.9}$$

$$\le C\|\varphi\|_{k+1,N_0}(1+t+r)^{-2}(1+|r-t|)^{-\nu}$$

for $(t, x) \in (0, \infty) \times \mathbf{R}^3$ satisfying $r \geq t/(2c_*) \geq 1$ with $r = |x|$ and $\omega = x|x|^{-1}$, provided that $\varphi \in \mathcal{S}(\mathbf{R}^3)$.

Next we derive a couple of estimates of the following integral operator for the latter sake:

$$T_\gamma[\varphi](t, x) = \int_{c_2 t}^{c_1 t} \tau^{-1} Q_\gamma[\varphi](\tau, x) d\tau, \quad (t, x) \in (0, \infty) \times \mathbf{R}^3. \tag{3.10}$$

Proposition 3.3. *Let k be a nonnegative integer, $\nu > 0$, γ be a multi-index, and $\varphi \in \mathcal{S}(\mathbf{R}^3)$. When $(t, x) \in (0, \infty) \times \mathbf{R}^3$ satisfies one of $r > 2c_1 t$, $r < c_2 t/2$ or $0 < t + r \leq 1$, we have*

$$|T_\gamma[\varphi](t, x)| \leq C\|\varphi\|_{0, \nu+2} (1 + t + r)^{-2-\nu}. \tag{3.11}$$

On the other hand, when $(t, x) \in (0, \infty) \times \mathbf{R}^3$ satisfies $c_2 t/2 < r < 2c_1 t$ and $t + r \geq 1$, we have

$$|T_\gamma[\varphi](t, x)| \leq C\|\varphi\|_{0, \nu+2} (1 + t + r)^{-3}, \tag{3.12}$$

provided $\nu > 1$. Moreover, if $k \geq 1$, then we have

$$|\partial_t^k T_\gamma[\varphi](t, x)| \leq C\|\varphi\|_{k, \nu+2} (1 + t)^{-1}(1 + t + r)^{-2} \max_{i=1,2}\{(1 + |r - c_i t|)^{-\nu}\} \tag{3.13}$$

for $(t, x) \in (0, \infty) \times \mathbf{R}^3$. Furthermore, we have

$$|T_\gamma[\partial_j \varphi](t, x)| \leq C\|\varphi\|_{2, N_0} (1 + t + r)^{-3} \max_{i=1,2}\{(1 + |r - c_i t|)^{-1}\}, \tag{3.14}$$

where N_0 is the number from Lemma 3.2.

Proof. First we prove (3.11). By (3.6) we have

$$|T_\gamma[\varphi](t, x)| \leq C\|\varphi\|_{0, \nu+2} \int_{c_2 t}^{c_1 t} \tau^{-1}(1 + \tau + r)^{-2}(1 + |r - \tau|)^{-\nu} d\tau$$

$$\leq C\|\varphi\|_{0, \nu+2} (1 + c_2 t + r)^{-2} \int_{c_2 t}^{c_1 t} \tau^{-1}(1 + |r - \tau|)^{-\nu} d\tau. \tag{3.15}$$

Observe that if $r \leq c_2 t/2$ and $\tau \geq c_2 t$ then $|\tau - r| \geq (c_2 t + r)/3$, and that if $r \geq 2c_1 t$ and $\tau \leq c_1 t$, then $|r - \tau| \geq (c_1 t + r)/3$. Thus we get (3.11) for $r \leq c_2 t/2$ or $r \geq 2c_1 t$. On the one hand, from (3.15) we have

$$|T_\gamma[\varphi](t, x)| \leq C\|\varphi\|_{0, \nu+2} \int_{c_2 t}^{c_1 t} \tau^{-1} d\tau \leq C\|\varphi\|_{0, \nu+2},$$

which yields (3.11) for $0 < t + r \leq 1$.

Next we prove (3.12). Since $\tau > C(1 + t + r)$ for $\tau > c_2 t$, $c_2 t/2 < r < 2c_1 t$, and $t + r \geq 1$, we get (3.12) from (3.15) by $\nu > 1$.

Next we prove (3.13). It follows from (3.10) that

$$\partial_t T_\gamma[\varphi](t, x) = t^{-1} \left(Q_\gamma[\varphi](c_1 t, x) - Q_\gamma[\varphi](c_2 t, x) \right). \tag{3.16}$$

When $t \geq 1$, we easily have (3.13) by (3.6). On the other hand, when $0 < t < 1$, we rewrite the right-hand side of (3.16) as

$$(c_1 - c_2) \int_0^1 (\partial_t Q_\gamma[\varphi]) (c_1 t\sigma + c_2 t(1 - \sigma), x) d\sigma.$$

Since $0 \leq c_1 t\sigma + c_2 t(1 - \sigma) \leq C$ for $0 < \sigma, t < 1$, we get from (3.6)

$$\left| \partial_t^k T_\gamma[\varphi](t, x) \right| \leq C\|\varphi\|_{k,\nu+2} (1 + r)^{-2-\nu},$$

which yields (3.13) for $0 < t \leq 1$.

Finally, we prove (3.14). When one of $r > 2c_1 t$, $r < c_2 t/2$ or $0 < t + r \leq 1$ holds, (3.11) with $\nu = 2$ yields (3.14). Therefore, we have only to consider the case where $c_2 t/2 \leq r \leq 2c_1 t$ and $t + r \geq 1$. We rewrite

$$T_\gamma[\partial_j \varphi](t, x) = (4\pi r)^{-1} \int_{c_2 t}^{c_1 t} \tau^{-2}(-\omega)^\gamma \mathcal{R}[\partial_j \varphi](r - \tau, \omega) d\tau$$

$$+ \int_{c_2 t}^{c_1 t} \tau^{-2} \left(\tau Q_\gamma[\partial_j \varphi](\tau, x) - (4\pi r)^{-1}(-\omega)^\gamma \mathcal{R}[\partial_j \varphi](r - \tau, \omega) \right) d\tau.$$

Let $\nu > 1$ in the following. Then, by (3.9) with $k = 0$ the second term on the right-hand side is estimated by

$$C\|\varphi\|_{2,N_0} \int_{c_2 t}^{c_1 t} \tau^{-2}(1 + \tau + r)^{-2}(1 + |r - \tau|)^{-\nu} d\tau$$

$$\leq C\|\varphi\|_{2,N_0} (1 + t + r)^{-4},$$

because $\tau \geq C(1 + t + r)$ in this case. Using (3.3), we can make integration by parts in τ in the first term. Then it is rewritten as

$$(4\pi r)^{-1} \int_{c_2 t}^{c_1 t} (-2\tau^{-3})\omega_j(-\omega)^\gamma \mathcal{R}[\varphi](r - \tau, \omega) d\tau$$

$$- (4\pi r)^{-1} \left((c_1 t)^{-2}\omega_j(-\omega)^\gamma \mathcal{R}[\varphi](r - c_1 t, \omega) \right.$$

$$\left. -(c_2 t)^{-2}\omega_j(-\omega)^\gamma \mathcal{R}[\varphi](r - c_2 t, \omega) \right).$$

By (3.4) we have $|\mathcal{R}[\varphi](s, \omega)| \leq C\|\varphi\|_{0,\nu+3} (1 + |s|)^{-\nu}$. Since $\nu > 1$, we thus find (3.14) in this case. This completes the proof. $\qquad \square$

4. The free elastic wave equation

In this section we consider the Cauchy problem:

$$(\partial_t^2 - L)u_0(t, x) = 0, \qquad\qquad (t, x) \in (0, \infty) \times \mathbf{R}^3, \qquad (4.1)$$

$$u_0(0, x) = f(x), \ (\partial_t u_0)(0, x) = g(x), \qquad\qquad x \in \mathbf{R}^3, \qquad (4.2)$$

where $f, g \in \left(\mathcal{S}(\mathbf{R}^3) \right)^3$. First we recall the explicit representation of the solution u_0. We define

$$E[g](t, x) = E_1[g](t, x) + E_2[g](t, x) + E_3[g](t, x), \qquad (4.3)$$

with

$$E_1[g](t,x) = \frac{t}{4\pi} \int_{\theta \in S^2} \Pi(\theta) g(x + c_1 t\theta) dS'_\theta, \tag{4.4}$$

$$E_2[g](t,x) = \frac{t}{4\pi} \int_{\theta \in S^2} (I - \Pi(\theta)) g(x + c_2 t\theta) dS'_\theta, \tag{4.5}$$

$$E_3[g](t,x) = -\frac{t}{4\pi} \int_{c_2 t}^{c_1 t} \tau^{-1} d\tau \int_{\theta \in S^2} (I - 3\Pi(\theta)) g(x + \tau\theta) dS'_\theta. \tag{4.6}$$

Here, for each fixed $\theta \in S^2$, $\Pi(\theta) : \mathbf{R}^3 \to \mathbf{R}^3$ is the projection defined by $\Pi(\theta)v = (\theta \cdot v)\theta$ for $v \in \mathbf{R}^3$. Then it is known that

$$u_0(t,x) = \partial_t E[f](t,x) + E[g](t,x), \quad (t,x) \in (0,\infty) \times \mathbf{R}^3 \tag{4.7}$$

holds (see, e.g., John [10]). The following estimates are extension of those in Theorem 1 in [10] in the sense that we can replace the decaying factor $1+r$ by $1+t+r$ and that the derivatives enjoy better decay property with respect to $1 + |r - c_i t|$ with $i = 1, 2$.

Proposition 4.1. *Let k be a nonnegative integer, f, $g \in (\mathcal{S}(\mathbf{R}^3))^3$, $\nu > 1$, and N_0 be the number from Proposition 3.2. Then, for $(t,x) \in (0,\infty) \times \mathbf{R}^3$, we have*

$$|u_0(t,x)|_k \le C\mathcal{A}_{k,\nu+2}[f,g] (1+t+r)^{-1} W_{-1}(t,r) \tag{4.8}$$

and

$$|\partial u_0(t,x)|_k \le C\mathcal{A}_{k+2,N_0}[f,g] (1+t+r)^{-1} W_{-2}(t,r), \tag{4.9}$$

where $\mathcal{A}_{k,\nu}[f,g]$ is defined by (2.4), and for $\nu \in \mathbf{R}$ we put

$$W_\nu(t,r) = \max_{i=1,2}\{(1 + |r - c_i t|)^\nu\}. \tag{4.10}$$

Proof. Suppose that all the assumptions of the proposition are fulfilled. Let $(t,x) \in (0,\infty) \times \mathbf{R}^3$ in the following. Since $[\widetilde{Z}_a, L] = 0$, we have

$$\partial_t^\ell \partial_x^\alpha \widetilde{O}^\beta E[g](t,x) = \partial_t^\ell E[\partial_x^\alpha \widetilde{O}^\beta g](t,x) \tag{4.11}$$

for any nonnegative integer ℓ and any multi-indices α, β. Therefore, it suffices to show that

$$|\partial_t^k E[g](t,x)| \le C\mathcal{A}_{k,\nu+2}[0,g] (1+t+r)^{-1} W_{-1}(t,x), \tag{4.12}$$

$$|\partial_t^k \partial E[g](t,x)| \le C\mathcal{A}_{k+2,N_0}[0,g] (1+t+r)^{-1} W_{-2}(t,x). \tag{4.13}$$

First we prove (4.12). We denote the ith component of $E_1[g]$ by $E_1^{(i)}[g]$. Then we see from (4.4), (3.5) that

$$E_1^{(i)}[g](t,x) = \sum_{j=1}^3 t Q_{\gamma_{ij}}[g_j](c_1 t, x) \quad (i = 1, 2, 3), \tag{4.14}$$

where $\gamma_{ij} = e_i + e_j$ with $e_1 = (1,0,0)$, $e_2 = (0,1,0)$, $e_3 = (0,0,1)$. Since

$$\partial_t^k \left(tQ_{\gamma_{ij}}[g_j](c_1 t, x) \right)$$
$$= c_1^k t \left(\partial_t^k Q_{\gamma_{ij}}[g_j] \right)(c_1 t, x) + k c_1^{k-1} \left(\partial_t^{k-1} Q_{\gamma_{ij}}[g_j] \right)(c_1 t, x) \tag{4.15}$$

for $k \geq 0$, applying (3.6) with t replaced by $c_1 t$, we get

$$\left| \partial_t^k \left(tQ_{\gamma_{ij}}[g_j](c_1 t, x) \right) \right| \leq C \|g_j\|_{k,\nu+2}(1+t+r)^{-1}(1+|r-c_1 t|)^{-\nu}$$

for $\nu > 0$. Thus we obtain

$$\left| \partial_t^k E_1[g](t,x) \right| \leq C \mathcal{A}_{k,\nu+2}[0,g] \, (1+t+r)^{-1}(1+|r-c_1 t|)^{-\nu}. \tag{4.16}$$

Similarly, we can handle $E_2[g]$ and find

$$\left| \partial_t^k E_2[g](t,x) \right| \leq C \mathcal{A}_{k,\nu+2}[0,g] \, (1+t+r)^{-1}(1+|r-c_2 t|)^{-\nu}. \tag{4.17}$$

Therefore, we have only to show

$$\left| \partial_t^k E_3[g](t,x) \right| \leq C \mathcal{A}_{k,\nu+2}[0,g] \, (1+t+r)^{-2}. \tag{4.18}$$

Observe that each component of $E_3[g]$ can be expressed as a linear combination of the following type of integrals:

$$t \int_{c_2 t}^{c_1 t} \tau^{-1} Q_\gamma[\varphi](\tau, x) d\tau = t \, T_\gamma[\varphi](t,x)$$

with a suitable $\varphi \in \mathcal{S}(\mathbf{R}^3)$ and multi-index γ. Here $T_\gamma[\varphi]$ is defined by (3.10). We see from (3.11), (3.12) that

$$|t \, T_\gamma[\varphi](t,x)| \leq C \|\varphi\|_{0,\nu+2} \, (1+t+r)^{-2}$$

holds, since $\nu > 1$. For $k \geq 1$, we have

$$\partial_t^k \left(t \, T_\gamma[\varphi](t,x) \right) = t \, \partial_t^k T_\gamma[\varphi](t,x) + k \, \partial_t^{k-1} T_\gamma[\varphi](t,x).$$

Therefore, using (3.13) together with (3.11) and (3.12), we obtain

$$\left| \partial_t^k \left(t \, T_\gamma[\varphi](t,x) \right) \right| \leq C \|\varphi\|_{k,\nu+2} \, (1+t+r)^{-2} \, W_{-1}(t,x) \tag{4.19}$$

since $\nu \geq 1$. Thus we get (4.18).

Next we prove (4.13). In view of (4.16) and (4.17) with $\nu = 2$, we see that it suffices to show

$$\left| \partial_t^k \partial E_3[g](t,x) \right| \leq C \|g\|_{k+2,N_0} \, (1+t+r)^{-2} \, W_{-1}(t,x). \tag{4.20}$$

When $\partial = \partial_t$, we immediately get (4.20) from (4.19). It remains to evaluate $\partial_t^k \partial_j (t \, T_\gamma[\varphi](t,x))$ for $k \geq 0$. Since $\partial_j T_\gamma[\varphi](t,x) = T_\gamma[\partial_j \varphi](t,x)$, it follows from (4.19) that

$$\left| \partial_t^k \partial_j \left(t \, T_\gamma[\varphi](t,x) \right) \right| \leq C \|\varphi\|_{k+1,\nu+2} \, (1+t+r)^{-2} \, W_{-1}(t,x)$$

for $k \geq 1$. Combining this estimate with (3.14), we find (4.20). This completes the proof. $\qquad\square$

Next we derive the radiation field for the free elastic wave (for the case of the scalar wave equation, see Friedlander [4], and also [11]). In view of (4.16), (4.17), (4.18), and (4.20), we find that

$$\lim_{t\to\infty} t u_0(t, (c_i t + s)\omega) = \lim_{t\to\infty} t\left(\partial_t E_i[f](t, (c_i t + s)\omega) + E_i[g](t, (c_i t + s)\omega)\right)$$

for each fixed $(s, \omega) \in \mathbf{R} \times S^2$ (recall $c_1 \neq c_2$). Having Proposition 3.2 in mind, we define the radiation field $\mathcal{F}_i[f, g]$ $(i = 1, 2)$ for u_0 associated with the propagation speed c_i by

$$\mathcal{F}_1[f, g](s, \theta) = \Pi(\theta)\widetilde{\mathcal{R}}_1[f, g](s, \theta), \tag{4.21}$$

$$\mathcal{F}_2[f, g](s, \theta) = (I - \Pi(\theta))\widetilde{\mathcal{R}}_2[f, g](s, \theta) \tag{4.22}$$

for $(s, \theta) \in \mathbf{R} \times S^2$, and $f, g \in \left(\mathcal{S}(\mathbf{R}^3)\right)^3$. Here, $\widetilde{\mathcal{R}}_i[f, g](s, \theta)$ is defined by (1.8), and $\Pi(\theta)v = (\theta \cdot v)\theta$ for $v \in \mathbf{R}^3$. We remark that (3.4) implies

$$|\partial_s^k o^\alpha \mathcal{F}_i[f, g](s, \theta)| \leq C(1 + |s|)^{-\nu}, \quad (s, \theta) \in \mathbf{R} \times S^2 \tag{4.23}$$

for any $\nu > 0$, nonnegative integer k, multi-index α, and $f, g \in \left(\mathcal{S}(\mathbf{R}^3)\right)^3$. One can express the leading term of u_0 in terms of the radiation fields as follows.

Proposition 4.2. *Let* $f, g \in \left(\mathcal{S}(\mathbf{R}^3)\right)^3$ *and let* u_0 *be the solution to the problem* (4.1)–(4.2). *Then for any nonnegative integer k and any multi-index α with $|\alpha| \geq 1$, there exists a positive constant C such that*

$$\left| u_0(t, x) - \sum_{m=1}^{2} r^{-1} \mathcal{F}_m[f, g](r - c_m t, \omega) \right|_k \leq C(1 + t + r)^{-2}, \tag{4.24}$$

and

$$\left| \partial_t u_0(t, x) - \sum_{m=1}^{2} (-c_m) r^{-1} (\partial_s \mathcal{F}_m[f, g])(r - c_m t, \omega) \right|_k \tag{4.25}$$

$$+ \left| \partial_x^\alpha u_0(t, x) - \sum_{m=1}^{2} \omega^\alpha r^{-1} (\partial_s^{|\alpha|} \mathcal{F}_m[f, g])(r - c_m t, \omega) \right|_k$$

$$\leq C(1 + t + r)^{-2} W_{-1}(t, r)$$

for $(t, x) \in (0, \infty) \times \mathbf{R}^3$ *with* $r \geq c_2 t/2 \geq 1$. *Here,* $\omega = (\omega_1, \omega_2, \omega_3) = r^{-1} x$.

Proof. First of all, recalling (4.7) and noting that

$$(\partial_s^{|\alpha|} \mathcal{F}_m[f, 0])(r - c_m t, \omega) = \partial_t \left((\partial_s^{|\alpha|} \mathcal{F}_m)[0, f](r - c_m t, \omega) \right)$$

for any multi-index α and $m = 1, 2$, we see that it suffices to show (4.24) and (4.25) with $u_0 = E[g]$ and $f \equiv 0$. In the following, we write $\mathcal{F}_m[g] \equiv \mathcal{F}_m[0, g]$ for simplicity.

Next we observe that the following property of $\mathcal{F}_m[g]$:

$$\left|\partial_x^\alpha \tilde{O}^\beta \left(\mathcal{F}_m[g](r - c_m t, \omega)\right) - \mathcal{F}_m[\partial_x^\alpha \tilde{O}^\beta g](r - c_m t, \omega)\right| \qquad (4.26)$$

$$\leq C\|g\|_{|\alpha|+|\beta|, \nu+3+|\alpha|+|\beta|} (1 + t + r)^{-1}(1 + |r - c_m t|)^{-\nu}$$

for $g \in (\mathcal{S}(\mathbf{R}^3))^3$ and multi-indices α, β. Since $O_i(\psi(\omega)) = (o_i\psi)(\omega)$ for any $\psi \in C^\infty(S^2)$, we find from (3.2) that

$$\tilde{O}_i(\mathcal{F}_m[g](r - c_m t, \omega)) = \mathcal{F}_m[\tilde{O}_i g](r - c_m t, \omega)$$

for $i = 1, 2, 3$ and $m = 1, 2$ (recall (2.2)). Thus, we may assume $\beta = 0$ in proving (4.26). By (2.1) and (3.3) we have

$$\partial_i(\mathcal{R}[g](r - c_m t, \omega)) = \mathcal{R}[\partial_i g](r - c_m t, \omega) - r^{-1}\Omega_i \mathcal{R}[g](r - c_m t, \omega),$$

where we put $\Omega = {}^t(\Omega_1, \Omega_2, \Omega_3) = \omega \wedge O$. Hence, noting that r is equivalent to $1 + t + r$ for $r \geq c_2 t/2 \geq 1$, we see from (3.4) that (4.26) with $\beta = 0$ holds.

Now, we are going to prove (4.24) with $u_0 = E[g]$ and $f \equiv 0$. In the following, we assume $r \geq c_2 t/2 \geq 1$ and $\nu \geq 0$. Then, by virtue of (4.11) and (4.26), the desired estimate follows from

$$\left|\partial_t^k \left(E[g](t, x) - \sum_{m=1}^2 r^{-1}\mathcal{F}_m[g](r - c_m t, \omega)\right)\right| \leq C(1 + t + r)^{-2}.$$

We see from (4.3) and (4.18) that the above estimate follows from

$$\sum_{m=1}^2 |\partial_t^k \left(E_m[g](t, x) - r^{-1}\mathcal{F}_m[g](r - c_m t, \omega)\right)| \qquad (4.27)$$

$$\leq C(1 + t + r)^{-2}W_{-\nu}(t, r).$$

Recalling (4.14) and (4.15), we get, for $i = 1, 2, 3$,

$$\partial_t^k E_1^{(i)}[g](t, x) = \sum_{j=1}^3 c_1^k t \left(\partial_t^k Q_{\gamma_{ij}}[g_j]\right)(c_1 t, x)$$

$$+ \sum_{j=1}^3 k c_1^{k-1} \left(\partial_t^{k-1} Q_{\gamma_{ij}}[g_j]\right)(c_1 t, x),$$

where $\gamma_{ij} = e_i + e_j$. By (3.6), the second term on the right-hand side is estimated by $\sum_{j=1}^3 C\|g_j\|_{k-1, \nu+2} (1 + t + r)^{-2}(1 + |r - c_1 t|)^{-\nu}$. Meanwhile, applying (3.9) with t and c_* by $c_1 t$ and $c_1/c_2(> 1)$ respectively, we get

$$\left|c_1 t \left(\partial_t^k Q_{\gamma_{ij}}[g_j]\right)(c_1 t, x) - (4\pi r)^{-1}(-\omega)^{\gamma_{ij}} \left((-\partial_s)^k \mathcal{R}[g_j]\right)(r - c_1 t, \omega)\right|$$

$$\leq C\|g_j\|_{k+1, N_0} (1 + t + r)^{-2}(1 + |r - c_1 t|)^{-\nu}.$$

Therefore, we have

$$\left| \partial_t^k \left(E_1^{(i)}[g](t,x) - (4\pi r)^{-1} \sum_{j=1}^{3} (-\omega)^{\gamma_{ij}} c_1^{-1} R[g_j](r - c_1 t, \omega) \right) \right| \tag{4.28}$$

$$\leq \sum_{j=1}^{3} C \|g_j\|_{k+1,N_0} (1+t+r)^{-2} (1+|r-c_1 t|)^{-\nu},$$

which implies

$$\left| \partial_t^k (E_1[g](t,x) - r^{-1} \mathcal{F}_1[g](r - c_1 t, \omega)) \right| \tag{4.29}$$
$$\leq C(1+t+r)^{-2}(1+|r-c_1 t|)^{-\nu},$$

since $(-\omega)^{\gamma_{ij}} = \omega^{\gamma_{ij}}$. Similarly, one can show

$$\left| \partial_t^k (E_2[g](t,x) - r^{-1} \mathcal{F}_2[g](r - c_2 t, \omega)) \right| \tag{4.30}$$
$$\leq C(1+t+r)^{-2}(1+|r-c_2 t|)^{-\nu},$$

and hence we obtain (4.27).

Finally, we prove (4.25) with $u_0 = E[g]$ and $f \equiv 0$. Proceeding as before, with the help of (4.20) instead of (4.18), we find that the desired estimate follows from

$$\sum_{m=1}^{2} \left| \partial_t^k \left(\partial_t E_m[g](t,x) - (-c_m) r^{-1} (\partial_s \mathcal{F}_m[g])(r - c_m t, \omega) \right) \right| \tag{4.31}$$

$$+ \sum_{m=1}^{2} \left| \partial_t^k \left(\partial_x^\alpha E_m[g](t,x) - \omega^\alpha r^{-1} (\partial_s^{|\alpha|} \mathcal{F}_m[g])(r - c_m t, \omega) \right) \right|$$
$$\leq C(1+t+r)^{-2} W_{-\nu}(t,r)$$

for $r \geq c_2 t/2 \geq 1$ and $\nu \geq 0$. Because of (3.3), it follows from (4.21) and (4.22) that $\mathcal{F}_m[\partial_i g](s, \theta) = \theta_i \partial_s \mathcal{F}_m[g](s, \theta)$. Therefore, recalling (4.11), we get (4.31) from (4.27). This completes the proof. $\qquad \square$

5. The inhomogeneous elastic wave equation

In this section we recall basic estimates for the solution of

$$\begin{cases} (\partial_t^2 - L)u(t,x) = h(t,x) & \text{for } (t,x) \in (0,T) \times \mathbf{R}^3, \\ u(0,x) = 0, \ (\partial_t u)(0,x) = 0 & \text{for } x \in \mathbf{R}^3. \end{cases} \tag{5.1}$$

The following estimate is a slight improvement of the corresponding estimate given by [1, Proposition 5.1] in the sense that $\left(\min_{i=0,1,2} \{1 + |c_i t - |x||\} \right)^{1+\mu}$ is replaced by $\left(\min_{i=0,1,2} \{1 + |c_i t - |x||\} \right)^{1-\mu}$ with being $c_0 = 0$. This kind of modification was well studied in the case of the scalar wave equation, and the detail of the proof of (5.2) will appear elsewhere.

Proposition 5.1. *Let u be the solution to (5.1) and let $\mu > 0$. Then we have*

$$|\partial u(t,x)| \le C(1+r)^{-1}W_{-1}(t,r) \tag{5.2}$$

$$\times \sup_{(s,x)\in[0,t]\times\mathbf{R}^3} |x|\,(1+s+|x|)^{1+\mu}\Big(\min_{i=0,1,2}\{1+|c_i s - |x||\}\Big)^{1-\mu}|h(s,x)|_1$$

for $(t,x) \in [0,T) \times \mathbf{R}^3$.

On the other hand, the following estimate was proved by [10, Theorem 3].

Proposition 5.2. *Let u be the solution to (5.1). Then we have*

$$|\partial u(t,x)| \le C(1+r)^{-1}W_{-1}(t,r) \tag{5.3}$$

$$\times \log(2+t+r) \sup_{s\in[0,t]} \int_{\mathbf{R}^3} \min_{i=1,2}\{1+|c_i s - |x||\}\,|h(s,x)|_7\,dy,$$

$$\int_{\mathbf{R}^3} |\partial u(t,x)| \frac{dx}{|x|} \le C\log(2+t) \tag{5.4}$$

$$\times \sup_{s\in[0,t]} \left(\int_{\mathbf{R}^3}\Big((1+|y|)\min_{i=1,2}\{1+|c_i s - |y||\}\,|h(s,x)|_1\Big)^2 dy\right)^{1/2}$$

for $t \in [0,T)$.

6. Approximate solutions

This section is the core of the present paper. We shall construct an approximate solution and derive important estimates given in Proposition 6.5 below in proving Theorem 1.1. Throughout this section we assume that $f, g \in \left(C_0^\infty(\mathbf{R}^3)\right)^3$ satisfy

$$f(x) = g(x) = 0 \text{ for } |x| \ge R \tag{6.1}$$

with some $R > 1$, and that $A_1 \ne 0$ and $p_0 \not\equiv 0$ on $\mathbf{R} \times S^2$, where p_0 is defined by (1.9).

Lemma 6.1. *Let $p(s,\theta,\tau)$ be the solution to (1.13)–(1.14) vanishing for $|s| \ge R$. Let $0 < \tau_0 < \tau_*$ with τ_* being defined by (1.10). Then for any $N > 0$, and for any multi-indices $\beta = (\beta_0, \ldots, \beta_4)$ and $\gamma = (\gamma_0, \ldots, \gamma_3)$, there exists a positive constant $C = C(\tau_0, \beta, \gamma, N)$ such that*

$$|\Lambda^\beta p(s,\theta,\tau)| \le C(1+|s|)^{-N}, \tag{6.2}$$

$$|\Lambda_*^\gamma\{p(s,\theta,\tau) - p_0(s,\theta)\}| \le C\tau(1+|s|)^{-N} \tag{6.3}$$

for all $(s,\theta,\tau) \in \mathbf{R} \times S^2 \times [0,\tau_0]$.

Proof. First of all, we note that (6.2) and (6.3) follow from

$$|\Lambda^\beta \partial_s p(s,\theta,\tau)| \le C(1+|s|)^{-N}, \tag{6.4}$$

$$|\Lambda_*^\gamma \partial_s\{p(s,\theta,\tau) - p_0(s,\theta)\}| \le C(1+|s|)^{-N} \tag{6.5}$$

for all $(s, \theta, \tau) \in \mathbf{R} \times S^2 \times [0, \tau_0]$ and $N > 0$, respectively, because both $p(s, \theta, \tau)$ and $p_0(s, \theta)$ vanish for $|s| \geq R$.

Next we prove (6.4). If we set $P = \partial_s p$, then it satisfies

$$c_1^2 \partial_\tau P + A_1 P \partial_s P = 0 \quad \text{in } \mathbf{R} \times S^2 \times [0, \tau_*), \tag{6.6}$$

$$P(s, \theta, 0) = \partial_s p_0(s, \theta) \quad \text{for } (s, \theta) \in \mathbf{R} \times S^2. \tag{6.7}$$

Observe that for $(s, s_0, \theta, \tau) \in \mathbf{R} \times \mathbf{R} \times S^2 \times [0, \tau_0)$, the equation

$$F(s, s_0, \theta, \tau) := c_1^2 (s_0 - s) + \partial_s p_0(s_0, \theta) A_1 \tau = 0 \tag{6.8}$$

determines the implicit function $s_0 = s_0(s, \theta, \tau)$, because

$$\partial_{s_0} F(s, s_0, \theta, \tau) = c_1^2 + \partial_s^2 p_0(s_0, \theta) A_1 \tau \geq c_1^2(1 - \tau/\tau_*) > 0.$$

Therefore, the solution to (6.6)–(6.7) is given by $P(s, \theta, \tau) = (\partial_s p_0)(s_0(s, \theta, \tau), \theta)$, and hence for $(s, \theta, \tau) \in \mathbf{R} \times S^2 \times [0, \tau_0)$, we have

$$\partial_s p(s, \theta, \tau) = (\partial_s p_0)(s_0(s, \theta, \tau), \theta). \tag{6.9}$$

Since (3.4) implies $|\Lambda^\beta p_0(s, \theta)| \leq C(1 + |s|)^{-N}$ for any $(s, \theta) \in \mathbf{R} \times S^2$ and $N > 0$, we see that $\Lambda^\beta s_0(s, \theta, \tau)$ is bounded for any $(s, \theta, \tau) \in \mathbf{R} \times S^2 \times [0, \tau_0)$, because we have

$$\partial_s s_0(s, \theta, \tau) = \frac{c_1^2}{c_1^2 + (\partial_s^2 p_0)(s_0(s, \theta, \tau), \theta) A_1 \tau},$$

$$\partial_\tau s_0(s, \theta, \tau) = \frac{-A_1(\partial_s p_0)(s_0(s, \theta, \tau), \theta)}{c_1^2 + (\partial_s^2 p_0)(s_0(s, \theta, \tau), \theta) A_1 \tau},$$

$$o_i s_0(s, \theta, \tau) = \frac{-A_1 \tau (o_i \partial_s p_0)(s_0(s, \theta, \tau), \theta)}{c_1^2 + (\partial_s^2 p_0)(s_0(s, \theta, \tau), \theta) A_1 \tau}.$$

Therefore, we get (6.4) by using (6.9).

Next we prove (6.5). Since $s_0(s, \theta, 0) = s$, we get

$$\partial_s p(s, \theta, \tau) - \partial_s p_0(s, \theta) = \tau \int_0^1 (\partial_s^2 p_0)(s_0(s, \theta, \sigma\tau), \theta) \partial_\tau s_0(s, \theta, \sigma\tau) d\sigma.$$

In view of (6.8), we see that $(1 + |s_0(s, \theta, \tau)|)^{-N}$ is equivalent to $(1 + |s|)^{-N}$ for $(s, \theta, \tau) \in \mathbf{R} \times S^2 \times [0, \tau_0)$, because $|\Lambda_*^\gamma(\partial_s p_0(s_0(s, \theta, \tau), \theta) A_1 \tau)|$ is bounded. Thus we find (6.5) holds. This completes the proof. □

For a real-valued function $\varphi(s, \theta, \tau)$ on $\mathbf{R} \times S^2 \times [0, \tau_*)$, we shall write

$$\widetilde{\varphi}(t, x) := \varphi(r - c_1 t, \omega, \varepsilon \log(\varepsilon t))$$

for $(t, x) \in [\varepsilon^{-1}, \varepsilon^{-1} \exp(\tau_* \varepsilon^{-1})) \times \mathbf{R}^3$ with $r = |x|$ and $\omega = r^{-1} x$. Then we have

$$\partial_t \widetilde{\varphi} = -c_1 \widetilde{\partial_s \varphi} + \varepsilon t^{-1} \widetilde{\partial_\tau \varphi}, \quad O_i \widetilde{\varphi} = \widetilde{o_i \varphi} \quad (i = 1, 2, 3), \tag{6.10}$$

$$\operatorname{grad} \widetilde{\varphi} = \omega \widetilde{\partial_s \varphi} - r^{-1} \omega \wedge \widetilde{o \varphi}, \tag{6.11}$$

where we have used (2.1) to get (6.11).

Let $p(s, \theta, \tau)$ be the solution to (1.13)–(1.14) vanishing for $|s| \geq R$. Using the above notation, we define

$$w_1(t, x) = \varepsilon r^{-1} \left(\widetilde{p}(t, x)\omega + \mathcal{F}_2[f, g](r - c_2 t, \omega) \right) \tag{6.12}$$

for $(t, x) \in [\varepsilon^{-1}, \varepsilon^{-1} \exp(\tau_* \varepsilon^{-1})) \times (\mathbf{R}^3 \setminus \{0\})$. Note that

$$w_1(t, x) = 0 \quad \text{for } |x| \geq c_1 t + R. \tag{6.13}$$

The following estimates, which shows that w_1 is a good approximation of u_0 near the characteristic cones $r = c_i t$ $(i = 1, 2)$, are reduced from Lemma 6.1.

Corollary 6.2. *Let* $0 < \tau_0 < \tau_*$ *and let* $0 < \varepsilon \leq 1$. *Then for any nonnegative integer* k, *there exists a positive constant* $C = C(\tau_0, k)$ *such that*

$$|w_1(t, x)|_k \leq C\varepsilon(1 + t + r)^{-1}, \tag{6.14}$$

$$|\partial w_1(t, x)|_k \leq C\varepsilon(1 + t + r)^{-1} W_{-1}(t, r), \tag{6.15}$$

$$|\operatorname{div} w_1(t, x)|_k \leq C\varepsilon(1 + t + r)^{-1}(1 + |r - c_1 t|)^{-1}, \tag{6.16}$$

$$|\operatorname{rot} w_1(t, x)|_k \leq C\varepsilon(1 + t + r)^{-1}(1 + |r - c_2 t|)^{-1} \tag{6.17}$$

for $c_2 t/2 \leq |x| \leq c_1 t + R$ *and* $t \in [\varepsilon^{+1}, \varepsilon^{-1} \exp(\tau_0 \varepsilon^{-1}))$. *Moreover, we have*

$$|w_1(t, x) - \varepsilon u_0(t, x)|_k \leq C\varepsilon(1 + t + r)^{-2}, \tag{6.18}$$

$$|\partial_t \{w_1(t, x) - \varepsilon u_0(t, x)\}|_k \leq C\varepsilon(1 + t + r)^{-2} W_{-1}(t, r), \tag{6.19}$$

for $c_2 t/2 \leq |x| \leq c_1 t + R$ *and* $1/\varepsilon \leq t \leq 2/\varepsilon$. *Here,* u_0 *is the solution of the Cauchy problem* (4.1)–(4.2).

Proof. We suppose that $c_2 t/2 \leq r \leq c_1 t + R$ and $t \in [\varepsilon^{-1}, \varepsilon^{-1} \exp(\tau_0 \varepsilon^{-1}))$ in what follows. Then we have

$$|t^{-1}|_k + |r^{-1}|_k + |(1 + t + r)^{-1}|_k \leq C(1 + t + r)^{-1}. \tag{6.20}$$

First we prove (6.14) and (6.15). It follows from (4.23), (2.1), and (6.20) that

$$|\mathcal{F}_2[f, g](r - c_2 t, \omega)|_k \leq C(1 + |r - c_2 t|)^{-1}.$$

Meanwhile, from (6.2), (6.4) with $N = 1$, (6.10), (6.11), and (6.20), we get

$$|\widetilde{p}(t, x)|_k \leq C \sum_{|\beta| \leq k} |\widetilde{\Lambda^\beta p}(t, x)| \leq C, \tag{6.21}$$

$$|\partial \widetilde{p}(t, x)|_k \leq C \sum_{|\beta| \leq k} |\widetilde{\Lambda^\beta \partial_s p}(t, x)| + C(1 + t + r)^{-1} \sum_{|\beta| \leq k+1} |\widetilde{\Lambda^\beta p}(t, x)| \tag{6.22}$$

$$\leq C(1 + |r - c_1 t|)^{-1}.$$

Thus we obtain (6.14) and (6.15) from (6.12).

Next we prove (6.16). A direct computation shows that

$$\operatorname{div}\left(r^{-1}\widetilde{p}(t,x)\omega\right) = r^{-1}\widetilde{\partial_s p}(t,x) + r^{-2}\widetilde{p}(t,x), \tag{6.23}$$

$$\operatorname{div}\left(r^{-1}\mathcal{F}_2[f,g](r-c_2t,\omega)\right) \tag{6.24}$$

$$= -r^{-2}(2\omega \cdot \widetilde{\mathcal{R}}_2[f,g](r-c_2t,\omega) + \Omega \cdot \widetilde{\mathcal{R}}_2[f,g](r-c_2t,\omega)),$$

where we put $\Omega \cdot f(x) = \sum_{j=1}^{3} \Omega_j f_j(x)$ with $\Omega = \omega \wedge O$ (recall also (4.22)). Therefore, by (6.2), (6.4) with $N = 1$, and (3.4), we get (6.16).

Next we prove (6.17). A direct computation shows that

$$\operatorname{rot}\left(r^{-1}\widetilde{p}(t,x)\omega\right) = -r^{-2}\Omega \wedge (\widetilde{p}(t,x)\omega), \tag{6.25}$$

$$\operatorname{rot}\left(r^{-1}\mathcal{F}_2[f,g](r-c_2t,\omega)\right) = r^{-1}\operatorname{rot}\widetilde{\mathcal{R}}_2[f,g](r-c_2t,\omega) \tag{6.26}$$

$$- r^{-2}(\omega \wedge \widetilde{\mathcal{R}}_2[f,g](r-c_2t,\omega) - \Omega \wedge \Pi(\omega)\widetilde{\mathcal{R}}_2[f,g](r-c_2t,\omega)).$$

Thus (6.2) and (3.4) yield (6.17).

Next we prove (6.19). Suppose that we also have $1/\varepsilon \leq t \leq 2/\varepsilon$ from now on. In view of (4.25), it suffices to show

$$\left|\partial_t\left\{w_1(t,x) - \sum_{m=1}^{2} \varepsilon r^{-1}\mathcal{F}_m[f,g](r-c_mt,\omega)\right\}\right|_k \leq C\varepsilon(1+t+r)^{-2}W_{-1}(t,r),$$

or

$$\left|\partial_t\{\widetilde{p}(t,x)\omega - \mathcal{F}_1[f,g](r-c_1t,\omega)\}\right|_k \leq C(1+t+r)^{-1}W_{-1}(t,r),$$

because of (6.12) and (6.20). We see from (4.21) and (1.9) that the above estimate follows from

$$\left|\partial_t\{\widetilde{p}(t,x) - p_0(r-c_1t,\omega)\}\right|_k \leq C(1+t+r)^{-1}W_{-1}(t,r). \tag{6.27}$$

It follows from (6.10), (6.11), (6.2), and (6.5) with $N = 1$ that the left-hand side of (6.27) is bounded by

$$C\sum_{|\gamma|\leq k} |\widetilde{\Lambda_*^\gamma \partial_s p}(t,x) - (\Lambda_*^\gamma \partial_s p_0)(r-c_1t,\omega)|$$

$$+ C\varepsilon(1+t+r)^{-1}\sum_{|\beta|\leq k} |\widetilde{\Lambda^\beta \partial_\tau p}(t,x)|$$

$$\leq C\varepsilon\left((\log(\varepsilon t))(1+|r-c_1t|)^{-1} + (1+t+r)^{-1}\right),$$

which yields (6.27), because $t \leq 2/\varepsilon$ implies $\varepsilon \leq C(1+t+r)^{-1}$.

Similarly, one can show (6.18) by using (4.24), (6.3) instead of (4.25), (6.5), respectively. This completes the proof. $\qquad\square$

Next we examine how well $w_1(t,x)$ satisfies the original equation (1.1) near the characteristic cones $r = c_it$ ($i = 1,2$). We set

$$E[u](t,x) = (\partial_t^2 - L)u(t,x) - F\left(\nabla u(t,x), \nabla^2 u(t,x)\right). \tag{6.28}$$

Lemma 6.3. *Let $0 < \tau_0 < \tau_*$ and let $0 < \varepsilon \le 1$. Then for any nonnegative integer k, there exists a positive constant $C = C(\tau_0, k)$ such that*

$$|E[w_1](t,x) - (c_1^2 - c_2^2)\varepsilon r^{-2}\{\omega \wedge o\,\widetilde{\partial_s p}(t,x) + (\partial_s Y)(r - c_2 t, \omega)\omega\} \qquad (6.29)$$
$$+ A_2\,\mathrm{grad}\,|\mathrm{rot}\,w_1(t,x)|^2|_k \le C\varepsilon(1 + t + r)^{-3},$$

for $c_2 t/2 \le |x| \le c_1 t + R$ and $t \in [\varepsilon^{-1}, \varepsilon^{-1}\exp(\tau_0\varepsilon^{-1}))$. Here we have set

$$Y(s,\omega) = 2\omega \cdot \widetilde{\mathcal{R}}_2[f,g](s,\omega) + \Omega \cdot \widetilde{\mathcal{R}}_2[f,g](s,\omega).$$

Proof. Let $c_2 t/2 \le |x| \le c_1 t + R$ and $t \in [\varepsilon^{-1}, \varepsilon^{-1}\exp(\tau_0\varepsilon^{-1}))$. Then, t and r are equivalent to $1 + t + r$.

It holds that

$$\partial_t^2 \widetilde{p}(t,x) = c_1^2 \widetilde{\partial_s^2 p}(t,x) - 2c_1\varepsilon t^{-1}\widetilde{\partial_r \partial_s p} + t^{-2}(\varepsilon^2 \widetilde{\partial_\tau^2 p} - \varepsilon\widetilde{\partial_\tau p}),$$

$$\Delta(r^{-1}\widetilde{p}(t,x)\omega) = r^{-1}\widetilde{\partial_s^2 p}(t,x)\omega + r^{-3}\Delta_\omega(\widetilde{p}(t,x)\omega),$$

$$\mathrm{grad}\,\mathrm{div}\,(r^{-1}\widetilde{p}(t,x)\omega) = r^{-1}\widetilde{\partial_s^2 p}(t,x)\omega - r^{-2}\omega \wedge o\,\widetilde{\partial_s p}(t,x)$$
$$- 2r^{-3}\widetilde{p}(t,x)\omega - r^{-3}\omega \wedge o\,\widetilde{p}(t,x),$$

where $\Delta_\omega = \sum_{j=1}^3 O_j^2$. Therefore, we have

$$\left|(\partial_t^2 - L)(\varepsilon r^{-1}\widetilde{p}(t,x)\omega) + 2c_1\varepsilon^2(tr)^{-1}\widetilde{\partial_r \partial_s p}(t,x)\omega \qquad (6.30)\right.$$
$$\left. -(c_1^2 - c_2^2)\varepsilon r^{-2}\omega \wedge o\,\widetilde{\partial_s p}(t,x)\right|_k \le C\varepsilon(1 + r + t)^{-3}.$$

Meanwhile, we have

$$(\partial_t^2 - c_2^2\Delta)(r^{-1}\mathcal{F}_2[f,g](r - c_2 t, \omega)) = -c_2^2 r^{-3}\Delta_\omega \mathcal{F}_2[f,g](r - c_2 t, \omega).$$

Hence, recalling (6.24), (6.10), and (6.11), we obtain

$$\left|(\partial_t^2 - L)(\varepsilon r^{-1}\mathcal{F}_2[f,g](r - c_2 t, \omega)) \qquad (6.31)\right.$$
$$\left. -(c_1^2 - c_2^2)\,\varepsilon r^{-2}(\partial_s Y)(r - c_2 t, \omega)\omega\right|_k \le C\varepsilon(1 + r + t)^{-3}.$$

Next we consider the nonlinear term F which is expressed as (1.3). It follows from (6.16), (6.17) that

$$|\mathrm{rot}\,((\mathrm{div}\,w_1(t,x))(\mathrm{rot}\,w_1(t,x)))|_k \le C\varepsilon^2(1 + t + r)^{-3}.$$

By using (2.1) together with the condition on B_{jkl}^{iab}, we get from (6.14) and (6.15)

$$|N(w_1(t,x))|_k \le C\varepsilon^2(1 + t + r)^{-3}.$$

We see from (6.23), (6.24) that

$$|\mathrm{grad}\,(\mathrm{div}\,w_1)^2 - \mathrm{grad}\,(\varepsilon r^{-1}\widetilde{\partial_s p}(t,x))^2|_k \le C\varepsilon^2(1 + t + r)^{-3},$$

and hence

$$|\mathrm{grad}\,(\mathrm{div}\,w_1)^2 - 2\varepsilon^2 r^{-2}\widetilde{\partial_s p}(t,x)\widetilde{\partial_s^2 p}(t,x)\,\omega|_k \le C\varepsilon^2(1 + t + r)^{-3}.$$

Thus we obtain

$$\left| F(\nabla w_1, \nabla^2 w_1) - A_2 \operatorname{grad} |\operatorname{rot} w_1|^2 \right. \tag{6.32}$$

$$\left. -2A_1 \varepsilon^2 r^{-2} \widetilde{\partial_s p}(t,x) \widetilde{\partial_s^2 p}(t,x) \, \omega \right|_k \leq C\varepsilon^2 (1+r+t)^{-3}.$$

Observe that (6.2) and (6.4) (with $N = 1/2$) yield

$$|\widetilde{\partial_s p}(t,x)|_k \leq C(1 + |c_1 t - r|)^{-1/2}. \tag{6.33}$$

By (6.6) with $P = \partial_s p$, and (6.33), we obtain

$$\left| 2c_1 \varepsilon^2 (tr)^{-1} \widetilde{\partial_\tau \partial_s p} + 2A_1 \varepsilon^2 r^{-2} \widetilde{\partial_s p} \, \widetilde{\partial_s^2 p} \right|_k \tag{6.34}$$

$$= \left| 2A_1 (r - c_1 t)\varepsilon^2 (c_1 t)^{-1} r^{-2} \widetilde{\partial_s p} \, \widetilde{\partial_s^2 p} \right|_k$$

$$\leq C\varepsilon^2 (1 + t + r)^{-3}.$$

Now (6.30), (6.31), (6.32), and (6.34) imply (6.29). This completes the proof. □

In order to eliminate $r^{-2}\{\omega \wedge o \, \widetilde{\partial_s p}(t,x) + (\partial_s Y)(r - c_2 t, \omega)\omega\}$ in the estimate (6.29), we need to construct a more precise approximation. For this reason, we set

$$q_1(s, \theta, \tau) = \int_s^\infty \theta \wedge (o \, p)(s', \theta, \tau) ds', \quad q_2(s, \theta) = \int_{-\infty}^s Y(s', \theta) \theta \, ds', \tag{6.35}$$

and define

$$w(t,x) = w_1(t,x) + \varepsilon r^{-2} \left(\widetilde{q_1}(t,x) + q_2(r - c_2 t, \omega) \right) \tag{6.36}$$

for $(t,x) \in [\varepsilon^{-1}, \varepsilon^{-1} \exp(\tau_0 \varepsilon^{-1})) \times (\mathbf{R}^3 \setminus \{0\})$. Then, w enjoys the same estimates as in Corollary 6.2 together with a suitable estimate for $E[w]$ as follows.

Lemma 6.4. *Let $0 < \tau_0 < \tau_*$. We assume that $0 < \varepsilon \leq 1$. Then for any nonnegative integer k, there exists a positive constant $C = C(\tau_0, k)$ such that*

$$|w(t,x)|_k \leq C\varepsilon(1 + t + r)^{-1}, \tag{6.37}$$

$$|\partial w(t,x)|_k \leq C\varepsilon(1 + t + r)^{-1} W_{-1}(t,r), \tag{6.38}$$

$$|\operatorname{div} w(t,x)|_k \leq C\varepsilon(1 + t + r)^{-1}(1 + |r - c_1 t|)^{-1}, \tag{6.39}$$

$$|\operatorname{rot} w(t,x)|_k \leq C\varepsilon(1 + t + r)^{-1}(1 + |r - c_2 t|)^{-1}, \tag{6.40}$$

$$|E[w](t,x) + A_2 \operatorname{grad} |\operatorname{rot} w_1(t,x)|^2|_k \leq C\varepsilon(1 + t + r)^{-3} \tag{6.41}$$

for $c_2 t/2 \leq |x| \leq c_1 t + R$ and $t \in [\varepsilon^{-1}, \varepsilon^{-1} \exp(\tau_0 \varepsilon^{-1}))$. Moreover, we have

$$|w(t,x) - \varepsilon u_0(t,x)|_k \leq C\varepsilon(1 + t + r)^{-2}, \tag{6.42}$$

$$|\partial_t\{w(t,x) - \varepsilon u_0(t,x)\}|_k \leq C\varepsilon(1 + t + r)^{-2} W_{-1}(t,r), \tag{6.43}$$

for $c_2 t/2 \leq |x| \leq c_1 t + R$ and $1/\varepsilon \leq t \leq 2/\varepsilon$. Here, u_0 is the solution of the Cauchy problem (4.1)–(4.2).

Proof. Since $p(s, \theta, \tau) = 0$ for $|s| \geq R$, we see from (6.2) and (3.4) that

$$|\Lambda^\beta q_1(s, \theta, \tau)| \leq C, \quad |\Lambda^\beta \partial_s q_1(s, \theta, \tau)| \leq C(1+s)^{-1}, \tag{6.44}$$

$$|\Lambda^\beta q_2(s, \theta)| \leq C, \quad |\Lambda^\beta \partial_s q_2(s, \theta)| \leq C(1+s)^{-1}, \tag{6.45}$$

for multi-indices β and $(s, \theta, \tau) \in \mathbf{R} \times S^2 \times [0, \tau_0]$. Therefore, if we set

$$w_2(t, x) = \varepsilon r^{-2} \left(\widetilde{q_1}(t, x) + q_2(r - c_2 t, \omega) \right),$$

then we get

$$|w_2(t, x)|_k \leq C\varepsilon(1 + t + r)^{-2}, \tag{6.46}$$

$$|\partial w_2(t, x)|_k \leq C\varepsilon(1 + t + r)^{-2} W_{-1}(t, r), \tag{6.47}$$

so that the estimates in Lemma 6.4 except for (6.41) immediately follow from Corollary 6.2.

In order to show (6.41), we write

$$E[w] + A_2 \operatorname{grad} |\operatorname{rot} w_1|^2 \tag{6.48}$$
$$= (E[w_1] + (c_1^2 - c_2^2)\varepsilon r^{-2}\{\widetilde{\partial_s^2 q_1}(t, x) - (\partial_s^2 q_2)(r - c_2 t, \omega)\}$$
$$+ A_2 \operatorname{grad} |\operatorname{rot} w_1|^2)$$
$$+ \left((\partial_t^2 - L)w_2 - (c_1^2 - c_2^2)\varepsilon r^{-2}\{\widetilde{\partial_s^2 q_1}(t, x) - (\partial_s^2 q_2)(r - c_2 t, \omega)\} \right)$$
$$+ \left(F(\nabla w_1, \nabla^2 w_1) - F(\nabla w, \nabla^2 w) \right).$$

By (6.15), (6.47) we get

$$|F(\nabla w_1, \nabla^2 w_1) - F(\nabla w, \nabla^2 w)|_k \leq C\varepsilon^2(1 + t + r)^{-3}. \tag{6.49}$$

Using (6.44), we find

$$|(\partial_t^2 - c_2^2 \Delta)(r^{-2}\widetilde{q_1}(t, x)) - (c_1^2 - c_2^2)r^{-2}\widetilde{\partial_s^2 q_1}(t, x)|_k \leq C(1 + t + r)^{-3}.$$

Since $\theta \cdot q_1(s, \theta) = 0$, we have $\operatorname{div}(r^{-2}\widetilde{q_1}(t, x)) = -r^{-3}\Omega \cdot \widetilde{q_1}(t, x)$ by (2.1). Therefore, we get

$$|(\partial_t^2 - L)(r^{-2}\widetilde{q_1}(t, x)) - (c_1^2 - c_2^2)r^{-2}\widetilde{\partial_s^2 q_1}(t, x)|_k \leq C(1 + t + r)^{-3}. \tag{6.50}$$

On the other hand, we have from (6.45)

$$|(\partial_t^2 - c_2^2 \Delta)(r^{-2}q_2(r - c_2 t, \omega))|_k \leq C(1 + t + r)^{-3}.$$

Since $\operatorname{div}(r^{-2}q_2(r - c_2 t, \omega)) = r^{-2}Y(r - c_2 t, \omega)$, we obtain

$$|(\partial_t^2 - L)(r^{-2}q_2(r - c_2 t, \omega)) \tag{6.51}$$
$$+ (c_1^2 - c_2^2)r^{-2}(\partial_s^2 q_2)(r - c_2 t, \omega)|_k \leq C(1 + t + r)^{-3}.$$

Now, in view of (6.48), we see from (6.49), (6.50), (6.51), and (6.29) that (6.41) holds, because $\widetilde{\partial_s^2 q_1}(t, x) - (\partial_s^2 q_2)(r - c_2 t, \omega) = -(\omega \wedge \widetilde{\partial_s p}(t, x) + (\partial_s Y)(r - c_2 t, \omega)\omega)$. This completes the proof. $\qquad \square$

Now we are in a position to construct an approximate solution u_1 for all $(t, x) \in [\varepsilon^{-1}, \varepsilon^{-1} \exp(\tau_0 \varepsilon^{-1})) \times (\mathbf{R}^3 \setminus \{0\})$: Let χ and ξ be smooth and nonnegative functions on $[0, \infty)$ such that

$$\chi(s) = \begin{cases} 1, & s \leq 1, \\ 0, & s \geq 2, \end{cases} \qquad \xi(s) = \begin{cases} 0, & s \leq c_2/2, \\ 1, & s \geq 3c_2/4. \end{cases}$$

Let $0 < \varepsilon \leq 1$ in the following. We put $\chi_\varepsilon(t) = \chi(\varepsilon t)$ and $\eta(t, x) = \xi(|x|/t)$. Since

$$\varepsilon \leq C(1+t)^{-1} \quad \text{if } 0 \leq \varepsilon t \leq 2, \tag{6.52}$$

we get

$$\left| \frac{d^m \chi_\varepsilon}{dt^m}(t) \right| = \varepsilon^m \left| \frac{d^m \chi}{dt^m}(\varepsilon t) \right| \leq C(1+t)^{-m} \quad \text{for } t \geq 0, \tag{6.53}$$

where m is a nonnegative integer. Meanwhile, we easily have $O_j \eta(t, x) = 0$ for $1 \leq j \leq 3$. Since $c_2 t/2 \leq r \leq 3c_2 t/4$ for $(t, x) \in \text{supp}\, \partial \eta$, we have

$$\sum_{|\alpha|=m} |\partial^\alpha \eta(t, x)| \leq C(1+t+r)^{-m} \quad \text{for } (t, x) \in [1, \infty) \times \mathbf{R}^3, \tag{6.54}$$

where m is a nonnegative integer, $\partial = (\partial_t, \nabla_x)$, and α is a multi-index. Besides, we get

$$W_{-1}(t, r) \leq C(1+t+r)^{-1} \quad \text{if } 0 \leq r \leq 3c_2 t/4. \tag{6.55}$$

Let u_0 be the solution of the Cauchy problem (4.1)–(4.2), and let w be given by (6.36). We define

$$u_1(t, x) = \chi_\varepsilon(t) \varepsilon u_0(t, x) + (1 - \chi_\varepsilon(t)) \eta(t, x) w(t, x) \tag{6.56}$$

for $(t, x) \in [0, \varepsilon^{-1} \exp(\tau_* \varepsilon^{-1})) \times \mathbf{R}^3$. By (6.1) and the property of finite propagation, we have $|x| \leq c_1 t + R$ in $\text{supp}\, u_0$. Hence, recalling (6.13), we find that

$$u_0(t, x) = w_1(t, x) = 0 \quad \text{for } |x| \geq c_1 t + R. \tag{6.57}$$

Then we have the following:

Proposition 6.5. *Let $0 < \tau_0 < \tau_*$, k be a nonnegative integer, $0 \leq \lambda \leq 1/2$, $0 < \mu \leq 1/4$, and $0 < \varepsilon \leq 1$. Then there exists a positive constant $C = C(\tau_0, k, \lambda, \mu)$ such that*

$$|u_1(t, x)|_k \leq C\varepsilon(1+t+r)^{-1}, \tag{6.58}$$

$$|\partial u_1(t, x)|_k \leq C\varepsilon(1+t+r)^{-1} W_{-1}(t, r), \tag{6.59}$$

$$|E[u_1](t, x) + A_2 \,\text{grad}\, |\text{rot}\, w_1(t, x)|^2|_k \tag{6.60}$$

$$\leq C\varepsilon^{1+\lambda}(1+t+r)^{-2+\lambda-\mu} W_{-1+\mu}(t, r)$$

for $(t, x) \in [0, \varepsilon^{-1} \exp(\tau_0 \varepsilon^{-1})] \times \mathbf{R}^3$, and

$$\left\| |E[u_1](t, \cdot) + A_2 \,\text{grad}\, |\text{rot}\, w_1(t, \cdot)|^2|_k \right\|_{L^2} \leq C\varepsilon^{1+\lambda}(1+t)^{-(3/2)+\lambda} \tag{6.61}$$

for $t \in [0, \varepsilon^{-1} \exp(\tau_0 \varepsilon^{-1})]$.

Proof. We write $x = r\omega$ with $r = |x|$ and $\omega \in S^2$. First we prove (6.58) and (6.59). It follows from (4.8) that

$$|u_0(t,x)|_k \leq C(1+t+r)^{-1} W_{-1}(t,r) \tag{6.62}$$

for $(t,x) \in [0,\infty) \times \mathbf{R}^3$. We see from (6.57) that $(1+t)^{-1} \leq C(1+t+r)^{-1}$ for $(t,x) \in \operatorname{supp} w_1$. Therefore, we get (6.58) and (6.59) from (6.37), (6.38), (6.53), (6.54), and (6.62).

Next we consider (6.60) and (6.61). If we set

$$v(t,x) = \eta(t,x)w(t,x) - \varepsilon u_0(t,x), \tag{6.63}$$

then we have $u_1 = \varepsilon u_0 + (1 - \chi_\varepsilon)v$ by (6.56). Therefore, it follows that

$$E[u_1] = I_0 + I_1 + I_2 + I_3, \tag{6.64}$$

where we put

$$
\begin{aligned}
I_0 &= -\chi_\varepsilon(t)F(\nabla u_1, \nabla^2 u_1), \\
I_1 &= -\chi_\varepsilon''(t)v(t,x), \\
I_2 &= -2\chi_\varepsilon'(t)\partial_t v(t,x), \\
I_3 &= \big(1 - \chi_\varepsilon(t)\big)\left\{\big(\partial_t^2 - L\big)\big(\eta(t,x)w(t,x)\big) - F(\nabla u_1, \nabla^2 u_1)\right\}.
\end{aligned}
$$

We will estimate I_j for $0 \leq j \leq 3$. Let $0 \leq \lambda \leq 1/2$ and $0 < \mu \leq 1/4$ in the following.

By (6.52) and (6.57), we have

$$\varepsilon \leq C(1+t+r)^{-1} \quad \text{for } (t,x) \in \operatorname{supp} u_0 \cup \operatorname{supp} w_1. \tag{6.65}$$

From (6.62), (6.15), (6.47), and (6.65) we get

$$|I_0|_k \leq C\varepsilon^{1+\lambda}(1+t+r)^{-3+\lambda}W_{-2}(t,r), \tag{6.66}$$

which yields

$$\big\||I_0|_k\big\|_{L^2} \leq C\varepsilon^{1+\lambda}(1+t)^{-2+\lambda}. \tag{6.67}$$

Next we estimate I_1. We may assume $t \geq 1$, because $\varepsilon t \geq 1$ in $\operatorname{supp} \chi_\varepsilon''$. Therefore, (6.54), (6.55) and (6.62) yield

$$\big|(1 - \eta(t,x))u_0(t,x)\big|_k \leq C(1+t+r)^{-2}. \tag{6.68}$$

Observe that we have $1/\varepsilon \leq t \leq 2/\varepsilon$ and $c_2 t/2 \leq r$ in $\operatorname{supp}(\chi_\varepsilon''\eta)$. Thus, writing $I_1 = -\varepsilon^2\chi''(\varepsilon t)\big(\eta(w - \varepsilon u_0) - \varepsilon(1 - \eta)u_0\big)$, by (6.42), (6.68), and (6.65), we get

$$|I_1|_k \leq C\varepsilon^3(1+t+r)^{-2} \leq C\varepsilon^2(1+t+r)^{-3}. \tag{6.69}$$

In order to evaluate I_2, we use

$$\big|(1 - \eta(t,x))\partial_t u_0(t,x)\big|_k \leq C(1+t+r)^{-3}, \tag{6.70}$$

which follows from (6.54), (6.55) and (4.9). Then, writing

$$
\begin{aligned}
I_2 = -\,2\varepsilon\chi'(\varepsilon t)\big((\partial_t\eta)(w - \varepsilon u_0) + (\partial_t\eta)\varepsilon u_0 \\
+ \eta(\partial_t w - \varepsilon\partial_t u_0) - (1 - \eta)\varepsilon\partial_t u_0\big),
\end{aligned}
$$

by (6.42), (6.43), (6.54), (6.62), (6.65), and (6.70) that

$$|I_2|_k \leq C\varepsilon^{1+\lambda}(1+t+r)^{-3+\lambda}W_{-1}(t,r). \tag{6.71}$$

By (6.69), (6.71), and (6.65) we get

$$|I_1 + I_2|_k \leq C\varepsilon^{1+\lambda}(1+t+r)^{-3+\lambda}W_{-1}(t,r), \tag{6.72}$$

$$\||I_1 + I_2|_k\|_{L^2} \leq C\varepsilon^{1+\lambda}(1+t)^{-2+\lambda}. \tag{6.73}$$

Next we consider I_3 by rewriting it as

$$I_3 = (1 - \chi_\varepsilon(t))(I_{31} + I_{32} + I_{33}), \tag{6.74}$$

where we have set

$$I_{31} = -F(\nabla u_1, \nabla^2 u_1) + \eta F(\nabla w, \nabla^2 w),$$
$$I_{32} = [\partial_t^2 - L, \eta]w$$
$$I_{33} = \eta((\partial_t^2 - L)w - F(\nabla w, \nabla^2 w)).$$

In the following, we assume $t \geq 1$, because $\varepsilon t \geq 1$ in $\mathrm{supp}(1 - \chi_\varepsilon)$.

We first estimate I_{31}. We may assume $\varepsilon t \leq 2$ or $r \leq 3c_2 t/4$, because $I_{31} = 0$ otherwise. If $0 \leq \varepsilon t \leq 2$, then we have

$$|(1 - \chi_\varepsilon)I_{31}|_k \leq C\varepsilon^{1+\lambda}(1+t+r)^{-3+\lambda}W_{-2}(t,r),$$

similarly to (6.66). Meanwhile, if $r \leq 3c_2 t/4$, then (6.38), (6.59), and (6.55) yield

$$|(1 - \chi_\varepsilon)I_{31}|_k \leq C\varepsilon^2(1+t+r)^{-4}.$$

Summing up, we have proved

$$|(1 - \chi_\varepsilon)I_{31}|_k \leq C\varepsilon^{1+\lambda}(1+t+r)^{-3+\lambda}W_{-2}(t,r) + C\varepsilon^2(1+t+r)^{-4}. \tag{6.75}$$

By (6.37), (6.38), (6.54) with $m = 1, 2$, and (6.55), we get

$$|(1 - \chi_\varepsilon)I_{32}|_k \leq C\varepsilon(1+t+r)^{-3}. \tag{6.76}$$

It follows that

$$(1 - \chi_\varepsilon)I_{33} + A_2 \, \mathrm{grad} \, |\mathrm{rot} \, w_1|^2 = (1 - \chi_\varepsilon)\eta(E[w] + A_2 \, \mathrm{grad} \, |\mathrm{rot} \, w_1|^2)$$
$$+ A_2 \left((1 - \chi_\varepsilon)(1 - \eta) + \chi_\varepsilon\right) \mathrm{grad} \, |\mathrm{rot} \, w_1|^2.$$

Therefore, by (6.41), (6.15), (6.55), and (6.65), we get

$$|(1 - \chi_\varepsilon)I_{33} + A_2 \, \mathrm{grad} \, |\mathrm{rot} \, w_1|^2|_k \tag{6.77}$$
$$\leq C\varepsilon(1+t+r)^{-3} + C\varepsilon^{1+\lambda}(1+t+r)^{-3+\lambda}W_{-2}(t,r).$$

Thus, (6.75), (6.76), and (6.77) lead to

$$|I_3 + A_2 \, \mathrm{grad} \, |\mathrm{rot} \, w_1|^2|_k \leq C\varepsilon(1+t+r)^{-3} + C\varepsilon^{1+\lambda}(1+t+r)^{-3+\lambda}.$$

Since $\varepsilon t \geq 1$ in $\mathrm{supp} \, I_3$, we have $\varepsilon \geq t^{-1} \geq (1+t+r)^{-1}$. Hence, we get

$$|I_3 + A_2 \, \mathrm{grad} \, |\mathrm{rot} \, w_1|^2|_k \leq C\varepsilon^{1+\lambda}(1+t+r)^{-3+\lambda}, \tag{6.78}$$

which yields

$$\left\| |I_3 + A_2 \operatorname{grad} |\operatorname{rot} w_1|^2|_k \right\|_{L^2} \leq C\varepsilon^{1+\lambda}(1+t)^{-(3/2)+\lambda}. \tag{6.79}$$

Finally (6.60) follows from (6.66), (6.72), and (6.78). We also obtain (6.61) from (6.67), (6.73), and (6.79). This completes the proof. □

7. Outline of the proof of Theorem 1.1

We assume that $0 < \varepsilon \leq 1$ and that (6.1) holds for some $R > 1$. Let $u_1(t, x)$ be the approximation defined by (6.56) for $(t, x) \in [0, \varepsilon^{-1} \exp(\tau_* \varepsilon^{-1})) \times \mathbf{R}^3$. If we set

$$u_2(t, x) = u(t, x) - u_1(t, x),$$

then (1.1)–(1.2) is reduced to

$$(\partial_t^2 - L)u_2 = H(u_1, u_2) - (E[u_1] + A_2 \operatorname{grad} |\operatorname{rot} w_1|^2) + A_2 \operatorname{grad} |\operatorname{rot} w_1|^2 \tag{7.1}$$
$$\text{in } [0, \varepsilon^{-1} \exp(\tau_* \varepsilon^{-1})) \times \mathbf{R}^3,$$

$$u_2(0, x) = (\partial_t u_2)(0, x) = 0 \quad \text{for } x \in \mathbf{R}^3, \tag{7.2}$$

where $E[u]$ is defined by (6.28), and $H(u_1, u_2)$ is given by

$$H(u_1, u_2) = F(\nabla(u_1 + u_2), \nabla^2(u_1 + u_2)) - F(\nabla u_1, \nabla^2 u_1).$$

Observe that for any nonnegative integer k, there exists a constant C_k such that

$$\sup_{x \in \mathbf{R}^3} |u_2(0, x)|_k \leq C_k \varepsilon^2, \tag{7.3}$$

because for $0 \leq t \leq \varepsilon^{-1}$ and $x \in \mathbf{R}^3$, we have

$$(\partial_t^2 - L)u_2 = F(\nabla(u_1 + u_2), \nabla^2(u_1 + u_2)),$$

$u_2(0, x) = \partial_t u_2(0, x) = 0$, and $u_1(t, x) = \varepsilon u_0(t, x)$ by (6.56). Therefore, by the local existence theorem (see [7]), what we need for proving Theorem 1.1 is to establish a suitable a priori estimate. More explicitly, for $0 < T < \max\{T_\varepsilon, \exp(\tau_0 \varepsilon^{-1})\}$ with $\tau_0 \in (0, \tau_*)$, we wish to evaluate the following quantity:

$$\sup_{(t,x) \in [0,T] \times \mathbf{R}^3} \{(1 + r)(W_{-1}(t, r))^{-1} |\partial u_2(t, x)|_K \tag{7.4}$$
$$+ (1 + r)(1 + |c_1 t - r|)|\operatorname{div} u_2(t, x)|_K$$
$$+ (1 + r)(1 + |c_2 t - r|)|\operatorname{rot} u_2(t, x)|_K\},$$

provided K is an integer large enough and ε is small enough.

In order to carry out this purpose, we employ (5.2) for estimating $E[u_1] + A_2 \operatorname{grad} |\operatorname{rot} w_1|^2$ and (5.3) for evaluating $H(u_1, u_2)$, respectively. Note that (6.60), (6.61) enable us to regard $E[u_1] + A_2 \operatorname{grad} |\operatorname{rot} w_1|^2$ as a harmless term. Moreover, one can handle $\operatorname{grad} |\operatorname{rot} w_1|^2$ by using the following lemma which is due to [1], because of (6.17) and the discrepancy of the propagation speeds c_1, c_2.

Lemma 7.1. *Let $\varphi = \varphi(t, x)$ be a smooth real-valued function. If u solves*

$$(\partial_t^2 - L)u = grad\,\varphi \qquad in\ [0, T) \times \mathbf{R}^3, \tag{7.5}$$

$$u(0, x) = (\partial_t u)(0, x) = 0 \quad for\ x \in \mathbf{R}^3, \tag{7.6}$$

then u also satisfies

$$(\partial_t^2 - c_1^2 \Delta)u = grad\,\varphi \qquad in\ [0, T) \times \mathbf{R}^3. \tag{7.7}$$

Proof. Let $\psi = \psi(t, x)$ be the solution of

$$(\partial_t^2 - c_1^2 \Delta)\psi = \varphi \qquad in\ [0, T) \times \mathbf{R}^3,$$

$$\psi(0, x) = (\partial_t \psi)(0, x) = 0 \quad for\ x \in \mathbf{R}^3.$$

Then we have

$$(\partial_t^2 - L)grad\,\psi = (\partial_t^2 - c_1^2 \Delta)grad\,\psi = grad\,\varphi.$$

Because of the uniqueness for the problem (7.5)–(7.6), we have $u = grad\,\psi$, so that u satisfies (7.7). This completes the proof. $\qquad\qquad\square$

Now, one can develop argument as in [1], because when $T < \varepsilon^{-1} \exp(\tau_0 \varepsilon^{-1})$, we see that $0 \le t \le T$ implies $\varepsilon \log(2 + t) \le C(1 + \tau_0)$ for ε small enough. Noting that $\lim_{\varepsilon \to 0} \varepsilon \log \varepsilon = 0$, we find that Theorem 1.1 is valid.

Acknowledgment

The author would like to express his gratitude to Prof. M. Kawashita for his useful comments. He is also grateful to the referee for careful reading of this paper and precise comments.

References

[1] R. Agemi, *Global existence of nonlinear elastic waves*, Invent. Math. **142** (2000), 225–250.

[2] S. Alinhac, *The null condition for quasilinear wave equations in two space dimensions II*, Amer. J. Math. **123** (2001), 1071–1101.

[3] D. Christodoulou, *Global solutions of nonlinear hyperbolic equations for small initial data*, Comm. Pure Appl. Math. **39** (1986), 267–282.

[4] F.G. Friedlander, *On the radiation field of pulse solutions of the wave equation*, Proc. Roy. Soc. A. **269** (1962), 53–65.

[5] L. Hörmander, *The lifespan of classical solutions of nonlinear hyperbolic equations*, Lecture Note in Math., **1256**, Springer, Berlin, (1987), 241–280.

[6] L. Hörmander, *"Lectures on nonlinear hyperbolic differential equations"*, Mathématiques & Applications, **26**, Springer-Verlag, Berlin, 1997.

[7] F. John, *Finite amplitude waves in a homogeneous isotropic elastic solid*, Comm. Pure Appl. Math. **30** (1977), 421–446.

[8] F. John, *Formation of singularities in elastic waves*, Lecture Notes in Phys., **195**, 194–210, Springer, Berlin, 1984.

212 H. Kubo

[9] F. John, *Existence for large times of strict solutions of nonlinear wave equations in three space dimensions for small initial data*, Comm. Pure Appl. Math. **40** (1987), 79–109.

[10] F. John, *Almost global existence of elastic waves of finite amplitude arising from small initial disturbances*, Comm. Pure Appl. Math. **41** (1988), 615–666.

[11] S. Katayama and H. Kubo, *The rate of convergence to the asymptotics for the wave equation in an exterior domain*, Funkcial. Ekvac. **53** (2010), 331–358.

[12] S. Klainerman, *The null condition and global existence to nonlinear wave equations*, Lectures in Appl. Math., **23** (1986) 293–326.

[13] S. Klainerman and T.C. Sideris, *On almost global existence for nonrelativistic wave equations in 3D*, Comm. Pure Appl. Math. **49** (1996), 307–321.

[14] H. Kubo and M. Ohta, *On the global behaviour of classical solutions to coupled systems of semilinear wave equations*, "New trends in the theory of hyperbolic equations" (M. Reissig and B.-W. Schulze eds.), Birkhäuser, 2005.

[15] T.C. Sideris, *Nonresonance and global existence of prestressed nonlinear elastic waves*, Ann. of Math. **151** (2000), 849–874.

Hideo Kubo
Division of Mathematics
Graduate School of Information Sciences
Tohoku University
Sendai 980-8579, Japan
e-mail: `kubo@math.is.tohoku.ac.jp`

Progress in Mathematics, Vol. 301, 213–225

Representation Formula of the Resolvent for Wave Equation with a Potential Outside the Convex Obstacle

Tokio Matsuyama

Abstract. The purpose of the present article is to provide the results on the perturbed resolvent of the Schrödinger operator in an exterior domain outside the convex obstacle. The representation formula for this resolvent via the free one in the whole space $\mathbb{R}^n$ will be given, which extends the result of [5] to higher-dimensional spaces.

Mathematics Subject Classification. Primary 35L05; Secondary 35L10.

Keywords. Resolvent, potential, exterior domain.

1. Introduction

Let $n \geq 3$ and Ω an exterior domain of $\mathbb{R}^n$ such that $\mathbb{R}^n \setminus \Omega$ is compact with smooth boundary $\partial\Omega$. We assume that $0 \notin \overline{\Omega}$. The present article is devoted to the investigation of the resolvent of the Schrödinger operator $-\Delta + V$, where the potential function V is real valued and measurable on Ω. More precisely, we shall derive the representation formula for the perturbed resolvent $R_V(z) = (A_V - z)^{-1}$ of $A_V = -\Delta + V$ via the free one $R_0(z) = (-\Delta - z)^{-1}$ over $\mathbb{R}^n$. It is proved in the author's recent paper [5] that when $n = 3$ and V decays quadratically, as will be stated later, zero is not an eigenvalue of A_V, and no resonances are present on $[0, \infty)$. In many articles on the dispersive estimates, the above fact is assumed and the spectral analysis is developed (see, e.g., Moulin [7], Vodev [10] and the references therein). When $0 \leq V(x) \leq C|x|^{-2}(|x|^\varepsilon + |x|^{-\varepsilon})^{-1}$ ($0 < \varepsilon \ll 1$) in 3-dimensional space $\mathbb{R}^3$, Georgiev & Visciglia studied the dispersive and Strichartz estimates for wave equation with the above potential by using the spectral analysis (see [4]), where the non-existence of eigenvalues and resonances on

The author was supported by Grant-in-Aid for Scientific Research (C) (No. 21540198), Japan Society for the Promotion of Science.

$[0, \infty)$ are discussed in detail. For the potentials of Kato class in $\mathbb{R}^3$, see D'Ancona & Pierfelice [3]. Quite recently, the author studied the dispersive and Strichartz estimates for wave equation with a polynomially decaying potential in an exterior domain outside the convex obstacle of $\mathbb{R}^3$ (see [5]); we assumed that

$$-c_0|x|^{-\delta_0} \leq V(x) \leq c_1|x|^{-\delta_0} \text{ for some } 0 < c_0 < 1/4, \, c_1 > 0 \text{ and } \delta_0 > 3.$$

Also, it is proved that zero is not an eigenvalue of A_V, and no resonances are present on $[0, \infty)$, provided $\delta_0 > 2$. However, it seems that *in exterior problems of higher-dimensional spaces, there are no results on the resolvent behaviour*. Thus, in this article we shall reveal the resolvent behaviour together with obtaining the representation formula of $R_V(\lambda^2 \pm i0)$ in higher-dimensional spaces. The main results will be stated in Theorem 3.2, which claims that no resonances are present on $[0, \infty)$. As a by-product, we will obtain the result in the whole space. This topic will be discuss in §4.

In the rest of this section we will review some known results on the perturbed resolvent and the free one. It is known from Mochizuki [6] that there exist the limits:

$$\text{s} - \lim_{\varepsilon \searrow 0} R_V(\lambda^2 \pm i\varepsilon) = R_V(\lambda^2 \pm i0) \quad \text{in } \mathscr{B}(L_s^2(\Omega), H_{-s}^2(\Omega)) \tag{1.1}$$

for any $s > 1/2$ and $\lambda > 0$, provided that the real-valued measurable function V on Ω satisfies

$$-c_0|x|^{-\delta_0} \leq V(x) \leq c_1|x|^{-\delta_0} \tag{1.2}$$

for some $0 < c_0 < (n-2)^2/4$, $c_1 > 0$ and $\delta_0 > 1$. Here $\mathscr{B}(L_s^2(\Omega), H_{-s}^2(\Omega))$ is the space of all bounded linear operators from $L_s^2(\Omega)$ to $H_{-s}^2(\Omega)$. The limits (1.1) are known as the limiting absorption principle. In particular case $V = 0$ and $\Omega = \mathbb{R}^n$, the limiting absorption principle for the free resolvent is known as

$$\text{s} - \lim_{\varepsilon \searrow 0} R_0(\lambda^2 \pm i\varepsilon) = R_0(\lambda^2 \pm i0) \quad \text{in } \mathscr{B}(L_s^2(\mathbb{R}^n), H_{-s}^2(\mathbb{R}^n)) \tag{1.3}$$

for any $s > 1/2$ and $\lambda > 0$ (see, e.g., Agmon [1]). We should also refer to the result of the limiting absorption principle in the critical case $s = 1/2$, which Ruzhansky and Sugimoto found (see [9]). Then, in an exterior domain outside the convex obstacle of 3-dimensional space it is proved in Proposition 3.2 from [5] that if the real-valued measurable function V on Ω satisfies (1.2) for some $\delta_0 > 2$, then

$$R_V(\lambda^2 \pm i0) = S_\lambda^{\pm} R_0(\lambda^2 \pm i0) J_*, \quad \forall \lambda \geq 0, \tag{1.4}$$

where $S_\lambda^{\pm}$ are the inverses of operators $J_* + R_0(\lambda^2 \pm i0)$ in a subspace of $L_{-s}^2(\Omega)$, J and J_* are identification operators, which will be defined in §3. In particular, when $\Omega = \mathbb{R}^3$, the operators $S_\lambda^{\pm}$ are coincide with the inverses of $I + R_0(\lambda^2 \pm i0)V$ in $L_{-s}^2(\mathbb{R}^n)$. Since we are in the exterior problem, the identity operator I will be perturbed as an identification operator J. For more details, see §3. The formula (1.4) provides several resolvent estimates, and hence, by developing the spectral analysis, we can get the dispersive and Strichartz estimates for wave equation with potential V satisfying (1.2) (see Theorems 1.1–1.2 in [5]). Thus the result in this article would be available to get these estimates in higher-dimensional spaces.

Now let us introduce the known uniform resolvent estimates. We will always assume that $n \geq 3$ and V is a real-valued measurable function on Ω satisfying

$$-c_0 |x|^{-\delta_0} \leq V(x) \leq c_1 |x|^{-\delta_0} \tag{1.5}$$

for some constants $0 < c_0 < (n-2)^2/4$, $c_1 > 0$ and $\delta_0 > (n+2)/2$. Without loss of generality, we may assume that the obstacle $\mathbb{R}^n \setminus \Omega$ contains the unit ball of $\mathbb{R}^n$. The uniform resolvent estimate for the perturbed version is known in Mochizuki [6]:

$$\left\| R_V(\lambda^2 \pm i0) f \right\|_{L^2_{-s}(\Omega)} \leq C\lambda^{-1} \|f\|_{L^2_s(\Omega)}, \quad \forall \lambda > 0 \tag{1.6}$$

for any $s > 1/2$, provided that $\mathbb{R}^n \setminus \Omega$ is star-shaped with respect to the origin. Actually, (1.6) holds for $\delta_0 > 1$. In particular case $V = 0$ and $\Omega = \mathbb{R}^n$ we have also the uniform resolvent estimate

$$\left\| R_0(\lambda^2 \pm i0) f \right\|_{L^2_{-s}(\mathbb{R}^n)} \leq C\lambda^{-1} \|f\|_{L^2_s(\mathbb{R}^n)}, \quad \forall \lambda > 0 \tag{1.7}$$

for any $s > 1/2$ (see also Ben-Artzi & Klainerman [2]). For small λ, it follows from Lemma 2.2 in Murata [8] that in $\mathscr{B}(L^2_s(\mathbb{R}^n), H^2_{-s}(\mathbb{R}^n))$ one has the expansion

$$R_0(\lambda^2 \pm i0) = \lambda^{n-2}(\log \lambda)^{\varepsilon(n)} F_1 + G_0 + o(1) \quad \text{as } \lambda \to +0 \tag{1.8}$$

for any $s > 1$, where $\varepsilon(n) = 0$ for n odd, and $\varepsilon(n) = 1$ for n even. Here the operators F_1 and G_0 belong to $\mathscr{B}(L^2_s(\mathbb{R}^n), H^2_{-s}(\mathbb{R}^n))$.

In this paper we denote by $\mathscr{B}(X, Y)$ the space of all bounded linear operators from a Banach space X to another Banach space Y. When $X = Y$, we denote $\mathscr{B}(X) = \mathscr{B}(X, X)$. The weighted Sobolev spaces are defined in the standard way: For $m \in \mathbb{N} \cup \{0\}$, $\kappa \in \mathbb{R}$ and a domain G of $\mathbb{R}^n$ we put

$$H^m_\kappa(G) = \left\{ f : \langle x \rangle^\kappa \partial_x^\alpha f \in L^2(G), \quad |\alpha| \leq m \right\},$$

and $H^m_{0,\kappa}(G)$ is the completion of $C_0^\infty(G)$ with respect to the norm $\| \cdot \|_{H^m_\kappa(G)}$. We put $L^2_\kappa(G) = H^0_\kappa(G)$. We will use the notation $\mathscr{R}(T)$ for the range of an operator T.

Finally, we state our plan in this paper. Section 2 is devoted to the proof of the assertion that zero is not an eigenvalue of the operator $-\Delta + V$. In Section 3 the main result will be stated as Theorem 3.2, and the complete proof will be given. In Section 4 the representation formula for the perturbed resolvent in the whole space will be given.

2. Uniqueness theorem for the Schrödinger equation

In this section we shall prove the uniqueness theorem for the Schrödinger equation. This means that zero is not an eigenvalue of A_V. The result will be used to obtain the representation formula for the perturbed resolvent in §3.

We shall prove here the following:

Lemma 2.1. *Let $n \geq 3$. Assume that the measurable potential V satisfies* (1.5) *for some $\delta_0 > (n+2)/2$. Suppose that $u \in H^1_{-s}(\Omega) \cap L^\infty(\Omega)$ for some $1/2 < s \leq \delta_0/2$ satisfies the boundary value problem for the Schrödinger equation:*

$$\begin{cases} -\Delta u + V(x)u = 0 & \text{in } \Omega, \\ \qquad\qquad\quad u = 0 & \text{on } \partial\Omega \end{cases} \tag{2.1}$$

and

$$\lim_{R \to \infty} \frac{1}{R^n} \int_{R < |x| < 2R} |u(x)|^2 \, dx = 0. \tag{2.2}$$

Then $u(x) = 0$ in Ω.

Actually, when $n = 3$, Lemma 2.1 is true for $\delta_0 > 2$ (see Lemma 2.1 from [5]).

To prove Lemma 2.1, we need the following lemma.

Lemma 2.2. *Let $n \geq 3$. Put $\tilde{u} = (-\Delta)^{-1} f$ in $\mathbb{R}^n$. If $f \in L^2_s(\mathbb{R}^n)$ for some $s > 1/2$, then*

$$\lim_{R \to \infty} \frac{1}{R^n} \int_{R < |x| < 2R} |\tilde{u}(x)|^2 \, dx = 0. \tag{2.3}$$

Proof. Taking the Fourier transform of the equation we have

$$\tilde{u}(x) = \mathscr{F}^{-1}\left[\frac{\hat{f}(\xi)}{|\xi|^2} \right](x),$$

where $\hat{f}$ is the Fourier transform of f and $\mathscr{F}^{-1}$ is the inverse Fourier transform. Inserting the cut-off function $\chi(\xi) \in C^\infty_0(\mathbb{R}^n)$ equal to one for $|\xi| \leq 1/2$ and zero for $|\xi| \geq 1$, we can write $\tilde{u} = u_1 + u_2$, where

$$u_1(x) = \mathscr{F}^{-1}\left[\frac{\chi(\xi)\hat{f}(\xi)}{|\xi|^2} \right](x), \quad u_2(x) = \mathscr{F}^{-1}\left[\frac{(1 - \chi(\xi))\hat{f}(\xi)}{|\xi|^2} \right](x).$$

We claim that u_2 fulfills (2.3). Indeed, by the assumption that $f \in L^2_s(\mathbb{R}^n)$, we readily see that $u_2 \in L^2_s(\mathbb{R}^n)$. Hence we conclude that

$$\frac{1}{R^n} \int_{R < |x| < 2R} |u_2(x)|^2 \, dx \leq C_s R^{-2s-n} \int_{R < |x| < 2R} \langle x \rangle^{2s} |u_2(x)|^2 \, dx$$

$$\leq C_s R^{-2s-n} \|u_2\|^2_{L^2_s(\mathbb{R}^n)} \to 0,$$

as $R \to \infty$.

In order to see that u_1 fulfills (2.3), we note that $(-\Delta)^{-1}$ is a convolution operator with a kernel $G_0(x - y)$ defined by

$$G_0(x) = \frac{\Gamma(n/2)}{4(n-2)\pi^{n/2}} \cdot \frac{1}{|x|^{n-2}}. \tag{2.4}$$

From (2.4) we observe that $\mathscr{F}^{-1}\left[|\xi|^{-2}\chi(\xi)\mathscr{F}\cdot\right]$ is a convolution operator with a C^{∞}-kernel majorized by $C\langle x-y\rangle^{-(n-2)}$, $C > 0$. Hence we have

$$\int_{R<|x|<2R}|u_1(x)|^2\,dx \leq C\int_{R<|x|<2R}\left(\int_{\mathbb{R}^n}\langle x-y\rangle^{-(n-2)}|f(y)|\,dy\right)^2 dx \qquad (2.5)$$

$$\leq C\int_{R<|x|<2R}\left(\int_{\mathbb{R}^n}\langle x-y\rangle^{-2(n-2)}\langle y\rangle^{-2s}\,dy\right)\|f\|_{L_s^2(\mathbb{R}^n)}^2\,dx$$

$$\leq C\int_{R<|x|<2R}\left(\int_{\mathbb{R}^n}|x-y|^{-(n-1)}\langle y\rangle^{-2s}\,dy\right)\|f\|_{L_s^2(\mathbb{R}^n)}^2\,dx,$$

where we used the fact that $n-1 \leq 2(n-2)$ in the last step, since $n \geq 3$. Here we can estimate

$$\int_{\mathbb{R}^n}|x-y|^{-(n-1)}\langle y\rangle^{-2s}\,dy \leq \begin{cases} C\langle x\rangle^{-(2s-1)} & \text{if } 1/2 < s < n/2, \\ C\langle x\rangle^{-(n-1)}\log(e+|x|) & \text{if } s = n/2, \\ C\langle x\rangle^{-(n-1)} & \text{if } s > n/2, \end{cases}$$

and it follows from (2.5) that, for $1/2 < s < n/2$,

$$\int_{R<|x|<2R}|u_1(x)|^2\,dx \leq C\left(\int_{R<|x|<2R}\langle x\rangle^{-(2s-1)}\,dx\right)\|f\|_{L_s^2(\mathbb{R}^n)}^2$$

$$\leq CR^{-(2s-1)+n}\|f\|_{L_s^2(\mathbb{R}^n)}^2.$$

Thus we conclude that u_1 satisfies (2.3), since $s > 1/2$. In a similar way, when $s \geq n/2$, we have

$$\int_{R<|x|<2R}|u_1(x)|^2\,dx \leq \begin{cases} CR\log(e+R)\|f\|_{L_s^2(\mathbb{R}^n)}^2, & \text{if } s = n/2, \\ CR\|f\|_{L_s^2(\mathbb{R}^n)}^2, & \text{if } s > n/2, \end{cases}$$

and hence, u_1 satisfies (2.3) also in these cases. The proof of Lemma 2.2 is now complete. $\qquad\square$

Proof of Lemma 2.1. First, we claim that

$$u(x) = \begin{cases} O\left(|x|^{-(\delta_0-2)}\right), & \text{if } 2 < \delta_0 < n, \\ O\left(|x|^{-(n-2)}\log|x|\right), & \text{if } \delta_0 = n, \qquad \text{as } |x| \to \infty, \qquad (2.6) \\ O\left(|x|^{-(n-2)}\right), & \text{if } \delta_0 > n, \end{cases}$$

$$|\nabla u(x)| = \begin{cases} O\left(|x|^{-(\delta_0-1)}\right), & \text{if } 2 < \delta_0 < n, \\ O\left(|x|^{-(n-1)}\log|x|\right), & \text{if } \delta_0 = n, \qquad \text{as } |x| \to \infty. \qquad (2.7) \\ O\left(|x|^{-(n-1)}\right), & \text{if } \delta_0 > n, \end{cases}$$

To show the asymptotic behaviour (2.6)–(2.7), let us consider the extension $\tilde{u}$ of u to $\mathbb{R}^n$. More precisely, we define $\tilde{u}$ to be $\tilde{u}(x) = \psi(x)u(x)$, where $\psi(x) \in C^{\infty}(\mathbb{R}^n)$ equals zero in a bounded open neighbourhood $\mathcal{O}$ of $\mathbb{R}^n \setminus \Omega$ and one in $\mathbb{R}^n \setminus \tilde{\mathcal{O}}$

such that the closure of $\widetilde{\mathcal{O}}$ is compact and $\mathcal{O} \Subset \widetilde{\mathcal{O}}$. This $\tilde{u}$ satisfies the equation $-\Delta \tilde{u} = f$ in $\mathbb{R}^n$, where

$$f = -\psi V u - 2\nabla\psi \cdot \nabla u - (\Delta\psi)u. \tag{2.8}$$

It is well known that the Poisson equation has a unique solution in $\mathscr{S}'(\mathbb{R}^n)$ (= the space of all tempered distributions) up to an additive polynomial. Hence $\tilde{u}$ can be represented as

$$\tilde{u}(x) = \frac{\Gamma(n/2)}{4(n-2)\pi^{n/2}} \int_{\mathbb{R}^n} \frac{f(y)}{|x-y|^{n-2}}\, dy + \text{polynomial}.$$

It can be readily checked that $f \in L_s^2(\mathbb{R}^n)$ provided $1/2 < s \le \delta_0/2$. In fact, by using the decay assumption (1.5) on V, we have

$$\|f\|_{L_s^2(\mathbb{R}^n)} \le C \left\| \langle \cdot \rangle^{s-\delta_0} u \right\|_{L^2(\mathbb{R}^n)} + \|2\nabla\psi \cdot \nabla u - (\Delta\psi)u\|_{L_s^2(\mathbb{R}^n)} \le C\|u\|_{H_{-s}^1(\Omega)} < \infty.$$

Hence we see from the asymptotic behaviour (2.2) in Lemma 2.2 that

$$\tilde{u}(x) = \frac{\Gamma(n/2)}{4(n-2)\pi^{n/2}} \int_{\mathbb{R}^n} \frac{f(y)}{|x-y|^{n-2}}\, dy,$$

and it follows from (2.8) that

$$\begin{aligned}
|\tilde{u}(x)| \le {} & \frac{\Gamma(n/2)}{4(n-2)\pi^{n/2}} \left(\int_{\mathbb{R}^n} \frac{|\psi(y)V(y)u(y)|}{|x-y|^{n-2}}\, dy \right. \\
& \left. + \int_{\mathbb{R}^n} \frac{2|\nabla\psi(y)||\nabla u(y)|}{|x-y|^{n-2}}\, dy + \int_{\mathbb{R}^n} \frac{|\Delta\psi(y)||u(y)|}{|x-y|^{n-2}}\, dy \right).
\end{aligned} \tag{2.9}$$

We shall estimate each term for large $|x| > 2R$. The second and third terms of (2.9) decay like $|x|^{-(n-2)}$ as $|x| \to \infty$. In fact, observing $|x-y| \ge |x| - R_0$ for $|x| > 2R$ and $y \in \mathrm{supp}|\nabla\psi|$, where R_0 is the diameter of $\widetilde{\mathcal{O}}$, we see that

$$\begin{aligned}
& \int_{\mathbb{R}^n} \left(\frac{|\nabla\psi(y)||\nabla u(y)|}{|x-y|^{n-2}} + \frac{|\Delta\psi(y)||u(y)|}{|x-y|^{n-2}} \right) dy \\
& \le C(|x| - R_0)^{-(n-2)} \left(\int_{\mathrm{supp}|\nabla\psi|} |\nabla u(y)|\, dy + \int_{\mathrm{supp}|\Delta\psi|} |u(y)|\, dy \right) \\
& \le C(|x| - R_0)^{-(n-2)} \|u\|_{H_{-s}^1(\Omega)}.
\end{aligned}$$

As to the term involving V, by using the decay assumption (1.5) and $u \in L^\infty(\Omega)$, we see that

$$\int_{\mathbb{R}^n} \frac{|\psi(y)V(y)u(y)|}{|x-y|^{n-2}}\, dy \le C \left(\int_{\mathbb{R}^n} \frac{dy}{|x-y|^{n-2}\langle y \rangle^{\delta_0}} \right) \|u\|_{L^\infty(\Omega)}.$$

Here the integral in the right-hand side behaves like

$$\int_{\mathbb{R}^n} \frac{dy}{|x-y|^{n-2}\langle y \rangle^{\delta_0}} = \begin{cases} O\left(|x|^{-(\delta_0-2)}\right) & \text{if } 2 < \delta_0 < n, \\ O\left(|x|^{-(n-2)} \log|x|\right) & \text{if } \delta_0 = n, \\ O\left(|x|^{-(n-2)}\right) & \text{if } \delta_0 > n, \end{cases} \quad \text{as } |x| \to \infty.$$

Summarizing this observation, we get (2.6). We now turn to prove (2.7). Writing

$$|\nabla \tilde{u}(x)| \leq c_n \int_{\mathbb{R}^n} \frac{|f(y)|}{|x - y|^{n-1}} \, dy$$

$$\leq c_n \left(\int_{\mathbb{R}^n} \frac{|\psi(y)V(y)u(y)|}{|x - y|^{n-1}} \, dy + \int_{\mathbb{R}^n} \frac{2|\nabla \psi(y)||\nabla u(y)|}{|x - y|^{n-1}} \, dy + \int_{\mathbb{R}^n} \frac{|\Delta \psi(y)||u(y)|}{|x - y|^{n-1}} \, dy \right),$$

we shall estimate each term. By the same calculation as above, we conclude that the second and third terms decay like $|x|^{-(n-1)}$ as $|x| \to \infty$. The first term is handled by using the decay assumption (1.5) on V and $u \in L^\infty(\Omega)$:

$$\int_{\mathbb{R}^n} \frac{|\psi(y)V(y)u(y)|}{|x - y|^{n-1}} \, dy \leq C \|u\|_{L^\infty(\Omega)} \int_{\mathbb{R}^n} \frac{dy}{|x - y|^{n-1} \langle y \rangle^{\delta_0}}$$

$$= \begin{cases} O(|x|^{-(\delta_0 - 1)}), & \text{if } 2 < \delta_0 < n, \\ O(|x|^{-(n-1)} \log |x|), & \text{if } \delta_0 = n, \\ O(|x|^{-(n-1)}), & \text{if } \delta_0 > n. \end{cases} \quad \text{as } |x| \to \infty.$$

Summarizing these estimates obtained now, we get (2.7).

Now, integrating by parts, we have

$$\int_{\Omega \cap \{|x| < R\}} \left\{ |\nabla u(x)|^2 + V(x)|u(x)|^2 \right\} dx = \int_{|x| = R} u_r(x) \overline{u(x)} \, dS_R, \tag{2.10}$$

where $u_r = \partial u / \partial r$ ($r = |x|$) and dS_R is the $(n-1)$-dimensional surface element. Since

$$(\delta_0 - 1) + (\delta_0 - 2) > n - 1$$

provided $\delta_0 > (n+2)/2$, by using the asymptotic behaviour (2.6)–(2.7) and taking the inferior limit in (2.10) as $R \to \infty$, we get

$$\int_\Omega \left\{ |\nabla u(x)|^2 + V(x)|u(x)|^2 \right\} dx = 0. \tag{2.11}$$

Here we may assume that $|x| \geq 1$ for all $x \in \Omega$ without loss of generality. By using the assumption (1.5) on V: $V(x) \geq -c_0 |x|^{-\delta_0}$, where $0 < c_0 < (n-2)^2/4$, we can estimate

$$\int_\Omega V(x)|u(x)|^2 \, dx \geq - \int_\Omega c_0 \frac{|u(x)|^2}{|x|^2} \, dx,$$

and hence, resorting to the Hardy inequality, we get

$$\int_\Omega \left\{ |\nabla u(x)|^2 + V(x)|u(x)|^2 \right\} dx \geq \int_\Omega \left\{ |\nabla u(x)|^2 - c_0 \frac{|u(x)|^2}{|x|^2} \right\} dx$$

$$\geq \left\{ 1 - \frac{4c_0}{(n-2)^2} \right\} \int_\Omega |\nabla u(x)|^2 \, dx.$$

Therefore we arrive at

$$\int_\Omega |\nabla u(x)|^2 \, dx = 0,$$

which implies that u is constant in Ω. Thus we conclude from the boundary condition that $u = 0$ in Ω. The proof of Lemma 2.1 is complete. $\qquad \square$

3. Representation formula for $R_V(\lambda^2 \pm i0)$

In this section let us investigate the relation among $R_V(\lambda^2 \pm i0)$ and $R_0(\lambda^2 \pm i0)$ for $n \geq 4$. For this purpose, we need identify the space $L^2(\mathbb{R}^n)$ that $-\Delta$ acts on, with $L^2(\Omega)$ that $-\Delta + V$ acts on. Following Mochizuki [6], we define *identification operators*

$$J \in \mathscr{B}(L^2(\mathbb{R}^n), L^2(\Omega)), \quad J_* \in \mathscr{B}(L^2(\Omega), L^2(\mathbb{R}^n)) \tag{3.1}$$

as follows. Let $j(x) = j(r\omega) \in C^\infty(\mathbb{R}^n)$ ($r = |x|, \omega = x/|x| \in \mathbb{S}^{n-1}$) be a non-decreasing function in r for each ω such that $j(x) = 0$ for $x \in \mathbb{R}^n \setminus \Omega$, and equal to 1 for $x \in \widetilde{\Omega} := \{x : \operatorname{dist}(x, \mathbb{R}^n \setminus \Omega) > 1\}$, which is possible if we assume that $\mathbb{R}^n \setminus \Omega$ is convex. In fact, for each $\omega \in \mathbb{S}^{n-1}$ taking $0 < r_1(\omega) < r_2(\omega)$ such that $r_1(\omega)\omega \in \partial(\mathbb{R}^n \setminus \Omega)$ and $r_2(\omega)\omega \in \partial\widetilde{\Omega}$, we let $j(r\omega) = 0$ for $r \leq r_1(\omega)$, $j(r\omega) = 1$ for $r \geq r_2(\omega)$, and $j(r\omega)$ is non-decreasing in $[r_1(\omega), r_2(\omega)]$. For $f \in L^2_{\mathrm{loc}}(\mathbb{R}^n)$ we define the operator J to be

$$(Jf)(x) = j(x)f(x), \quad x \in \Omega,$$

and for $g \in L^2_{\mathrm{loc}}(\overline{\Omega})$ define

$$(J_* g)(x) = \begin{cases} j(x)g(x), & x \in \Omega, \\ 0, & x \in \mathbb{R}^n \setminus \Omega. \end{cases}$$

Then $J \in \mathscr{B}(L^2_s(\mathbb{R}^n), L^2_s(\Omega))$ and $J_* \in \mathscr{B}(L^2_s(\Omega), L^2_s(\mathbb{R}^n))$ for any $s \in \mathbb{R}$, and hence, (3.1) holds. It is readily checked that J_* is the adjoint operator of J: For any $f \in L^2(\mathbb{R}^n)$ and $g \in L^2(\Omega)$,

$$(Jf, g)_{L^2(\Omega)} = (f, J_* g)_{L^2(\mathbb{R}^n)}.$$

Put

$$W_* = J_*(-\Delta + V(x)) + \Delta J_*.$$

By elementary calculus, we can write

$$W_* g = j(x)V(x)g - (\Delta j)g - 2\nabla j \cdot \nabla g. \tag{3.2}$$

Then

$$W_* \in \mathscr{B}(H^1_s(\Omega), L^2_{s+\delta_0}(\mathbb{R}^n))$$

for any $s \in \mathbb{R}$ on account of the decay assumption (1.5) on V, and we have the "resolvent equations":

$$J_* R_V(\lambda^2 \pm i0) = R_0(\lambda^2 \pm i0)\left\{J_* - W_* R_V(\lambda^2 \pm i0)\right\},$$

which implies that

$$\left\{J_* + R_0(\lambda^2 \pm i0)W_*\right\} R_V(\lambda^2 \pm i0) = R_0(\lambda^2 \pm i0)J_*. \tag{3.3}$$

Recalling that $\mathscr{R}(T)$ is the range of an operator T, we prepare the following lemma:

Lemma 3.1. *Let $n \geq 4$. Assume that the measurable potential V satisfies (1.5) for some $\delta_0 > (n+2)/2$. Let $1 < s \leq \delta_0/2$. Then the inclusion*

$$\mathscr{R}(R_0(\lambda^2 \pm i0)J_*) \subset \mathscr{R}(J_* + R_0(\lambda^2 \pm i0)W_*) \subset L^2_{-s}(\mathbb{R}^n) \tag{3.4}$$

holds for any $\lambda \geq 0$.

Proof. The first inclusion in (3.4) for $\lambda \geq 0$ is proved by using the resolvent equations (3.3):

$$\mathscr{R}(R_0(\lambda^2 \pm i0)J_*) = \mathscr{R}((J_* + R_0(\lambda^2 \pm i0)W_*)R_V(\lambda^2 \pm i0)) \tag{3.5}$$
$$\subset \mathscr{R}(J_* + R_0(\lambda^2 \pm i0)W_*).$$

When $\lambda = 0$, we used the existence of $R_V(0)$, which is assured by Lemma 2.1. As to the second inclusion in (3.4), we have only to show that

$$\left\| R_0(\lambda^2 \pm i0)W_* f \right\|_{L^2_{-s}(\mathbb{R}^n)} \leq C \| f \|_{H^1_{-s}(\Omega)}, \quad f \in H^1_{-s}(\Omega) \tag{3.6}$$

for any $\lambda \geq 0$. For any $\lambda > 0$, by using the resolvent estimate (1.7) and the decay assumption (1.5) on V, we can estimate

$$\left\| R_0(\lambda^2 \pm i0)W_* f \right\|_{L^2_{-s}(\mathbb{R}^n)} \tag{3.7}$$
$$\leq \left\| R_0(\lambda^2 \pm i0)(jVf) \right\|_{L^2_{-s}(\mathbb{R}^n)} + \left\| R_0(\lambda^2 \pm i0)\{(\Delta j)f\} \right\|_{L^2_{-s}(\mathbb{R}^n)}$$
$$+ 2 \left\| R_0(\lambda^2 \pm i0)(\nabla j \cdot \nabla f) \right\|_{L^2_{-s}(\mathbb{R}^n)}$$
$$\leq C\lambda^{-1}\left\{ \|\langle \cdot \rangle^{s-\delta_0} f \|_{L^2(\Omega)} + \|(\Delta j)f\|_{L^2_s(\Omega)} + \|\nabla j \cdot \nabla f\|_{L^2_s(\Omega)} \right\} \leq C\lambda^{-1}\|f\|_{H^1_{-s}(\Omega)},$$

provided $1/2 < s \leq \delta_0/2$. For small λ, by using the asymptotic expansion (1.8) and the decay assumption (1.5) on V, we conclude from the same calculation as in (3.7) that

$$\left\| R_0(\lambda^2 \pm i0)W_* f \right\|_{L^2_{-s}(\mathbb{R}^n)} \leq C\|f\|_{H^1_{-s}(\Omega)}, \tag{3.8}$$

provided $1 < s \leq \delta_0/2$. Thus (3.7)–(3.8) imply (3.6). The proof of Lemma 3.1 is now finished. $\qquad\square$

To state the main result, let us introduce a subspace of $H^2_{-s}(\Omega)$:

$$\hat{H}^m_{0,-s}(\Omega) = \left\{ u \in H^m_{0,-s}(\Omega) : \lim_{R \to 0} \frac{1}{R^n} \int_{R < |x| < 2R} |u(x)|^2 \, dx = 0 \right\}, \quad (m = 1, 2),$$

which comes from Lemma 2.1. Then we have:

Theorem 3.2. *Let $n \geq 4$ and $\mathbb{R}^n \setminus \Omega$ is convex. Assume that the measurable potential V satisfies (1.5) for some $\delta_0 > (n+2)/2$. Let $1 < s \leq \delta_0/2$. Then there exists a family of operators $S^{\pm}_\lambda$ from $\mathscr{R}(J_* + R_0(\lambda^2 \pm i0)W_*)$ to $H^2_{0,-s}(\Omega)$ such that*

$$S^{\pm}_\lambda \circ \left\{ J_* + R_0(\lambda^2 \pm i0)W_* \right\} = I, \tag{3.9}$$

$$R_V(\lambda^2 \pm i0) = S^{\pm}_\lambda R_0(\lambda^2 \pm i0)J_* \tag{3.10}$$

for any $\lambda \geq 0$. When $\lambda = 0$, $\mathscr{R}(S^{\pm}_0)$ equals $\hat{H}^2_{0,-s}(\Omega) \cap L^\infty(\Omega)$. Furthermore, $S^{\pm}_\lambda$ are bounded from $\mathscr{R}(R_0(\lambda^2 \pm i0)J_*)$ to $H^2_{0,-s}(\Omega)$ for any $\lambda > 0$ and there exists a *constant $C > 0$ such that*

$$\left\| S_\lambda^\pm R_0(\lambda^2 \pm i0) J_* f \right\|_{H^1_{-s}(\Omega)} \le C \left\| R_0(\lambda^2 \pm i0) J_* f \right\|_{L^2_{-s}(\mathbb{R}^n)}, \quad \forall \lambda > 0 \qquad (3.11)$$

for $f \in L^2_s(\Omega)$.

Proof. First, we claim that $J_* + R_0(\lambda^2 \pm i0) W_*$ are injective from $H^2_{0,-s}(\Omega)$ (resp. $\hat{H}^2_{0,-s}(\Omega) \cap L^\infty(\Omega)$) to $L^2_{-s}(\mathbb{R}^n)$ for any $1 < s \le \delta_0/2$ and $\lambda > 0$ (resp. $\lambda = 0$). Let $f_\pm \in H^2_{0,-s}(\Omega)$ be solutions to the following integral equations:

$$\{J_* + R_0(\lambda^2 \pm i0) W_*\} f_\pm = 0. \qquad (3.12)$$

Taking a sequence $\{f_{\pm,j}\}_j \subset C_0^\infty(\overline{\Omega})$ in such a way that

$$f_{\pm,j} \to f_\pm \quad \text{in } H^2_{0,-s}(\Omega) \text{ as } j \to \infty, \qquad (3.13)$$

we find from the resolvent equations (3.3) that

$$
\begin{aligned}
u_{\pm,j} &:= R_0(\lambda^2 \pm i0) J_* \left\{ -\Delta + V(x) - \lambda^2 \right\} f_{\pm,j} \\
&= \left\{ J_* + R_0(\lambda^2 \pm i0) W_* \right\} R_V(\lambda^2 \pm i0) \left\{ -\Delta + V(x) - \lambda^2 \right\} f_{\pm,j} \\
&= \left\{ J_* + R_0(\lambda^2 \pm i0) W_* \right\} f_{\pm,j}.
\end{aligned}
$$

Then it follows from (3.12)–(3.13) that $u_{\pm,j} \to 0$ in $H^2_{0,-s}(\mathbb{R}^n)$ as $j \to \infty$, which implies that

$$J_* \left\{ -\Delta + V(x) - \lambda^2 \right\} f_{\pm,j} = \left(-\Delta - \lambda^2 \right) u_{\pm,j} \to 0 \quad \text{in } L^2_{-s}(\mathbb{R}^n) \text{ as } j \to \infty.$$

Hence, by using the definition of J_* and (3.13), we see that $\{-\Delta + V(x) - \lambda^2\} f_\pm = 0$ a.e. in Ω. Denote by Ω_0 a sub-domain of Ω such that $\{-\Delta + V(x) - \lambda^2\} f_\pm$ vanish. Then $f_\pm$ satisfy the following boundary value problems:

$$
\begin{cases}
\{-\Delta + V(x) - \lambda^2\} f_\pm = 0 & \text{in } \Omega_0, \\
\qquad\qquad\qquad\quad f_\pm = 0 & \text{on } \partial\Omega_0.
\end{cases}
$$

When $\lambda > 0$, since the operator $-\Delta + V(x)$ has no positive eigenvalues (see Theorem 2.1 from [6, Mochizuki]), it follows that $f_\pm = 0$ in Ω_0, and hence, $f_\pm = 0$ in $H^2_{0,-s}(\Omega)$. This means that the operators $J_* + R_0(\lambda^2 \pm i0) W_*$ are injective for $\lambda > 0$. When $\lambda = 0$, if $H^2_{0,-s}(\Omega)$ is replaced by $\hat{H}^2_{0,-s}(\Omega) \cap L^\infty(\Omega)$, then it follows from Lemma 2.1 that $f_\pm = 0$ in Ω_0. In any case, the injection of $J_* + R_0(\lambda^2 \pm i0) W_*$ is proved for all $\lambda \ge 0$.

Next, by the injection of $J_* + R_0(\lambda^2 \pm i0) W_*$, the inverses $S_\lambda^\pm$ of $J_* + R_0(\lambda^2 \pm i0) W_*$ exist and map from $\mathscr{R}(J_* + R_0(\lambda^2 \pm i0) W_*)$ to $H^2_{0,-s}(\Omega)$ for any $\lambda > 0$, which proves (3.9). When $\lambda = 0$, the range of $S_0^\pm$ is $\hat{H}^2_{0,-s}(\Omega) \cap L^\infty(\Omega)$, and (3.9) is also valid in this case. The equations (3.10) readily follows from (3.9) and the resolvent equations (3.3). In particular, $S_\lambda^\pm$ map from $\mathscr{R}(R_0(\lambda^2 \pm i0) J_*)$ to $H^2_{0,-s}(\Omega)$ for any $\lambda \ge 0$ by the inclusion (3.4) from Lemma 3.1. In particular, $S_\lambda^\pm$ map from $\mathscr{R}(R_0(\lambda^2 \pm i0) J_*)$ to $H^2_{0,-s}(\Omega)$ for any $\lambda \ge 0$ by the inclusion (3.4) from Lemma 3.1. In particular, $S_\lambda^\pm$ map from $\mathscr{R}(R_0(\lambda^2 \pm i0) J_*)$ to $H^2_{0,-s}(\Omega)$ for any $\lambda \ge 0$ by the inclusion (3.4) from Lemma 3.1.

We now turn to prove that $S_\lambda^\pm \in \mathscr{B}(\mathscr{R}(R_0(\lambda^2 \pm i0)J_*), H_{0,-s}^1(\Omega))$ for each $\lambda > 0$. We note that $\mathscr{R}(R_0(\lambda^2 \pm i0)J_*)$ are closed subspaces in $L_{-s}^2(\mathbb{R}^3)$ for $s > 1$, since $R_0(\lambda^2 \pm i0)$ are compact operators from $L_s^2(\Omega)$ to $L_{-s}^2(\Omega)$ for $s > 1$ and J_* is bounded from $L_{-s}^2(\Omega)$ to $L_{-s}^2(\mathbb{R}^3)$ for $s > 0$. For this purpose, we claim that the spaces $\mathscr{R}(J_* + R_0(\lambda^2 \pm i0)W_*)$ are closed subspaces in $L_{-s}^2(\mathbb{R}^3)$. In fact, let $\{f_{\pm,j}\}_{j=1}^\infty$ be sequences in $H_{0,-s}^2(\Omega)$ such that

$$\{J_* + R_0(\lambda^2 \pm i0)W_*)\}f_{\pm,j} \to g \in H_{0,-s}^2(\Omega) \quad \text{as } j \to \infty.$$

Then the sequences $\{\{J_* + R_0(\lambda^2 \pm i0)W_*)\}f_{\pm,j}\}_{j=1}^\infty$ are Cauchy sequences, and hence, by the same argument as in the proof of injection argument, we see that

$$(-\Delta - V(x) - \lambda^2)(f_{\pm,j} - f_{\pm,k}) \to 0 \quad \text{in } L_{-s}^2(\Omega) \text{ as } j, k \to \infty.$$

Hence it follows from the limiting absorption principle (1.1) that

$$f_{\pm,j} - f_{\pm,k} \to 0 \quad \text{in } H_{0,-s}^2(\Omega) \text{ as } j, k \to \infty,$$

which implies that there exist the limits $f_\pm$:

$$f_{\pm,j} \to f_\pm \quad \text{in } H_{0,-s}^2(\Omega) \text{ as } j \to \infty.$$

Since $J_* + R_0(\lambda^2 \pm i0)W_*$ are continuous from $H_{0,-s}^2(\Omega)$ to itself, we conclude that $g = \{J_* + R_0(\lambda^2 \pm i0)\}f_\pm \in \mathscr{R}(J_* + R_0(\lambda^2 \pm i0))$, which proves that $\mathscr{R}(J_* + R_0(\lambda^2 \pm i0)W_*)$ are closed. Since $J_* + R_0(\lambda^2 \pm i0)W_*$ are closed operators, it follows from the closed range theorem that the inverses $S_\lambda^\pm$ are in $\mathscr{B}(\mathscr{R}(J_* + R_0(\lambda^2 \pm i0)W_*), H_{0,-s}^2(\Omega))$ for each $\lambda > 0$. Thus we conclude from the first inclusion in Lemma 3.1 that $S_\lambda^\pm \in \mathscr{B}(\mathscr{R}(R_0(\lambda^2 \pm i0)J_*), H_{0,-s}^2(\Omega))$ for each $\lambda > 0$.

It remains to prove (3.11) for any $\lambda > 0$. Since $\mathscr{X}_\pm^\lambda(\mathbb{R}^n) := \mathscr{R}(J_* + R_0(\lambda^2 \pm i0)W_*)) \subset L_{-s}^2(\mathbb{R}^n)$, it follows that

$$\left\|J_* + R_0(\lambda^2 \pm i0)W_*\right\|_{\mathscr{B}(H_{-s}^1(\Omega), \mathscr{X}_\pm^\lambda(\mathbb{R}^n))} \geq C\left\|J_* + R_0(\lambda^2 \pm i0)W_*\right\|_{\mathscr{B}(H_{-s}^1(\Omega), L_{-s}^2(\mathbb{R}^n))}$$

for any $\lambda > 0$ and $1/2 < s \leq \delta_0/2$, where the constant $C > 0$ is independent of λ. If we can prove that there exists a constant $C > 0$ such that

$$\left\|J_* + R_0(\lambda^2 \pm i0)W_*\right\|_{\mathscr{B}(H_{-s}^1(\Omega), L_{-s}^2(\mathbb{R}^n))} \geq C \tag{3.14}$$

for any $\lambda > 0$, then by the above estimate and (3.14), we get

$$\left\|J_* + R_0(\lambda^2 \pm i0)W_*\right\|_{\mathscr{B}(H_{-s}^1(\Omega), \mathscr{X}_\pm^\lambda(\mathbb{R}^n))} \geq C \tag{3.15}$$

for any $\lambda > 0$, which proves the estimates (3.11) for $N = 0$ and $\lambda > 0$. Hence all we have to do is to prove (3.14). However, we have proved the estimates (3.7) in the proof of Lemma 3.1:

$$\left\|R_0(\lambda^2 \pm i0)W_* f\right\|_{L_{-s}^2(\mathbb{R}^n)} \leq C\lambda^{-1}\|f\|_{H_{-s}^1(\Omega)}, \quad \forall \lambda > 0.$$

Thus these estimates imply that there exists $\lambda_1 > 1$ such that

$$\left\|R_0(\lambda^2 \pm i0)W_*\right\|_{\mathscr{B}(H_{-s}^2(\Omega), L_{-s}^2(\mathbb{R}^n))} \leq \frac{1}{2}\|J_*\|_{\mathscr{B}(H_{-s}^1(\Omega), L_{-s}^2(\mathbb{R}^n))}$$

for any $\lambda > \lambda_1$, and hence

$$\left\|J_* + R_0(\lambda^2 \pm i0)W_*\right\|_{\mathscr{B}(H^2_{-s}(\Omega), L^2_{-s}(\mathbb{R}^n))} \geq \frac{1}{2}\|J_*\|_{\mathscr{B}(H^1_{-s}(\Omega), L^2_{-s}(\mathbb{R}^n))}$$

for any $\lambda > \lambda_1$. Hence we get (3.14) for $\lambda > \lambda_1$. We now turn to the estimates for $\lambda \in (0, \lambda_1]$. If we suppose that there is a number $\lambda_* \in (0, \lambda_1]$ such that

$$\left\|J_* + R_0(\lambda_*^2 \pm i0)W_*\right\|_{\mathscr{B}(H^1_{-s}(\Omega), L^2_{-s}(\mathbb{R}^n))} = 0,$$

then this contradicts the injection of $J_* + R_0(\lambda_*^2 \pm i0)W_*$ in $\mathscr{B}(H^2_{-s}(\Omega), L^2_{-s}(\mathbb{R}^n))$. Thus there exists a constant $C > 0$ such that

$$\left\|J_* + R_0(\lambda^2 \pm i0)W_*\right\|_{\mathscr{B}(H^1_{-s}(\Omega), L^2_{-s}(\mathbb{R}^n))} \geq C$$

for any $\lambda \in (0, \lambda_1]$, which implies (3.14) for $0 < \lambda \leq \lambda_1$. Summarizing the estimates obtained now, we conclude (3.14) for any $\lambda > 0$. The proof of Theorem 3.2 is complete. □

4. Final remark

As an immediate consequence of Theorem 3.2 for $n \geq 4$, and Proposition 3.2 for $n = 3$ from [5], we have the result on the whole space $\mathbb{R}^n$. For, in particular case $\Omega = \mathbb{R}^n$, we can choose $j(x)$ as $j(x) = 1$ on $\mathbb{R}^n$, and hence, the identification operator J becomes the identity operator I. As a result of this observation, we conclude that the operator W_* is reduced to the multiplication operator V.

Summarizing the above argument, we have the following:

Theorem 4.1. *Let $n \geq 4$. Assume that the measurable potential V satisfies*

$$-c_0 \langle x \rangle^{-\delta_0} \leq V(x) \leq c_1 \langle x \rangle^{-\delta_0}$$

for some $0 < c_0 < (n-2)^2/4$, $c_1 > 0$ and $\delta_0 > (n+2)/2$. Let $1 < s \leq \delta_0/2$. Then there exists a family of bounded operators $S_\lambda^\pm \in \mathscr{B}(L^2_{-s}(\mathbb{R}^n))$ such that

$$S_\lambda^\pm \circ \{I + R_0(\lambda^2 \pm i0)V\} = I,$$

$$R_V(\lambda^2 \pm i0) = S_\lambda^\pm R_0(\lambda^2 \pm i0)$$

for any $\lambda \geq 0$. Furthermore, there exists a constant $C > 0$ such that

$$\left\|S_\lambda^\pm f\right\|_{L^2_{-s}(\mathbb{R}^n)} \leq C \|f\|_{L^2_{-s}(\mathbb{R}^n)}, \quad \forall \lambda \geq 0$$

for $f \in L^2_{-s}(\mathbb{R}^n)$. When $n = 3$, we may take $\delta_0 > 2$, and the same results hold true.

Let us give a remark on Theorem 4.1. Thanks to the argument of [4], observing that $R_0(\lambda^2 \pm i0)V$ are the compact perturbations of the identity operator I, we can then apply Fredholm alternative theorem to deduce the existence of the inverses $S_\pm^\lambda$ of operators $I + R_0(\lambda^2 \pm i0)V$.

Acknowledgment

The author would like to express his sincere gratitude to Professor Michael Ruzhansky for giving him a chance to write this article. The author would like to thank also Professor Yasuo Komori-Furuya for providing him the basic properties of Fourier transforms.

References

[1] S. Agmon, *Spectral properties of Schrödinger operators and scattering theory,* Ann. Scuola Norm. Sup. Pisa, Ser. IV, **2** (1975), 151–218.

[2] M. Ben-Artzi and S. Klainerman, *Decay and regularity for the Schrödinger equation,* J. Analyse Math. **58** (1992), 25–37.

[3] P. D'Ancona and V. Pierfelice, On the wave equation with a large potential, J. Functional Analysis **227** (2005), 30–77.

[4] V. Georgiev and N. Visciglia, *Decay estimates for the wave equation with potential,* Comm. Partial Differential Equations **29** (2003), 101–153.

[5] T. Matsuyama, *Dispersion for 3D wave equation with a potential in an exterior domain,* submitted.

[6] K. Mochizuki, *Spectral and Scattering Theory for Second-Order Elliptic Differential Operators in an Exterior Domain,* Lecture Notes Univ. Utah, Winter and Spring 1972

[7] S. Moulin, *Low-frequency dispersive estimates for the wave equation in higher dimensions,* Asymptot. Anal. **60** (2008), 15–27.

[8] M. Murata, *Asymptotic expansions in time for solutions of Schrödinger-type equations,* J. Functional Analysis **49** (1982), 10–56.

[9] M. Ruzhansky and M. Sugimoto, *A smoothing property of Schrödinger equations in the critical case,* Math. Ann. **335** (2006), 645–673.

[10] G. Vodev, Dispersive estimates of solutions to the wave equation with a potential in dimension $n \geq 4$, Comm. Partial Differential Equations **31** (2006), 1709–1733.

Tokio Matsuyama
Department of Mathematics
Faculty of Science and Engineering
Chuo University
1-13-27 Kasuga, Bunkyo-ku
Tokyo 112-8551, Japan
e-mail: tokio@math.chuo-u.ac.jp

Progress in Mathematics, Vol. 301, 227–245

On the Scattering on a Loop-shaped Graph

K. Mochizuki and I.Yu. Trooshin

Abstract. We treat an inverse scattering problem on a graph with an infinite ray and a loop joined at one point. Our problem amounts to the reconstruction of potential on the basis of the scattering data of operator.

Mathematics Subject Classification. 78A46; 34L25.

Keywords. Scattering, graph.

1. Introduction

Differential equations on graphs arise as simplified models in mathematics, physics, chemistry and engineering (nanotechnology), when one considers the propagation of waves of different natures in thin, tube-like domains (for more details see P. Exner, P. Seba [3], Yu. Pokornyi et al. [16], the papers of P. Kuchment [10]–[11] and the references within). Among several problems in this field, the scattering problems have been studied by many authors (e.g., B. Pavlov [5], N. Gerasimenko [5], [6], V. Kostrykin, R. Schrader [8], M. Harmer [7], P. Kurasov-F. Stenberg [9], J. Boman-P. Kurasov [2], V. Pivovarchik [15], Y. Latushkin-V. Pivovarchik [12]) because of the general importance of their applications.

In our paper we investigate the case of a relatively simple looking "loop-shaped" graph, i.e., a graph consisting of an infinite ray and a loop joined at one point.

As we have already mentioned, very interesting and deep results have already been obtained in the case of some general and complex graphs. However, the problem of characterizing scattering data, i.e., S-function, eigenvalues and normalizing constants (weight numbers), appears to be rather complicated and in this paper we investigate the spectral properties more precisely, allowing us to progress deep investigation of the corresponding inverse scattering problem.

Let Γ be a graph which consists of a half-line $\gamma = \{x \mid 0 < x < \infty\}$ and a loop $\kappa = \{z \mid 0 < z < 2\pi\}$, joint at the point $\vartheta = \{\{x = 0\} = \{z = 0\} = \{z = 2\pi\}\}$ (we call this point the vertex of the graph). We consider on Γ the following spectral problem describing the one-dimensional scattering of a quantum particle:

$$-u'' + \{q(X) - \lambda^2\} u = 0, \quad X \in \Gamma \setminus \{\vartheta\}, \tag{1.1}$$

$$u(x = 0) = u(z = 0) = u(z = 2\pi), \tag{1.2}$$

$$u'(x = 0 + 0) + u'(z = 0 + 0) - u'(z = 2\pi - 0) = 0. \tag{1.3}$$

Here differentiation with respect to the variable X is understood as differentiation with respect to x, when $X \in \gamma$, and as differentiation with respect to z, when $X \in \kappa$. Differentiation is not defined at the vertex. The potential $q(X)$ is real valued, and is required to satisfy $q(X) \in L^2_{\text{loc}}(\Gamma)$ (i.e., $q(X) \in L^2(\delta)$ for any part $\delta \subset \Gamma$) and also $(1 + x)q(x) \in L^1(\gamma)$. Parameter λ is a complex number such that $\text{Im}\lambda \geq 0$.

The matching conditions (1.2), (1.3) at the vertex are called the Kirchhoff conditions. For electric circuits they express the Kirchhoff law.

The Kirchhoff conditions guarantee the selfadjointness of the resulting Schrödinger operator

$$\mathcal{L}(u) = u''(X) + q(X)u(X), \quad X \in \Gamma. \tag{1.4}$$

More precisely, on the Hilbert space $L^2(\Gamma)$ with norm

$$\|f\|_{L^2(\Gamma)} = \left\{ \|f\|^2_{L^2(\gamma)} + \|f\|^2_{L^2(\kappa)} \right\}^{1/2} \tag{1.5}$$

we introduce the operator $\mathcal{L}$ with domain

$$\mathcal{D}(\mathcal{L}) = \{u(X), X \in \Gamma \mid u(X) \in H^2(\gamma) \cup H^2(\kappa)$$
$$\text{and satisfies the Kirchhoff conditions}\}$$

where H^k ($k = 1, 2, \dots$) are the usual Sobolev spaces. Then under the above assumptions on $q(X)$, $\mathcal{L}$ is selfadjoint in $L^2(\Gamma)$ and bounded from below. Moreover, the essential spectrum of $\mathcal{L}$ consists of the half-line $[0, \infty)$.

Theorem 1.1. *For any real $\lambda \neq 0$ there exists a solution $\Phi(X, \lambda)$ of problem (1.1)–(1.3) which is represented on γ uniquely as follows:*

$$\Phi(x, \lambda) = e(x, -\lambda) - S(\lambda)e(x, \lambda). \tag{1.6}$$

(More precise formulation of Theorem 1.1 is presented in §2.)

Here $e(x, \lambda)$ are so-called Jost functions, which behave on the closed upper half-plane of the spectral parameter λ as

$$e(x, \lambda) = e^{i\lambda x} \{1 + o(1)\}. \tag{1.7}$$

Function $S(\lambda)$ is continuous on the whole line $-\infty < \lambda < \infty$ (except, possibly, at the point $\lambda = 0$), $|S(\lambda)| = 1$ and

$$S(\lambda) - S_0(\lambda) = O\left(\frac{1}{\lambda}\right), \quad |\lambda| \to \infty, \tag{1.8}$$

$$S_0(\lambda) = \frac{2\sin\lambda\pi - i\cos\lambda\pi}{2\sin\lambda\pi + i\cos\lambda\pi} = \frac{3e^{i\lambda\pi} - e^{-i\lambda\pi}}{e^{i\lambda\pi} - 3e^{-i\lambda\pi}}, \tag{1.9}$$

by drawing an analogy with classical scattering on a half-line (e.g., [1]), the function $S(\lambda)$ is called the scattering function for the boundary value problem (1.1)–(1.3). Function $S_0(\lambda)$ is the scattering function of (1.1)–(1.3) in the case $q(X) \equiv 0$, $X \in \Gamma$.

Precise formulas for $S(\lambda)$ are given by (2.3)–(2.4).

Next we investigate a point spectrum of the boundary value problem (1.1)–(1.3).

Definition 1.2. Eigenvalues of $\mathcal{L}$ are called "visible at infinity" if there exist corresponding eigenfunctions which are not identically vanishing on γ.

Remark 1.3. The boundary value problem (1.1)–(1.3) can possess eigenvalues, which are "invisible at infinity", i.e., corresponding eigenfunctions vanish identically on γ. For example, in the case $q(X) \equiv 0$, $X \in \Gamma$, any number k^2, $k = 1, 2, \ldots$ is an "invisible at infinity" eigenvalue with eigenfunction $\sin kz$, $z \in \kappa$.

Theorem 1.4. *Operator $\mathcal{L}$ possesses at most a finite number of "visible at infinity" eigenvalues $-\lambda_j^2$, $0 < \lambda_1 < \lambda_2 < \cdots < \lambda_n$, which are all negative and simple.*

Remark 1.5. "Invisible at infinity" eigenvalues are also simple and can not coincide with eigenvalues "visible at infinity".

Now we are ready to state the inverse scattering problem.
We call
$$m_j = \|E(\cdot, i\lambda_j)\|_{L^2(\Gamma)}^{-1}, \ j = 1, \ldots, n \tag{1.10}$$
the "weight numbers" of problem (1.1)–(1.3). Here $E(X, i\lambda_j)$ is an eigenfunction of operator L, corresponding to the eigenvalue "visible at infinity" $-\lambda_j^2$, which is normalized as $E(x, i\lambda_j) = e(x, i\lambda_j)$ for $x \in \gamma$.

The scattering data is then given by $\{S(\lambda), \lambda_j, m_j | \lambda \in \Re \backslash \{0\}, j = 1, \ldots, n\}$, where $-\lambda_j^2$, $j = 1, \ldots, n$ are eigenvalues "visible at infinity", and m_j, $j = 1, \ldots, n$ are corresponding "weight numbers".

Our inverse scattering problem is the following:

IScP: Given scattering data $\{S(\lambda), \lambda_j, m_j | \lambda \in \Re \backslash \{0\}, j = 1, \ldots, n\}$, recover potential $q(X) \in \Gamma$.

As is well known (see for instance [13], Chapter 3), the Jost solution $e(x, \lambda)$ of equation (1.1) in γ can be represented as
$$e(x, \lambda) = e^{i\lambda x} + \int_x^\infty K(x, t) e^{i\lambda t} dt, \tag{1.11}$$
where the kernel $K(x, t)$ is continuous on $0 \le x \le t < \infty$ and satisfies the equation
$$K(x, x) = \frac{1}{2} \int_x^\infty q(t) dt, \quad x > 0. \tag{1.12}$$
As a function of variable λ, $e(x, \lambda)$ is analytic in the open half-plane $\text{Im}\lambda > 0$ and continuous on $\text{Im}\lambda \ge 0$.

The following theorem allows us to reconstruct potential $q(X)$ on the semi-line γ and at the vertex.

We define the function

$$F(x) = \sum_{k=1}^{n} m_k^2 e^{-\lambda_k x} + F_S(x), \qquad (1.13)$$

$$F_S(x) = \frac{1}{2\pi} \int_{-\infty}^{\infty} (S_0(\lambda) - S(\lambda)) e^{i\lambda x} d\lambda. \qquad (1.14)$$

Function $F_S(x)$ is understood as the Fourier transform of the function from $L^2(-\infty, \infty)$.

Theorem 1.6.

(i) *For any fixed $x \in \gamma$ the kernel $K(x,t)$ of transformation operator (1.11) satisfies the equation*

$$F(x+t) + K(x,t) + \int_x^{\infty} K(x,y) F(t+y) dy = 0, \quad 0 < x < t < \infty. \qquad (1.15)$$

(ii) *If function $F(t)$ is constructed from the scattering data of problem (1.1)–(1.3) according to (1.13)–(1.14), then equation (1.15) has a unique solution $K(x,t)$, belonging to $L^1(x, \infty)$ for each fixed $x \geq 0$.*

Theorem 1.6 allows us to prove the uniqueness of the solution to the inverse scattering problem (IScP) on the semi-line γ in the following sense.

Let us consider a second boundary value problem

$$-u'' + \{\tilde{q}(X) - \lambda^2\} u = 0, \quad X \in \Gamma, \qquad (1.16)$$

subject to Kirchhoff boundary conditions (1.2)–(1.3). Here function $\tilde{q}(X)$ is real valued, required to satisfy $\tilde{q}(X) \in L^2_{loc}(\Gamma)$ and $(1+x)\tilde{q}(x) \in L^1(\gamma)$

This second boundary value problem possesses the scattering function $\tilde{S}(\lambda)$, "visible at infinity" eigenvalues $\tilde{\lambda}_j$ and corresponding weight numbers $\tilde{m}_j$.

Theorem 1.7. *Let us suppose that scattering data of problems (1.1)–(1.3) and (1.16), (1.2)–(1.3) coincide, i.e., $S(\lambda) = \tilde{S}(\lambda)$, $\lambda \in \Re \setminus \{0\}$, $\lambda_j = \tilde{\lambda}_j$, $m_j = \tilde{m}_j$, $1 \leq j \leq n = \tilde{n}$. Then $q(X) = \tilde{q}(X)$, $X \in \gamma$.*

Theorem 1.6 allows us also to reconstruct potential $q(X)$ for any $X \in \gamma$.

Reconstruction procedure. Given scattering data $\{S(\lambda), \lambda_j, m_j | \lambda \in \Re \setminus \{0\}, j = 1, \ldots, n\}$

Step 1. Construct $F(x)$ via formula (1.13)

Step 2. Find $K(x,t)$, $0 \leq x \leq t < \infty$ by solving main equation (1.15).

Step 3. Recover the potential according to formula

$$q(x) = -2\frac{d}{dx} K(x,x), \ x > 0. \qquad (1.17)$$

Remark. The knowledge of scattering data $\{S(\lambda), \lambda_j, m_j | \lambda \in \Re \setminus \{0\}, j = 1, \ldots, n\}$ allows us to reconstruct potential $q(X)$ only on a semiinfinite line γ. However, given some additional information we can reconstruct potential $q(X)$ on the whole graph Γ. For example, in the case of a potential which is constant on the loop and continuous on γ in the neighborhood of the vertex, we can extend the reconstruction procedure to the whole graph.

The inverse scattering problem, investigated in this paper, was initially suggested to us by Prof. V.A. Marchenko and brief results are reported in Marchenko-Mochizuki-Trooshin [14] subject to the requirement that potentials have compact support. The present paper generalizes the results to non-compactly supported potentials and includes full details of the proof.

We would like to mention some results closely related to ours.

B. Pavlov, N. Gerasimenko [5], [6] started the rigorous investigation of scattering problems on graphs. M. Harmer had deduced the Marchenko equation in the case of a star-shaped graph, i.e., a graph consisting of a finite set of semiinfinite lines, joined at one point. P. Kurasov, J. Boman [2] had proved the existence of graphs equipped with different potentials, which possesses the same scattering matrix. (Their results do not mean a non-uniqueness in our case under consideration.) V. Pivovarchik, Y. Latushkin [15], [12] had investigated cases of loop-shaped and fork-shaped graphs with potentials vanishing on semi-lines. They used a connection with the Regge-type spectral problem on a finite interval to investigate spectral and scattering properties of a problem on graphs and they proved the existence of loop-shaped graphs equipped with different potentials on the loop, which possesses the same scattering function.

The rest of the paper will be organized as follows. Theorem 1.1 is proved in §2, and Theorem 1.4 is proved in §3. After preparing some other lemmas, we prove Theorems 1.6 and 1.7 in §4.

2. Proof of Theorem 1.1

A precise formulation of Theorem 1.1 is given by Theorem 2.1.

Let $c(z, \lambda)$, $s(z, \lambda)$ be linear independent solutions of equation (1.1) on the loop κ, satisfying the initial conditions

$$c(0, \lambda) = s'(0, \lambda) = 1, \quad c'(0, \lambda) = s(0, \lambda) = 0.$$

Note that their Wronskian is

$$\langle c(z, \lambda), s(z, \lambda) \rangle = 1. \tag{2.1}$$

Theorem 2.1. *For any real $\lambda \neq 0$ there exists a solution $\Phi(X, \lambda)$ of problem* (1.1)–(1.3) *which is represented as follows: On γ we have*

$$\Phi(x, \lambda) = e(x, -\lambda) - S(\lambda)e(x, \lambda); \tag{2.2}$$

$$S(\lambda) = \frac{e(0, -\lambda)[c(2\pi, \lambda) + s'(2\pi, \lambda) - 2] - e'(0, -\lambda)s(2\pi, \lambda)}{e(0, \lambda)[c(2\pi, \lambda) + s'(2\pi, \lambda) - 2] - e'(0, \lambda)s(2\pi, \lambda)}, \quad \lambda \neq \lambda_0, \tag{2.3}$$

$$S(\lambda_0) = \frac{e(0, -\lambda_0)c'(2\pi, \lambda_0) - e'(0, -\lambda_0)}{e(0, \lambda_0)c'(2\pi, \lambda_0) - e'(0, \lambda_0)}. \tag{2.4}$$

Here we denote by λ_0 such a value of parameter λ that

$$s(2\pi, \lambda_0) = 0 \quad and \quad c(2\pi, \lambda_0) = 1.$$

On the loop κ we have

$$\Phi(z, \lambda) = \frac{-2i\lambda[s(2\pi, \lambda)c(z, \lambda) + (1 - c(2\pi, \lambda))s(z, \lambda)]}{e(0, \lambda)[c(2\pi, \lambda) + s'(2\pi, \lambda) - 2] - e'(0, \lambda)s(2\pi, \lambda)}, \quad \lambda \neq \lambda_0 \qquad (2.5)$$

$$\Phi(z, \lambda_0) = \frac{-2i\lambda_0 c(z, \lambda_0)}{e(0, \lambda_0)c'(2\pi, \lambda_0) - e'(0, \lambda_0)} + b(\lambda_0)s(z, \lambda_0). \qquad (2.6)$$

Here $b(\lambda_0)$ is an arbitrary constant.

Proof. For any real $\lambda \neq 0$, the Wronskian of functions $e(x, \lambda)$, $e(x, -\lambda)$ is

$$\langle e(x, \lambda), e(x, -\lambda) \rangle = e(x, \lambda)e'(x, -\lambda) - e'(x, \lambda)e(x, -\lambda) = -2i\lambda. \qquad (2.7)$$

This implies that these two functions form a fundamental system of solutions of equation (1.1) in γ, and thus, if u satisfies (1.1) for any real $\lambda \neq 0$, then we have for some constants $m(\lambda)$, $n(\lambda)$,

$$u(x, \lambda) = m(\lambda)e(x, \lambda) + n(\lambda)e(x, -\lambda). \qquad (2.8)$$

On the other hand, any solution of equation (1.1) on the loop κ can be represented as

$$y(z, \lambda) = a(\lambda)c(z, \lambda) + b(\lambda)s(z, \lambda). \qquad (2.9)$$

Let us find $S(\lambda)$ on such a way that the solution in the form (2.2) would satisfy boundary conditions (1.2)–(1.3). According to representation (2.9) of the solution to equation (1.1) on the "loop" κ, the following equations hold

$$e(0, -\lambda) - S(\lambda)e(0, \lambda) = a(\lambda) = a(\lambda)c(2\pi, \lambda) + b(\lambda)s(2\pi, \lambda), \qquad (2.10)$$

$$e'(0, -\lambda) - S(\lambda)e'(0, \lambda) = a(\lambda)c'(2\pi, \lambda) + b(\lambda)(s'(2\pi, \lambda) - 1) \qquad (2.11)$$

As result, in the case $s(2\pi, \lambda) \neq 0$ the following equation takes place

$$b(\lambda) = a(\lambda)\frac{1 - c(2\pi, \lambda)}{s(2\pi, \lambda)}.$$

It follows from equation (2.11) that

$$e'(0, -\lambda) - S(\lambda)e'(0, \lambda) = [e(0, -\lambda) - S(\lambda)e(0, \lambda)]$$
$$\times \left(c'(2\pi, \lambda) + \frac{(1 - c(2\pi, \lambda))(s'(2\pi, \lambda) - 1)}{s(2\pi, \lambda)} \right),$$

and as result,

$$S(\lambda)\left[e(0, \lambda)\left(c'(2\pi, \lambda) + \frac{(1 - c(2\pi, \lambda))(s'(2\pi, \lambda) - 1)}{s(2\pi, \lambda)} \right) - e'(0, \lambda) \right]$$
$$= e(0, -\lambda)\left(c'(2\pi, \lambda) + \frac{(1 - c(2\pi, \lambda))(s'(2\pi, \lambda) - 1)}{s(2\pi, \lambda)} \right) - e'(0, -\lambda). \qquad (2.12)$$

If we take into account the Wronskian (2.1), we will come to (2.3). The solution to equation (1.1) in the case $s(2\pi, \lambda) \neq 0$ will be represented on the "loop" κ as

$$\Phi(z, \lambda) = a(z)\left[c(z, \lambda) + \frac{1 - c(2\pi, \lambda)}{s(2\pi, \lambda)} s(z, \lambda)\right].$$

If we put

$$a(\lambda) = e(0, -\lambda) - S(\lambda)e(0, \lambda)$$

$$= \frac{-2i\lambda s(2\pi, \lambda)}{e(0, \lambda)[c(2\pi, \lambda) + s'(2\pi, \lambda) - 2] - e'(0, \lambda)s(2\pi, \lambda)},$$

then we come to representation (2.5).

Next we examine such a parameter λ that $s(2\pi, \lambda) = 0$, $c(2\pi, \lambda) \neq 1$. In that case it has to be $a(\lambda) = 0$ and it follows from (2.10) that

$$S(\lambda) = \frac{e(0, -\lambda)}{e(0, \lambda)}.$$

If we put it into formula (2.11) and take into account the Wronskian (2.7) then

$$b(\lambda) = \frac{2i\lambda}{(1 - s'(2\pi, \lambda))e(0, \lambda)}$$

and as result,

$$\Phi(z, \lambda) = \frac{2i\lambda s(z, \lambda)}{(1 - s'(2\pi, \lambda))e(0, \lambda)}.$$

We came to formulas (2.3), (2.5) with $s(2\pi, \lambda) = 0$.

Lastly we investigate the case $\lambda = \lambda_0$. In this case it follows from equation (2.1) that $s'(2\pi, \lambda_0) = 1$. Taking also into account (2.10), (2.11) and (2.7) we come to (2.4) and formula

$$a(\lambda_0) = \frac{-2i\lambda_0}{e(0, \lambda_0)c'(2\pi, \lambda_0) - e'(0, \lambda_0)}.$$

As result we obtain equation (2.6).   □

Lemma 2.2. *The solution* $\Phi(X, \lambda)$ *to the boundary value problem* (1.1)–(1.3)*, defined by* (2.2)–(2.6) *can be analytically extended, as a function of variable* λ*, to the function, which is meromorphic in the half-plane* $\mathrm{Im}\lambda > 0$ *and continuous on the closed half-plane* $\mathrm{Im}\lambda \geq 0$*, except for its poles and, probably, zero.*

Such an analytically extended function is the solution to the boundary value problem (1.1)–(1.3) *for any* $\mathrm{Im}\lambda \geq 0$*,* $\lambda \neq 0$ *and can be represented on the "semi-infinite string"* γ *as*

$$\Phi(x, \lambda) = -\frac{2i\lambda}{e(0, \lambda)}w(x, \lambda) - \frac{2i\lambda}{\Delta(\lambda)e(0, \lambda)}e(x, \lambda), \quad \lambda \neq \tilde{\lambda}, \tag{2.13}$$

$$\Phi(x, \tilde{\lambda}) = -\frac{2i\tilde{\lambda}}{e(0, \tilde{\lambda})}w(x, \tilde{\lambda}), \tag{2.14}$$

and on the "loop" $\kappa = \{0 < z < 2\pi\}$ according to formulas (2.5)–(2.6). Here function $\omega(x, \lambda)$ is the solution to equation (1.1) on "string" γ, satisfying initial conditions $\omega(0, \lambda) = 0$, $\omega'(0, \lambda) = 1$,

$$\Delta(\lambda) = \frac{e(0, \lambda)[c(2\pi, \lambda) + s'(2\pi, \lambda) - 2]}{s(2\pi, \lambda)} - e'(0, \lambda), \ \lambda \neq \lambda_0 \qquad (2.15)$$

$$\Delta(\lambda_0) = e(0, \lambda_0)c'(2\pi, \lambda_0) - e'(0, \lambda_0), \qquad (2.16)$$

and $\tilde{\lambda}$ is a value of parameter λ, satisfying $s(2\pi, \tilde{\lambda}) = 0$ and $c(2\pi, \tilde{\lambda}) \neq 1$.

Proof. In the case $s(2\pi, \lambda_0) = 0$ and $c(2\pi, \lambda_0) = 1$ it follows from equation (2.1) that $s'(2\pi, \lambda_0) = 1$. Zeros of function $s(2\pi, \lambda)$ are simple and, as result,

$$\frac{c(2\pi, \lambda) + s'(2\pi, \lambda) - 2}{s(2\pi, \lambda)}$$
$$= \frac{(c(2\pi, \lambda) - 1)(1 - s'(2\pi, \lambda))}{s(2\pi, \lambda)} + c'(2\pi, \lambda) \to c'(2\pi, \lambda_0) \qquad (2.17)$$

as $\lambda \to \lambda_0$. It means that function $\Phi(z, \lambda)$, $z \in \kappa$, which is defined by formulae (2.5)–(2.6), is analytic for any $\Im\lambda > 0$ and continuous on the closed half-plane $\mathrm{Im}\lambda \geq 0$, except for zeros of function $\Delta(\lambda)$.

Let us consider now the solution of problems (1.1)–(1.3) on the "semi-infinite string" γ, defined by formula (2.2) in the case of real $\lambda \neq 0$.

We can represent this solution as

$$\Phi(x, \lambda) = e(x, -\lambda) - \frac{e(0, -\lambda)}{e(0, \lambda)}e(x, \lambda) - \left(S(\lambda) - \frac{e(0, -\lambda)}{e(0, \lambda)}\right)e(x, \lambda).$$

We come to (2.14) immediately in the case $\lambda = \tilde{\lambda}$. Otherwise we use the expression

$$\omega(x, \lambda) = -\frac{e(0, \lambda)e(x, -\lambda) - e(0, -\lambda)e(x, \lambda)}{2i\lambda} \qquad (2.18)$$

to obtain

$$\Phi(x, \lambda) = -\frac{2i\lambda}{e(0, \lambda)}\omega(x, \lambda) - \frac{2i\lambda}{\Delta(\lambda)e(0, \lambda)}e(x, \lambda). \qquad (2.19)$$

This function can be analytically extended in the whole half-plane $\Im\lambda > 0$ except either for zeros of function $e(\lambda, 0)$ or for zeros of function $\Delta(\lambda)$. It could be verified directly, that function $\Phi(x, \lambda)$, so-defined on the upper half-plane $\mathrm{Im}\lambda \geq 0$, satisfies the boundary value problem (1.1)–(1.3). $\qquad\qquad\square$

3. Proof of Theorem 1.4

First we prove several lemmas which will be used to prove Theorem 1.4.

Lemma 3.1. *Any solution $y(X, \lambda)$ of equation (1.1) for any $a, b \in \Gamma$ satisfies the formula*

$$2\lambda \int_a^b y^2(X, \lambda)dX + (y(X, \lambda)\dot{y}'(X, \lambda) - \dot{y}(X, \lambda)y'(X, \lambda))\,|_{x=a}^b = 0. \qquad (3.1)$$

Here $\dot{y} = \dfrac{\partial}{\partial \lambda} y$.

Proof. If we differentiate equation (1.1) with respect to λ, we obtain the following equation

$$\dot{y}''(X, \lambda) - (q(X) - \lambda^2)\,\dot{y}(X, \lambda) + 2\lambda y(X, \lambda) = 0,\ X \in \Gamma. \qquad (3.2)$$

We multiply equation (3.2) by $y(X, \lambda)$ and subtract equation (1.1) multiplied by $\dot{y}(X, \lambda)$. We obtain the equation:

$$2\lambda y^2(X, \lambda) + \dot{y}''(X, \lambda)y(X, \lambda) - y''(X, \lambda)\dot{y}(X, \lambda) = 0,\ X \in \Gamma. \qquad (3.3)$$

After integrating this equation, we obtain equality (3.1). □

Lemma 3.2. *Zeros of functions $\Delta(\lambda)$ and $e(0, \lambda)$ do not coincide in the upper half-plane $\mathrm{Im}\lambda > 0$.*

Proof. We first mention that functions $e(0, \lambda)$ and $e'(0, \lambda)$ can not become 0 simultaneously. We fix any $\tilde{\lambda}$ from the upper half-plane $\mathrm{Im}\tilde{\lambda} > 0$.

1) In the case $s(2\pi, \tilde{\lambda}) \neq 0$ it follows from definition (2.15) that $e(0, \tilde{\lambda})$ and $\Delta(\tilde{\lambda})$ can not become 0 simultaneously.

2) In the case $s(2\pi, \tilde{\lambda}) = 0$ and $c(2\pi, \tilde{\lambda}) = 1$ it follows from equation (2.1) that $s'(2\pi, \tilde{\lambda}) = 1$. Zeros of function $s(2\pi, \lambda)$ are simple and, as result,

$$\frac{c(2\pi, \lambda) + s'(2\pi, \lambda) - 2}{s(2\pi, \lambda)} = \frac{(c(2\pi, \lambda) - 1)(1 - s'(2\pi, \lambda))}{s(2\pi, \lambda)} + c'(2\pi, \lambda) \to c'(2\pi, \tilde{\lambda})$$

as $\lambda \to \tilde{\lambda}$. It means that in the case under consideration $\Delta(\tilde{\lambda}) = 0$ and $e(0, \tilde{\lambda}) = 0$ implies $e'(0, \tilde{\lambda}) = 0$ which is impossible.

3) In the case $s(2\pi, \tilde{\lambda}) = 0$ and $c(2\pi, \tilde{\lambda}) \neq 1$ function $\Delta(\tilde{\lambda})$ can be finite only if $e(0, \tilde{\lambda}) = 0$. Such a point $\tilde{\lambda}$ belongs to imaginary axis.

We choose $\lambda = \tilde{\lambda} + \epsilon$, $\epsilon \in \mathbf{R}\backslash\{0\}$ and multiply formula (2.15) by $\overline{e(0, \lambda)}$ to obtain an equality

$$2i\mathrm{Im}(\Delta(\lambda)\overline{e(0, \lambda)}) = 2i\mathrm{Im}\left\{\frac{(c(2\pi, \lambda) - 1)(1 - s'(2\pi, \lambda))}{s(2\pi, \lambda)} + c'(2\pi, \lambda)\right\}|e(0, \lambda)|^2$$
$$- \{e'(0, \lambda)\overline{e(0, \lambda)} - \overline{e'(0, \lambda)}e(0, \lambda)\}$$

Since $c(2\pi, \tilde{\lambda})s'(2\pi, \tilde{\lambda}) = 1$ by (2.1), it follows that

$$(c(2\pi, \lambda) - 1)(1 - s'(2\pi, \lambda)) = \frac{(c(2\pi, \tilde{\lambda}) - 1)^2 \xi_1(\epsilon)}{c(2\pi, \tilde{\lambda})}, \quad \xi_1(\epsilon) = 1 + O(\epsilon)$$

as $\epsilon \to 0$. Next we take $y = s(z, \tilde{\lambda})$, $(a, b) = \kappa$ in Lemma 3.1. Then the condition $s(2\pi, \tilde{\lambda}) = 0$, $s(0, \lambda) = \dot{s}(0, \lambda) = 0$ imply

$$\dot{s}(2\pi, \tilde{\lambda}) = 2\tilde{\lambda}c(2\pi, \tilde{\lambda}) \int_\kappa s(z, \tilde{\lambda})^2 dz.$$

Substituting this into the formula

$$s(2\pi, \lambda) = \epsilon \dot{s}(2\pi, \tilde{\lambda}) \xi_2(\epsilon), \quad \xi_2(\epsilon) = 1 + O(\epsilon),$$

and taking into account that functions $c(z, \tilde{\lambda})$, $s(z, \tilde{\lambda})$ are real valued we come to the following equation

$$2i\mathrm{Im}\left\{ \frac{(c(2\pi, \lambda) - 1)(1 - s'(2\pi, \lambda))}{s(2\pi, \lambda)} + c'(2\pi, \lambda) \right\}$$

$$= 2i\mathrm{Im}\left\{ \frac{(c(2\pi, \tilde{\lambda}) - 1)^2 \xi_1(\epsilon)\overline{\xi_2(\epsilon)}}{2\epsilon \tilde{\lambda} c(2\pi, \tilde{\lambda})^2 \|s(\cdot, \tilde{\lambda})\|^2_{L^2(\kappa)} |\xi_2(\epsilon)|^2} \right\} + 2i\mathrm{Im}c'(2\pi, \lambda)$$

$$= -\frac{i}{\epsilon}B(1 + \delta(\epsilon)) + 2i\delta_1(\epsilon).$$

Here

$$B = \frac{(c(2\pi, \tilde{\lambda}) - 1)^2}{|\tilde{\lambda}|c(2\pi, \tilde{\lambda})^2 \|s(\cdot, \tilde{\lambda})\|^2_{L^2(\kappa)}} > 0$$

and $\delta(\epsilon)$, $\delta_1(\epsilon) \to 0$ as $\epsilon \to 0$. On the other hand, since $\tilde{\lambda}$ is pure imaginary we come to the formula

$$e'(0, \lambda)\overline{e(0, \lambda)} - \overline{e'(0, \lambda)}e(0, \lambda) = 4i\epsilon|\tilde{\lambda}|\|e(\cdot, \lambda)\|^2_{L^2(\gamma)}.$$

Taking into account

$$e(0, \lambda) = \epsilon \dot{e}(0, \tilde{\lambda}) \xi_3(\epsilon), \quad \xi_3(\epsilon) = 1 + O(\epsilon), \quad \epsilon \to 0,$$

we conclude

$$\mathrm{Im}(\Delta(\lambda)\overline{e(0, \lambda)}) = -2\epsilon|\tilde{\lambda}|\|e(\cdot, \lambda)\|^2_{L^2(\gamma)} - \frac{\epsilon}{2}B|\dot{e}(0, \tilde{\lambda}|^2(1 + \delta(\epsilon)). \quad (3.4)$$

If $\Delta(\tilde{\lambda}) = 0$, function $\Delta(\lambda)e(0, \lambda)$ (and function $\Delta(\lambda)\overline{e(0, \lambda)}$) has to have a zero of second order at point $\tilde{\lambda}$, which is impossible according to formula (3.4).

This implies, that $\Delta(\tilde{\lambda}) \neq 0$. $\square$

Lemma 3.3. *Eigenvalues of boundary problem* (1.1)–(1.3), *visible at infinity, are all negative and compose the following set*

$$\sigma_d(\mathcal{L}) = \{\tilde{\lambda}^2 | \; \tilde{\lambda} \in U_1 \cup U_2 \cup U_3\};$$
$$U_1 = \{\tilde{\lambda}| \; \text{Im}\tilde{\lambda} > 0, \; s(2\pi, \tilde{\lambda}) \neq 0, \; \Delta(\tilde{\lambda}) = 0\},$$
$$U_2 = \{\tilde{\lambda}| \; \text{Im}\tilde{\lambda} > 0, \; s(2\pi, \tilde{\lambda}) = 0, \; c(2\pi, \tilde{\lambda}) = 1, \; \Delta(\tilde{\lambda}) = 0\},$$
$$U_3 = \{\tilde{\lambda}| \; \text{Im}\tilde{\lambda} > 0, \; s(2\pi, \tilde{\lambda}) = 0, \; c(2\pi, \tilde{\lambda}) \neq 1, \; e(0, \tilde{\lambda}) = 0\}.$$

Proof. It follows from the selfadjointness of problem (1.1)–(1.3), that all eigenvalues of the problem are real.

For any real $\lambda \neq 0$ there exists a fundamental system of solutions $e(x, \lambda)$, $e(x, -\lambda)$ to equation (1.1) on the semiinfinite string γ, which behaves like $\exp(i\lambda x)$ and $\exp(-i\lambda x)$ for sufficiently large x. It means, that any solution to equation (1.1) does not decrease for any positive value of parameter λ and, as result, problem (1.1)–(1.3) can not possess positive eigenvalues, visible at infinity. In the case $\lambda = 0$ there exists a fundamental system of solutions, which behaves as 1 and x for sufficiently large x (see §5, Chapter I, [1]), which means that problem (1.1)–(1.3) also cannot possess zero eigenvalue, visible at infinity.

If $\tilde{\lambda}^2$ is an eigenvalue visible at infinity, then the corresponding eigenfunction has to be proportional to function $e(x, \tilde{\lambda})$ on γ. This eigenfunction has to satisfy the boundary conditions (1.2)–(1.3). As result, the following relation has to hold:

$$e(0, \tilde{\lambda}) = a(\tilde{\lambda}) = a(\tilde{\lambda})c(2\pi, \tilde{\lambda}) + b(\tilde{\lambda})s(2\pi, \tilde{\lambda}), \tag{3.5}$$
$$e'(0, \tilde{\lambda}) = a(\tilde{\lambda})c'(2\pi, \tilde{\lambda}) + b(\tilde{\lambda})(s'(2\pi, \tilde{\lambda}) - 1) \tag{3.6}$$

Consequently, in the case $s(2\pi, \tilde{\lambda}) \neq 0$, $e(0, \tilde{\lambda}) \neq 0$ it should be

$$e'(0, \tilde{\lambda}) = e(0, \tilde{\lambda})\left[c'(2\pi, \tilde{\lambda}) + \frac{(1 - c(2\pi, \tilde{\lambda}))(s'(2\pi, \tilde{\lambda}) - 1)}{s(2\pi, \tilde{\lambda})}\right]. \tag{3.7}$$

In the case $s(2\pi, \tilde{\lambda}) = 0$, $c(2\pi, \tilde{\lambda}) = 1$ we have $s'(2\pi, \tilde{\lambda}) = c(2\pi, \tilde{\lambda})^{-1} = 1$ and consequently

$$e'(0, \tilde{\lambda}) = e(0, \tilde{\lambda})c'(2\pi, \tilde{\lambda}). \tag{3.8}$$

These equations show that $\Delta(\tilde{\lambda}) = 0$, and as result $\tilde{\lambda} \in U_1 \cup U_2$.

Lastly, in the case $s(2\pi, \tilde{\lambda}) = 0$, $c(2\pi, \tilde{\lambda}) \neq 1$ parameter $\tilde{\lambda}^2$ will become an eigenvalue, visible at infinity, only if $e(0, \lambda) = 0$ and, as result, $\tilde{\lambda} \in U_3$. □

Lemma 3.4. *All the zeros $i\lambda_k$ of function $\Delta(\lambda)$ are simple. Moreover, the following equality holds*

$$\frac{\dot{\Delta}(i\lambda_k)e(0, i\lambda_k)}{2i\lambda_k} = -m_k^{-2}, \tag{3.9}$$

where m_k are weight numbers defined by (1.10).

Proof. Let us denote by $h(X, \lambda)$ a solution to (1.1)–(1.3), satisfying the initial condition $h(0, \lambda) = 1$. Such a solution exists in the neighbourhood of $i\lambda_k$ and we have

$$h(X, i\lambda_k) = \frac{E(X, i\lambda_k)}{e(0, i\lambda_k)}, \qquad (3.10)$$

where $E(X, i\lambda_k)$ is an eigenfunction normalized as $E(x, i\lambda_k) = e(x, i\lambda_k)$, $x \in \gamma$.

We employ Lemma 3.1 to obtain the following equality

$$\|h(i\lambda_k, x)\|_{L^2(\gamma)}^2 = \frac{\dot{e}'(0, i\lambda_k)e(0, i\lambda_k) - e'(0, i\lambda_k)\dot{e}(0, i\lambda_k)}{2i\lambda_k e^2(0, i\lambda_k)}. \qquad (3.11)$$

Solution $h(x, \lambda)$ satisfies the initial condition $h(0, \lambda) = 1$ for any λ in the neighbourhood of $i\lambda_k$, and, as result, $\dot{h}(0, \lambda) = 0$ there. Then it follows from Lemma 3.1 that

$$\|h(z, i\lambda_k)\|_{L^2(\kappa)}^2 = -\frac{1}{2i\lambda_k}\left(\dot{h}'(2\pi, i\lambda_k) - \dot{h}'(0, i\lambda_k)\right). \qquad (3.12)$$

Function $h(X, i\lambda_k)$ satisfies boundary conditions (1.2)–(1.3). As result,

$$h'(x = 0, i\lambda_k) = h'(z = 2\pi, i\lambda_k) - h'(z = 0, i\lambda_k) = A(i\lambda_k), \qquad (3.13)$$

where

$$A(\lambda) = \frac{c(2\pi, \lambda) + s'(2\pi, \lambda) - 2}{s(2\pi, \lambda)}, \ \lambda \neq \lambda_0,$$

and $A(\lambda_0) = c'(2\pi, \lambda_0)$. Here λ_0 is such a value of parameter λ, that $s(2\pi, \lambda_0) = 0$ and $c(2\pi, \lambda_0) = 1$.

As result we obtained the formula

$$\|h(z, i\lambda_k)\|_{L^2(\kappa)}^2 = -\frac{\dot{A}(i\lambda_k)}{2i\lambda_k}. \qquad (3.14)$$

Combining (3.11) and (3.14), we come to

$$\|h(X, \lambda)\|_{L^2(\Gamma)}^2 = \frac{\dot{e}'(0, i\lambda_k)e(0, i\lambda_k) - e'(0, i\lambda_k)\dot{e}(0, i\lambda_k) - \dot{A}(i\lambda_k)e^2(0, i\lambda_k)}{2i\lambda_k e^2(0, i\lambda_k)}. \qquad (3.15)$$

From definitions (2.15)–(2.16) of $\Delta(\lambda)$, differentiating with respect to λ, we obtain the formula

$$\dot{\Delta}(\lambda) = -\dot{e}'(0, \lambda) + \dot{A}(\lambda)e(0, \lambda) + A(\lambda)\dot{e}(0, \lambda). \qquad (3.16)$$

It follows from (3.10)–(3.13), that

$$\dot{\Delta}(i\lambda_k)e(0, i\lambda_k) = -\dot{e}'(0, i\lambda_k)e(0, i\lambda_k) + e'(0, i\lambda_k)\dot{e}(0, i\lambda_k) + \dot{A}(i\lambda_k)e^2(0, i\lambda_k). \qquad (3.17)$$

Comparing this formula with formula (3.15), we obtain the equality

$$m_k^{-2} = e^2(0, i\lambda_k)\|h(i\lambda_k, x)\|_{L^2(\Gamma)}^2 = -\frac{\dot{\Delta}(i\lambda_k)e(0, i\lambda_k)}{2i\lambda_k}. \qquad (3.18)$$

$\square$

Proof of Theorem 1.4. According to Lemma 3.3, we have only to show that the sets U_1, U_2, U_3 are at most finite.

Function $s(2\pi, \lambda)$ can have only a finite number of imaginary zeros. So, U_2 and U_3 are at most finite. We now show that the set of zeros of function $\Delta(\lambda)$ is also finite.

We can show that $\Delta(\lambda) \geq C > 0$ for sufficiently large $|\lambda|$ and, consequently, the zeros of $\Delta(\lambda)$ form at most a countable set having 0 as the only possible limit point.

If $\Delta(0) \neq 0$, then the set of zeros cannot have a limit point and, as result, the number of zeros of $\Delta(\lambda)$ is finite. We now investigate the case $\Delta(0) = 0$. If 0 is the limiting point of zeros $\{i\mu_k\}$ then there exists

$$M > \mu_1 > \mu_2 > \cdots > \mu_k > \cdots \to 0.$$

Eigenfunctions $E(X, i\mu_k)$ satisfy the following equation:

$$
\begin{aligned}
0 &= \int_\Gamma E(X, i\mu_k)E(X, i\mu_{k+1})dX \\
&= \int_{\Gamma_A} E(X, i\mu_k)\{E(X, i\mu_{k+1}) - E(X, i\mu_k)\}dX \\
&\quad + \int_{\Gamma_A} E(X, i\mu_k)^2 dX + \int_A^\infty e(x, i\mu_k)e(x, i\mu_{k+1})dx.
\end{aligned}
$$

Here $\Gamma_A = (0, A) \cup \kappa$ for any $A > 0$ and we can show that

$$\lim_{k\to\infty} \{E(X, i\mu_{k+1}) - E(X, i\mu_k)\} = 0$$

uniformly for $X \in \Gamma_A$. It implies, that

$$\liminf_{k\to\infty} \int_A^\infty e(x, i\mu_k)e(x, i\mu_{k+1})dx \leq 0.$$

But for sufficiently large $A > 0$ there is a uniform estimate $e(x, i\mu) > \dfrac{1}{2}e^{-\mu x}$ on $(x, \mu) \in [A, \infty) \times [0, \infty)$ and, consequently,

$$\int_A^\infty e(x, i\mu_k)e(x, i\mu_{k+1})dx > \frac{e^{-2AM}}{8M}.$$

We come to a contradiction. $\square$

4. Proof of Theorem 1.6 and Theorem 1.7

To prove Theorem 1.6 we need some more lemmas.

Lemma 4.1. *Scattering function $S(\lambda)$, defined in Theorem 1.1 (Theorem 2.1), is continuous on the whole line (except, probably, at the point $\lambda = 0$), $|S(\lambda)| = 1$ and*

$$S(\lambda) - S_0(\lambda) = O\left(\frac{1}{\lambda}\right), \quad |\lambda| \to \infty. \tag{4.1}$$

Proof. We can represent function $S(\lambda)$ as

$$S(\lambda) = \frac{\overline{\Delta(\lambda)}s(2\pi,\lambda)}{\Delta(\lambda)s(2\pi,\lambda)}.$$

Assertions of the lemma follow from this representation, formula (2.17) and the asymptotics of corresponding functions:

$$e(0,\lambda) - 1 = O\left(\frac{1}{\lambda}\right), \quad s(2\pi,\lambda) - \frac{\sin 2\pi\lambda}{\lambda} = O\left(\frac{1}{\lambda}\right), \quad \text{etc.} \qquad \square$$

Lemma 4.2. *Function $\Phi(X,\lambda)$, defined by (2.5)–(2.6), (2.13)–(2.14) possesses the following properties:*

(i) *Function $\Phi(X,\lambda)$ is bounded in a vicinity of $\lambda = 0$.*

(ii) *Let us define by $\Lambda_0 > 0$ such a positive number, that all the poles of $\Phi(X,\lambda)$ belong to the ball $|\lambda| < \Lambda_0$. Then there are exists a positive number $C > 0$ such that*

$$|\Phi(x,\lambda)| \le Ce^{|\lambda|x}, \quad |\lambda| \ge \Lambda_0. \qquad (4.2)$$

Proof. (i) If $\Delta(0)e(0,0) \ne 0$, then the statement follows immediately from the definition of function $\Phi(X,\lambda)$.

Let us investigate now the case $\Delta(0)e(0,0) = 0$, using the argument of [13], Chapter 3.

We introduce the family of auxiliary problems $-u'' + q_\beta(X)u$ with potentials

$$q_\beta(X) = q(X), \ X \in \Gamma_\beta, \ = 0, \ x > \beta.$$

Here $\Gamma_\beta = (0,\beta) \cup \kappa$.

In the case of compact supported potentials $q_\beta(X)$ the corresponding functions $\Delta_\beta(\lambda)e_\beta(x,\lambda)$ are meromorphic on the whole plane $\lambda \in \mathbf{C}$.

Kernel $K(x,t)$ of transformation operator (1.11) is integrable with respect to both variables and satisfies the following estimates (see, for instance [13] Chapter 3 or [4] Chapter 2):

$$|K(x,t)| \le \frac{1}{2}Q_0\left(\frac{x+t}{2}\right)\exp\left(Q_1(x) - Q_1\left(\frac{x+t}{2}\right)\right), \qquad (4.3)$$

$$\left|\frac{\partial K(x_1,x_2)}{\partial x_i} + \frac{1}{4}q\left(\frac{x_1+x_2}{2}\right)\right|$$
$$\le \frac{1}{2}Q_0(x_1)Q_0\left(\frac{x_1+x_2}{2}\right)\exp\left(Q_1(x_1) - Q_1\left(\frac{x_1+x_2}{2}\right)\right), \ i = 1,2. \qquad (4.4)$$

Here

$$Q_0(x) = \int_x^\infty |q(t)|dt, \ Q_1(x) = \int_x^\infty Q_0(t)dt. \qquad (4.5)$$

We can choose sufficiently small $\rho > 0$ such, that functions $\Delta_\beta(\lambda)e_\beta(x,\lambda)$ have no more than one zero in the half-disk $D_\rho = \{\lambda; \ |\lambda| \le \rho, \ \text{Im}\lambda \ge 0\}$ for all sufficiently large β. We also suppose that $\rho > 0$ is small enough and $s(2\pi,\lambda)$ does not have any zeros in D_ρ except probably 0.

We denote by λ_β a zero of $\Delta_\beta(\lambda)e_\beta(x,\lambda)$, belonging to D_ρ (if there is no such zero, we put $\lambda_\beta = 0$).

It follows from Lemma 3.4 that all the zeros λ_β are simple and it implies that functions

$$\Phi_\beta(x,\lambda)\frac{\lambda - \lambda_\beta}{\lambda + \lambda_\beta}$$

are analytic in D_ρ and bounded on arc $\{\lambda; |\lambda| = \rho, \mathrm{Im}\lambda \geq 0\}$ uniformly on β for all sufficiently large β. Also it follows from Lemma 4.1 that $\Phi_\beta(x,\lambda)$ are bounded on interval $-\rho \leq \lambda \leq \rho$ uniformly on β for all sufficiently large β. It means that

$$\sup_{\lambda \in D_\rho} \left| \Phi_\beta(x,\lambda)\frac{\lambda - \lambda_\beta}{\lambda + \lambda_\beta} \right| = C(x) < \infty$$

for all sufficiently large β.

We mention that

$$\lim_{\beta \to \infty} \frac{\lambda - \lambda_\beta}{\lambda + \lambda_\beta} = 1, \quad \lim_{\beta \to \infty} \Phi_\beta(x,\lambda) = \Phi(x,\lambda)$$

for any $\lambda \in D_\rho\backslash\{0\}$. As result, we come to the following estimate:

$$\left| \Phi(x,\lambda) \right| \leq 2C(x)$$

for any $\lambda \in D_\rho\backslash\{0\}$.

(ii) Estimate (4.2) follows easily from definitions (2.5)–(2.6), (2.13)–(2.14) and the asymptotics of the corresponding functions. □

Lemma 4.3. *For every eigenvalue $i\mu_j$, visible at infinity, the following equality takes place:*

$$\mathrm{Res}_{\lambda = i\mu_j} e^{i\lambda t} \Phi(x,\lambda) = im_j^2 e(x, i\mu_j) e^{-\mu_j t}. \tag{4.6}$$

Proof. In the case $i\mu_j \in U_1 \cup U_2$ we have

$$\mathrm{Res}_{\lambda = i\mu_j} e^{i\lambda t} \Phi(x,\lambda) = -\frac{2i(i\mu_j)e^{-\mu_j t}e(x, i\mu_j)}{\dot{\Delta}(i\mu_j)e(0, i\mu_j)}$$

and formula (4.6) follows immediately from Lemma 3.4.

Next, in the case $i\mu_j \in U_3$ it can be easily seen that

$$\frac{e(0,\lambda)}{s(2\pi,\lambda)} \to \frac{\dot{e}(0, i\mu_j)}{\dot{s}(2\pi, i\mu_j)}, \quad \lambda \to i\mu_j,$$

$$c(2\pi, i\mu_j) + s'(2\pi, i\mu_j) - 2 = \frac{(c(2\pi, i\mu_j) - 1)^2}{c(2\pi, i\mu_j)}.$$

Then

$$\mathrm{Res}_{\lambda = i\mu_j} e^{i\lambda t} \Phi(x,\lambda) = \frac{2\mu_j e^{-\mu_j t}e(x, i\mu_j)}{\dot{e}(0, i\mu_j)}\left[\frac{1}{e'(0, i\mu_j)} \right.$$
$$\left. + \frac{\dot{s}(2\pi, i\mu_j)}{\dot{e}(0, i\mu_k)(c + s' - 2)(2\pi, i\mu_j) - e'(0, i\mu_k)\dot{s}(2\pi, i\mu_j)} \right]$$

$$= \frac{2\mu_j e^{-\mu_j t} e(x, i\mu_j)}{\dot{e}(0, i\mu_j)}$$

$$\times \frac{\dot{e}(0, i\mu_j)(c(2\pi, i\mu_j) - 1)^2}{e'(0, i\mu_j)[\dot{e}(0, i\mu_j)(c(2\pi, i\mu_j) - 1)^2 - c(2\pi, i\mu_j)\dot{s}(2\pi, i\mu_j)e'(0, i\mu_k)]}.$$

By definition $E(x, i\mu_j) = e(x, i\mu_j)$, $x \in \gamma$, and

$$E(z, i\mu_j) = \frac{e'(0, i\mu_j)c(2\pi, i\mu_j)}{1 - c(2\pi, i\mu_j)} s(z, i\mu_j), \quad z \in \kappa.$$

According to Lemma 3.1 we can calculate

$$2i\mu_j \|E(\cdot, i\mu_j)\|^2_{L^2(\gamma)} = -\dot{e}(0, i\mu_j)e'(0, i\mu_j),$$

$$2i\mu_j \|E(\cdot, i\mu_j)\|^2_{L^2(\kappa)} = \left(\frac{e'(0, i\mu_j)c(2\pi, i\mu_j)}{1 - c(2\pi, i\mu_j)}\right)^2 \dot{s}(2\pi, i\mu_j)s'(2\pi, i\mu_j),$$

and as result,

$$2i\mu_j \|E(\cdot, i\mu_j)\|^2_{L^2(\Gamma)}$$

$$= -\dot{e}(0, i\mu_j)e'(0, i\mu_j) + \frac{e'(0, i\mu_j)^2 c^2(2\pi, i\mu_j)}{(c(2\pi, i\mu_j) - 1)^2} \dot{s}(2\pi, i\mu_j)s'(2\pi, i\mu_j)$$

$$= \frac{-(c(2\pi, i\mu_j) - 1)^2 \dot{e}(0, i\mu_j)e'(0, i\mu_j) + e'(0, i\mu_j)^2 c^2(2\pi, i\mu_j)\dot{s}(2\pi, i\mu_j)s'(2\pi, i\mu_j)}{(c(2\pi, i\mu_j) - 1)^2}.$$

Finally we use the equality $c(2\pi, i\mu_j)s'(2\pi, i\mu_j) = 1$, which takes place in the case $i\mu_k \in U_3$, to finish the proof of the lemma. $\qquad\square$

Proof of Theorem 1.6. (i) Solution $\Phi(x, \lambda)$, $x \in \gamma$, to the boundary problem (1.1)–(1.3) can be represented as

$$\Phi(x, \lambda) = e(x, -\lambda) - S_0(\lambda)e(x, \lambda) + (S_0(\lambda) - S(\lambda))e(x, \lambda).$$

Using formula (1.11) we can rewrite it as

$$\Phi(x, \lambda) = e^{-i\lambda x} - S_0(\lambda)e^{i\lambda x} + \int_x^\infty K(x, t)(e^{-i\lambda t} - S_0(\lambda)e^{i\lambda t})dt$$

$$+ (S_0(\lambda) - S(\lambda))\left(e^{i\lambda x} + \int_x^\infty K(x, t)e^{i\lambda t}dt\right). \quad (4.7)$$

We fix $x > 0$. For any arbitrary chosen $y > x$ we let $\chi(t)$ be an infinitely differentiable function such that $\chi(t) = 0$, $t \in (-\infty, y)$ and tends rapidly to zero for $t \to \infty$.

We denote by $\tilde{\chi}(\lambda)$ a Fourier transform of $\chi(t)$:

$$\tilde{\chi}(\lambda) = \int_y^\infty \chi(t)e^{i\lambda t}dt. \quad (4.8)$$

We multiply both sides of (4.7) by $\frac{1}{2\pi}\hat{\chi}(\lambda)$ and integrate on $(-\infty, \infty)$. Function $e^{-i\lambda x} - S_0(\lambda)e^{i\lambda x}$ is analytic in the upper half-plane $\operatorname{Im}\lambda > 0$ and function $S_0(\lambda)$

is uniformly bounded there, we can use the Jordan lemma to see that

$$\frac{1}{2\pi}\int_{-\infty}^{\infty}(e^{-i\lambda x}-S_0(\lambda)e^{i\lambda x})\hat{\chi}(\lambda)d\lambda$$

$$=\int_y^{\infty}\chi(t)\left[\frac{1}{2\pi}\int_{-\infty}^{\infty}(e^{-i\lambda x}-S_0(\lambda)e^{i\lambda x})e^{i\lambda t}d\lambda\right]dt=0.$$

As $S_0(\lambda)-S(\lambda)=\int_{-\infty}^{\infty}F_S(t)e^{-i\lambda t}dt$, we get from formula (4.7) that

$$\frac{1}{2\pi}\int_{-\infty}^{\infty}\Phi(x,\lambda)\hat{\chi}(\lambda)d\lambda$$

$$=\int_y^{\infty}\chi(t)\left[F_S(x+t)+\int_{-\infty}^{\infty}F_S(t-\tau)K(x,-\tau)d\tau+K(x,t)\right]dt.$$

Function $\Phi(x,\lambda)$ is analytic in the half-plane $\mathrm{Im}\lambda>0$ except zeros of $\Delta(\lambda)$ and $e(0,\lambda)$. Then it follows from Lemma 4.3 that

$$\frac{1}{2\pi}\int_{-\infty}^{\infty}\Phi(x,\lambda)\hat{\chi}(\lambda)d\lambda=\int_y^{\infty}\chi(t)\left[-\sum_{j=1}^{n}m_j^2e(x,i\mu_j)e^{-\mu_jt}\right]dt$$

$$-\int_y^{\infty}\chi(t)\left[\sum_{j=1}^{m}\left(\frac{2\nu_j\omega(x,i\nu_j)e^{-\nu_jt}}{\dot{e}(0,i\nu_j)}+\frac{2\nu_je(x,i\nu_j)e^{-\nu_jt}}{\Delta(i\nu_j)\dot{e}(0,i\nu_j)}\right)\right]dt,$$

where $\{i\nu_j\}$, $j=1,\ldots,m$, are zeros of $e(0,\lambda)$, which are not included in U_3. Taking into account the equality $\Delta(i\nu_j)=-e'(0,i\nu_j)$ we see that the second integral becomes zero. Hence

$$\frac{1}{2\pi}\int_{-\infty}^{\infty}\Phi(x,\lambda)\hat{\chi}(\lambda)d\lambda=\int_y^{\infty}\chi(t)\left[-\sum_{j=1}^{n}m_j^2e(x,i\lambda_j)e^{-\lambda_jt}\right]dt$$

$$=\int_y^{\infty}\chi(t)\left[-\sum_{j=1}^{n}m_j^2\left\{e^{-\lambda_j(x+t)}+\int_x^{\infty}K(x,\tau)e^{-\lambda_j(\tau+t)}d\tau\right\}\right]dt.$$

Since function $\chi(t)$ is arbitrary, the assertion (i) follows from these equations.

(ii) In the case under consideration there is only a finite number of eigenvalues visible at infinity, which are simple, as in the case of the scattering problem on the semi-axis. According to Lemma 4.1 function $F(X)$ possesses the same properties as the corresponding function in the case of the scattering problem on the semi-axis, which allows us to repeat arguments [1], Chapter III. to prove the unique solvability of the Marchenko equation (1.15). $\square$

Proof of Theorem 1.7. In the case under consideration $F(x)=\tilde{F}(x)$ and, as result of unique solvability of the main equation (1.15), $K(x,t)=\tilde{K}(x,t)$, $0<x<t<\infty$ and, by continuity, $K(x,x)=\tilde{K}(x,x)$, $0\leq x<\infty$. Then, according to (1.17), $q(x)=\tilde{q}(x)$, $x\in\gamma$. $\square$

Acknowledgment

The authors wish to express their heartfelt thanks to Prof. V.A. Marchenko for the formulation of this problem and for his extremely valuable advice in helping to solve it. The first author was partly supported by Grant-in-Aid for Scientific Research (C) (No.22540204, No.23540181), Japan Society for the Promotion of Science. The second author was partly supported by Grant-in-Aid for Scientific Research (C) (No.23540181), Japan Society for the Promotion of Science.

References

[1] Z.S. Agranovich and V.A. Marchenko, *The Inverse Problem of Scattering Theory*, New York and London: Gordon and Breach, 1963

[2] J. Boman and P. Kurasov, *Symmetries of quantum graphs and the inverse scattering problem* Advances in Applied Mathematics **32** (2005), 58–70

[3] P. Exner and P. Seba *Free quantum motion on a branching graph* Rep. Math. Phys. (1989), 7–26

[4] G. Freiling and V.A. Yurko *Inverse Sturm-Liouville Problems and their Applications*, New York: NOVA Science Publishers, 2001

[5] N.I. Gerasimenko and B.S. Pavlov *Scattering problems on noncompacy graphs* Theor. Math. Phys. **75** (1988), 230–240

[6] N.I. Gerasimenko *Inverse scattering problem on a noncompact graphs* Theor. Math. Phys. **75** (1988), 460–470

[7] M. Harmer 2002 *Inverse scattering for the matrix Schrödinger operator and Schrödinger operator on graphs with general self-adjoint boundary conditions* ANZIAM J. **44** (2002), 161–168

[8] V. Kostrykin and R. Schrader R 2000 *Kirchhoff's rule for quantum wires II: the inverse problem with possible applications to quantum computers* Fortschr. Phys. **48** (2000), 703–716

[9] P. Kurasov and F. Stenberg *On the inverse scattering problem on branching graphs* J. Phys. A **35** (2002), 101–121

[10] P. Kuchment *Quantum graphs. I. Some basic structures* Waves Random Media **14** no. 1 (2004), S107–S128

[11] P. Kuchment *Quantum graphs. II. Some spectral properties of quantum and combinatorial graphs* J. Phys. A **38** no. 22 (2005), 4887–4900

[12] Y. Latushkin and V. Pivovarchik V *Scattering in a forked-shaped waveguide* Integral Equations Operator Theory **61** no. 3 (2008), 365–399

[13] V.A. Marchenko *Sturm-Liouville Operators and Applications* Operator Theory: Advances and Applications vol. 22), Birkhäuser, 1986

[14] V. Marchenko, K. Mochizuki and I. Trooshin *Inverse scattering on a graph containing circle* Analytic methods of analysis and differential equations: AMADE 2006 (Cambridge: Camb. Sci. Publ.) (2008), 237–243

[15] V. Pivovarchik *Scattering in a loop-shaped waveguide* Recent advances in operator theory (Groningen, 1998) (Oper. Theory Adv. Appl. vol. 124), Birkhäuser (2001), 527–543

[16] Yu.V. Pokornyi, O.M. Penkin, V.I. Pryadiev, A.V. Borovskikh, K.P. Lazarev and S.A. Shabrov *Differential equations on geometric graphs* (in Russian) Moscow: Fiziko-Matematicheskaya Literatura, 2005

K. Mochizuki
Emeritus, Tokyo Metropolitan University
8315-7, Hotaka, Azumino-shi, Nagano
399-8303, Japan
e-mail: mochizuk@math.chuo-u.ac.jp

I.Yu. Trooshin
Institute for International Education
Tohoku University
41 Kawauchi, Aoba-ku, Sendai
Miyagi 980-8576, Japan

   and

Institute of Problems of Precise Mechanics and Control
Russian Academy of Sciences
Rabochaya 24, Saratov
410028, Russia
e-mail: trushin@m.tohoku.ac.jp

Progress in Mathematics, Vol. 301, 247–266

On the Cauchy Problem for Hyperbolic Operators with Double Characteristics

Cesare Parenti and Alberto Parmeggiani

Abstract. The aim of this work is to provide a survey, along with new recent results, of what is known about the C^∞ well-posedness of the Cauchy problem for hyperbolic operators with double characteristics. Some conjectures and open problems will also be presented.

Mathematics Subject Classification. Primary 35A05; Secondary 35L15, 35L80.

Keywords. Cauchy problem, weakly hyperbolic operator.

1. Introduction and state of the art

We are concerned here with the classical problem of the C^∞ well-posedness of the Cauchy problem (C.p., for short) for (linear) hyperbolic operators with double characteristics. This argument has been intensely studied in past years, but in our opinion much remains yet to be done to have a reasonably complete picture. Our hope is mainly to convince the reader about the still actual interest of this field and, along the path, the fundamental role played by symplectic geometry for obtaining invariant results.

Although most of what we will be saying may be formulated for hyperbolic operators of order ≥ 2, with at most double characteristics, for the sake of clarity, and simplicity, we will restrict ourselves to considering only the second-order case.

In $\mathbb{R}^{n+1}_x = \mathbb{R}_{x_0} \times \mathbb{R}^n_{x'}$ $(n \geq 2)$ consider the operator

$$P = -D_{x_0}^2 + \sum_{\substack{j+|\alpha|\leq 2 \\ j=0,1}} a_{j\alpha}(x) D_{x_0}^j D_{x'}^\alpha \quad \left(D = \frac{1}{i}\partial\right)$$

with smooth coefficients. Denote by

$$p(x, \xi = (\xi_0, \xi')) = -\xi_0^2 + 2b(x, \xi')\xi_0 + c(x, \xi')$$

the *principal symbol* of P.

Suppose that P is *hyperbolic* with respect to x_0 (the time variable), that is

- $b(x, \xi')$, $c(x, \xi')$ are *real* (homogeneous of degree 1, 2, respectively, in ξ');
- $\Delta(x, \xi') := b(x, \xi')^2 + c(x, \xi') \geq 0$, $\forall x, \forall \xi' \neq 0$.

Then, the *characteristic set* of P is

$$p^{-1}(0) = \{(x, \xi);\ \xi' \neq 0,\ \xi_0 = b(x, \xi') \pm \sqrt{\Delta(x, \xi')}\}.$$

If $\Delta(x, \xi') > 0$ for all x and all $\xi' \neq 0$, then P would be *strictly hyperbolic*, a case for which the C^∞ well-posedness has no secrets (see, e.g., Hörmander [7]).

Since we are interested in the *weakly hyperbolic* case, we will assume throughout that Δ *vanishes* **exactly** *to second order on some smooth (connected) submanifold*

$$\Sigma' = \{(x, \xi) \in T^*\mathbb{R}^{n+1};\ \xi' \neq 0,\ \Delta(x, \xi') = 0\}$$

of fixed codimension d ($1 \leq d < 2n + 1$).

Recall that the vanishing condition on Δ means that for all compact $K \subset \mathbb{R}^{n+1}$ there exists $C_K > 0$ such that

$$C_K^{-1}|\xi'|^2 \mathrm{dist}_{\Sigma'}(x, \xi')^2 \leq \Delta(x, \xi') \leq C_K|\xi'|^2 \mathrm{dist}_{\Sigma'}(x, \xi')^2, \quad \forall x \in K,\ \forall \xi' \neq 0,$$

where $\mathrm{dist}_{\Sigma'}(x, \xi')$ denotes the distance of $(x, \xi_0 = 0, \xi'/|\xi'|)$ to Σ'. We shall write

$$\Delta(x, \xi') \approx |\xi'|^2 \mathrm{dist}_{\Sigma'}(x, \xi'/|\xi'|)^2.$$

Define next

$$\Sigma = p^{-1}(0) \cap \Sigma' \quad (\mathrm{codim}\, \Sigma = d + 1),$$

so that $p|_\Sigma = 0$, $dp|_\Sigma = 0$ and $p^{-1}(0) \setminus \Sigma \neq \emptyset$.

Problem: *The well-posedness of the C.p. for P in C^∞.*

The C^∞ well-posedness we aim at is that as defined in Hörmander [6]:

(i) *For any given $t \in \mathbb{R}$, for every $u \in \mathscr{E}'(\mathbb{R}^{n+1})$ with $Pu = 0$ in $\{x_0 > t\}$ we have $u = 0$ in $\{x_0 > t\}$;*

(ii) *For any given $t \in \mathbb{R}$, for every $f \in C_0^\infty(\mathbb{R}^{n+1})$ there exists $u \in \mathscr{E}'(\mathbb{R}^{n+1})$ satisfying $Pu = f$ in $\{x_0 > t\}$.*

It is worth noting that if we *do not* assume the above vanishing condition on Δ, then very little is known, and mainly in classes of examples.

A main concern in the study of degenerate operators is to have a coordinate-independent approach, and for this reason geometry (namely, symplectic geometry) plays a fundamental role.

To start with, we recall one of the main invariant objects.

At any given $\rho \in \Sigma$ and for every $v \in T_\rho T^*\mathbb{R}^{n+1}$ we have

$$p(\rho + v) = \frac{1}{2}\langle \mathrm{Hess}\, p(\rho)v, v \rangle + O(|v|^3).$$

Using the symplectic form $\sigma = \sum_{j=0}^n d\xi_j \wedge dx_j$, we can write

$$\frac{1}{2}\langle \mathrm{Hess}\, p(\rho)v, w \rangle = \sigma(v, F(\rho)w), \quad \forall v, w \in T_\rho T^*\mathbb{R}^{n+1},$$

where the linear map $F(\rho)\colon T_\rho T^*\mathbb{R}^{n+1} \longrightarrow T_\rho T^*\mathbb{R}^{n+1}$ is called the *Hamilton map* or *fundamental matrix* of P at ρ.

Note that $2F(\rho)$ is the linearization at ρ of the Hamilton vector-field H_p associated with the principal symbol p of P. In the coordinates (x, ξ) we thus have

$$H_p = \sum_{j=0}^{n} \Big(\frac{\partial p}{\partial \xi_j} \frac{\partial}{\partial x_j} - \frac{\partial p}{\partial x_j} \frac{\partial}{\partial \xi_j} \Big),$$

and

$$F(\rho) = \frac{1}{2} \left[\begin{array}{cc} p''_{x\xi}(\rho) & p''_{\xi\xi}(\rho) \\ -p''_{xx}(\rho) & -p''_{\xi x}(\rho) \end{array} \right], \quad \rho \in \Sigma.$$

It is important also to note that one rephrases the vanishing condition on Δ in terms of $F(\rho)$ by saying that

$$\operatorname{Ker} F(\rho) = T_\rho \Sigma, \quad \forall \rho \in \Sigma.$$

From Hörmander [7] one knows that

$$z \in \operatorname{Spec}(F(\rho)) \iff -z, \bar{z} \in \operatorname{Spec}(F(\rho)),$$

and only three possible cases can occur:

(i) There exists $\lambda > 0$ such that

$$\{-\lambda, \lambda\} \subset \operatorname{Spec}(F(\rho)), \quad \operatorname{Spec}(F(\rho)) \setminus \{-\lambda, \lambda\} \subset i\mathbb{R}.$$

In this case one says that P is *effectively hyperbolic at ρ*. From now on we shall put

$$\Sigma_+ := \{\rho \in \Sigma;\ P \text{ is effectively hyperbolic at } \rho\}.$$

Note that Σ_+ is always a (possibly empty) open subset of Σ.

(ii) $\operatorname{Spec}(F(\rho)) \subset i\mathbb{R}$ and in the Jordan normal form of $F(\rho)$ corresponding to the eigenvalue 0, there are only Jordan blocks of dimension 2, i.e., $\operatorname{Ker} F(\rho)^2 \cap \operatorname{Im} F(\rho)^2 = \{0\}$. In this case one says that P is *non-effectively hyperbolic of type 1 at ρ*. From now on we shall put

$$\Sigma_- := \{\rho \in \Sigma;\ P \text{ is non-effectively hyperbolic of type 1 at } \rho\}.$$

(iii) $\operatorname{Spec}(F(\rho)) \subset i\mathbb{R}$ and in the Jordan normal form of $F(\rho)$ corresponding to the eigenvalue 0, there is only a Jordan block of dimension 4 and no block of dimension 3, i.e., $\operatorname{Ker} F(\rho)^2 \cap \operatorname{Im} F(\rho)^2$ is two-dimensional. In this case one says that P is *non-effectively hyperbolic of type 2 at ρ*. From now on we shall put

$$\Sigma_0 := \{\rho \in \Sigma;\ P \text{ is non-effectively hyperbolic of type 2 at } \rho\}.$$

Obviously,

$$\Sigma = \Sigma_- \sqcup \Sigma_0 \sqcup \Sigma_+ \quad \text{(disjoint union)}$$

and we say that *we have a transition* exactly when *at least two among the above sets are non-empty*.

At this point we recall one of the most important results due to Ivrii and Petkov [9], and Hörmander [6].

Theorem 1.1. *In order that the C.p. for P be well posed in C^∞ it is* necessary *that at any given $\rho \in \Sigma_- \sqcup \Sigma_0$ we have*

$$\operatorname{Im} p_1^s(\rho) = 0, \quad |\operatorname{Re} p_1^s(\rho)| \leq \operatorname{Tr}^+ F(\rho), \tag{1.1}$$

where

$$p_1^s(x, \xi) = p_1(x, \xi) + \frac{i}{2} \sum_{j=0}^n \frac{\partial^2 p}{\partial x_j \partial \xi_j}(x, \xi)$$

is the subprincipal symbol of P (p_1 being the symbol of the 1st-order part of P), and

$$\operatorname{Tr}^+ F(\rho) := \sum_{\substack{\mu \geq 0 \\ i\mu \in \operatorname{Spec}(F(\rho))}} \mu,$$

is the positive trace of $F(\rho)$.

Remark 1.2. Note that in case $\operatorname{Tr}^+ F(\rho) = 0$ (which is equivalent to saying that $\operatorname{Spec}(F(\rho)) = \{0\}$, since $\rho \notin \Sigma_+$), condition (1.1) reduces to the celebrated *Levi condition* $p_1^s(\rho) = 0$.

1.1. Known results for the C.p.

On the side of positive results for the C^∞ well-posedness of the C.p. we have that:

- When $\Sigma = \Sigma_+$, the C.p. is well posed independently of the lower-order terms of P. This result is due to Iwasaki [10] and [11], and Nishitani [14] (see also Melrose [12]).
- When $\Sigma = \Sigma_-$, and $\operatorname{rk} \sigma|_\Sigma$ is constant (recall that, for $\rho \in \Sigma$, $\operatorname{rk} \sigma_\rho = \dim\left(\frac{T_\rho \Sigma}{T_\rho \Sigma \cap T_\rho \Sigma^\sigma}\right)$), the C.p. is well posed if the following *strict Ivrii-Petkov-Hörmander* condition is satisfied:

$$\operatorname{Im} p_1^s(\rho) = 0, \quad |\operatorname{Re} p_1^s(\rho)| < \operatorname{Tr}^+ F(\rho), \quad \rho \in \Sigma. \tag{1.2}$$

This result is due to Ivrii [8] and Hörmander [6]. Notice that the condition reduces to $p_1^s|_\Sigma = 0$ when $\operatorname{Tr}^+ F|_\Sigma = 0$.

A natural case to consider is now when $\Sigma = \Sigma_0$. Supposing that $\operatorname{rk} \sigma|_\Sigma$ is constant and that (1.2) holds, can we conclude that the C.p. is C^∞ well posed?

The answer is in general *in the negative*.

The reason why this is the case, is related to the behavior of the *null-bicharacteristics* of p, that is of the integral curves $t \mapsto \gamma(t)$ of the Hamilton vector field H_p that by definition solve

$$\begin{cases} \dot{\gamma}(t) = H_p(\gamma(t)) \\ \gamma(0) \in p^{-1}(0) \setminus \Sigma. \end{cases}$$

Of course, one has that $\gamma \subset p^{-1}(0) \setminus \Sigma$, but the problem is: *Can γ have limit-points in Σ?*

- The answer is *no* when $\Sigma = \Sigma_-$ (in this case, Σ is roughly a *center* for H_p).

- When $\Sigma = \Sigma_+$, for any given $\rho \in \Sigma$ there are exactly four half-bicharacteristics that enter Σ at ρ (as $t \to \pm\infty$) *transversally* (the directions being given by the eigenvectors belonging to the real non-zero eigenvalues of $F(\rho)$). In this case Σ behaves like a *saddle* for H_p.
- What happens hence when $\Sigma = \Sigma_0$?

Consider the example $(n = 2)$

$$p(x,\xi) = -(\xi_0 - x_1\xi_2)^2 + \xi_1^2 + (x_1^2 + \beta x_1^3)\xi_2^2, \quad \beta \in \mathbb{R}.$$

An explicit computation gives (see Bernardi-Bove [1]) that

- If $\beta = 0$, *no* null-bicharacteristic of H_p can have limit-points in Σ.
- If $\beta \neq 0$, *there are* indeed null-bicharacteristics entering Σ *tangentially*.

A precise explanation to this phenomenon was given (long time ago) by Bernardi-Bove [1], and by Nishitani [15]. Recall that at every $\rho \in \Sigma$ we have

$$T_\rho\Sigma = \operatorname{Ker} F(\rho) \subsetneq \operatorname{Ker} F(\rho)^2 \subsetneq \operatorname{Ker} F(\rho)^3 \subsetneq \operatorname{Ker} F(\rho)^4 = \operatorname{Ker} F(\rho)^5,$$

with a one-dimensional gap between $\operatorname{Ker} F^3$ and $\operatorname{Ker} F^4$, and between $\operatorname{Ker} F^2$ and $\operatorname{Ker} F^3$. Fix next a smooth non-zero vector field $\Sigma \ni \rho \longmapsto z(\rho) \in T_\rho T^*\mathbb{R}^{n+1}$ (positively homogeneous of degree 0 in the fibers; from now on we shall refer to positive homogeneity of degree d in the fibers for short as "p.h. of deg. d") such that

$$\operatorname{Ker} F(\rho)^4 = \operatorname{Ker} F(\rho)^3 \oplus \operatorname{Span}\{z(\rho)\}, \quad \operatorname{Ker} F(\rho)^3 = \operatorname{Ker} F(\rho)^2 \oplus \operatorname{Span}\{F(\rho)z(\rho)\}$$

(so that $\operatorname{Ker} F(\rho)^2 \cap \operatorname{Im} F(\rho)^2 = \operatorname{Span}\{F(\rho)^2 z(\rho), F(\rho)^3 z(\rho)\}$), and fix a real symbol $S(x,\xi)$ (p.h. of deg. 1), vanishing on Σ, such that

$$0 \neq H_S(\rho) /\!\!/ F(\rho)^2 z(\rho), \quad \rho \in \Sigma.$$

One then has the following theorems (see [1], [3] and [15]).

Theorem 1.3. *No null-bicharacteristic γ of p has limit-points in Σ iff*

$$H_S^3 p\big|_\Sigma = 0.$$

Theorem 1.4. *If $H_S^3 p\big|_\Sigma = 0$ and (1.2) holds, the C.p. is microlocally well posed in C^∞.*

Theorem 1.4 has been recently improved by Nishitani [16], who proved the *local* well-posedness in C^∞, and made more precise by Bernardi-Nishitani [4], who proved that *if $(H_S^3 p)(\rho) \neq 0$ for some point $\rho \in \Sigma$ and if (1.2) holds, then the C.p. is **not** well posed in C^∞*.

1.2. New problems: transition

From Subsection 1.1 we therefore may conclude that we have a quite complete picture when the *spectrum of $F(\rho)$ is of one and the same type at every point ρ of Σ*. We note in passing that, however, there is still the obvious gap between (1.1) and its strict counterpart (1.2), and we address the reader to the work of Parenti-Parmeggiani [17] for results in that direction.

Now, very simple examples show that $\mathrm{Spec}(F(\rho))$ *can* indeed change at different points of Σ.

Consider for example

$$p(x,\xi) = -\xi_0^2 + \xi_1^2 + (x_0 - x_1 - f(x_1, x'))^2 |\xi'|^2, \qquad (1.3)$$

where $x' = (x_2, \ldots, x_n)$, $n \geq 2$, f is smooth and real, with $1 + f'_{x_1} > 0$. Then a point $\rho = (x_0 = x_1 + f(x_1, x'), x_1, x'; \xi_0 = 0, \xi_1 = 0, \xi' \neq 0) \in \Sigma$ belongs to

- Σ_+ iff $f'_{x_1} < 0$;
- Σ_- iff $f'_{x_1} > 0$;
- Σ_0 iff $f'_{x_1} = 0$.

Hence, *we have a transition iff $f'_{x_1} \not\equiv 0$ and f'_{x_1} vanishes somewhere.*

The previous example is very simple, but one should bear in mind that transition may occur in a very complicated way (see, e.g., Bernardi-Bove [2]).

Our purpose is next that of studying the C^∞ well-posedness of the C.p. when transitions are present. Cases where $\Sigma = \Sigma_+ \sqcup \Sigma_-$ were already treated in [2], and we will consider here a rather different class of examples.

2. The case of transition: a geometric class of operators

The first step in our program is that of selecting a geometric-definite class of operators. We proceed as follows. Near any given $\rho \in \Sigma$, by performing a suitable canonical transformation which preserves the "time" direction x_0, and by using the Morse Lemma, we may microlocally write

$$p(x,\xi) = -\xi_0^2 + \sum_{j=1}^{d} f_j(x, \xi')^2,$$

for some smooth real functions f_j, p.h. of deg. 1, with independent Hamilton vector fields at all points of

$$\Sigma = \{(x,\xi);\ \xi' \neq 0,\ \xi_0 = 0,\ f_1(x,\xi') = \cdots = f_d(x,\xi') = 0\}.$$

For $\rho \in \Sigma$ define (recall that for functions f, g the Poisson bracket $\{f, g\} = \sum (\partial_{\xi_j} f \partial_{x_j} g - \partial_{x_j} f \partial_{\xi_j} g)$)

$$\alpha(\rho) = \begin{bmatrix} \{\xi_0, f_1\}(\rho) \\ \vdots \\ \{\xi_0, f_d\}(\rho) \end{bmatrix}, \quad M(\rho) = \Big[\{f_j, f_k\}(\rho)\Big]_{1 \leq j, k \leq d}.$$

The matrix $M(\rho)$ is therefore a $d \times d$ skew-symmetric matrix. Suppose from now on the following

Hypothesis (H1):

- $\mathrm{rk}\, M(\rho)$ is constant as ρ varies in Σ;
- $\alpha(\rho) \in \mathrm{Im}\, M(\rho)$ for all $\rho \in \Sigma$.

It can be shown (see Bernardi-Parenti-Parmeggiani [5]) that **(H1)** has a symplectically invariant meaning. Namely, **(H1)** is equivalent to requiring
  (i) $\mathrm{rk}\,\sigma|_{\Sigma'}$ is a constant ≥ 2;
  (ii) $\mathrm{rk}\,\sigma|_{\Sigma} = \mathrm{rk}\,\sigma|_{\Sigma'} - 2$.

Denoting by $M(\rho)^{-1}$ the inverse of $M(\rho)|_{\mathrm{Im}\,M(\rho)}$, one can show that

  (i) $\rho \in \Sigma_+$ iff $|M(\rho)^{-1}\alpha(\rho)| > 1$;
  (ii) $\rho \in \Sigma_-$ iff $|M(\rho)^{-1}\alpha(\rho)| < 1$;
  (iii) $\rho \in \Sigma_0$ iff $|M(\rho)^{-1}\alpha(\rho)| = 1$.

Hence, to have a transition we must have $|M(\rho)^{-1}\alpha(\rho)| = 1$ somewhere and have $\mathrm{codim}\,\Sigma \geq 3$. (Note that $M(\rho)^{-1}\alpha(\rho)$ is p.h. of deg. 0.)

Next, suppose also from now on the following

Hypothesis (H2): $\Sigma_0 = \{\rho \in \Sigma;\ |M(\rho)^{-1}\alpha(\rho)| = 1\} \subsetneq \Sigma$ *is a smooth submanifold of Σ of fixed codimension ≥ 1.*

Correspondingly, we have

$$\Sigma_\pm = \{\rho \in \Sigma;\ |M(\rho)^{-1}\alpha(\rho)| - 1 \gtrless 0\}.$$

Remark 2.1. The quantity $|M(\rho)^{-1}\alpha(\rho)|$ is *invariant*: it is independent of the Morse normal form of p.

We therefore have, in principle, three possible cases of transition:
(T1) $\Sigma = \Sigma_- \sqcup \Sigma_0$;
(T2) $\Sigma = \Sigma_+ \sqcup \Sigma_0$;
(T3) $\Sigma = \Sigma_- \sqcup \Sigma_0 \sqcup \Sigma_+$.

Notice that in cases **(T1)** and **(T2)** the submanifold Σ_0 can have any codimension in Σ, but in case **(T3)** one must have $\mathrm{codim}_\Sigma \Sigma_0 = 1$.

We next turn to examine what happens to the spectrum of $F(\rho)$. This is what one can prove (see [5]):
(S1) exactly two eigenvalues $\pm i\mu(\rho)$ collapse to 0 as ρ approaches Σ_0 from within Σ_-;
(S2) the two real eigenvalues $\pm\lambda(\rho)$ collapse to 0 as ρ approaches Σ_0 from within Σ_+;
(S3) the picture is either **(S1)** or **(S2)**, depending on whether we approach Σ_0 from within Σ_- or Σ_+.

It is important to note that the order of vanishing on Σ_0 of the eigenvalues that collapse to 0 is that of

$$\sqrt{1 - |M(\rho)^{-1}\alpha(\rho)|^2} \quad (\rho \in \Sigma_-), \quad \sqrt{|M(\rho)^{-1}\alpha(\rho)|^2 - 1} \quad (\rho \in \Sigma_+).$$

Hence, in general we *do not* have a smooth collapse.

Having selected a geometric class of operators, for which a transition occurs, we now turn to the study of the C.p. for these operators. As a preliminary remark motivated by the case $\Sigma = \Sigma_0$, we expect that the behavior of the null-bicharacteristics of P play now a crucial role in all the above cases. Moreover, once the dynamics of H_p has been understood, we still have to consider the Ivrii-Petkov-Hörmander condition (1.1) that must be satisfied, at least on Σ_0.

At the moment, we have some (positive) results in what we think to be the simplest case, namely $\Sigma = \Sigma_- \sqcup \Sigma_0$, and some conjectures in the case $\Sigma = \Sigma_+ \sqcup \Sigma_0$.

3. Results and conjectures when $\Sigma = \Sigma_- \sqcup \Sigma_0$

Since p is *non-effectively hyperbolic* everywhere, for simplicity we shall suppose that $\mathrm{Tr}^+ F(\rho) > 0$ for all $\rho \in \Sigma$, and that the strict Ivrii-Petkov-Hörmander condition (1.2) holds on Σ. Note that, geometrically, we necessarily have $\mathrm{Tr}^+ F\big|_{\Sigma \backslash \Sigma_0} > 0$, but that we could also very well have $\mathrm{Tr}^+ F\big|_{\Sigma_0} = 0$. When this happens, the situation becomes even more involved, as shown in Parenti-Parmeggiani [17].

Consider now the dynamical system associated with H_p on $p^{-1}(0) \backslash \Sigma$. As in Section 1, select a smooth non-zero vector field $\Sigma_0 \ni \rho \longmapsto v(\rho) \in T_\rho T^* \mathbb{R}^{n+1}$ such that

$$\mathrm{Ker}\, F(\rho)^2 \cap \mathrm{Im}\, F(\rho)^2 = \mathrm{Span}\{v(\rho), F(\rho)v(\rho)\}, \quad \rho \in \Sigma_0.$$

Recall that $v(\rho)$ is obtained as the image through $F(\rho)^2$ of the one-dimensional gap between $\mathrm{Ker}\, F^3$ and $\mathrm{Ker}\, F^4$.

Let now $S(x, \xi)$ be a smooth real symbol (p.h. of deg. 1) vanishing on Σ such that

$$0 \neq H_S(\rho) /\!\!/ v(\rho), \quad \rho \in \Sigma_0.$$

We have, at this point, the following conjectures.

Conjecture 3.1. No null-bicharacteristic of p can have limit-points in Σ_0 iff

$$H_S^3 p\big|_{\Sigma_0} = 0.$$

Conjecture 3.2. If $H_S^3 p\big|_{\Sigma_0} = 0$ and (1.2) holds, then the C.p. for P is (at least microlocally) well posed in C^∞.

Conjecture 3.1 sounds reasonable, but we are still unable to prove it in general. We can prove it in case hypothesis **(H3)** below is satisfied.

As for Conjecture 3.2, again, we do not have a proof in general, but we do have a proof (see [5]) under the following two additional assumptions, that is under the following

Hypothesis (H3):

- Recall that $1 - |M(\rho)^{-1}\alpha(\rho)|^2$ is positive when $\rho \in \Sigma \backslash \Sigma_0$ and vanishes when $\rho \in \Sigma_0$. We assume that

$$1 - |M(\rho)^{-1}\alpha(\rho)|^2 \quad \textit{vanishes exactly to second order on } \Sigma_0. \tag{3.1}$$

Equivalently, in a more invariant form, the vanishing eigenvalues of $F(\rho)$, as we approach Σ_0, vanish *exactly* as the distance of ρ to Σ_0.

- The second assumption concerns $F(\rho)v(\rho)$. Because of its definition

$$F(\rho)v(\rho) \in T_\rho\Sigma \cap T_\rho\Sigma^\sigma, \quad \forall \rho \in \Sigma_0.$$

We assume that

$$F(\rho)v(\rho) \in T_\rho\Sigma_0 \subsetneq T_\rho\Sigma, \quad \forall \rho \in \Sigma_0. \tag{3.2}$$

We wish to comment on the latter condition. First of all, the condition is *invariant*, because for all $\rho \in \Sigma_0$ the vector field $v(\rho)$ is in the image through $F(\rho)^2$ of the one-dimensional gap between $\operatorname{Ker} F(\rho)^4$ and $\operatorname{Ker} F(\rho)^3$. The second observation is that at a first glance one might think there is no correlation between the symplectic nature of Σ_0 and the tangency of $F(\rho)v(\rho)$ to Σ_0. On the contrary, there is indeed a strong link between these two.

Lemma 3.3. *Write $z(\rho) := F(\rho)v(\rho)$, $\rho \in \Sigma_0$. Then, for any given $\rho \in \Sigma_0$ the following conditions are equivalent:*

(i) $z(\rho) \notin T_\rho\Sigma_0$;

(ii) *There exists a symplectic subspace $V \subset T_\rho T^*\mathbb{R}^{1+n}$ such that*

$$T_\rho\Sigma_0 \subset V \subset \operatorname{Span}\{z(\rho)\}^\sigma.$$

Proof. (ii) $\Rightarrow$ (i): by contradiction. If $z(\rho) \in T_\rho\Sigma_0$, then $z(\rho) \in V \subset \operatorname{Span}\{z(\rho)\}^\sigma$, whence $z(\rho) \in V \cap V^\sigma = \{0\}$, which is impossible, for we have $z(\rho) \neq 0$.

(i) $\Rightarrow$ (ii): as $\operatorname{Span}\{z(\rho)\} \not\subset T_\rho\Sigma_0$, we have $T_\rho\Sigma_0^\sigma \not\subset \operatorname{Span}\{z(\rho)\}^\sigma$, i.e., there exists $v \in T_\rho\Sigma_0^\sigma$ such that $\sigma(v, z(\rho)) \neq 0$. Set $W = \operatorname{Span}\{v, z(\rho)\}$. Then W is symplectic and so is W^σ. Obviously, $W^\sigma \subset \operatorname{Span}\{z(\rho)\}^\sigma$ and $W \subset T_\rho\Sigma_0^\sigma$, because $z(\rho) \in \operatorname{Im} F(\rho) = T_\rho\Sigma^\sigma \subset T_\rho\Sigma_0^\sigma$. Hence $T_\rho\Sigma_0 \subset W^\sigma$, and it suffices to take $V = W^\sigma$. $\square$

Note, in particular, that when Σ_0 is symplectic, then $z(\rho)$ is *never* tangent to Σ_0.

To make an example, consider (1.3) with $x' = (y', y'')$ and

$$f(x_1, x') = \frac{\gamma_1}{3}x_1^3 + \gamma_2 x_1 |y'|^2, \quad \gamma_1, \gamma_2 \geq 0, \ \gamma_1 + \gamma_2 > 0.$$

A direct computation shows that

$$\Sigma_0 = \begin{cases} \xi_0 = x_0 = 0, \ \xi_1 = x_1 = 0, \ y' = 0, & \text{when } \gamma_1, \gamma_2 > 0, \\ \xi_0 = x_0 = 0, \ \xi_1 = x_1 = 0, & \text{when } \gamma_1 > 0, \ \gamma_2 = 0, \\ \xi_0 = \xi_1 = 0, \ x_0 = x_1, \ y' = 0, & \text{when } \gamma_1 = 0, \ \gamma_2 > 0. \end{cases}$$

Moreover, on Σ_0, the vector field $z(\rho)$ is a (non-zero) smooth multiple of the vector field

$$\frac{\partial}{\partial x_0} + \frac{\partial}{\partial x_1},$$

so that $z(\rho) \in T_\rho\Sigma_0$, $\rho \in \Sigma_0$, iff $\gamma_1 = 0$ and $\gamma_2 > 0$. Note that, consistently with Lemma 3.3, in all the other cases Σ_0 is either symplectic or contained in a symplectic submanifold. Note also that condition $H_S^3 p\big|_{\Sigma_0} = 0$ is satisfied, for H_S is a multiple of the vector field

$$\frac{\partial}{\partial \xi_0} - (1 + f'_{x_1})\frac{\partial}{\partial \xi_1} - \left\langle \nabla_{x'} f, \frac{\partial}{\partial \xi'} \right\rangle.$$

To prove at the same time microlocal Carleman estimates and Conjecture 3.1, the standard way (see Ivrii [8]) is to find a microlocal factorization of the principal symbol p of P of the kind

$$p(x, \xi) = - \underbrace{(-\xi_0 + \ell(x, \xi'))}_{N(x,\xi)} \underbrace{(-\xi_0 - \ell(x, \xi'))}_{\Lambda(x,\xi)} + q(x, \xi'), \qquad (3.3)$$

with $q \geq 0$ vanishing on Σ and

  (i) $|\{N, \Lambda\}| \lesssim |\Lambda - N| + \sqrt{q}$,
  (ii) $|\{\Lambda, q\}| \lesssim q$.

From now on we shall refer to such a factorization as to *an Ivrii-factorization*.

An Ivrii-factorization of p near points of $\Sigma \setminus \Sigma_0$ was already proved by Hörmander in [6] and Ivrii in [8]. Hence we concentrate on finding a Ivrii-factorization near any arbitrarily fixed point of Σ_0, the main point being the choice of the symbol ℓ (appearing in N and Λ in (3.3)), which is *forced* on Σ. When p is written as

$$p(x, \xi) = -\xi_0^2 + \sum_{j=1}^{d} f_j(x, \xi')^2,$$

then, for $\rho \in \Sigma$,

$$H_\Lambda(\rho) = -H_{\xi_0}(\rho) - \sum_{j=1}^{d} (M(\rho)^{-1}\alpha(\rho))_j H_{f_j}(\rho) =: -H_{\xi_0}(\rho) - \langle M(\rho)^{-1}\alpha(\rho), H_f(\rho) \rangle.$$

Note that $H_\Lambda(\rho)$ is collinear to $z(\rho)$. One then starts by taking (see [5])

$$\ell(x, \xi') = \sum_{j=1}^{d} \lambda_j(x, \xi') f_j(x, \xi') = \langle \lambda(x, \xi'), f(x, \xi') \rangle,$$

where

$$\lambda(\rho) = \begin{bmatrix} \lambda_1(\rho) \\ \vdots \\ \lambda_d(\rho) \end{bmatrix} = M(\rho)^{-1}\alpha(\rho), \quad \rho \in \Sigma,$$

with $|\lambda(x, \xi')| \leq 1$ and $|\lambda(x, \xi')| = 1$ only on Σ_0, with $1 - |\lambda|$ *vanishing exactly to second order* on Σ_0. Then, necessarily,

$$q(x, \xi') = |f(x, \xi')|^2 - \langle \lambda(x, \xi'), f(x, \xi') \rangle^2,$$

so that q is transversally elliptic with respect to Σ only if we stay *away* from Σ_0.

Unfortunately, without any further condition the above factorization *does not* satisfy Ivrii's condition (ii). To cope with this problem, we first of all microlocally choose new equations for Σ', $\psi = 0$, $\zeta_1 = \cdots = \zeta_{d-1} = 0$, so that $\psi = \langle \lambda, f \rangle$, and $q = h\psi^2 + \sum_j \zeta_j^2 = h\psi^2 + |\zeta|^2$, where $h = (1 - |\lambda|^2)/|\lambda|^2 \geq 0$ vanishes *exactly to second order* on Σ_0.

In order to control the Poisson bracket

$$-\{\Lambda, q\} = \{\xi_0 + \psi, h\psi^2 + |\zeta|^2\}$$

by q, we have to control the two terms $\{\xi_0 + \psi, h\psi^2\}$ and $\{\xi_0 + \psi, |\zeta|^2\}$, which give different kinds of problems.

As for the former, we write

$$\{\xi_0 + \psi, h\psi^2\} = 2h\psi\{\xi_0 + \psi, \psi\} + \psi^2\{\xi_0 + \psi, h\}.$$

The first term is harmless, because H_Λ is tangent to Σ, so that

$$|h\psi\{\xi_0 + \psi, \psi\}| \leq C(h\psi^2 + |\zeta|^2),$$

and the second term is under control, for $H_\Lambda(\rho) \in T_\rho\Sigma_0$ on Σ_0, so that, h vanishing exactly to second order on Σ_0, we have

$$|\{\xi_0 + \psi, h\}|\psi^2 \leq Ch\psi^2 \quad \text{near} \quad \Sigma_0.$$

As for the latter term, $\{\xi_0 + \psi, |\zeta|^2\}$, this requires taking into account the condition $H_g^3 p|_{\Sigma_0} = 0$, which is at this point used to deform Λ (and accordingly N) by adding to $-\xi_0 - \psi$ terms that vanish to higher order on Σ' (see [5] for the non-trivial details).

In the end, we get

$$\begin{cases} \Lambda = -\xi_0 - \psi - (\text{higher-order terms in } \psi, \zeta) \\ N = -\xi_0 + \psi + (\text{the same higher-order terms}) \\ q \approx \tilde{q} = h\psi^2 + |\zeta|^2 + \psi^4/|\xi'|^2. \end{cases}$$

The next step is to use the factorization to prove microlocal (forward) Carleman estimates: *For any given $\tau \in \mathbb{R}$ and any given compact $K \subset \mathbb{R}^{n+1}$, for any given (properly supported) microlocalizer $\chi(x, \xi')$ supported in a conic neighborhood (with compact base), there exists $C > 0$ and $\lambda_0 \gg 1$ such that for all $v \in C_0^\infty(K)$ and $u = \chi(x, D')v$ one has*

$$\lambda^4 \int_0^\infty e^{2\lambda x_0}|u|_\tau^2 dx_0 + \lambda^2 \int_0^\infty e^{2\lambda x_0}|u|_{\tau+1/2}^2 dx_0 \leq C \int_0^\infty e^{2\lambda x_0}|Pu|_\tau^2 dx_0, \quad (3.4)$$

for all $\lambda \geq \lambda_0$.
Here $|u|_\tau$ is the Sobolev norm of order τ with respect to the space variables x'.

By Hörmander [6], things are OK when χ is supported away from Σ_0. So, we need only prove (3.4) when $\operatorname{supp}\chi \cap \Sigma_0 \neq \emptyset$. We shall sketch the proof when $\tau = 0$.

Remark 3.4. It is useful to take into account that estimates (3.4) are invariant

- through conjugation by a unitary, zeroth-order Fourier integral operator in x', depending smoothly on x_0;
- by perturbations of P by zeroth-order terms.

To simplify notation, we shall denote by OPS^m the class of all *classical*, properly supported ψdos $A(x, D')$ which depend smoothly on x_0 and are pseudo-differential of order m in x' (the total symbol $a(x, \xi')$ of $A(x, D')$ will be written as $a_m(x, \xi') + a_{m-1}(x, \xi') + \cdots$).

By Remark 3.4, we may therefore suppose P of the form

$$P = -D_0^2 + A(x, D') + iB(x, D'),$$

where $A = A^* \in \mathrm{OPS}^2$, with principal symbol $a_2 \geq 0$ vanishing exactly to second order on Σ', $B = B^* \in \mathrm{OPS}^1$. Notice that

$$\mathrm{Re}\, p_1^s = \mathrm{Re}\, a_1, \quad \mathrm{Im}\, p_1^s = b_1.$$

Hence, condition (1.2) reads

$$\mathrm{Re}\, a_1(\rho) + \mathrm{Tr}^+ F(\rho) > 0, \quad b_1(\rho) = 0, \quad \forall \rho \in \Sigma.$$

It is at this point that we have to exploit the factorization of p that we have constructed before.

On defining

$$\Lambda(x, D) := -D_0 - L(x, D'), \quad N(x, D) = -D_0 + L(x, D'),$$

with $L = L^* \in \mathrm{OPS}^1$, $\ell_1 = \ell$, we have

$$P = -N\Lambda + Q + iR,$$

with $Q = Q^* \in \mathrm{OPS}^2$, $q_2 = q$, $R = R^* \in \mathrm{OPS}^1$, and

$$q_1^s\big|_\Sigma = \mathrm{Re}\, p_1^s\big|_\Sigma, \quad r_1 = b_1 + \{\xi_0 + \ell_1, \ell_1\}.$$

Remark 3.5. We have $H_\Lambda(\rho) = -H_{\xi_0 + \ell_1}(\rho)$, with $\ell_1\big|_\Sigma = 0$, so that by the tangency we get $\{\xi_0 + \ell_1, \ell_1\}\big|_\Sigma = 0$. Hence $r_1\big|_\Sigma = 0$.

Now, following Hörmander, estimates (3.4) are obtained through an *energy form* of the kind

$$E(x_0; u) = |\Lambda(x, D)u|^2 + (Qu, u).$$

One next computes dE/dx_0, considers $-\displaystyle\int_0^\infty e^{2\lambda x_0}(dE/dx_0)dx_0$ and integrates by parts, and runs into a problem:

Problem: *One has to control* $|(R^2 u, u)|$ *by* (Qu, u). *Since we are in a situation where* Q *is not transversally elliptic, the usual approach does not work.*

Luckily, there is a remedy to this. We modify $\Lambda(x, D)$ by a zeroth-order term as follows:

$$\tilde\Lambda(x, D) := \Lambda(x, D) - iS(x, D'), \quad S = S^* \in \mathrm{OPS}^0,$$

so that, modulo OPS^0,

$$P = -N\tilde\Lambda + Q + i\tilde R + iS\tilde\Lambda,$$

and we may choose S in such a way that

$$\tilde r_1 = O(\sqrt{q}).$$

The term $S\tilde{\Lambda}$ will be under control, provided we choose the right energy:

$$E(x_0; u) = |\tilde{\Lambda}(x, D)u|^2 + (Qu, u).$$

Then

$$\frac{dE}{dx_0} = -2\mathsf{Im}(Pu, \tilde{\Lambda}u) + i([Q, \Lambda]u, u) + 2\mathsf{Re}(SQu, u)$$

$$+ 2\mathsf{Re}(\tilde{R}u, \tilde{\Lambda}u) + 2(S\tilde{\Lambda}u, \tilde{\Lambda}u),$$

and, through an integration by parts, from

$$-\int_0^\infty e^{2\lambda x_0} \frac{dE}{dx_0} dx_0 = E(0) + 2\lambda \int_0^\infty e^{2\lambda x_0} E(x_0) dx_0,$$

one arrives at, for $\lambda \geq \lambda_0 \gg 1$,

$$\lambda E(0) + \lambda^2 \int_0^\infty e^{2\lambda x_0} |\tilde{\Lambda}u|^2 dx_0 + \lambda^2 \int_0^\infty e^{2\lambda x_0}(Q_\lambda u, u) dx_0 \leq C \int_0^\infty e^{2\lambda x_0} |Pu|^2 dx_0,$$

with Q_λ given by

$$Q_\lambda = Q + \frac{1}{2\lambda}(SQ + QS) - \frac{1}{2\lambda i}[Q, \Lambda] - \frac{1}{\lambda^2}\tilde{R}^2. \tag{3.5}$$

It is elementary to control the term containing $|\tilde{\Lambda}u|^2$. In fact one has

$$\frac{\lambda^3}{2}|u(0)|^2 + \frac{\lambda^4}{2} \int_0^\infty e^{2\lambda x_0} |u|^2 dx_0 \leq \lambda^2 \int_0^\infty e^{2\lambda x_0} |\tilde{\Lambda}u|^2 dx_0.$$

To see this, write $-\tilde{\Lambda} = D_0 + Z(x, D')$, where Z has order 1 and *real* principal symbol. Then

$$\frac{d}{dx_0}|u(x_0)|^2 = -2\mathsf{Im}(\tilde{\Lambda}u, u) - i(\underbrace{(Z - Z^*)}_{\text{order } 0} u, u),$$

$$-\int_0^\infty e^{2\lambda x_0} \frac{d}{dx_0}|u|^2 dx_0 = |u(0)|^2 + 2\lambda \int_0^\infty e^{2\lambda x_0}|u|^2 dx_0,$$

whence, by the Cauchy-Schwarz inequality,

$$\frac{1}{\lambda} \int_0^\infty e^{2\lambda x_0}|\tilde{\Lambda}u|^2 dx_0 \geq |u(0)|^2 + \underbrace{(\lambda - C)}_{\geq \lambda/2} \int_0^\infty e^{2\lambda x_0}|u|^2 dx_0,$$

which proves the claim when λ_0 is chosen sufficiently large.

Next, to control the term $(Q_\lambda u, u)$, we have to use *Melin's inequality* (see, e.g., Hörmander [7]).

Theorem 3.6. *Suppose that $A = A^*$ is a ψdo with symbol $a \sim a_m + a_{m-1} + \cdots$ satisfying*

$$a_m \geq 0, \quad \text{and} \quad a_m = 0 \Longrightarrow a_{m-1}^s + \mathrm{Tr}^+ F > 0.$$

Then for any given compact K there exist $c_K, C_K > 0$ such that

$$(Au, u) \geq c_K |u|^2_{(m-1)/2} - C_K |u|^2_{(m-2)/2}, \quad \forall u \in C_0^\infty(K).$$

To be in a position to use Melin's theorem, we have to examine the principal and subprincipal symbols, $q_{\lambda,2}$ and $q_{\lambda,1}^s$, respectively, and the Hamilton map $F_{q_{\lambda,2}}$ of Q_λ. One has

$$q_{\lambda,2} = \left(1 + \frac{s_0}{\lambda}\right)q + \frac{1}{2\lambda}\{\xi_0 + \ell_1, q\} - \frac{1}{\lambda^2}\tilde{r}_1^2 \gtrsim q, \quad \text{provided } \lambda \geq \lambda_0 \gg 1,$$

$$q_{\lambda,1}^s(\rho) = q_1^s(\rho) + \frac{1}{2\lambda}(H_\Lambda q)(\rho) = q_1^s(\rho), \quad \rho \in \Sigma,$$

and

$$F_{q_{\lambda,2}}(\rho) = F_q(\rho) + \frac{1}{\lambda}s_0(\rho)F_q(\rho) + \frac{1}{2\lambda}F_{\{\xi_0+\ell_1,q\}}(\rho) - \frac{1}{\lambda^2}F_{\tilde{r}_1^2}(\rho).$$

Hence,

$$F_{q_{\lambda,2}}(\rho) \longrightarrow F_q(\rho), \quad \text{as } \lambda \to +\infty, \text{ locally uniformly in } \rho \in \Sigma.$$

Since

$$\mathrm{Tr}^+ F_q(\rho) = \mathrm{Tr}^+ F(\rho), \quad \rho \in \Sigma,$$

Melin's inequality yields the inequality

$$\lambda^2 \int_0^\infty e^{2\lambda x_0}(Q_\lambda u, u)dx_0 \geq c_1\lambda^2 \int_0^\infty e^{2\lambda x_0}|u|_{1/2}^2 dx_0 - c_2\lambda^2 \int_0^\infty e^{2\lambda x_0}|u|^2 dx_0.$$

$$(3.6)$$

Melin's inequality is also used to control from below $(Q(0, x', D')u(0), u(0))$ by $-C|u(0)|^2$, and hence to get rid of the term $\lambda E(0)$ by using the term $\lambda^3|u(0)|^2$ with λ large.

Hence we arrive at

$$\lambda^4 \int_0^\infty e^{2\lambda x_0}|u|^2 dx_0 + \lambda^2 \int_0^\infty e^{2\lambda x_0}|u|_{1/2}^2 dx_0 \leq C \int_0^\infty e^{2\lambda x_0}|Pu|^2 dx_0, \quad (3.7)$$

for all $\lambda \geq \lambda_0 \gg 1$, and we are done.

To pass from the Sobolev order $\tau = 0$ to the Sobolev order $\tau \neq 0$, we take $M_\tau = M_\tau(x', D')$ with principal symbol $|\xi'|^\tau$ and parametrix $M'_{-\tau}$, and see at once that the Carleman estimate (3.4) with $\tau = 0$ gives the Carleman estimate (3.4) with $\tau \neq 0$, just by the replacements

$$u \rightsquigarrow M_\tau u, \quad P \rightsquigarrow M_\tau P M'_{-\tau}.$$

Since P and $M_\tau P M'_{-\tau}$ have equal principal symbols and on Σ equal subprincipal symbols, we are done.

Remark 3.7. One gets in the same way estimates when $\int_0^\infty$ is replaced by $\int_t^\infty$, any given $t \in \mathbb{R}$, and also when $\int_0^\infty e^{2\lambda x_0}$ is replaced by $\int_{-\infty}^t e^{-2\lambda x_0}$.

The natural step at this point would be passing from microlocal Carleman estimates to local Carleman estimates. This requires taking under control the commutators $[P, \chi]$, which is no problem when χ is supported away from Σ_0, but that in fact seems to be out of control near Σ_0, for it involves control from below of

$$(Q_\lambda u, u) - \varepsilon(Z^* Z u, u), \quad 0 < \varepsilon \ll 1,$$

for an arbitrary first-order operator $Z(x, D')$, the principal symbol of which we *only* know that it vanishes on Σ. Hence, the alluded to control is in general impossible regardless the size of λ and ε, because Q_λ is not transversally elliptic.

Henceforth, we just have microlocal energy estimates, and a plan for proving C^∞ well-posedness seems to go through a combination of microlocal estimates and results on the propagation of singularities.

We now wish to give an idea of this, at least in the case $p(x, \xi) = -\xi_0^2 + |f(x, \xi')|^2$. Recall that when $\rho \in \Sigma$, the *hyperbolicity cone of the localized polynomial* $\sigma(v, F(\rho)v)$ is defined to be

$$\Gamma_\rho = \{v \in T_\rho T^* \mathbb{R}^{n+1}; \ \sigma(v, H_{\xi_0}(\rho)) > |\sigma(v, H_f(\rho))|\},$$

and its *polar* by

$$\Gamma_\rho^\sigma := \{w \in T_\rho T^* \mathbb{R}^{n+1}; \ \sigma(w, v) \geq 0, \ \forall v \in \Gamma_\rho\}.$$

(Recall, see, e.g., Wakabayashi [18], the "semicontinuity property" of the cone-field $\rho \longmapsto \Gamma_\rho$: Given $\rho_0 \in \Sigma$, and given K, a compact set in Γ_{ρ_0}, there exists a neighborhood U of ρ_0 in Σ such that $K \subset \Gamma_\rho$ for all $\rho \in U$.)

As is well known (see, e.g., Hörmander [6]),

$$\rho \in \Sigma_- \Longleftrightarrow \Gamma_\rho \cap \operatorname{Ker} F(\rho)^2 \neq \emptyset,$$

and the proper *time-directions* to derive energy estimates are chosen in

$$F(\rho)\Big(\Gamma_\rho \cap \operatorname{Ker} F(\rho)^2\Big).$$

Lemma 3.8. *Let* $\rho \in \Sigma_-$. *Denote by* $\dot{\Gamma}_\rho^\sigma := \{w \in T_\rho T^* \mathbb{R}^{n+1}; \ \sigma(w, v) > 0, \forall v \in \Gamma_\rho\}$, *the "interior" of* Γ_ρ^σ. *Then*

$$F(\rho)\Big(\Gamma_\rho \cap \operatorname{Ker} F(\rho)^2\Big) = \dot{\Gamma}_\rho^\sigma \cap \Big(T_\rho \Sigma \cap T_\rho \Sigma^\sigma\Big).$$

Proof. We start by defining the linear maps

$$\gamma = \gamma_\rho \colon T_\rho T^* \mathbb{R}^{n+1} \ni v \longmapsto \begin{bmatrix} \sigma(v, H_{\xi_0}(\rho)) \\ \sigma(v, H_f(\rho)) \end{bmatrix} \in \mathbb{R} \times \mathbb{R}^d,$$

$$\theta = \theta_\rho \colon \mathbb{R} \times \mathbb{R}^d \ni \begin{bmatrix} \zeta_0 \\ \zeta' \end{bmatrix} \longmapsto -\zeta_0 H_{\xi_0}(\rho) + \langle \zeta', H_f(\rho) \rangle \in T_\rho T^* \mathbb{R}^{n+1}.$$

Using

$$F(\rho)v = -\sigma(v, H_{\xi_0}(\rho))H_{\xi_0}(\rho) + \langle \sigma(v, H_f(\rho)), H_f(\rho) \rangle,$$

one then computes

$$\operatorname{Ker} F(\rho)^2 = \left\{ v; \ \gamma(v) = \begin{bmatrix} \zeta_0 \\ \zeta' - \zeta_0 M(\rho)^{-1} \alpha(\rho) \end{bmatrix}, \zeta_0 \in \mathbb{R}, \zeta' \in \operatorname{Ker} M(\rho) \right\},$$

$$\Gamma_\rho = \left\{ v; \ \gamma(v) = \begin{bmatrix} \eta_0 \\ \eta' \end{bmatrix}, \ \eta_0 > |\eta'| \right\},$$

and

$$\operatorname{Ker} F(\rho)^2 \cap \Gamma_\rho = \left\{ v; \ \gamma(v) = \left[\begin{array}{c} \zeta_0 \\ \zeta' - \zeta_0 M(\rho)^{-1}\alpha(\rho) \end{array} \right], \right.$$
$$\left. \zeta_0 > 0, \zeta_0 > \frac{|\zeta'|}{\sqrt{1 - |M(\rho)^{-1}\alpha(\rho)|^2}}, \zeta' \in \operatorname{Ker} M(\rho) \right\}.$$

Let now $v \in \operatorname{Ker} F(\rho)^2 \cap \Gamma_\rho$, with $\gamma(v) = \left[\begin{array}{c} \zeta_0 \\ \zeta' - \zeta_0 M(\rho)^{-1}\alpha(\rho) \end{array} \right]$. Then

$$w = F(\rho)v = \theta(\gamma(v)) = -\zeta_0 H_{\xi_0} + \langle \zeta' - \zeta_0 M(\rho)^{-1}\alpha(\rho), H_f(\rho) \rangle,$$

with $\zeta_0 > |\zeta'|/\sqrt{1 - |M(\rho)^{-1}\alpha(\rho)|^2}$. Hence $w \neq 0$. Notice that since $T_\rho \Sigma = \operatorname{Ker} F(\rho)$ and $\operatorname{Im} F(\rho) = \left(\operatorname{Ker} F(\rho) \right)^\sigma = T_\rho \Sigma^\sigma$, it is immediate that $w \in T_\rho \Sigma \cap T_\rho \Sigma^\sigma$. If we prove that $w \in \dot{\Gamma}_\rho^\sigma$, we then have

$$F(\rho) \left(\operatorname{Ker} F(\rho)^2 \cap \Gamma_\rho \right) \subset \dot{\Gamma}_\rho^\sigma \cap \left(T_\rho \Sigma \cap T_\rho \Sigma^\sigma \right).$$

Now, for any given $v \in \Gamma_\rho$ with $\gamma(v) = \left[\begin{smallmatrix} \eta_0 \\ \eta' \end{smallmatrix} \right]$,

$$\sigma(w, v) = \zeta_0 \left(\eta_0 + \left\langle M(\rho)^{-1}\alpha(\rho) - \frac{\zeta'}{\zeta_0}, \eta' \right\rangle \right),$$

and since

$$\left| M(\rho)^{-1}\alpha(\rho) - \frac{\zeta'}{\zeta_0} \right|^2 < 1, \quad \text{we get} \quad \eta_0 + \left\langle M(\rho)^{-1}\alpha(\rho) - \frac{\zeta'}{\zeta_0}, \eta' \right\rangle \geq \eta_0 - |\eta'| > 0,$$

which gives the claimed inclusion.

Conversely, let $w \in \dot{\Gamma}_\rho^\sigma \cap \left(T_\rho \Sigma \cap T_\rho \Sigma^\sigma \right)$. Then $w \neq 0$ and $w \in \operatorname{Im} F(\rho)$, so that we may write $w = -\lambda_0 H_{\xi_0}(\rho) + \langle \lambda', H_f(\rho) \rangle$. Since, on the other hand, we also have $F(\rho)w = 0$, it then follows that

$$0 = F(\rho)w = \theta(\gamma(w)) = \theta \left[\begin{array}{c} \sigma(w, H_{\xi_0}(\rho)) \\ \sigma(w, H_f(\rho)) \end{array} \right] = \theta \left[\begin{array}{c} -\langle \alpha(\rho), \lambda' \rangle \\ -\lambda_0 \alpha(\rho) - M(\rho)\lambda' \end{array} \right],$$

whence

$$M(\rho)(\lambda' + \lambda_0 M(\rho)^{-1}\alpha(\rho)) = 0 \Longrightarrow \lambda' = -\lambda_0 M(\rho)^{-1}\alpha(\rho) + \theta', \ \theta' \in \operatorname{Ker} M(\rho).$$

Therefore we have

$$0 \neq w = -\lambda_0 H_{\xi_0}(\rho) + \langle \theta' - \lambda_0 M(\rho)^{-1}\alpha(\rho), H_f(\rho) \rangle,$$

and, for $v \in \Gamma_\rho$ with $\gamma(v) = \left[\begin{smallmatrix} \eta_0 \\ \eta' \end{smallmatrix} \right]$,

$$0 < \sigma(w, v) = \lambda_0 \eta_0 + \langle \lambda_0 M(\rho)^{-1}\alpha(\rho) - \theta', \eta' \rangle,$$

for all $\left[\begin{smallmatrix} \eta_0 \\ \eta' \end{smallmatrix} \right]$ with $\eta_0 > |\eta'|$. It thus follows that

$$\lambda_0 > |\lambda_0 M(\rho)^{-1}\alpha(\rho) - \theta'|,$$

and since $|\lambda_0 M(\rho)^{-1}\alpha(\rho) - \theta'|^2 = \lambda_0^2 |M(\rho)^{-1}\alpha(\rho)|^2 + |\theta'|^2$, we obtain

$$\lambda_0 > |\theta'| \Big/ \sqrt{1 - |M(\rho)^{-1}\alpha(\rho)|^2}\,,$$

that is, $w = F(\rho)v_0$ with $\gamma(v_0) = \left[\begin{smallmatrix}\lambda_0\\\lambda'\end{smallmatrix}\right]$, whence

$$w \in F(\rho)\Big(\operatorname{Ker} F(\rho)^2 \cap \Gamma_\rho\Big),$$

which shows the remaining inclusion and proves the lemma. □

This observation allows us to define the following family of sets:

$$\Sigma \ni \rho \longmapsto V_\rho := \begin{cases} \dot{\Gamma}_\rho^\sigma \cap (T_\rho\Sigma \cap T_\rho\Sigma^\sigma), & \text{for } \rho \in \Sigma_-, \\ \mathbb{R}_+ z(\rho)\,(\subset T_\rho\Sigma \cap T_\rho\Sigma^\sigma), & \text{for } \rho \in \Sigma_0, \end{cases}$$

where $z(\rho)$ is the vector field of Lemma 3.3.

Remark 3.9. Notice that $V_\rho \subset T_\rho\Sigma \cap T_\rho\Sigma^\sigma$ for all $\rho \in \Sigma$. Since $z(\rho) \in T_\rho\Sigma_0$, when $\rho \in \Sigma_0$, we therefore have

$$V_\rho \subset T_\rho\Sigma_0 \cap T_\rho\Sigma_0^\sigma \neq \{0\}, \quad \forall \rho \in \Sigma_0,$$

because Σ_0 *is not symplectic*.

We next define the *microlocal influence domain* (*in the future*) by

$$K_\rho^+ := \{\gamma(t);\ \gamma\colon [0, T_\gamma] \overset{\text{Lipschitz}}{\longrightarrow} \Sigma,\ \gamma(0) = \rho,\ \dot{\gamma}(t) \in V_{\gamma(t)}\ t\text{--a.e.}\}.$$

The theorem we expect to prove (which is work in progress) is the following.

Theorem 3.10. *If $\rho \in \Sigma \setminus \operatorname{WF}(Pu)$ and, for some neighborhood $U \subset T^*\mathbb{R}^{n+1}$ of ρ,*

$$\operatorname{WF}(u) \cap U \cap \Big(K_\rho^+ \setminus \{\rho\}\Big) = \emptyset,$$

then $\rho \notin \operatorname{WF}(u)$.

We expect that an analogous result holds in the case of the (Sobolev-based) wave-front sets $\operatorname{WF}_\tau(Pu)$ and $\operatorname{WF}_{\tau+1/2}(u)$.

We hope that the combination of microlocal estimates and results on the propagation of singularities give hold on the whole C^∞ well-posedness of the C.p., in the spirit of work done by Kajitani and Wakabayashi.

4. Conjectures when $\Sigma = \Sigma_+ \sqcup \Sigma_0$.

This case is much more difficult to treat, in that, to start with, one should find the correct Ivrii-Petkov-Hörmander condition (1.1). One should bear in mind that we *must* have (1.1) on Σ_0, but no condition at all on $p_1^s(\rho)$ is a priori expected as long as ρ stays away from Σ_0.

To simplify things a little, let us suppose that $\operatorname{Tr}^+ F(\rho) > 0$ at every $\rho \in \Sigma$. Let $\lambda(\rho)$, $\rho \in \Sigma \setminus \Sigma_0$, be the real positive (smooth) eigenvalue of $F(\rho)$, and extend

λ to Σ by putting $\lambda(\rho) = 0$ when $\rho \in \Sigma_0$. Recall also that

$$\lambda(\rho) \approx \sqrt{|M(\rho)^{-1}\alpha(\rho)|^2 - 1}.$$

Now, for $\rho \in \mathbb{S}^*\Sigma$ (where the *cosphere* $\mathbb{S}^*\Sigma$ of Σ is defined to be the set of those $\rho \in \Sigma$ which have fiber ξ of length 1) and $0 < \varepsilon \ll 1$, consider the closed set $I_\varepsilon(\rho) \subset \mathbb{C}$ given by

$$I_\varepsilon(\rho) = \{z \in \mathbb{C}; \ \mathrm{Im}\, z = 0, \ |\mathrm{Re}\, z| \leq \mathrm{Tr}^+ F(\rho) - \varepsilon\}.$$

Let us consider the following analog of the Ivrii-Petkov-Hörmander condition:

For any given $\rho_0 \in \mathbb{S}^*\Sigma_0$ there is a neighborhood $V \subset \mathbb{S}^*\Sigma$ of ρ_0 and some $0 < \varepsilon \ll 1$ such that

$$\mathrm{dist}(p_1^s(\rho), I_\varepsilon(\rho)) \leq C\lambda(\rho)^2, \quad \rho \in V, \tag{4.1}$$

for some $C > 0$ independent of ρ and ε.

Condition (4.1) obviously implies

$$\mathrm{Im}\, p_1^s(\rho) = 0, \quad |\mathrm{Re}\, p_1^s(\rho)| < \mathrm{Tr}^+ F(\rho), \quad \forall \rho \in \Sigma_0,$$

but is very demanding on the behavior of $p_1^s(\rho)$ as ρ approaches Σ_0 from within Σ_+.

Condition (4.1) should be compared with *Melrose's conjecture* (see Melrose [13]), which states the following: *If the C.p. for P is well posed in $\Omega \cap \{t > 0\}$, where $\Omega \ni 0$ is an open set of $\mathbb{R}^{n+1}$, then $s(\rho)/\lambda(\rho)$ is uniformly bounded in Σ_+ for all $\rho \in \Sigma_+ \cap (T^*V_0 \setminus 0)$, where V_0 is some neighborhood of $0 \in \mathbb{R}^{n+1}$ and where*

$$s(\rho) := |\mathrm{Im}\, p_{m-1}^s(\rho)| + \inf\{|\mathrm{Re}\, p_1^s(\rho) - s|; \ |s| \leq \mathrm{Tr}^+ F(\rho)\}.$$

Now, once more, we are faced with another difficult geometric problem related to the null-bicharacteristics of p. As in the preceding cases, we can consider $H_{S}^3 p|_{\Sigma_0}$.

Problem: *Is condition $H_{S}^3 p|_{\Sigma_0} = 0$ necessary and/or sufficient to ensure that no null-bicharacteristic of p can* tangentially *enter Σ_0?*

The problem is open.

On the side of the C.p. we have the following conjecture.

Conjecture 4.1. Suppose that Σ_0 is non-symplectic and that $H_{S}^3 p|_{\Sigma_0} = 0$. If $\lambda(\rho)^2$ vanishes exactly to second order on Σ_0 and $F(\rho)v(\rho) \in T_\rho\Sigma_0$, and if (4.1) holds, then the C.p. is C^∞ well posed.

5. A final example

We end this paper by giving an example, which puts at work the approach we explained in Setions 2 and 3, and for which the C.p. is well posed, *not only microlocally*, in C^∞, this being due to the fact that the Ivrii-factorization that is constructed is global (and not only microlocal).

Let $\mu > 0$ and let

$$p(x, \xi) = -\xi_0^2 + \xi_1^2 + \xi_2^2 + \mu^2 x_2^2 |\xi''|^2 + (x_0 - x_1 - f(x_1, x''))^2 |\xi''|^2,$$

where $\xi'' = (\xi_3, \ldots, \xi_n)$, $x'' = (y', y'')$, $f(x_1, x'') = \gamma x_1 |y'|^2$ with $\gamma > 0$. Then

$$\Sigma = \{\xi_0 = \xi_1 = \xi_2 = 0,\ x_2 = 0,\ x_0 = x_1 + f(x_1, x''),\ \xi'' \neq 0\},$$

$$\alpha(\rho) = \begin{bmatrix} 0 \\ 0 \\ 0 \\ |\xi''| \end{bmatrix}, \quad M(\rho) = \begin{bmatrix} 0 & 0 & 0 & -f'_{x_1}|\xi''| \\ 0 & 0 & \mu|\xi''| & 0 \\ 0 & -\mu|\xi''| & 0 & 0 \\ f'_{x_1}|\xi''| & 0 & 0 & 0 \end{bmatrix},$$

$$M(\rho)^{-1}\alpha(\rho) = \begin{bmatrix} 1/(1 + f'_{x_1}) \\ 0 \\ 0 \\ 0 \end{bmatrix},$$

so that $|M(\rho)^{-1}\alpha(\rho)| = 1/(1 + \gamma|y'|^2)$ and

$$\Sigma_0 = \{\xi_0 = \xi_1 = \xi_2 = 0,\ x_0 = x_1, x_2 = 0, y' = 0, \xi'' \neq 0\},$$

which is *non-symplectic*, and we have a Ivirii-factorization by taking

$$\Lambda = -\xi_0 - \frac{1}{1 + \gamma|y'|^2}\xi_1.$$

Hence at $\rho \in \Sigma_0$

$$H_\Lambda(\rho) = \frac{\partial}{\partial x_0} + \frac{\partial}{\partial x_1} \in T_\rho \Sigma_0,$$

and as H_S one may choose a multiple of

$$\frac{\partial}{\partial \xi_0} - (1 + f'_{x_1})\frac{\partial}{\partial \xi_1} - \left\langle \nabla_{x''}f, \frac{\partial}{\partial \xi''} \right\rangle,$$

and check that $H_S^3 p\big|_{\Sigma_0} = 0$. Finally, the eigenvalues of $F(\rho)$ are given by $\pm i\mu|\xi''|$ and $\pm i|\xi''|\sqrt{f'_{x_1}(2 + f'_{x_1})}$, so that

$$\mathrm{Tr}^+ F(\rho) = \mu|\xi''| + |\xi''|\sqrt{\gamma|y'|^2(2 + \gamma|y'|^2)}.$$

The admissible lower-order terms are of the form

$$p_1(x, \xi) = a\xi_0 + b\xi_1 + c\xi_2 + \langle d, \xi'' \rangle,$$

where $\mathrm{Im}\, d\big|_\Sigma = 0$, and $|\mathrm{Re}\, d\big|_\Sigma| < \mu + \sqrt{\gamma|y'|^2(2 + \gamma|y'|^2)}$.

For the relative operator P, the C.p. is (locally) well posed in C^∞.

References

[1] E. Bernardi and A. Bove. Geometric results for a class of hyperbolic operators with double characteristics. *Comm. Partial Differential Equations* **13** (1988), 61–86.

[2] E. Bernardi and A. Bove. Geometric transition for a class of hyperbolic operators with double characteristics. *Japan. J. Math. (N.S.)* **23** (1997), 1–87.

[3] E. Bernardi, A. Bove and C. Parenti. Geometric results for a class of hyperbolic operators with double characteristics. II. *J. Funct. Anal.* **116** (1993), 62–82.

[4] E. Bernardi and T. Nishitani. On the Cauchy problem for non-effectively hyperbolic operators, the Gevrey 5 well-posedness. *J. Anal. Math.* **105** (2008), 197–240.

[5] E. Bernardi, C. Parenti and A. Parmeggiani. The Cauchy problem for hyperbolic operators with double characteristics in presence of transition. To appear in Comm. Partial Differential Equations.

[6] L. Hörmander. The Cauchy problem for differential equations with double characteristics. *J. Analyse Math.* **32** (1977), 118–196.

[7] L. Hörmander. The analysis of linear partial differential operators. III. Pseudodifferential operators. Grundlehren der Mathematischen Wissenschaften, 274. *Springer-Verlag, Berlin*, 1985. viii+525 pp.

[8] V. Ivrii. The well-posedness of the Cauchy problem for nonstrictly hyperbolic operators. III. The energy integral. *Trans. Moscow Math. Soc.* **34** (1978), 149–168.

[9] V. Ivrii and V. Petkov. Necessary conditions for the correctness of the Cauchy problem for non-strictly hyperbolic equations. *Uspehi Mat. Nauk.* **29** (1974), no. 5(179), 3–70.

[10] N. Iwasaki. The Cauchy problem for effectively hyperbolic equations. A standard type. *Publ. Res. Inst. Math. Sci.* **20** (1984), no. 3, 543–584.

[11] N. Iwasaki. The Cauchy problem for effectively hyperbolic equations (general cases). *J. Math. Kyoto Univ.* **25** (1985), no. 4, 727–743.

[12] R.B. Melrose. The Cauchy problem for effectively hyperbolic operators. *Hokkaido Math. J.*, **12** (1983), no. 3, part 2, 371–391.

[13] R.B. Melrose. The Cauchy problem and propagation of singularities. *Seminar on nonlinear partial differential equations (Berkeley, Calif., 1983)*, 185–201, Math. Sci. Res. Inst. Publ., 2, Springer, New York, 1984.

[14] T. Nishitani. The Cauchy problem for effectively hyperbolic operators. *Nonlinear variational problems (Isola d'Elba, 1983)*, 9–23, Res. Notes in Math., 127, *Pitman, Boston, MA*, 1985.

[15] T. Nishitani. Non-effectively hyperbolic operators, Hamilton map and bicharacteristics. *J. Math. Kyoto Univ.* **44** (2004), 55–98.

[16] T. Nishitani. Cauchy problem for noneffectively hyperbolic operators. To appear in the Memoirs of the Mathematical Society of Japan.

[17] C. Parenti and A. Parmeggiani. On the Cauchy problem for hyperbolic operators with double characteristics. *Comm. Partial Differential Equations* **34** (2009), 837–888.

[18] S. Wakabayashi. Singularities of solutions of the Cauchy problem for hyperbolic systems in Gevrey classes. *Japan J. Math.* **11** (1985), 157–201.

Cesare Parenti
Department of Computer Science, University of Bologna
Via Mura Anteo Zamboni, 7
I-40126 Bologna, Italy
e-mail: `parenti@cs.unibo.it`

Alberto Parmeggiani
Department of Mathematics, University of Bologna
Piazza di Porta San Donato, 5
I-40126 Bologna, Italy
e-mail: `parmeggi@dm.unibo.it`

Progress in Mathematics, Vol. 301, 267–283
© 2012 Springer Basel

Modulation Spaces and Nonlinear Evolution Equations

Michael Ruzhansky, Mitsuru Sugimoto and Baoxiang Wang

Abstract. We survey some recent progress on modulation spaces and the well-posedness results for a class of nonlinear evolution equations by using the frequency-uniform localization techniques.

Mathematics Subject Classification. Primary 42B35, 35Q55;
Secondary 46E35, 35L05.

Keywords. Frequency-uniform localization, modulation spaces, nonlinear Schrödinger equation.

1. Definitions of modulation spaces

The modulation space was introduced by Feichtinger [15] in 1983. We denote by $\mathfrak{S}$ the Schwartz space and by $\mathfrak{S}'$ its dual space. Let f, g be Schwartz functions, and set

$$V_g f(x, \omega) = \int_{\mathbb{R}^n} e^{-it\omega} \overline{g(t-x)} f(t) dt.$$

$V_g f$ is said to be the short-time Fourier transform of f. In the following, if there is no explanation, we will always assume that

$$-\infty < s < \infty, \quad 0 < p, q \leq \infty, \quad \langle \cdot \rangle = (1 + |\cdot|^2)^{1/2}.$$

The norm on modulation spaces is given by

$$\|f\|_{M_{p,q}^s}^{\circ} = \left(\int_{\mathbb{R}^n} \left(\int_{\mathbb{R}^n} |V_g f(x, \omega)|^p dx \right)^{q/p} \langle \omega \rangle^{sq} d\omega \right)^{1/q}, \qquad (1.1)$$

with a natural modification for $p, q = \infty$. Applying the frequency-uniform localization techniques, one can give an equivalent norm on modulation spaces. Let Q_k be the unit cube with its center at k, $\{Q_k\}_{k \in \mathbb{Z}^n}$ constitutes a decomposition of

This work was completed with the support of NSFC grant. M.R. is supported by the EPSRC Leadership Fellowship.

$\mathbb{R}^n$. Such kind of decomposition goes back to the work of N. Wiener [57], and we say that it is the Wiener decomposition of $\mathbb{R}^n$. We can roughly write

$$\Box_k \sim \mathfrak{F}^{-1}\chi_{Q_k}\mathfrak{F}, \quad k \in \mathbb{Z}^n, \tag{1.2}$$

where χ_E denotes the characteristic function on the set E, and $\mathfrak{F}$ is the Fourier transform. Since Q_k is a translation of Q_0, $\Box_k$ ($k \in \mathbb{Z}^n$) have the same localized structures in the frequency space, which are said to be the *frequency-uniform decomposition operators*. Similar to Besov spaces, one can use $\{\Box_k\}_{k\in\mathbb{Z}^n}$ and $\ell^q(L^p)$ to generate a class of function spaces, so-called modulation spaces for which the norm is defined by

$$\|f\|_{M_{p,q}^s} = \left(\sum_{k\in\mathbb{Z}^n} \langle k\rangle^{sq}\|\Box_k f\|_p^q\right)^{1/q},$$

with the natural modifications for $p, q = \infty$. We now give an exact definition on frequency-uniform decomposition operators. Since χ_{Q_k} can not be differentiated, it is convenient to replace χ_{Q_k} in (1.2) by a smooth cut-off function. Let $\rho \in \mathfrak{S}(\mathbb{R}^n)$, $\rho : \mathbb{R}^n \to [0,1]$ be a smooth function verifying $\rho(\xi) = 1$ for $|\xi|_\infty \le 1/2$ and $\rho(\xi) = 0$ for $|\xi|_\infty \ge 1$[1]. Let ρ_k be a translation of ρ,

$$\rho_k(\xi) = \rho(\xi - k), \quad k \in \mathbb{Z}^n. \tag{1.3}$$

We see that $\rho_k(\xi) = 1$ in Q_k and so, $\sum_{k\in\mathbb{Z}^n} \rho_k(\xi) \ge 1$ for all $\xi \in \mathbb{R}^n$. Denote

$$\sigma_k(\xi) = \rho_k(\xi)\left(\sum_{\ell\in\mathbb{Z}^n} \rho_\ell(\xi)\right)^{-1}, \quad k \in \mathbb{Z}^n. \tag{1.4}$$

Then we have

$$\begin{cases} |\sigma_k(\xi)| \ge c, & \forall\, \xi \in Q_k, \\ \operatorname{supp} \sigma_k \subset \{\xi : |\xi - k|_\infty \le 1\}, & \\ \sum_{k\in\mathbb{Z}^n} \sigma_k(\xi) \equiv 1, & \forall\, \xi \in \mathbb{R}^n, \\ |D^\alpha \sigma_k(\xi)| \le C_{|\alpha|}, & \forall\, \xi \in \mathbb{R}^n, \ \alpha \in (\mathbb{N}\cup\{0\})^n. \end{cases} \tag{1.5}$$

Hence, the set

$$\Upsilon_n = \{\{\sigma_k\}_{k\in\mathbb{Z}^n} : \{\sigma_k\}_{k\in\mathbb{Z}^n} \text{ satisfies } (1.5)\} \tag{1.6}$$

is non-void. If there is no confusion, in the sequel we will write $\Upsilon = \Upsilon_n$. Let $\{\sigma_k\}_{k\in\mathbb{Z}^n} \in \Upsilon$. The frequency-uniform decomposition operators can be exactly defined by

$$\Box_k := \mathfrak{F}^{-1}\sigma_k\mathfrak{F}, \quad k \in \mathbb{Z}^n. \tag{1.7}$$

For simplicity, we write $M_{p,q}^0 = M_{p,q}$. The space $M_{p,q}^s$ is said to be the modulation space. One can prove that $\|\cdot\|_{M_{p,q}^s}^0$ and $\|\cdot\|_{M_{p,q}^s}$ are equivalent norms; cf. [15] for a proof on modulation spaces defined on Abelian groups, and [55] for a straightforward proof.

---

[1] For $\xi = (\zeta_1,\ldots,\zeta_n)$, $|\zeta|_\infty := \max_{i=1,\ldots,n} |\zeta_i|$.

2. Questions on the Schrödinger equation

Let $S(t) = e^{it\Delta}$ be the Schrödinger semi-group. It is known that $S(t) = e^{it\Delta}$: $L^p \to L^p$ is bounded if and only if $p = 2$. Indeed, one can show that

$$\sup_{\varphi \in \mathfrak{S} \setminus \{0\}} \frac{\|S(t)\varphi\|_{L^p(\mathbb{R}^n)}}{\|\varphi\|_{L^p(\mathbb{R}^n)}} = \infty, \quad p \neq 2, \tag{2.1}$$

by taking $\varphi = e^{-(a+ib)|x|^2}$. So, a natural question is the following

Question 2.1. Is there any Banach function space X satisfying

$$\sup_{\varphi \in X \setminus \{0\}} \frac{\|S(t)\varphi\|_{X(\mathbb{R}^n)}}{\|\varphi\|_{X(\mathbb{R}^n)}} < \infty ? \tag{2.2}$$

Recall that (cf. [5, 6])

$$\|S(t)f\|_p \leq C|t|^{-n(1/2-1/p)}\|f\|_{p'}, \tag{2.3}$$

where $2 \leq p \leq \infty$, $1/p + 1/p' = 1$. Estimate (2.3) contains singularity at $t = 0$, and another natural question is

Question 2.2. Is there any Banach function space X satisfying the following truncated decay

$$\|S(t)\varphi\|_{X(\mathbb{R}^n)} \leq C(1 + |t|)^{-\delta(X)}\|\varphi\|_{X'(\mathbb{R}^n)}? \tag{2.4}$$

where $\delta > 0$, X' denotes the dual space of X, X and X' have the same regularity.

Consider the Cauchy problem for the nonlinear Schrödinger equation (NLS)

$$iu_t + \Delta u = |u|^{2\kappa}u, \quad u(0, x) = u_0(x), \quad \kappa \in \mathbb{N}. \tag{2.5}$$

It is known that if u is a solution of (2.5), so is $u_\mu(t, x) = \mu^{1/\kappa}u(\mu^2 t, \mu x)$ with the initial datum $\mu^{1/\kappa}u_0(\mu x)$. Taking notice of

$$\|u_\mu(0)\|_{\dot{H}^s} = \mu^{s-n/2+1/\kappa}\|u_0\|_{\dot{H}^s}, \quad \mu > 0,$$

one sees that $s = s_\kappa := n/2 - 1/\kappa$ is the index such that $\|u_\mu(0)\|_{\dot{H}^s}$ is invariant for all $\mu > 0$. $\dot{H}^{s_\kappa}$ is said to be a critical space for NLS (2.5); cf. [6]. Up to now, we can solve (2.5) in H^s for $s \geq s_\kappa \geq 0$, which correspond to the critical and sub-critical cases in H^s. However, we cannot solve (2.5) in H^s in the case $s < s_\kappa$, where H^s is said to be the super-critical space. Our question is

Question 2.3. Let $0 \leq s < s_\kappa$. Are there any initial data $u_0 \in H^s \setminus H^{s_\kappa}$ so that NLS (2.5) is still locally and globally well posed?

In order to answer the above questions, modulation spaces play important roles. Roughly speaking, Schrödinger semi-group is bounded in modulation spaces and it satisfies a truncated decay in modulation spaces. Moreover, the modulation space $M_{2,1}$ enjoys lower derivative regularity and we can solve NLS (2.5) in $M_{2,1}$ for all $\kappa \in \mathbb{N}$. $M_{2,1}$ contains a class of data which are out of the control of H^{s_κ} if $s_\kappa > 0$.

Finally, we mention that all the questions above make sense for other equations as well. For example, the propagators $W_\pm(t) = e^{\pm it\sqrt{-\Delta}}$ for the wave equation display the loss of regularity for $p \neq 2$. More precisely, we have that for $1 < p < \infty$ and $t > 0$, the operator $W_+(t) : W^{s,p}(\mathbb{R}^n) \to L^p(\mathbb{R}^n)$ is (locally) bounded if and only if $s \geq (n-1)|1/p - 1/2|$. The same holds for $W_-(t)$, and Question 2.1 is valid. Also, a variant (2.3) holds[2], so Question 2.2 is also valid, as well as the corresponding version of Question 2.3.

We note that the dependence of estimates and large time asymptotics in Sobolev spaces for the propagator $e^{ita(D)}$ on $a(\xi)$ is known, especially in the first-order case. For example, let $a \in C^\infty(\mathbb{R}^n \backslash 0)$ be a real-valued positively homogeneous function, satisfying $\nabla a(\xi) \neq 0$ for all $\xi \neq 0$. If we denote

$$k := \max_{|\xi|=1} \mathrm{rank} \nabla_\xi^2 a(\xi),$$

then there is a loss of derivatives in L^p depending on k. More precisely, as a special case of estimates for Fourier integral operators for $1 < p < \infty$, it was shown in [37] (see also [38]) that if $e^{ia(D)} : W^{s,p} \to L^p$ is locally bounded then $s \geq k|1/p - 1/2|$. The dispersive estimates[3]

$$\|e^{ita(D)}f\|_p \leq C|t|^{-\kappa(1/2-1/p)}\|f\|_{p'},$$

where $2 \leq p \leq \infty$, $1/p + 1/p' = 1$, depend on the convexity of the level sets $\Sigma = \{\xi : a(\xi) = 1\}$ as well as on the maximal orders of contacts between Σ and its tangent lines. For a comprehensive analysis of such types of estimates and their applications to the Strichartz estimates we refer to [39].

3. Some results on modulation spaces

Roughly speaking, frequency-uniform decomposition operators combined with function spaces $\ell^q(L^p)$ produce modulation spaces. During the past thirty years, the importance of the frequency-uniform decomposition in applications seems to be not fully recognized, and it is even not mentioned in Gröchenig's famous book [18]. However, from PDE point of view, the combination of frequency-uniform decomposition operators and Banach function spaces $\ell^q(X(\mathbb{R}^n))$[4] seems to be important in making nonlinear estimates, which contains an automatic decomposition on high-low frequencies.

3.1. Basic properties on modulation spaces

Proposition 3.1 (Completeness). *Let* $0 < p, q \leq \infty$ *and* $s \in \mathbb{R}$.

(1) $M_{p,q}^s$ *is a (quasi-) Banach space. Moreover, if* $1 \leq p, q \leq \infty$, *then* $M_{p,q}^s$ *is a Banach space.*

(2) $\mathfrak{S}(\mathbb{R}^n) \subset M_{p,q}^s \subset \mathfrak{S}'(\mathbb{R}^n)$.

(3) *Let* $0 < p, q < \infty$, *then* $\mathfrak{S}(\mathbb{R}^n)$ *is dense in* $M_{p,q}^s$.

---

[2]with n replaced by $n - 1$.

[3]or rather the value of κ.

[4]X is a Banach function space defined in $\mathbb{R}^n$.

Proposition 3.2 (Equivalent norm). *Let $\{\sigma_k\}_{k\in\mathbb{Z}^n}$, $\{\varphi_k\}_{k\in\mathbb{Z}^n} \in \Upsilon$. Then $\{\sigma_k\}_{k\in\mathbb{Z}^n}$ and $\{\varphi_k\}_{k\in\mathbb{Z}^n}$ generate equivalent norms on $M_{p,q}^s$.*

Proposition 3.2 indicates that one can choose $\{\sigma_k\}_{k\in\mathbb{Z}^n} \in \Upsilon_n$ according to our requirement. In applications of PDE, it is convenient to use the following $\{\sigma_k\}_{k\in\mathbb{Z}^n} \in \Upsilon_n$. Let $\{\eta_k\}_{k\in\mathbb{Z}} \in \Upsilon_1$, we denote

$$\sigma_k(\xi) := \eta_{k_1}(\xi_1)\ldots\eta_{k_n}(\xi_n), \tag{3.1}$$

then we have $\{\sigma_k\}_{k\in\mathbb{Z}^n} \in \Upsilon_n$. the above $\sigma_k(\xi)$ realizes the separation of different variables.

Proposition 3.3 (Embedding). *Let $s_1, s_2 \in \mathbb{R}$ and $0 < p_1, p_2, q_1, q_2 \leq \infty$.*

(1) *If $s_2 \leq s_1$, $p_1 \leq p_2$ and $q_1 \leq q_2$, then $M_{p_1,q_1}^{s_1} \subset M_{p_2,q_2}^{s_2}$.*
(2) *If $q_2 < q_1$ and $s_1 - s_2 > n/q_2 - n/q_1$, then $M_{p,q_1}^{s_1} \subset M_{p,q_2}^{s_2}$.*

Proposition 3.4 (Dual space). *Let $s \in \mathbb{R}$ and $0 < p, q < \infty$. If $p \geq 1$, we denote $1/p + 1/p' = 1$; If $0 < p < 1$, we write $p' = \infty$. Then*

$$(M_{p,q}^s)^* = M_{p',q'}^{-s}. \tag{3.2}$$

If $p \geq 1$, Proposition 3.4 is similar to that of Besov spaces, however, if $0 < p < 1$, the result is quite different from that of Besov spaces. The details of the proof of Proposition 3.4 can be found in [55] by following the proof of the relevant result in Besov spaces.

Remark 3.5. If $p, q \in [1, \infty]$, Propositions 3.1 and 3.4 were obtained by Feichtinger [15]. In [55, 56], the cases $0 < p < 1$ and $0 < q < 1$ were considered.

Soon after the work [56], Kobayashi [29] independently defined $M_{p,q}$ for all $0 < p, q \leq \infty$ and obtained Proposition 3.1. Almost at the same time as [55], Kobayashi [30] discussed the dual space of $M_{p,q}$ and obtained partial results of Proposition 3.4: if $0 < p < 1$ or $1 < q < \infty$, he obtained $M_{p',q'} \subset (M_{p,q})^* \subset M_{\infty,\infty}$. For the other cases, he showed $(M_{p,q})^* = M_{p',q'}$. Recently, by using the molecular decomposition techniques of modulation spaces, Kobayashi and Sawano [31] reconsidered the dual space of $M_{p,q}^s$ and they also obtained the result of Proposition 3.4. It is worth to mention that Triebel [50] introduced a class of generalized modulation spaces for all indices $0 < p, q \leq \infty$, however, those spaces have no complete norms, which seems harder to use in the study of PDEs.

3.2. Inclusions between Besov and modulation spaces

From the definitions, we see that Besov spaces and modulation spaces are rather similar, both of them are the combinations of frequency decomposition operators and function spaces $\ell^q(L^p)$. In fact, we have the following inclusion results.

Theorem 3.6 (Embedding). *Let $0 < p, q \leq \infty$ and $s_1, s_2 \in \mathbb{R}$. We have the following results.*

(1) $B^{s_1}_{p,q} \subset M^{s_2}_{p,q}$ if and only if $s_1 \geq s_2 + \tau(p,q)$, where

$$\tau(p,q) = \max\left\{0,\ n\left(\frac{1}{q} - \frac{1}{p}\right),\ n\left(\frac{1}{q} + \frac{1}{p} - 1\right)\right\};$$

(2) $M^{s_1}_{p,q} \subset B^{s_2}_{p,q}$ if and only if $s_1 \geq s_2 + \sigma(p,q)$, where

$$\sigma(p,q) = \max\left\{0,\ n\left(\frac{1}{p} - \frac{1}{q}\right),\ n\left(1 - \frac{1}{p} - \frac{1}{q}\right)\right\}.$$

The inclusions between Besov and modulation spaces in the cases $(1/p, 1/q) \in [0,1]^2$ were first discussed by Gröbner [17] and he has never published his results. When $(1/p, 1/q)$ is in the vertices of the square $[0,1]^2$, Gröbner's results are optimal. Afterwards, Toft [48] obtained the sufficiency of Theorem 3.6 in the cases $(1/p, 1/q) \in [0,1]^2$. Sugimoto and Tomita [46] showed the necessity of the first inclusion of Theorem 3.6 in the cases $(1/p, 1/q) \in [0,1]^2$, and by duality the second inclusion is also sharp if $(1/p, 1/q) \in [0,1]^2$ and $p, q \neq \infty$. Sugimoto and Tomita's idea is to use Feichtinger's norm and the dilation property of modulation spaces. In [54, 55, 56] the authors proved the conclusions of Theorem 3.6 by using frequency-uniform decomposition techniques.

Corollary 3.7. *We have the following inclusions.*

$$B^{s+n/2}_{2,1} \subset M^s_{2,1} \subset B^s_{2,1}, \quad B^{s+n}_{\infty,1} \subset M^s_{\infty,1} \subset B^s_{\infty,1}.$$

The above embedding theorem is of importance for the study of nonlinear PDEs. As for the inclusions between L^p-Sobolev spaces and modulation spaces, it has been explicitly determined by a recent work of Sugimoto and Kobayashi [32]:

Theorem 3.8. *Let $1 \leq p, q \leq \infty$ and $s_1, s_2 \in \mathbb{R}$. Then $H^{s_1}_p \subset M^{s_2}_{p,q}$ if and only if one of the following conditions is satisfied:*

(1) $q \geq p > 1$, $s_1 \geq s_2 + \tau(p,q)$;   (2) $p > q$, $s_1 > s_2 + \tau(p,q)$;
(3) $p = 1, q = \infty$, $s_1 \geq s_2 + \tau(1,\infty)$;   (4) $p = 1, q \neq \infty$, $s_1 > s_2 + \tau(1,q)$,

and $M^{s_1}_{p,q} \subset H^{s_2}_p$ if and only if one of the following conditions is satisfied:

(1) $q \leq p < \infty$, $s_1 \geq s_2 + \sigma(p,q)$;   (2) $p < q$, $s_1 > s_2 + \sigma(p,q)$;
(3) $p = \infty, q = 1$, $s_1 \geq s_2 + \sigma(\infty,1)$;   (4) $p = \infty, q \neq 1$, $s_1 > s_2 + \sigma(\infty,q)$,

where $\tau(p,q)$ and $\sigma(p,q)$ are the same indices as in Theorem 3.6.

3.3. Dilation property of modulation spaces

Roughly speaking, homogeneous Sobolev spaces and their generalizations including homogeneous Besov and Triebel spaces have scales like

$$\|f(\lambda \cdot)\|_X \sim \lambda^{\theta(X)} \|f\|_X.$$

However, the scaling properties of modulation spaces are very complicated, which is quite different from the classical Sobolev spaces. The following result is due to Sugimoto and Tomita. Let $\tau(p,q)$ and $\sigma(p,q)$ be as in Theorem 3.6. Denote

$$\mu_1(p,q) = \tau(p,q) - n/p, \quad \mu_2(p,q) = -\sigma(p,q) - n/p$$

Theorem 3.9 ([46]). *Let $1 \leq p, q \leq \infty$. The following are true.*

(1) *There exists a constant $C > 0$ such that*

$$C^{-1}\lambda^{\mu_2(p,q)}\|f\|_{M_{p,q}} \le \|f(\lambda \cdot)\|_{M_{p,q}} \le C\lambda^{\mu_1(p,q)}\|f\|_{M_{p,q}}$$

for all $f \in M_{p,q}$ and $\lambda \ge 1$. Conversely, if there is a constant $C > 0$ such that

$$C^{-1}\lambda^{\beta}\|f\|_{M_{p,q}} \le \|f(\lambda \cdot)\|_{M_{p,q}} \le C\lambda^{\alpha}\|f\|_{M_{p,q}}$$

for all $f \in M_{p,q}$ and $\lambda \ge 1$, then $\alpha \ge \mu_1(p,q)$ and $\beta \le \mu_2(p,q)$.

(2) *There exists a constant $C > 0$ such that*

$$C^{-1}\lambda^{\mu_1(p,q)}\|f\|_{M_{p,q}} \le \|f(\lambda \cdot)\|_{M_{p,q}} \le C\lambda^{\mu_2(p,q)}\|f\|_{M_{p,q}}$$

for all $f \in M_{p,q}$ and $0 < \lambda \le 1$. Conversely, if there is a constant $C > 0$ such that

$$C^{-1}\lambda^{\alpha}\|f\|_{M_{p,q}} \le \|f(\lambda \cdot)\|_{M_{p,q}} \le C\lambda^{\beta}\|f\|_{M_{p,q}}$$

for all $f \in M_{p,q}$ and $0 < \lambda \le 1$, then $\alpha \ge \mu_1(p,q)$ and $\beta \le \mu_2(p,q)$.

4. NLS and NLKG in modulation spaces

As indicated in §2, the dispersive semi-group combined with the frequency-uniform decomposition operator has some advantages and we discuss them in this section. The results of this section can be found in [2, 3, 11, 56, 55].

4.1. Schrödinger and Klein-Gordon semigroup in modulation spaces

Let $S(t) = e^{it\Delta}$ denote the Schrödinger semi-group. In [56], Wang, Zhao and Guo obtained the uniform boundedness for the Ginzburg-Landau semi-group $L(t) = e^{(a+i)t\Delta}$ ($a > 0$) in modulation spaces and their proof is also adapted to the Schrödinger semi-group ($a = 0$ in $L(t)$).

Proposition 4.1 (Uniform boundedness of $S(t)$ in $M_{p,q}^s$). *Let $s \in \mathbb{R}$, $1 \le p \le \infty$ and $0 < q < \infty$. Then we have*

$$\|S(t)f\|_{M_{p,q}^s} \le C(1 + |t|)^{n|1/2 - 1/p|}\|f\|_{M_{p,q}^s}. \tag{4.1}$$

Shortly after the work [56], Proposition 4.1 is independently obtained by Bényi, Gröchenig, Okoudjou and Rogers in [3] and their result contains more general semi-group $e^{it(-\Delta)^{\alpha}}$ with $\alpha \le 1$, whose proof is based on the short-time frequency analysis technique. Miyachi, Nicola, Riveti, Taracco and Tomita [34] were able to consider the case $\alpha > 1$, Chen, Fan and Sun [7] obtained some refined estimates for $e^{it(-\Delta)^{\alpha}}$ with any $\alpha > 0$.

Now we consider the truncated decay of $S(t)$.

Proposition 4.2. *Let $s \in \mathbb{R}$, $2 \le p < \infty$, $1/p + 1/p' = 1$ and $0 < q < \infty$. Then we have*

$$\|S(t)f\|_{M_{p,q}^s} \le C(1 + |t|)^{-n(1/2 - 1/p)}\|f\|_{M_{p',q}^s}. \tag{4.2}$$

Propositions 4.1 and 4.2 answer Questions 2.1 and 2.2, and moreover, they are optimal in the sense that the powers of time variable are sharp, cf. [12]. Now we consider the truncated decay estimate for the Klein-Gordon semi-group $G(t) = e^{it\omega^{1/2}}$ where $\omega = I - \Delta$.

Proposition 4.3. *Let* $s \in \mathbb{R}$, $1 \le p \le \infty$ *and* $0 < q < \infty$. *Then we have*

$$\|G(t)f\|_{M_{p,q}^s} \le C(1 + |t|)^{n|1/2-1/p|}\|f\|_{M_{p,q}^s}. \tag{4.3}$$

It is known that $G(t)$ satisfies the following $L^p - L^{p'}$ estimate

$$\|G(t)f\|_{H_p^{-2\sigma(p)}} \le C|t|^{-n(1/2-1/p)}\|f\|_{p'}, \tag{4.4}$$

where

$$2 \le p < \infty, \quad 2\sigma(p) = (n+2)\left(\frac{1}{2} - \frac{1}{p}\right). \tag{4.5}$$

From (4.4) it follows that

Proposition 4.4. *Let* $s \in \mathbb{R}$, $2 \le p < \infty$, $1/p + 1/p' = 1$, $0 < q < \infty$, $\theta \in [0, 1]$ *and* $\sigma(p)$ *is as in* (4.5). *Then we have*

$$\|G(t)f\|_{M_{p,q}^s} \le C(1 + |t|)^{-n\theta(1/2-1/p)}\|f\|_{M_{p',q}^{s+2\sigma(p)\theta}}. \tag{4.6}$$

4.2. Strichartz estimates in modulation spaces

For convenience, we write

$$\|f\|_{\ell_{\Box}^{s,q}(L^\gamma(I,L^p))} = \left(\sum_{k \in \mathbb{Z}^n} \langle k \rangle^{sq}\|\Box_k f\|_{L^\gamma(I,L^p)}^q\right)^{1/q}, \tag{4.7}$$

$\ell_{\Box}^q(L^\gamma(I, L^p)) := \ell_{\Box}^{0,q}(L^\gamma(I, L^p))$, $\ell_{\Box}^q(L_{x,t\in I}^p) := \ell_{\Box}^q(L^p(I, L^p))$. Recall that the truncated decay can be generalized to the following estimate

$$\|U(t)f\|_{M_{p,q}^\alpha} \le C(1 + |t|)^{-\delta}\|f\|_{M_{p',q}}, \tag{4.8}$$

where $2 \le p < \infty$, $1 \le q < \infty$, $\alpha = \alpha(p) \in \mathbb{R}$, $\delta = \delta(p) > 0$, α and δ are independent of $t \in \mathbb{R}$, $U(t)$ is a dispersive semi-group,

$$U(t) = \mathfrak{F}^{-1}e^{itP(\xi)}\mathfrak{F}, \tag{4.9}$$

and $P(\cdot) : \mathbb{R}^n \to \mathbb{R}$ is a real-valued function. In the sequel we will assume that $U(t)$ satisfies conditions (4.8) and (4.9), from which we can get some Strichartz inequalities for $U(t)$ in modulation spaces (cf. [55]).

Proposition 4.5 (Strichartz inequalities). *Let* $U(t)$ *satisfy* (4.8) *and* (4.9). *For any* $\gamma \ge 2 \vee (2/\delta)$, *we have*

$$\|U(t)f\|_{\ell_{\Box}^{\alpha/2,q}(L^\gamma(\mathbb{R},L^p))} \le C\|f\|_{M_{2,q}}. \tag{4.10}$$

In addition, if $\gamma \ge q$, *then we have*

$$\|U(t)f\|_{L^\gamma(\mathbb{R},M_{p,q}^{\alpha/2})} \le C\|f\|_{M_{2,q}}. \tag{4.11}$$

Denote

$$(\mathfrak{U}f)(t) = \int_0^t U(t-s)f(s,\cdot)ds. \tag{4.12}$$

Proposition 4.6. *Let $U(t)$ satisfy (4.8) and (4.9). For any $\gamma \geq 2 \vee (2/\delta)$, we have*

$$\|\mathfrak{U}f\|_{\ell_\square^q(L^\infty(\mathbb{R},L^2))} \leq C\|f\|_{\ell_\square^{-\alpha/2,q}(L^{\gamma'}(\mathbb{R},L^{p'}))}. \tag{4.13}$$

In addition, if $\gamma' \leq q$, then

$$\|\mathfrak{U}f\|_{L^\infty(\mathbb{R},M_{2,q})} \leq C\|f\|_{L^{\gamma'}(\mathbb{R},M_{p',q}^{-\alpha/2})}. \tag{4.14}$$

Proposition 4.7. *Assume that $U(t)$ satisfies (4.8) and (4.9), $\gamma \geq \max(2/\delta, 2)$. Then we have*

$$\|\mathfrak{U}f\|_{\ell_\square^{\alpha/2,q}(L^\gamma(\mathbb{R},L^p))} \leq C\|f\|_{\ell_\square^q(L^1(\mathbb{R},L^2))}. \tag{4.15}$$

In addition, if $\gamma \geq q$, then

$$\|\mathfrak{U}f\|_{L^\gamma(\mathbb{R},M_{p,q}^{\alpha/2})} \leq C\|f\|_{L^1(\mathbb{R},M_{2,q})}. \tag{4.16}$$

The Schrödinger semi-group corresponds to the cases $\alpha = 0$, $\delta = n(1/2-1/p)$ and $2 \leq p < \infty$. Taking $q = 1$ in Propositions 4.5–4.7, we immediately have

Corollary 4.8. *Let $2 \leq p < \infty$, $\gamma \geq 2 \vee \gamma(p)$, and*

$$\frac{2}{\gamma(p)} = n\left(\frac{1}{2} - \frac{1}{p}\right). \tag{4.17}$$

Let $S(t) = e^{it\Delta}$, $\mathfrak{A} = \int_0^t S(t-s)\cdot ds$. Then

$$\|S(t)\varphi\|_{\ell_\square^1(L^\gamma(\mathbb{R},L^p))} \leq C\|\varphi\|_{M_{2,1}}, \tag{4.18}$$

$$\|\mathfrak{A}f\|_{\ell_\square^1(L^\gamma(\mathbb{R},L^p))\cap\ell_\square^1(L^\infty(\mathbb{R},L^2))} \leq C\|f\|_{\ell_\square^1(L^{\gamma'}(\mathbb{R},L^{p'}))}. \tag{4.19}$$

Similar to Corollary 4.8, we have

Corollary 4.9. *Let $2 \leq p < \infty$, $\theta \in (0,1]$, $1 \leq q < \infty$,*

$$\frac{2}{\gamma_\theta(p)} = n\theta\left(\frac{1}{2} - \frac{1}{p}\right), \quad 2\sigma = (n+2)\theta\left(\frac{1}{2} - \frac{1}{p}\right). \tag{4.20}$$

Let $G(t)$ be as in (4.4), $\mathfrak{G} = \int_0^t G(t-s)\cdot ds$. Then for any $\gamma \geq 2 \vee \gamma_\theta(p)$, we have

$$\|G(t)\varphi\|_{\ell_\square^{-\sigma,q}(L^\gamma(\mathbb{R},L^p))} \leq C\|\varphi\|_{M_{2,q}}, \tag{4.21}$$

$$\|\mathfrak{G}f\|_{\ell_\square^{-\sigma,q}(L^\gamma(\mathbb{R},L^p))\cap\ell_\square^q(L^\infty(\mathbb{R},L^2))} \leq C\|f\|_{\ell_\square^{\sigma,q}(L^{\gamma'}(\mathbb{R},L^{p'}))}. \tag{4.22}$$

Related Strichartz estimates in Wiener amalgam spaces for the Schrödinger equation were obtained by Cordero and Nicola [10].

4.3. Wellposedness for NLS and NLKG

We study the Cauchy problem for NLS and give partial answers to Question 2.3. Let us consider

$$iu_t + \Delta u = f(u), \quad u(0, x) = u_0(x). \tag{4.23}$$

Noticing that $B_{\infty,1}^n \subset M_{\infty,1} \subset B_{\infty,1}^0 \subset L^\infty$ are sharp embeddings, up to now, we can not get the wellposedness of NLS in L^∞ or in $B_{\infty,1}^0$. However, we can obtain the local wellposedness of NLS in $M_{\infty,1}$. We have

Theorem 4.10 ([2, 11]). Let $n \geq 1$, $f(u) = \lambda |u|^\kappa u$, $\kappa \in 2\mathbb{N}$, $\lambda \in \mathbb{R}$, $u_0 \in M_{p,1}$ and $1 \leq p \leq \infty$. Then there exists a $T > 0$ such that (4.23) has a unique solution $u \in C([0, T), M_{p,1})$. Moreover, if $T < \infty$, then $\limsup_{t \nearrow T} \|u(t)\|_{M_{p,1}} = \infty$.

If the nonlinearity has an exponential growth, say $f(u) = \lambda(e^{|u|^2} - 1)u$, the result in Theorem 4.10 also holds. Noticing that $B_{2,1}^{n/2} \subset M_{2,1} \subset B_{2,1}^0 \cap C(\mathbb{R}^n)$ are sharp embeddings, we can get that NLS is globally well posed in $M_{2,1}$ if initial data are sufficiently small.

Theorem 4.11 ([55]). Let $n \geq 1$, $f(u) = \lambda |u|^\kappa u$, $\kappa \in 2\mathbb{N}$, $\lambda \in \mathbb{R}$, $\kappa \geq 4/n$, $u_0 \in M_{2,1}$ and there exists a sufficiently small $\delta > 0$ such that $\|u_0\|_{M_{2,1}} \leq \delta$. Then (4.23) has a unique solution

$$u \in C(\mathbb{R}, M_{2,1}) \cap \ell_\square^1(L_{x,t\in\mathbb{R}}^p), \tag{4.24}$$

where $p \in [2 + 4/n, 2 + \kappa] \cap \mathbb{N}$, $\ell_\square^1(L_{x,t\in\mathbb{R}}^p)$ is as in (4.7).

Theorem 4.12 ([55]). Let $n \geq 2$, $f(u) = \lambda(e^{\varrho|u|^2} - 1)u$, $\lambda \in \mathbb{C}$ and $\varrho > 0$. Assume that $u_0 \in M_{2,1}$ and there exists a sufficiently small $\delta > 0$ such that $\|u_0\|_{M_{2,1}} \leq \delta$. Then (4.23) has a unique solution

$$u \in C(\mathbb{R}, M_{2,1}) \cap \ell_\square^1(L_{x,t\in\mathbb{R}}^4). \tag{4.25}$$

We now consider the initial value problem for NLKG,

$$u_{tt} + (I - \Delta)u + f(u) = 0, \quad u(0) = u_0, \, u_t(0) = u_1. \tag{4.26}$$

Analogous to NLS, we have

Theorem 4.13 ([2, 11]). Let $n \geq 1$, $f(u) = \lambda |u|^\kappa u$, $\kappa \in 2\mathbb{N}$, $\lambda \in \mathbb{R}$, $(u_0, u_1) \in M_{p,1} \times M_{p,1}^{-1}$ and $1 \leq p \leq \infty$. Then there exists a $T > 0$ such that (4.23) has a unique solution $(u, u_t) \in C([0, T), M_{p,1}) \times C([0, T), M_{p,1}^{-1})$. Moreover, if $T < \infty$, then $\limsup_{t \nearrow T}(\|u(t)\|_{M_{p,1}} + \|u_t(t)\|_{M_{p,1}^{-1}}) = \infty$.

If the nonlinearity has an exponential growth, the corresponding results as in Theorem 4.13 also hold.

Theorem 4.14 ([55]). Let $n \geq 1$, $f(u) = \lambda u^{1+\kappa}$, $\kappa \in \mathbb{N}$ and $\kappa \geq 4/n$. Put

$$\sigma = \frac{n+2}{n(2+\kappa)}. \tag{4.27}$$

Assume that $(u_0, u_1) \in M_{2,1}^\sigma \times M_{2,1}^{\sigma-1}$ and there exists a sufficiently small $\delta > 0$ such that $\|u_0\|_{M_{2,1}^\sigma} + \|u_1\|_{M_{2,1}^{\sigma-1}} \leq \delta$. Then (4.26) has a unique solution

$$u \in C(\mathbb{R}, M_{2,1}^\sigma) \cap \ell_\square^1(L_{x,t\in\mathbb{R}}^{2+\kappa}). \tag{4.28}$$

Theorem 4.15 ([55]). *Let $n \geq 2$, $f(u) = \sinh u - u$ and $\sigma = (n+2)/4n$. Assume that $(u_0, u_1) \in M_{2,1}^\sigma \times M_{2,1}^{\sigma-1}$ and there exists a sufficiently small $\delta > 0$ such that $\|u_0\|_{M_{2,1}^\sigma} + \|u_1\|_{M_{2,1}^{\sigma-1}} \leq \delta$. Then (4.26) has a unique solution*

$$u \in C(\mathbb{R}, M_{2,1}^\sigma) \cap \ell_\square^1(L_{x,t\in\mathbb{R}}^4). \tag{4.29}$$

5. Derivative nonlinear Schrödinger equations

We study the initial value problem for the derivative nonlinear Schrödinger equation (gDNLS)

$$iu_t + \Delta_\pm u = F(u, \bar{u}, \nabla u, \nabla \bar{u}), \quad u(0,x) = u_0(x), \tag{5.1}$$

where u is a complex-valued function of $(t,x) \in \mathbb{R} \times \mathbb{R}^n$,

$$\Delta_\pm u = \sum_{i=1}^n \varepsilon_i \partial_{x_i}^2, \quad \varepsilon_i \in \{1, -1\}, \quad i = 1, \ldots, n, \tag{5.2}$$

$\nabla = (\partial_{x_1}, \ldots, \partial_{x_n})$, $F : \mathbb{C}^{2n+2} \to \mathbb{C}$ is a series of $z \in \mathbb{C}^{2n+2}$,

$$F(z) = F(z_1, \ldots, z_{2n+2}) = \sum_{3 \leq |\beta| < \infty} c_\beta z^\beta, \quad c_\beta \in \mathbb{C}, \tag{5.3}$$

$|c_\beta| \leq C^{|\beta|}$. The typical nonlinear term is

$$F(u, \bar{u}, \nabla u, \nabla \bar{u}) = |u|^2 \vec{\lambda} \cdot \nabla u + u^2 \vec{\mu} \cdot \nabla \bar{u} + |u|^2 u,$$

see [9, 14, 51]. Another model is

$$F(u, \bar{u}, \nabla u, \nabla \bar{u}) = (1 + |u|^2)^{-1}(\nabla u)^2 \bar{u} = \sum_{k=0}^\infty (-1)^k |u|^{2k}(\nabla u)^2 \bar{u}, \quad |u| < 1,$$

which is an equivalent version of the Schrödinger flow [20]. The non-elliptic gDNLS arises in the strongly interacting many-body systems near the criticality, where anisotropic interactions are manifested by the presence of the non-elliptic case, as well as additional residual terms which involve cross derivatives of the independent variables [9, 14, 51]. Some water wave and completely integrable system models in higher spatial dimensions are also non-elliptic, cf. [1, 58, 59]. A large amount of work has been devoted to the study of gDNLS, see [19, 20, 22, 23, 24, 25, 28, 36, 42, 43, 45].

Since the nonlinearity in gDNLS contains derivative terms and the Strichartz inequalities can not absorb any derivatives, gDNLS can not be solved if we use only the Strichartz estimate. One needs to look for some other ways to handle the derivative terms in the nonlinearity. Up to now, three kinds of methods seem to be very useful for gDNLS. One is to use the energy estimate to deal with the derivatives in the nonlinearity, the second way is to use Bourgain's space $X^{s,b}$ and the third technique is Kato's smooth effect estimates. Of course, there are some connections between these methods.

In this survey paper we only discuss the smooth effect estimates together with the frequency-uniform decomposition techniques for gDNLS and we show that it is globally well posed and scattering in a class of modulation spaces.

For convenience, we denote

$$S(t) = e^{it\Delta_\pm} = \mathfrak{F}^{-1} e^{it\sum_{j=1}^n \varepsilon_j \xi_j^2} \mathfrak{F}, \quad \mathfrak{A}f(t,x) = \int_0^t S(t-\tau)f(\tau,x)d\tau.$$

We now state a global wellposedness and scattering result for gDNLS in modulation spaces. We denote by $L_{x_i}^{p_1} L_{(x_j)_{j\neq i}}^{p_2} L_t^{p_2} := L_{x_i}^{p_1} L_{(x_j)_{j\neq i}}^{p_2} L_t^{p_2}(\mathbb{R}^{1+n})$ the anisotropic Lebesgue space for which the norm is defined by

$$\|f\|_{L_{x_i}^{p_1} L_{(x_j)_{j\neq i}}^{p_2} L_t^{p_2}} = \left\| \|f\|_{L_{x_1,\ldots,x_{j-1},x_{j+1},\ldots,x_n}^{p_2} L_t^{p_2}(\mathbb{R}\times\mathbb{R}^{n-1})} \right\|_{L_{x_i}^{p_1}(\mathbb{R})}. \tag{5.4}$$

For $k = (k_1,\ldots,k_n)$, we write

$$\|u\|_{X_\alpha^s} = \sum_{i,\ell=1}^n \sum_{k\in\mathbb{Z}^n, |k_i|>10\vee\max_{j\neq i}|k_j|} \langle k_i \rangle^{s-1/2} \left\| \partial_{x_\ell}^\alpha \Box_k u \right\|_{L_{x_i}^\infty L_{(x_j)_{j\neq i}}^2 L_t^2}$$

$$+ \sum_{i,\ell=1}^n \sum_{k\in\mathbb{Z}^n} \left\| \partial_{x_\ell}^\alpha \Box_k u \right\|_{L_{x_i}^2 L_{(x_j)_{j\neq i}}^\infty L_t^\infty}, \tag{5.5}$$

$$\|u\|_{S_\alpha^s} = \sum_{\ell=1}^n \sum_{k\in\mathbb{Z}^n} \langle k \rangle^{s-1} \left\| \partial_{x_\ell}^\alpha \Box_k u \right\|_{L_t^\infty L_x^2 \cap L_t^3 L_x^6}, \tag{5.6}$$

$$\|u\|_{X^s} = \sum_{\alpha=0,1} \|u\|_{X_\alpha^s}, \quad \|u\|_{S^s} = \sum_{\alpha=0,1} \|u\|_{S_\alpha^s}. \tag{5.7}$$

Theorem 5.1 ([53, 52]). *Let $n \geq 3$, $u_0 \in M_{2,1}^{3/2}$ and there exists a suitably small $\delta > 0$ such that $\|u_0\|_{M_{2,1}^{3/2}} \leq \delta$. Then (5.1) has a unique solution $u \in C(\mathbb{R}, M_{2,1}^{3/2}) \cap X^{3/2} \cap S^{3/2}$, $\|u\|_{X^{3/2}\cap S^{3/2}} \leq C\delta$. Moreover, the scattering operator S of (5.1) carries a whole zero neighborhood in $C(\mathbb{R}, M_{2,1}^{3/2})$ into $C(\mathbb{R}, M_{2,1}^{3/2})$.*

Remark 5.2. Recently, this technique was also developed for the Navier-Stokes equation and the dissipative nonlinear electrohydrodynamic system [56, 21, 13].

6. Canonical transformations

It is interesting to generalize results in previous two sections to the case of dispersive operators $a(D)$ instead of $-\Delta$. Recently Ruzhansky and Sugimoto have introduced a new idea to establish fundamental estimates for dispersive equations based on the idea of canonical transformations, and this attempt is quite successful for smoothing estimates ([41]).

Let $\psi : \mathbb{R}^n \setminus 0 \to \mathbb{R}^n \setminus 0$ be C^∞-maps satisfying $\psi(\lambda\xi) = \lambda\psi(\xi)$ for all $\lambda > 0$ and $\xi \in \mathbb{R}^n \setminus 0$, and let

$$I_\psi u(x) = \mathfrak{F}^{-1}[(\mathfrak{F}u)(\psi(\xi))](x)$$
$$= (2\pi)^{-n} \int_{\mathbb{R}^n} \int_{\mathbb{R}^n} e^{i(x\cdot\xi - y\cdot\psi(\xi))} u(y)\,dy d\xi. \tag{6.1}$$

We remark that we have the formula

$$a(D) \cdot I_\psi = I_\psi \cdot \sigma(D), \quad a(\xi) = (\sigma \circ \psi)(\xi). \tag{6.2}$$

For example, for a positive function $a(\xi)$ satisfying $a(\lambda\xi) = \lambda^2 a(\xi)$ for all $\lambda > 0$ and $\xi \in \mathbb{R}^n \setminus 0$, we have

$$a(D) \cdot I_\psi = I_\psi \cdot (-\Delta)$$

if we take

$$\sigma(\eta) = |\eta|^2, \quad \psi(\xi) = \sqrt{a(\xi)}\frac{\nabla a(\xi)}{|\nabla a(\xi)|},$$

provided that $\nabla a(\xi) \neq 0$ for $\xi \neq 0$. The latter is achieved if we assume that the Gaussian curvature of the hypersurface $\Sigma = \{\xi : a(\xi) = 1\}$ never vanishes, in which case also I_ψ has the inverse $I_{\psi^{-1}} = I_\psi^{-1}$ because the Gauss map $\Sigma \ni \xi \mapsto \frac{\nabla a(\xi)}{|\nabla a(\xi)|} \in \mathbb{S}^{n-1}$ is a diffeomorphism.

Thus we can induce the same estimate for $a(D)$ from the estimates for $-\Delta$ if we establish the boundedness of operators I_ψ and $I_{\psi^{-1}}$ on modulation spaces. Ruzhansky, Sugimoto, Toft and Tomita [40] discuss such boundedness properties, and we have a positive result for the local boundedness. Let L_s^q be the space of functions such that $\langle x \rangle^s f \in L^q$:

Theorem 6.1. *Let $s \in \mathbb{R}$, $1 \leq p, q \leq \infty$, and let $\psi : \mathbb{R}^n \to \mathbb{R}^n$ be such that the pullback $\psi^* : f \mapsto f \circ \psi$ is bounded on $L_s^q(\mathbb{R}^n)$. Then I_ψ is locally bounded on $M_{p,q}^s(\mathbb{R}^n)$.*

By Theorem 6.1 and Propositions 4.1, 4.2, we straightforwardly obtain estimates (4.1) and (4.2) for $S(t) = \chi(x)e^{-ita(D)}\tilde{\chi}(x)$ (and we have to also restrict the case to $1 \leq q \leq \infty$) form the estimates for $S(t) = e^{it\Delta}$, where $\chi, \tilde{\chi}$ are cut-off functions.

As for the global boundedness of I_φ, we have unfortunately a negative result:

Theorem 6.2. *Let $1 \leq p, q \leq \infty$, $2 \neq p < \infty$, and let $\psi : \mathbb{R}^n \to \mathbb{R}^n$ be a C^1-function. Assume that operator I_ψ is bounded on $M_{p,q}(\mathbb{R}^n)$. Then ψ is an affine mapping.*

7. Open questions

It is known that the algebra property of function spaces is of importance for PDE. Up to now the following question is not clear for us:

Question 7.1. Let $\alpha \in (0, \infty)$. Does

$$\||u|^\alpha u\|_{M_{p,1}} \leq C\|u\|_{M_{p,1}}^{\alpha+1}$$

hold for all $u \in M_{p,1}$?

It is known that if $\alpha \in 2\mathbb{N}$, the answer is affirmative. If α is not an even integer, the question seems very difficult, cf. [47].

The global well posedness of NLS in modulation spaces for large initial data seems open.

Question 7.2. Can we show that NLS (2.5) is global well posed if the initial data $u_0 \in M_{2,1}$ is large?

We should also discuss non-affine transforms which induce the globally bounded canonical transformations. Note that such transforms must not be C^1-mappings in view of Theorem 6.2.

Question 7.3. Can we show the global boundedness of the operator I_ψ in (6.1) on modulation spaces for a homogeneous change of variables ψ?

Some partial answers to this question appeared in [40].

References

[1] M.J. Ablowitz, R. Haberman, Nonlinear evolution equations in two and three dimensions, Phys. Rev. Lett., **35** (1975), 1185–1188.

[2] A. Bényi, K.A. Okoudjou, Local well-posedness of nonlinear dispersive equations on modulation spaces, Bull. London Math. Soc., **41** (2009), 549–558.

[3] A. Bényi, K. Gröchenig, K.A. Okoudjou and L.G. Rogers, Unimodular Fourier multiplier for modulation spaces, J. Funct. Anal., **246** (2007), 366–384.

[4] J. Bergh and J. Löfström, *Interpolation Spaces*, Springer-Verlag, 1976.

[5] T. Cazenave and F.B. Weissler, The Cauchy problem for the critical nonlinear Schrödinger equation in H^s, Nonlinear Anal. TMA, **14** (1990), 807–836.

[6] T. Cazenave, *Semilinear Schrödinger equations*, Courant Lecture Notes in Mathematics, Vol. **10**, 2003.

[7] J.C. Chen, D.S. Fan and L. Sun, Asymptotic estimates For unimodular Fourier multipliers on modulation spaces, preprint.

[8] J. Colliander, *Nonlinear Schrödinger equations*, Lecture Notes, 2004.

[9] P.A. Clarkson and J.A. Tuszyriski, Exact solutions of the multidimensional derivative nonlinear Schrödinger equation for many-body systems near criticality, J. Phys. A: Math. Gen. **23** (1990), 4269–4288.

[10] E. Cordero, F. Nicola, Some new Strichartz estimates for the Schrödinger equation. J. Differential Equations, **245** (2008), 1945–1974.

[11] E. Cordero, F. Nicola, Remarks on Fourier multipliers and applications to the wave equation, J. Math. Anal. Appl., **353** (2009), 583–591.

[12] E. Cordero, F. Nicola, Sharpness of some properties of Wiener amalgam and modulation spaces. Bull. Aust. Math. Soc., **80** (2009), 105–116.

[13] C. Deng, J.H. Zhao, S.B. Cui, Well-posedness of a dissipative nonlinear electrohydro-dynamic system in modulation spaces, Nonlinear Anal., TMA, **73** (2010), 2088–2100.

[14] J.M. Dixon and J.A. Tuszynski, Coherent structures in strongly interacting many-body systems: II, Classical solutions and quantum fluctuations, J. Phys. A: Math. Gen., **22** (1989), 4895–4920.

[15] H.G. Feichtinger, Modulation spaces on locally compact Abelian group, Technical Report, University of Vienna, 1983. Published in: "Proc. Internat. Conf. on Wavelet and Applications", 99–140. New Delhi Allied Publishers, India, 2003.

[16] L. Grafakos, *Classical and modern Fourier analysis*, Pearson/Prentice Hall, 2004.

[17] P. Gröbner, *Banachräume Glatter Funktionen und Zerlegungsmethoden*, Doctoral thesis, University of Vienna, 1992.

[18] K. Gröchenig, *Foundations of Time-Frequency Analysis*, Birkhäuser, Boston, MA, 2001.

[19] N. Hayashi and T. Ozawa, Finite energy solutions of nonlinear Schrödinger equations of derivative type, SIAM J. Math. Anal., **25** (1994), 1488–1503.

[20] A. Ionescu and C.E. Kenig, Low-regularity Schrödinger maps, II: Global well-posedness in dimensions $d \geq 3$, Commun. Math. Phys., **271** (2007), 523–559.

[21] T. Iwabuchi, Navier-Stokes equations and nonlinear heat equations in modulation spaces with negative derivative indices. J. Differential Equations, **248** (2010), 1972–2002.

[22] C.E. Kenig, G. Ponce, C. Rolvent, L. Vega, The general quasilinear ultrahyperbolic Schrodinger equation, Adv. Math., **206** (2006), 402–433.

[23] C.E. Kenig, G. Ponce, L. Vega, Small solutions to nonlinear Schrödinger equation, Ann. Inst. Henri Poincaré, Sect C, **10** (1993), 255–288.

[24] C.E. Kenig, G. Ponce and L. Vega, Smoothing effects and local existence theory for the generalized nonlinear Schrödinger equations, Invent. Math., **134** (1998), 489–545.

[25] S. Klainerman, Long-time behavior of solutions to nonlinear evolution equations, Arch. Rational Mech. Anal., **78** (1982), 73–98.

[26] S. Klainerman and M. Machedon, Space-time estimates for null forms and the local existence theorem, Comm. Pure Appl. Math., **46** (1993), no. 9, 1221–1268.

[27] S. Klainerman and M. Machedon, Smoothing estimates for null forms and applications, Duke Math. J., **81** (1995), 99–133.

[28] S. Klainerman, G. Ponce, Global small amplitude solutions to nonlinear evolution equations, Commun. Pure Appl. Math., **36** (1983), 133–141.

[29] M. Kobayashi, Modulation spaces $M^{p,q}$ for $0 < p, q \leq \infty$, J. of Funct. Spaces and Appl., **4** (2006), no.3, 329–341.

[30] M. Kobayashi, Dual of modulation spaces, J. of Funct. Spaces and Appl., **5** (2007), 1–8.

[31] M. Kobayashi, Y. Sawano, Molecular decomposition of the modulation spaces, Osaka J. Math., **47** (2010), 1029–1053.

[32] M. Kobayashi, M. Sugimoto, The inclusion relation between Sobolev and modulation spaces, J. Funct. Anal. **260** (2011), 3189–3208.

[33] F. Linares and G. Ponce, On the Davey–Stewartson systems, Ann. Inst. H. Poincaré, Anal. Non Linéaire, **10** (1993), 523–548.

[34] A. Miyachi, F. Nicola, S. Riveti, A. Taracco and N. Tomita, Estimates for unimodular Fourier multipliers on modulation spaces, Proc. Amer. Math. Soc., **137** (2009), 3869–3883.

[35] T. Ozawa, On the nonlinear Schrödinger equations of derivative type, Indiana Univ. Math. J., **45** (1996), 137–163.

[36] T. Ozawa and J. Zhai, Global existence of small classical solutions to nonlinear Schrödinger equations, Ann. I. H. Poincaré, Anal. Non Linéaire, **25** (2008), 303–311.

[37] M. Ruzhansky, On the sharpness of Seeger-Sogge-Stein orders, Hokkaido Math. J., **28** (1999), 357–362.

[38] M.V. Ruzhansky, Singularities of affine fibrations in the regularity theory of Fourier integral operators, Russian Math. Surveys, **55** (2000), 93–161.

[39] M. Ruzhansky, J. Smith, *Dispersive and Strichartz estimates for hyperbolic equations with constant coefficients*, MSJ Memoirs, **22**, Mathematical Society of Japan, Tokyo, 2010.

[40] M. Ruzhansky, M. Sugimoto, J. Toft and N. Tomita *Changes of variables in modulation and Wiener amalgam spaces*, Math. Nachr., **284** (2011), 2078–2092.

[41] M. Ruzhansky and M. Sugimoto, *Smoothing properties of evolution equations via canonical transforms and comparison principle*, Proc. London Math. Soc. (2012), doi: 10.1112/plms/pds006.

[42] M. Ruzhansky and M. Sugimoto, *Structural resolvent estimates and derivative nonlinear Schrödinger equations*, arXiv:1101.5026v1, to appear in Comm. Math. Phys.

[43] J. Shatah, Global existence of small classical solutions to nonlinear evolution equations, J. Differential Equations, **46** (1982), 409–423.

[44] J. Shatah, Normal forms and quadratic nonlinear Klein–Gordon equations, Comm. Pure Appl. Math., **38** (1985), 685–696.

[45] M. Sugimoto, M. Ruzhansky, A smoothing property of Schrödinger equations and a global existence result for derivative nonlinear equations, Advances in analysis, 315–320, World Sci. Publ., Hackensack, NJ, 2005.

[46] M. Sugimoto amd N. Tomita, The dilation property of modulation spaces and their inclusion relation with Besov spaces, J. Funct. Anal., **248** (2007), 79–106.

[47] M. Sugimoto; N. Tomita; B. X. Wang, Remarks on nonlinear operations on modulation spaces, Integral Transforms and Special Functions, **22** (2011), 351–358

[48] J. Toft, Continuity properties for modulation spaces, with applications to pseudodifferential calculus, I, J. Funct. Anal., **207** (2004), 399–429.

[49] H. Triebel, *Theory of Function Spaces,* Birkhäuser Verlag, 1983.

[50] H. Triebel, Modulation spaces on the Euclidean *n*-spaces, Z. Anal. Anwendungen, **2** (1983), 443–457.

[51] J.A. Tuszynski and J.M. Dixon, Coherent structures in strongly interacting manybody systems: I, Derivation of dynamics, J. Phys. A: Math. Gen., **22** (1989), 4877–4894.

[52] B.X. Wang, Sharp global well-posedness for non-elliptic derivative Schrödinger equations with small rough data, arXiv:1012.0370.

[53] B.X. Wang, L.J. Han, C.Y. Huang, Global well-Posedness and scattering for the derivative nonlinear Schrödinger equation with small rough data, Ann. I. H. Poincaré, Anal. Non Linéaire, **26** (2009), 2253–2281.

[54] B.X. Wang and C.Y. Huang, Frequency-uniform decomposition method for the generalized BO, KdV and NLS equations, J. Differential Equations, **239** (2007), 213–250.

[55] B.X. Wang and H. Hudzik, The global Cauchy problem for the NLS and NLKG with small rough data, J. Differential Equations, **231** (2007), 36–73.

[56] B.X. Wang, L.F. Zhao, B.L. Guo, Isometric decomposition operators, function spaces $E_{p,q}^\lambda$ and their applications to nonlinear evolution equations, J. Funct. Anal., **233** (2006), 1–39.

[57] N. Wiener, Tauberian theorems, Ann. of Math., **33** (1932), 1–100.

[58] V.E. Zakharov, E.A. Kuznetson, Multi-scale expansions in the theory of systems integrable by inverse scattering method, Physica D, **18** (1986), 455–463.

[59] V.E. Zakharov, E.I. Schulman, Degenerated dispersion laws, motion invariant and kinetic equations, Physica D, **1** (1980), 185–250.

Michael Ruzhansky
Department of Mathematics
Imperial College London
180 Queens Gate
London SW7 2AZ, UK
e-mail: m.ruzhansky@imperial.ac.uk

Mitsuru Sugimoto
Graduate School of Mathematics
Nagoya University
Furocho, Chikusa-ku
Nagoya 464-8602, Japan
e-mail: sugimoto@math.nagoya-u.ac.jp

Baoxiang Wang
LMAM
School of Mathematical Sciences
Peking University
Beijing 100871, PR of China
e-mail: wbx@math.pku.edu.cn

Progress in Mathematics, Vol. 301, 285–300

An Optimal Control Problem for a Nonlinear Hyperbolic Equation with an Infinite Time Horizon

Simon Serovajsky and Kanat Shakenov

Abstract. An optimization control problem for a nonlinear hyperbolic equation with non-smooth nonlinearity and infinite time horizon without global solvability of the boundary problem is considered. This problem is solved using an approximation. The convergence of the approximation is proved. Necessary conditions of optimality are obtained.

Mathematics Subject Classification. 49K20.

Keywords. Optimization, hyperbolic equation, singularity, non-smoothness, infinite time horizon, approximation.

1. Introduction

We consider an optimization control problem for a system described by a nonlinear hyperbolic equation. The existence and uniqueness of the boundary problem is not guaranteed for arbitrary control data. Furthermore, the nonlinear term of the equation is non-smooth and the problem is considered on an infinite time interval.

Optimization methods for systems described by nonlinear parabolic and elliptic equations are well known. There exist a lot of results for control systems characterized by Goursat–Darboux problems. Some results for optimization problems for usual boundary problems are obtained by Matveev and Yakubovich [1], Tiba [2], Fursikov [3]. They prove the existence of the optimal control and necessary conditions for optimality in the following case. The boundary problem has a unique solution for all admissible control data, nonlinear terms are smooth, and the time interval is finite. Relaxation methods for these problems are used by Tiba [4] and Sumin [5]. Banks and Kunisch [6] apply numerical methods for its solution. Kuliev and Gasanov (see [7]) consider optimization problems for nonlinear hyperbolic equations with a control in coefficients and with state constraints. Optimization methods for systems described by regular boundary problems for nonlinear hyperbolic equations under smoothness assumptions and with finite time horizon are well known.

Singular control systems may be not solvable or may have non-unique solutions for admissible control data. The use of variational methods or gradient methods for functional minimization are a matter of serious difficulty in this situation. The control is the primary object, and the state function is second for standard optimization methods. It is determined by the state equation for the given value of the control. However, the unique solvability of the problem can be violated when varying the control in the singular case. Then the control and the state function should be interpreted as an equal in rights pair. The state equation is interpreted as a constraint in this situation. The cost functional is minimized here on the admissible set of pairs, that is the set of control-state pairs such that the state equation holds true. This conditional extremal problem can be solved by means of the infinite-dimensional Lagrange multipliers method (see Fursikov [3]) or penalty method (see Lions [8]). Optimization problems for nonlinear hyperbolic equations are considered for the singular case in these papers. However the considered systems are smooth and the corresponding time interval is finite.

Two types of non-smooth optimization problems are known. Non-smooth terms can be included either in the cost functional or in the state equation. If the state operator is smooth and there is non-smoothness only in the functional the problem can be solved using nonsmooth analysis methods (see, for example, Rockafellar [9] and Clarke [10]). The classical derivatives (Gataux, Fréchet, some other) can be replaced by its non-smooth extension, for example, sub-gradient or Clarke derivative. Using these methods for problems with non-smooth terms in the equation is very difficult because of the absence of the effective non-smooth analogues of the inverse function theorem and the implicit function theorem. They are used for proving the differentiability of the control-state mapping. However, such optimization problems can be solved by means of smooth approximation of the state equation (see Barbu [11]). This idea is used for nonlinear singular elliptic equation in [12].

The additional difficulty of our optimization problem is the non-compactness of the time interval. Optimization problems with infinite time horizon are well known for systems described by ordinary differential equations (see, for example, Seierstad [13] or Aseev and Kryazhimskiy [14]). The analogous problems for distributed systems are seldom considered. However, we note the result of Lions [15] for systems described by linear parabolic equations. He proposes the approximation of the initial system by the analogous system on the finite time interval. The conditions of optimality for the given problem are obtained after passing to the limit in the necessary conditions of optimality for the approximate problem. But the linearity of the system is used substantially in this case. Optimization problems for nonlinear parabolic equations with infinite time horizon are considered by [16] and Cannarsa and Da Prato [17]. However they solve only feedback problems by means of Hamilton–Jacobi equations.

Optimization problems for nonlinear hyperbolic equations without smoothness and regularity, and with infinite time horizon are not solved yet. We will use some ideas and technical methods for its resolution.

Our system is singular. So we will interpret the state equation as a constraint. The control data and the state function are equal in rights in this situation (see [3] and [8]). The considered problem can be solved using Lagrange multipliers method or a penalty method. These methods are equivalent for the optimization problems of [3] and [8]. However we have additional difficulties because of non-smoothness and non-compactness of the time interval. So we prefer to use the Penalty method because it is an approximation method as opposed to the Lagrange multipliers one.

The peculiarity of our problem is the existence of non-smooth terms in the state equation but not in the cost functional. This difficulty will be overcome by means of smooth approximations of the equation. This idea was used in [11], [18] for regular systems. It will be natural to use two forms of approximation (penalty method and smooth approximation) simultaneously. The approximation of the non-smooth term will be realized in the penalty functional. The corresponding smooth penalty approximation method was used in [12] for an optimization problem for a singular elliptic equation with non-smooth nonlinearity. However, it is not sufficient for obtaining the effective results for the system with infinite time horizon.

We know that an optimization problem for linear parabolic equations with infinite time horizon was solved in [15] by means of the finite time approximation method. The corresponding approximate optimization problem has a finite time interval. We propose to use this idea in our problem. The approximation will be realized here at two stages. At first we will use finite time approximation. The obtained optimization problem will be solved with using of the smooth penalty approximation method.

In known results based on approximation techniques (penalty method [8], smooth approximation method [11], and finite time approximation method [15]) necessary conditions of optimality for the initial problem are obtained by passing to the limit in the optimality conditions for the approximate problems. Here we have several difficulties. The high order of difficulty of the given problem does not allow to obtain an analogous statement. But this peculiarity is not an obstacle for solving the problem.

We will find an approximate solution of the problem, it is necessary to define exactly the notion of approximate solution. It will be best to find an admissible control, which is close enough to the optimal one. However, it is only possible for simple enough optimization problems to obtain this form of approximate solution. There is another notion of approximate solution, which is often used. For weak approximate solutions the aim is to find an admissible control where the value of the minimizing functional is close enough to its minimum on the admissible control set. These two forms of approximate solutions are equivalent for optimization problems well posed in the sense of Tikhonov [19]. But the values of the functionals can be close for controls which are not close if the optimization problem is ill posed.

It is known (see [20]), that the majority of optimization problems are ill posed. So the weak approximate solution is used in the practical solution of optimization problems as a rule.

Unfortunately, it may be hard to find a weak approximate solution if the problem is very difficult. We will define a weaker approximate solution. Both, strong and weak approximate solutions are admissible controls. We determine the weaker approximate solution as a control, which is close enough to some admissible control, and the corresponding value of the minimizing functional, which is close enough to its minimum on the admissible control set. This object is weaker because we permit the realization of the given constraint with some small error and do not require it exactly. So the class of solvable problems is extended by weakening the requirement of the approximate solution. The analogical idea was be realized in [21] for an optimization problem for a singular elliptic equation. Our problem is difficult enough. So we will try to find its weaker approximate solution.

There are different methods of practical solution of optimization problems. The first class includes direct methods. The practical algorithm is determined by the problem statement directly. This is true for example for gradient methods (see [22]).

However, using direct practical methods can be very hard for a difficult problem. The given problem may be transformed to another form (condition of optimality) in this situation and the obtained problem may be easier for using numerical methods. The methods of the second class are realized at two steps, obtaining of the optimality conditions and its immediate resolution. However, using the optimality conditions can be very hard too for very difficult optimization problems. In these situations we can approximate the initial problem. So we obtain a third class of practical optimization methods, which are realized in three steps. At first we approximate the given problem. Then we obtain the conditions of optimality for the approximate problem. The last step is the resolution of optimality conditions for the approximate problem. Our problem is very difficult. So we will use a method of the third class.

2. Statement of the problem

Let Ω be an bounded open n-dimensional set with smooth boundary S and let $Q = \Omega \times (0, \infty)$ and $\Sigma = S \times (0, \infty)$. The state function $y = y(x, t)$ is the solution of the initial boundary value problem

$$y'' - \Delta y + f(y) = v, \qquad (x, t) \in Q, \tag{2.1}$$

$$y = 0, \qquad (x, t) \in \Sigma, \tag{2.2}$$

$$y(x, 0) = \varphi(x), \quad y'(x, 0) = \psi(x), \qquad x \in \Omega. \tag{2.3}$$

The functions in the right side of the equations (2.3) are known. They satisfy the inclusions

$$\varphi \in H_0^1(\Omega), \quad \psi \in L_2(\Omega).$$

The function f belongs to the set

$$F = \left\{ f \in C(\mathbb{R}) : \ |f(\eta)| \le c|\eta|^3 \ \forall \eta \right\},$$

where $c > 0$.

The function $v = v(x, t)$ is the control. It is an element of the set

$$V = \left\{ x \in L_2(Q) : \ v(x, t) \in G(x), \ (x, t) \in Q \right\},$$

where $G(x)$ is closed and convex for all x and $0 \in G(x)$. The solution of the boundary problem (2.1)–(2.3) will be found from the space

$$Y = \left\{ y : \ y \in L_\infty(0, \infty; H_0^1), \ y' \in L_\infty(0, \infty; L_2) \right\}.$$

It is important that we cannot guarantee the existence of the solution of this problem for arbitrary control (see [8]). However, we can determine the set U of admissible pairs for the system (2.1)–(2.3) (see [3] and [8]).

Definition 1. *The pair (v, y) from the set $V \times Y$ is called* **admissible***, if it satisfies the equations (2.1)–(2.3).*

The state functional is determine by

$$I(v, y) = \frac{1}{6}\|y - z\|_{L_6(Q)}^6 + \frac{\alpha}{2}\|v\|_{L_2(Q)}^2,$$

where $\alpha > 0$ and z is a given function from the space $L_6(Q)$. We consider the following optimization problem.

Problem P. *Find an admissible pair (v, y) that minimizes the state functional I on the set U.*

The existence of its solution is guaranteed by the following result.

Theorem 1. *If the set U is nonempty, then the Problem **P** is solvable.*

Proof. The functional I is bounded from below. Therefore, there exists a minimizing sequence $\{u_n\}$ for this problem. Let $u_n = (v_n, y_n)$, where $v_n \in V, \ y_n \in Y$. It satisfies the equations

$$y_n'' - \Delta y_n + f(y_n) = v_n, \qquad (x, t) \in Q, \tag{2.4}$$

$$y_n = 0, \qquad (x, t) \in \Sigma, \tag{2.5}$$

$$y_n(x, 0) = \varphi(x), \qquad y_n'(x, 0) = \psi(x), \quad x \in \Omega. \tag{2.6}$$

Furthermore, we have the convergence

$$I(u_n) \to \inf_{u \in U} I(u). \tag{2.7}$$

The sequences $\{v_n\}$ and $\{y_n\}$ are bounded in the spaces $L_2(Q)$ and $L_6(Q)$ because of the coercitivity of the functional. Therefore, the sequence

$$f_n = v_n - f(y_n)$$

is bounded in $L_2(Q)$. The function y_n is the solution of the equation

$$y_n'' - \Delta y_n = f_n, \quad (x, t) \in Q$$

with boundary conditions (2.5), (2.6). By the classical theory of linear hyperbolic equations (see, for example, [15], Chapter 4) it follows that the sequence $\{y_n\}$ is bounded in the space Y. Choosing a suitable subsequence, we obtain the convergence $v_n \to v$ weakly in $L_2(Q)$ and $y_n \to y$ weakly in Y, in particular $v \in V$. So $y_n \to y$ strongly in $L_2(Q)$ and a.e. in Q by compactness of the embedding from Y into $L_2(Q)$ (see [23], Chapter 1, Theorem 5.1). Thus, $f(y_n) \to f(y)$ a.e. in Q. We get the convergence $f(y_n) \to f(y)$ weakly in $L_2(Q)$ (see [23], Chapter 1, Lemma 1.3). We pass to the limit in the equations (2.4)–(2.6). Then the function y satisfies (2.1)–(2.3). Hence, for $u = (v, y)$. we obtain $u \in U$.

Powers of the norms of the considered spaces are lower semicontinuous. So we get the inequality

$$I(u) \le \liminf I(u_n).$$

Using (2.7) we obtain, that the pair u is a solution of the Problem **P**. □

Our next step is a solving this problem.

3. Finite time approximation

We will find an approximate solution of the given problem. If the control space is normed, then it is naturally to define an approximate solution as an element of the admissible control set U such that

$$\|u - u_0\| \le \varepsilon$$

with small enough $\varepsilon > 0$, where u_0 is an exact solution of the given problem.

Unfortunately finding such an approximate solution can be very hard for difficult enough optimization problem. So one often defines a weak approximate solution as a control $u \in U$, that satisfies the inequality

$$\left| I(u) - \inf_{u \in U} I(u) \right| \le \varepsilon$$

for small enough $\varepsilon > 0$. In reality it is sufficient that

$$I(u) \le \inf_{u \in U} I(u) + \varepsilon,$$

because the value of the functional in the admissible control cannot be less than its lower bound. The closeness of functional values is a corollary of the closeness of controls if the functional is continuous. Then this approximate solution is weaker than first one. Obviously these notions are equivalent if the optimization problem is well posed in the sense of Tikhonov [19]. But the class of ill-posed optimization problem is much larger. Thus, usually only weak approximate solutions will be found practical optimization problems. Unfortunately, finding a weak approximate solution can be hard too if the optimization problem is very difficult. Then it is necessary to try to determine weaker forms of approximate solutions. Both of the above-defined approximate solutions are elements of the set U. We permit small errors of the optimal control and of the minimum of the functional, but constraints are satisfied exactly. However all objects of the problem statement are only known in approximate form. Therefore it is naturally to require the approximate realization of the given constraints. Of course, the corresponding error should be small. So we have the following weaker form of approximate solutions of optimization problems.

Definition 2. *The control u denotes a* **weaker approximate solution** *of the minimization problem of the functional I on the set U, if $u \in O$ and*

$$I(u) \leq \inf_{u \in U} I(u) + \varepsilon$$

for a small enough neighborhood O of the set U and a small enough value $\varepsilon > 0$.

The weaker approximate solution may be not admissible. But it is close enough to a point of the set U. Besides this, the corresponding value of the functional may exceed its lower bound only by a small value. It is obvious, that the state functional in the weaker approximate solution must be approximated too. If the weaker approximate solution belongs to the set U, it is the weak approximate solution of the given problem.

We have three difficulties of the optimization problem. It is the singularity of the state equation, the smoothness of the nonlinear term, and the non-compactness of the time interval. Therefore, we will use finite time approximation [15], the penalty method [8], and a smooth approximation [16] for finding the approximate solution of the given problem.

Our first step is the finite time approximation. We fix the value $T > 0$. Let $Q_T = \Omega \times (0, T)$ and denote by V_T and Y_T the set of restrictions of functions from V and Y to Q_T. The set U_T of pairs (v, y) is determined from the product $V_T \times Y_T$ such that (2.1) is satisfied on the set Q_T and the initial conditions (2.3) hold true. It is obvious that the restriction of an admissible pair of the system (2.1)–(2.3) to the set Q_T is an element of the set U_T. Furthermore, for all pair $u = (v, y)$ from U_T its trivial extension $\bar{u} = (\bar{v}, \bar{y})$ by zero outside of the set U_T is an admissible pair of the system (2.1)–(2.3). We define the functional

$$I_T(v, y) = \frac{1}{6}\|y - z\|^6_{L_6(Q_T)} + \frac{\alpha}{2}\|v\|^2_{L_2(Q_T)}$$

and consider the finite time approximation problem of the given optimization problem.

Problem $\mathbf{P}_T$. *Find the control data from U_T that minimize the functional I_T on this set.*

Theorem 2. *The problem $\mathbf{P}_T$ is solvable.*

Proof. By lower boundedness of the functional I_T there exists a sequence $\{u_n\}$ from U_T such that

$$I(u_n) \to \inf_{u \in U} I(u). \tag{3.1}$$

We denote $u_n = (v_n, y_n)$. Then we get the equations

$$y_n'' - \Delta y_n + f(y_n) = v_n, \qquad (x,t) \in Q_T, \tag{3.2}$$

$$y_n = 0, \quad x \in S, \qquad t \in (0,T), \tag{3.3}$$

$$y_n(x,0) = \varphi(x), \quad y_n'(x,0) = \psi(x), \qquad x \in \Omega \tag{3.4}$$

from the definition of the set U_T. The sequence $\{y_n\}$ is bounded in the space $L_6(Q_T)$ and $\{v_n\}$ is bounded in $L_2(Q_T)$ because of the coercitivity of the functional I_T. Therefore, the sequence is bounded in the space $L_2(Q_T)$ by definition of the set F. It is obvious that the function y_n is a solution of the equation

$$y_n'' - \Delta y_n = f(y_n), \qquad (x,t) \in Q_T,$$

where

$$f_n = v_n - f(y_n).$$

The sequence $\{f_n\}$ is also bounded in the space $L_2(Q_T)$ and the sequence $\{y_n\}$ is bounded in Y_T by standard theory of linear hyperbolic equations. After passing to subsequences we get $v_n \to v$ weakly in $L_2(Q_T)$, and $y_n \to y$ weakly in Y_T together with $v \in V_T$. We repeat reasoning from the proof of Theorem 1 to obtain $f(y_n) \to f(y)$ weakly in $L_2(Q_T)$. After passing to the limit in the equation (3.2)–(3.4), we obtain that the function y is a solution to the equation (2.1) within Q_T. Thus the pair $u = (v, y)$ belongs to the set U_T. Thus

$$I_T(u) \le \liminf I_T(u_n)$$

and the pair u is a solution of the problem $\mathbf{P}_T$ because of (3.1). $\qquad\square$

We denote the solution of the approximate problem $\mathbf{P}_T$ by $u_T = (v_T, y_T)$. Now we prove the convergence of the approximation scheme.

Theorem 3. *If $T \to \infty$ then $I(\overline{u}_T) \to \min\limits_{u \in U} I(u)$.*

Proof. By the lower boundedness of the functional I on the set U there exists for all $\delta > 0$ a pair u^δ such that

$$I(u^\delta) \le \min_{u \in U} I(u) + \delta.$$

Therefore, we get the inequality

$$I_T(u_T) \le I_T(u^\delta) \le I(u^\delta) \le \min_{u \in U} I(u) + \delta. \tag{3.5}$$

Besides this, we obtain

$$
\min_{u \in U} I(u) \leq I(\bar{u}_T) = \frac{\alpha}{2} \int_Q |\bar{v}_T|^2 dQ + \frac{1}{6} \int_Q |\bar{y}_T - z|^6 dQ
$$

$$
= \frac{\alpha}{2} \int_{Q_T} |\bar{v}_T|^2 dQ_T + \frac{1}{6} \int_{Q_T} |\bar{y}_T - z|^6 dQ_T + \int_T^\infty \int_\Omega |z|^6 d\Omega dt
$$

$$
= I_T(u_T) + \int_T^\infty \int_\Omega |z|^6 d\Omega dt.
$$

By (3.5) we get

$$
\min_{u \in U} I(u) \leq I(\bar{u}_T) \leq \min_{u \in U} I(u) + \delta + \int_T^\infty \int_\Omega |z|^6 d\Omega dt.
$$

After the passing to the limit we obtain

$$
\min_{u \in U} I(u) \leq \lim_{T \to \infty} I(\bar{u}_T) \leq \min_{u \in U} I(u) + \delta.
$$

Then $I(\bar{u}_T) \to \min_{u \in U} I(u)$ because of arbitrariness of δ.   □

By the proved assertion the extension $\bar{u}_T$ of the solution u_T of problem $\mathbf{P}_T$ can be chosen as the weak solution of the initial optimization problem for a small enough value T. It is an admissible pair with the value of the minimizing functional close enough to its minimum. Our next step is an analysis of the problem $\mathbf{P}_T$.

4. Smooth penalty approximation

We use the penalty method with a smooth approximation for solving $\mathbf{P}_T$. For this we define the functional

$$
I_T^k(v, y) = \frac{1}{6} \|y - z\|_{L_6(Q_T)}^6 + \frac{\alpha}{2} \|v\|_{L_2(Q_T)}^2 + \frac{1}{2\varepsilon^k} \|y'' - \Delta y + f^k(y) - v\|_{L_2(Q_T)}^2,
$$

where $\varepsilon^k > 0$ and $\varepsilon^k \to 0$ as $k \to \infty$ and where f^k are continuous functions with

$$
(\varepsilon^k)^{-1/2} \|f^k(y) - f(y)\|_{L_2(Q_T)} \to 0 \tag{4.1}
$$

uniformly with respect to $y \in Y_T$. We denote further by W_T the set of pairs (v, y) from $(V_T \times Y_T)$ such that the function y satisfies the initial conditions (2.3).

Problem $\mathbf{P}_T^k$ *Find the pair that minimizes the functional I_T^k over the set W_T.*

Theorem 4. *The problem $\mathbf{P}_T^k$ is solvable.*

Proof. Obviously, there exists a sequence $\{u_n\}$ from W_T, where $u_n = (v_n, y_n)$, such that

$$I_T^k(u_n) \to \inf_{u \in U} I_T^k(u). \tag{4.2}$$

By coercitivity of the functional I_k the sequence $\{y_n\}$ is bounded in the space $L_6(Q_T)$ and the sequences $\{v_n\}$ and $\{g_n\}$ are bounded in $L_2(Q_T)$, where

$$g_n = y_n'' - \Delta y_n + f^k(y_n) - v_n.$$

Furthermore, y_n is the solution of the equation

$$y_n'' - \Delta y_n = h_n, \quad (x, t) \in Q_T$$

with boundary conditions

$$y_n = 0, \quad x \in S, \ t \in (0, T),$$

$$y_n(x, 0) = \varphi(x), \quad y_n'(x, 0) = \psi(x), \quad x \in \Omega,$$

where

$$h_n = g_n + v_n - f^k(y_n).$$

The sequence $\{h_n\}$ is bounded in the space $L_2(Q_T)$. So the sequence $\{y_n\}$ is bounded in the space Y_T. After passing to subsequences we get $v_n \to v$ weakly in $L_2(Q_T)$, $g_n \to g$ weakly in $L_2(Q_T)$, and $y_n \to y$ weakly in Y_T, besides $v \in V_T$. Thus the pair $u = (v, y)$ is an element of U_T. We repeat the reasoning from the proof of Theorem 1 and obtain $f^k(y_n) \to f^k(y)$ weakly in $L_2(Q_T)$. Then $h_n \to h$, where

$$h = g + v - f^k(y).$$

Hence we get

$$I_T^k(u) \leq \liminf I_T^k(u_n).$$

Therefore, the pair $u = (v, y)$ is a solution of the problem $\mathbf{P}_T^k$. $\qquad\square$

Next, we prove the convergence of the approximation scheme. We denote by $u_T^k = (v_T^k, y_T^k)$ the solution of the problem $\mathbf{P}_T^k$.

Theorem 5. *If $k \to \infty$ then*

$$\liminf I_T(u_T^k) \leq \min_{u \in U_T} I_T(u),$$

and $u_T^k \to u_T$ weakly in $L_2(Q_T) \times Y_T$ where u_T is the solution of the problem $\mathbf{P}_T$.

Proof. By lower boundedness of the functional I_T on the set U_T there exists for all $\delta > 0$ a pair $u^\delta = (v^\delta, y^\delta)$ from U_T such that

$$I_T(u^\delta) \leq \inf_{u \in U_T} I_T(u) + \delta.$$

Using the optimality of the pair u_T^k for the problem $\mathbf{P}_T^k$, we get

$$I_T^k(u_T^k) = \min_{u \in W_T} I_T^k(u) \le I_T^k(u^\delta)$$

$$= I_T(u^\delta) + \frac{1}{2\varepsilon k} \left\| (y^\delta)'' - \Delta y^\delta + f^k(y^\delta) - v^\delta \right\|_{L_2(Q_T)}^2$$

$$\le \min_{u \in U_T} I_T(u) + \delta + \frac{1}{2\varepsilon k} \left\| f^k(y^\delta) - f(y^\delta) \right\|_{L_2(Q_T)}^2$$

because of the previous inequality and the inclusion $y^\delta \in Y_T$. Using (4.1), we obtain after passing to the limit $k \to \infty$

$$\liminf I_T^k(u_T^k) \le \min_{u \in U_T} I_T(u) + \delta.$$

After passing to the limit $\delta \to 0$, we obtain

$$\liminf I_T^k(u_T^k) \le \min_{u \in U_T} I_T(u). \tag{4.3}$$

Then by definition of the functional I_T^k we obtain the boundedness of the sequences $\{v_T^k\}$ and $\{y_T^k\}$ in the spaces $L_2(Q_T)$ and $L_6(Q_T)$, besides y_T^k is a solution of the equation

$$(y_T^k)'' - \Delta y_T^k + f^k(y_T^k) - v_T^k = \sqrt{\varepsilon^k} g_T^k, \quad (x.t) \in Q_T, \tag{4.4}$$

where the sequence $\{g_T^k\}$ is bounded in $L_2(Q_T)$. So the sequence $\{f^k(y_T^k)\}$ is bounded in this space. Thus the function y_T^k satisfies the equation

$$(y_T^k)'' - \Delta y_T^k = h_T^k,$$

where

$$h_T^k = v_T^k + \sqrt{\varepsilon^k} g_T^k - f^k(y_T^k).$$

It is obvious that the sequence $\{h_T^k\}$ is bounded in the space $L_2(Q_T)$. Using the theory of linear hyperbolic equations, we prove the boundedness of the sequences $\{y_T^k\}$ in the space Y_T.

After passing to subsequences we obtain convergence $v_T^k \to v_T$ weakly in $L_2(Q_T)$, $y_T^k \to y_T$ weakly in Y_T, $g_T^k \to g_T$ weakly in $L_2(Q_T)$, and in particular $v_T \in V_T$. Using the standard method, we get $f^k(y_T^k) \to f(y_T)$ weakly in $L_2(Q_T)$. Then we obtain

$$\left| \int_{Q_T} \left(f^k(y_T^k) - f(y_T) \right) \lambda dQ_T \right|$$

$$\le \left| \int_{Q_T} \left(f^k(y_T^k) - f(y_T^k) \right) \lambda dQ_T \right| + \left| \int_{Q_T} \left(f(y_T^k) - f(y_T) \right) \lambda dQ_T \right|$$

$$\le \sup_{y \in Y} \left\| f^k(y) - f(y) \right\|_{L_2(Q_T)} + \left| \int_{Q_T} \left(f(y_T^k) - f(y_T) \right) \lambda dQ_T \right|$$

for all $\lambda \in L_2(Q_T)$.

Using the uniform convergence of the sequence $\{f^k\}$, we get $f^k(y_T^k) \to f(y_T)$ weakly in $L_2(Q_T)$. After the passing to the limit in (4.4), we obtain, that the function y_T satisfies the equation (2.1) in Q_T. Thus we get the inclusion $y_T \in Y_T$. Then $u_T \in U_T$, where $u_T = (v_T, y_T)$. The following inequality

$$I_T\left(u_T^k\right) \le I_T^k\left(u_T^k\right)$$

is true because of the definition of the functional I_T^k. Using 4.3, we get (if $k \to \infty$)

$$\liminf I_T\left(u_T^k\right) \le \liminf I_T^k\left(u_T^k\right) \le \min_{u \in U_T} I_T(u).$$

Thus, we obtain

$$I_T(u_T) \le \liminf_{k \to \infty} I_T\left(u_T^k\right) \le \min_{u \in U_T} I_T(u)$$

from the weak lower semicontinuity of the powers of norms. Therefore u_T is a solution of the problem $\mathbf{P}_T$. □

Therefore, a weaker approximate solution can be found from the obtained results.

Theorem 6. *The extension $\overline{u}_T^k$ of the solution u_T^k of the problem $\mathbf{P}_T^k$ is a weaker approximate solution of the problem $\mathbf{P}$ for large enough k and T.*

Proof. Using Theorem 3, we obtain that for all $\varepsilon > 0$ there exists a value T such that the extension $\overline{u}_T$ of the solution u_T of the problem $\mathbf{P}_T$ satisfies the inequality

$$I(\overline{u}_T) \le \min_{u \in U} I(u) + \varepsilon/2. \tag{4.5}$$

By Theorem 5, we obtain the convergence $u_T^k \to u_T$ weakly in $L_2(Q_T) \times Y_T$. Therefore,

$$\overline{u}_T^k \to \overline{u}_T \text{ weakly in } L_2(Q_T) \times Y. \tag{4.6}$$

Furthermore,

$$\liminf_{k \to \infty} I_T\left(u_T^k\right) \le \min_{u \in U_T} I_T(u). \tag{4.7}$$

It is obvious, that $I_T(u) = I(\overline{u})$ for all $u \in L_2(Q_T) \times Y_T$. So we get

$$\min_{u \in U_T} I_T(u) = I_T(u_T) = I(\overline{u}_T)) \le \min_{u \in U} I(u) + \varepsilon/2 \tag{4.8}$$

from inequality (4.5). Based on (4.6) and (4.7), we obtain for given $\varepsilon > 0$, corresponding value T and neighbourhood O of U in the sense of the weak topology of the product $L_2(Q_T) \times Y_T$ a sufficiently large number k such that $\overline{u}_T \in O$ and

$$I(\overline{u}_T^k) = I_T\left(u_T^k\right) \le I_T(U_T) + \varepsilon/2.$$

Using (4.8), we get

$$I(\overline{u}_T^k) \le I(U) + \varepsilon/2.$$

Hence, the assertions of the theorem are true. □

Thus we can choose the extension $\overline{u}_T^k$ of the solution to the problem $\mathbf{P}_T^k$ as the weaker approximate solution of the initial optimization problem. Our last step is to obtain necessary conditions of optimality for the problem $\mathbf{P}_T^k$.

5. Solving the approximate optimization problem

We consider the problem $\mathbf{P}_T^k$ of minimizing the smooth functional I_T^k on the convex set W_T. This problem can be solved by means of standard methods.

Theorem 7. *The solution* $u_T^k = (v_T^k, y_T^k)$ *to the problem* $\mathbf{P}_T^k$ *satisfies the variational inequality*

$$\int_{Q_T} \left(\alpha v_T^k + p_T^k\right)\left(v - v_T^k\right)dQ_T \geq 0 \qquad \forall v \in V_T, \tag{5.1}$$

where p_T^k *is the solution of the initial boundary value problem*

$$\left(p_T^k\right)'' - \Delta p_T^k + \left(f^k\right)'\left(y_T^k\right)p_T^k = \left(y_T^k - z\right)^5, \qquad (x,t) \in Q_T, \tag{5.2}$$

$$p_T^k = 0, \qquad x \in \Omega, \ t \in (0,T), \tag{5.3}$$

$$p_T^k(x,T) = 0, \qquad \left(p_T^k\right)'(x,T) = 0, \ x \in \Omega, \tag{5.4}$$

and y_T^k *is the solution of the initial boundary value problem*

$$\left(y_T^k\right)'' - \Delta y_T^k + f^k\left(y_T^k\right) = v_T^k + \varepsilon^k p_T^k, \qquad (x,t) \in Q_T, \tag{5.5}$$

$$y_T^k = 0, \qquad x \in \Omega, \ t \in (0,T), \tag{5.6}$$

$$y_T^k(x,0) = \varphi(x), \quad \left(y_T^k\right)'(x,0) = \psi(x), \qquad x \in \Omega. \tag{5.7}$$

Proof. By Theorem 3.1 (of [15], Chapter 1) the minimal point u_T^k of the functional I_T^k on the convex set W_T satisfies the variational inequality

$$\left\langle \left(I_T^k\right)'\left(u_T^k\right), \ u - u_T^k \right\rangle \geq 0 \qquad \forall u \in W_T$$

where $\langle \lambda, u \rangle$ is the value of the linear continuous functional λ in the point u, and $\left(I_T^k\right)'\left(u_T^k\right)$ is the Gataux derivative of the functional I_T^k in the point u_T^k. Using the definition of the set W_T, we obtain, that the last formula can be transformed to the variational inequality

$$\left\langle \left(I_{T_v}^k\right)\left(u_T^k\right), \ v - v_T^k \right\rangle \geq 0 \qquad \forall v \in V_T \tag{5.8}$$

and stationarity condition

$$I_{T_y}^k\left(u_T^k\right) = 0, \tag{5.9}$$

where $I_{T_v}^k\left(u_T^k\right)$ and $I_{T_y}^k\left(u_T^k\right)$ are the partial derivatives of the functional I_T^k in the considered point. Using the definition of the functional I_T^k we find the values of its partial derivatives from the formula

$$I_{T_v}^k\left(u_T^k\right) = \alpha v_T^k - \frac{1}{\varepsilon^k}\left(\left(y_T^k\right)'' - \Delta y_T^k + f^k\left(y_T^k\right) - v_T^k\right),$$

such that

$$\left\langle \left(I_{T_y}^k\right)\left(u_T^k\right), h \right\rangle = \int\limits_{Q_T} \left(y_T^k - z\right)^5 dQ_T$$

$$+ \frac{1}{\varepsilon^k} \int\limits_{Q_T} \left(\left(y_T^k\right)'' - \Delta y_T^k + f^k\left(y_T^k\right) - v_T^k\right)\left(h'' - \Delta h + \left(f^k\right)'\left(y_T^k\right)h\right) dQ_T$$

holds true for all $h \in Y_T$. We define

$$p_T^k = \frac{1}{\varepsilon^k}\left(\left(y_T^k\right)'' - \Delta y_T^k + f^k\left(y_T^k\right) - v_T^k\right).$$

Then the function y_T^k satisfies the equation (5.5). The boundary conditions (5.6), (5.7) can be obtained by using of the definition of the set W_T. We find the value of the partial derivative

$$I_{T_v}^k\left(u_T^k\right) = \alpha v_T^k - p_T^k.$$

Thus the variational inequality (5.8) is transformed to (5.1). We obtain in an analogous way the partial derivative from the equation

$$\left\langle \left(I_{T_y}^k\right)\left(u_T^k\right), h \right\rangle = \int\limits_{Q_T} \left(\left(y_T^k - z\right)^5 h + p_T^k\left(h'' - \Delta h + \left(f^k\right)'\left(y_T^k\right)h\right)\right) dQ_T$$

$$= \int\limits_{Q_T} \left(\left(y_T^k - z\right)^5 + \left(p_T^k\right)'' - \Delta p_T^k + \left(f^k\right)'\left(y_T^k\right)p_T^k\right) h dQ_T$$

$$+ \int\limits_{\Omega} \left(p_T^k(x,T)h'(x,T) - \left(p_T^k\right)'(x,T)h(x,T)\right) dx$$

$$+ \int\limits_0^T \int\limits_S p_T^k \frac{\partial h}{\partial \vec{n}} dSdt \qquad \forall h \in Y_T,$$

where $\vec{n}$ is the outward normal of S. Using (5.9), we obtain

$$\int\limits_{Q_T} \left(\left(y_T^k - z\right)^5 + \left(p_T^k\right)'' - \Delta p_T^k + \left(f^k\right)'\left(y_T^k\right)p_T^k\right) h dQ_T$$

$$+ \int\limits_S \left(p_T^k(x,T)h'(x,T) - \left(p_T^k\right)'(x,T)h(x,T)\right) dx + \int\limits_0^T \int\limits_S p_T^k \frac{\partial h}{\partial \vec{n}} dSdt = 0$$

for all $h \in Y_T$ and the function p_T^k solves the boundary problem (5.2)–(5.4). $\square$

Thus we have obtained a system including the variational inequality (5.1), the state equations (5.5)–(5.7), and the adjoint system (5.2)–(5.4) for solving the problem $\mathbf{P}_T^k$. It can be computed by using standard iterative methods, see [24]. By Theorem 6 the extension of the solution $u_T^k = \left(v_T^k, y_T^k\right)$ of the problem $\mathbf{P}_T^k$

can be chosen as the weaker approximate solution of the initial problem for large enough value of k and T. Note, that the equation (5.5) implies that the initial state equation is satisfied approximately but not exactly because of the second term in its right-hand side. Therefore, the pair $\bar{u}_T^k$ is not admissible and we have indeed a weaker approximate solution of the problem **P**.

References

[1] A. Matveev and V. Jakubovich, *Optimal control for a system with distributed parameters.* J. Sib. Math. **19** (1978), 5, 1109–1140.

[2] D. Tiba, *Optimality conditions for nonlinear distributed control problems.* Proc. 22 IEEE Conf. N.Y. Dec. Contr., 1983, 1251–1252.

[3] A.V. Fursikov, *Optimal Control of Distributed Systems, Theory and Applications.* Providence, Amer. Math. Soc., 1999.

[4] D. Tiba, *Optimal control for second-order semilinear hyperbolic equations.* Contr. Theory Adv. Techn. **3** (1987), 1, 33–46.

[5] V. Sumin, *About relaxation of optimization problems for functional equations in spaces of essential bounded functions.* Vestnik Nizhegorod. Univ. 1998, 1, 126–133.

[6] H. Banks, K. Kunisch, *An approximation theory for nonlinear partial differential equations with applications to identification and control.* SIAM J. Contr. Optim. **20** (1982), 6, 815–849.

[7] H. Kuliev, K. Gasanov, *Necessary conditions of optimality for system with distributed parameters and control in the coefficients of high derivatives.* J. Differ. Equat. **18** (1982), 6, 1028–1036.

[8] J.L. Lions, *Contrôle de Systèmes Distribués Singuliers.* Paris, Gauthiers–Villars, 1983.

[9] R. Rockafellar, *Convex Analysis.* Princeton Univ. Press, 1996.

[10] F.H. Clarke, *Optimization and Nonsmooth Analysis.* N.Y., Wiley, 1983.

[11] V. Barbu, *Boundary control problems with nonlinear state equation.* SIAM J. Control and Optim. **20** (1982), 1, 125–143.

[12] S. Serovajsky, *Optimal control for the elliptic equation with nonsmooth nonlinearity.* J. Differ. Equat. **39** (2003), 10, 1497–1502.

[13] A. Seierstad, *Necessary conditions for nonsmooth, infinite–horizon, optimal control problems.* J. Optim. Theory Appl. **103** (1999), 1, 201–209.

[14] S. Aseev and A. Kryazhimskiy, *The Pontryagin maximum principle and transversality conditions for a class of optimal control problems with infinite time horizons.* SIAM J. Control Optim. **43** (2004), 3, 1094–1119.

[15] J.L. Lions, *Contrôle Optimal de Systèmes Gouvernés par des Equations aux Dérivées Partielles.* Paris, Gauthiers–Villars, 1983.

[16] V. Barbu, *Optimal feedback control for a class of nonlinear distributed parameter systems.* SIAM J. Control and Optim. **21** (1983), 6, 871–894.

[17] P. Cannarsa, G. Da Prato, *Nonlinearity of control with infinite horizon for distributed parameter systems and stationary Hamilton–Jacobi equations.* SIAM J. Contr. Optim. **27** (1989), 4, 861–875.

[18] D. Tiba, *Optimal control of nonsmooth distributed parameter systems*. Lecture Notes in Mathematics. **1459** (1990), Berlin, WJ Springer-Verlag, 159 p.

[19] F.P. Vasiliev, *Optimization Metods*. Moscow, Factorial, 2002.

[20] T.A. Zolezzi, *A characterization of well-posed optimal control systems*. SIAM J. Contr. Optim. **19** (1981), 5, 605–616.

[21] S. Serovajsky, *Approximate Solution of Singular Optimization Problems*. J. Mathematical Notes, **74** (2003), 5, 685–694.

[22] J. Sea, *Optimisation. Théorie et algorithmes*. Paris, Dunod, 1971.

[23] J.L. Lions, *Quelques Méthodes de Résolution des Problèmes aux Limites non Linéaires*. Dunod, Gauthiers–Villars, Paris, 1969.

[24] F.L. Chernous'ko and V.V. Kolmanovsky, *Computing and Approximate Optimization Methods*. In "Mathematical Analysis. Results of Science and Technique". **14** (1977), 101–166.

Simon Serovajsky and Kanat Shakenov
Masanchi str. 39/47
al-Farabi Kazakh National University
050012, Almaty, Kazakhstan

e-mail: serovajskys@mail.ru
        shakenov2000@mail.ru

Progress in Mathematics, Vol. 301, 301–313

Local in Space Energy Estimates for Second-order Hyperbolic Equations

Sergio Spagnolo and Giovanni Taglialatela

Abstract. In the first part of the paper we review some recent results concerning the propagation of analytic regularity for the s-Gevrey solutions, with $s < \overline{m}/(\overline{m} - 1)$, to the semilinear (weakly) hyperbolic equations with characteristics of multiplicity $\leq \overline{m}$. The main results are concerning two special classes of equations: the equations with coefficients depending only on time, and those in space dimension one. These results rely on suitable a priori estimates for the corresponding linearized equations, based on the theory of quasi-symmetrizer. In view of these a priori estimates, the case of several space variables is quite different from that of one space variable: in the latter the quasi-symmetrizer provides an energy integral on each open subset of $\mathbb{R}^1$, and this ensures the propagation of regularity along the cones of determinacy, whereas in the multidimensional case we can only define the energy on the whole $\mathbb{R}^n$.

The second part of the paper is devoted to the Cauchy problem for the second-order linear hyperbolic equations. For these equations, we are in the position to get an energy estimate along the cones of determinacy, which will imply, for the corresponding semilinear equations, the analytic propagation along these cones for all the s-Gevrey solution with $s < 2$. The proof of this energy estimate is only sketched.

Mathematics Subject Classification. Primary 35L15; Secondary 35B65.

Keywords. Energy estimates, weakly hyperbolic equations, analytic propagation.

1. The analytic propagation

The propagation (in time) of the analyticity (in space variables) for solutions to nonlinear hyperbolic equations has been the object of many investigations. Roughly speaking, the propagation of analyticity can be described as follows:

Let $u(t, x)$ be a sufficiently smooth solution of an hyperbolic equation

$$\mathcal{P}u \equiv \partial_t^m u + \sum_{h=1}^{m} p_h(t, x, \partial_x)\partial_t^{m-h}u = f(u) \quad \text{in } [0, T] \times \mathbb{R}^n, \qquad (1)$$

$p_h(t, x, \xi)$ being polynomials of degree $\leq h$ in $\xi \in \mathbb{R}^n$, with analytic coefficients.

If $u(t,x)$, as well as its time derivatives $\partial_t^j u(t,x)$ of order $j \leq m$, are analytic in x at $t = 0$, can we conclude that $u(t, \cdot)$ remains analytic for all $t \geq 0$?

Denote by

$$p^0(t,x,\tau,\xi) \equiv \tau^m + \sum_{h=1}^m p_h^0(t,x,\xi)\tau^{m-h} = \prod_{j=0}^m (\tau - \tau_j(t,x,\xi)), \qquad (2)$$

where p_h^0 is the homogeneous part of degree h of p_h, the *principal symbol* of the operator $\mathcal{P}$:

We say that $\mathcal{P}$ is *hyperbolic* when the characteristic roots $\{\tau_j(t,x,\xi)\}_{1 \leq j \leq m}$ are *real*.

Thus, *hyperbolic* is here synonymous of *weakly hyperbolic*.

In the following, we shall assume that $p_h^0(t,x,\xi)$ is *bounded* on $[0,T] \times \mathbb{R}^n$, for $|\xi| = 1$, $h = 1, \ldots, n$, hence

$$\Lambda_0 \equiv \sup_{(t,x,\xi) \in [0,T] \times \mathbb{R}^n \times (\mathbb{R}^n \setminus \{0\})} \left| \tau_h(t,x,\xi/|\xi|) \right| < +\infty. \qquad (3)$$

We shall refer to the cones

$$\Gamma \equiv \Gamma(x_0, r_0, \Lambda) = \left\{ (t,x) \in [0,T] \times \mathbb{R}^n \,\middle|\, |x - x_0| \leq r_0 - \Lambda t \right\},$$

with $\Lambda \geq \Lambda_0$, as to the *determinacy cones* of (1).

We consider two different versions of analytic propagation for the equation (1): either

$$\partial_t^j u(0, \cdot) \in \mathcal{C}^\omega(\Gamma_0) \qquad \text{for all } j < m \qquad \Longrightarrow \qquad u \in \mathcal{C}^\omega(\Gamma), \qquad (\text{PA})_{\text{loc}}$$

for each determinacy cone Γ with base $\Gamma_0 = \Gamma \cap \{t = 0\}$ at the initial time; or

$$\partial_t^j u(0, \cdot) \in \mathcal{C}_{L^2}^\omega(\mathbb{R}^n) \qquad \text{for all } j < m \qquad \Longrightarrow \qquad u(t, \cdot) \in \mathcal{C}_{L^2}^\omega(\mathbb{R}^n). \qquad (\text{PA})_{\text{glob}}$$

By (3) it follows that

$$(\text{PA})_{\text{loc}} \implies (\text{PA})_{\text{glob}} .$$

The *analytic propagation property* has a rather long history. The case of *strictly hyperbolic* equations (which means that the characteristic roots τ_j are *simple*) has been considered by Lax [L] in space dimension $n = 1$, and by Alinhac and Métivier [AM] in several space dimensions. They proved $(\text{PA})_{\text{loc}}$ for those solutions which are *a priori* assumed to belong to $\mathcal{C}^\infty$, or, more generally, to a Sobolev class of a sufficiently large order. In the linear case the analytic propagation had been previously stated by Mizohata [Mi].

In the *nonstrictly hyperbolic case*, it is natural to think that we need to strengthen the degree of regularity which should be *a priori* required for the given solution. More precisely, we conjecture that the propagation of analyticity holds for the solutions which belong to a functional class $\mathcal{X}$ in which the Cauchy problem for the linearized equation is well posed:

Ansatz. *If the Cauchy Problem for the linearized equation $\mathcal{P}u = f(t,x)$ is well posed in some class $\mathcal{X}$ ($\mathcal{C}^\omega \subseteq \mathcal{X} \subseteq \mathcal{C}^\infty$), regardless to lower-order terms, then a solution $u(t,\cdot)$ enjoys the analytic propagation as long as it remains bounded in $\mathcal{X}$.*

In this direction, in [S1], some answers were given, limited to the class of semilinear second-order equations. Subsequently, various other results have been obtained: [Ma] and [RY] for second-order equations, [CZ] for equations with constant multiplicity, [KY] for uniformly symmetrizable systems. It should be noted that for a (hyperbolic) *linear system*, the analytic propagation is a direct consequence of the global wellposedness in $\mathcal{C}^\omega$ (Theorem of Bony-Shapira [BS]) together with Holmgren uniqueness.

If $\mu(\tau_j)$ is the multiplicity of the characteristic root τ_j, the number (among 1 and m)

$$\overline{m} \equiv \max_j \max_{t,x,\xi} \mu\big(\tau_j(t,x,\xi)\big),$$

is called the *multiplicity* of the equation (1). For $\overline{m} = 1$, the equation is *strictly hyperbolic*.

Now, after Bronšteĭn [B] we know that the sharp wellposedness spaces for a linear hyperbolic equation with multiplicity $\leq \overline{m}$, are the Gevrey classes γ^s with exponent $s < \overline{m}/(\overline{m}-1)$; thus we expect that every solution which is bounded in one of these classes enjoys the analytic propagation. More precisely we think that the following statements holds true:

Conjecture 1. *A solution $u(t,\cdot)$ to an analytic hyperbolic equation (1) enjoys* $(\mathrm{PA})_{\mathrm{loc}}$ *as long as it remains bounded in some Gevrey class γ^s with $s < \overline{s} \equiv \overline{m}/(\overline{m}-1)$.*

GLOBAL VERSION: $u(t,\cdot)$ *enjoys* $(\mathrm{PA})_{\mathrm{glob}}$ *as long as it remains bounded in some* $\gamma^s_{L^2}$ *with $s < \overline{s}$.*

Notations. If Ω is an open set of $\mathbb{R}^n$, the Gevrey class $\gamma^s \equiv \gamma^s(\Omega)$ is formed by the functions $\varphi \in \mathcal{C}^\infty(\Omega)$ satisfying

$$\sup_{x \in K} \big|\partial_x^\alpha \varphi(x)\big| \leq C_K \Lambda_K^{|\alpha|} \alpha!^s \quad \text{for all compact } K \subset \Omega, \quad \forall \alpha,$$

while

$$\gamma^s_{L^2} = \Big\{ \varphi \in \mathcal{C}^\infty(\mathbb{R}^n) : \|\partial_x^\alpha \varphi\|_{L^2} \leq C\Lambda^{|\alpha|} \alpha!^s, \ \forall \alpha \in \mathbb{N}^n \Big\}.$$

Note that $\mathcal{C}^\omega = \gamma^1 \subset \gamma^s \subset \gamma^{s'} \subset \mathcal{C}^\infty$ for $1 < s < s'$.

In [DS], Conjecture 1 was proved, in the global version, for the special case of space-independent leading coefficients. More recently, in [ST], Conjecture 1 was proved for any equation of the type (1), but limited to one space variable.

The case of hyperbolic systems was also considered, and it was proved:

Theorem 1 ([DS]). *A solution $U(t,x)$ to an analytic hyperbolic system of the form*

$$U_t + A_1(t)\, U_{x_1} + \cdots + A_n(t)\, U_{x_n} = f(t,x,U) \quad in \quad [0,T] \times \mathbb{R}^n,$$

enjoys $(PA)_{\mathrm{glob}}$ *whenever it is bounded in $\gamma^s_{L^2}$ for some $s < \overline{s} \equiv \overline{m}/(\overline{m}-1)$.*

Theorem 2 ([ST]). *A solution $U(t,x)$ to an analytic hyperbolic system of the form*

$$U_t = f(t,x,U,U_x) \quad in \quad [0,T] \times \mathbb{R}, \tag{4}$$

enjoys $(PA)_{\mathrm{loc}}$ *in a determinacy cone Γ, whenever, in a neighborhood of Γ, it is bounded in γ^s for some $s < \overline{s}$.*

Here, the nonlinear terms $f(t,x,U)$ and $f(t,x,U,U_x)$ are assumed to be analytic in (t,x) and *entire analytic* in U and (U,U_x) respectively.

Roughly speaking, the proofs of these theorems can be splitted in two parts: an a priori energy estimate (in Gevrey classes) for the linearized operators, and an estimate of the *analytic norm* of the nonlinear terms. The techniques employed for proving the linear estimates are elaborate, and apply also to other questions on hyperbolic systems and on parameter-depending polynomials. The main tool is the *quasi-symmetrizer* introduced by Jannelli in 1989 [J1], and by D'Ancona-Spagnolo in 1998 [DS] (see also [J2]).

Open Problem. *To extend Theorem 2 to the case of several space variables, i.e., to the systems*

$$U_t = f(t,x,U,U_{x_1},\ldots,U_{x_n}) \quad in \quad [0,T] \times \mathbb{R}^n. \tag{5}$$

The main difficulty is concerning the *linear* estimates. We need to find a *Gevrey energy function* for the linear hyperbolic system

$$U_t + A_1(t,x)\, U_{x_1} + \cdots + A_n(t,x)\, U_{x_n} = f(t,x), \tag{6}$$

and to prove an a priori estimate ensuring the γ^s-wellposedness for $s < \overline{m}/(\overline{m}-1)$. This wellposedness was proved in 1982 by Bronšteĭn [B], but his proof is not based on an energy estimate, and so does no seem to be directly exploitable for the nonlinear equations.

Actually, a *global* (Gevrey) energy function for the solutions to (6), leading to the global analytic propagation for (5), could be constructed by setting up a *pseudo-differential version* of the quasi-symmetrizer, suitable to be applied to linear equations with space-dependent coefficients. On the other hand, the construction of a *local* energy function for (6) is much harder, since to this end we cannot use the pseudodifferential operators. As a matter of facts, we are able to construct such a local energy only in the special case of second-order scalar equations. This construction will be sketched in the next section.

We remark that, in space dimension $n = 1$, the construction of a local energy does not present any additional difficulty w.r. to that of a global energy (see Subsection 2.9).

2. Local energy for second-order equations

Let us now restrict our investigation to the second-order equations in n space variables, i.e.,

$$\mathcal{P}u \equiv \partial_t^2 u - 2\sum_{j=1}^n b_j(t,x)\partial_j\partial_t u - \sum_{i,j=1}^n a_{ij}(t,x)\partial_i\partial_j u + l.o.t. = f(t,x), \quad (7)$$

where the leading coefficients are smooth real functions on a strip $[0,T] \times \mathbb{R}^n$.

We assume the hyperbolicity condition:

$$\left(\sum_{j=1}^n b_j(t,x)\xi_j\right)^2 + \sum_{i,j=1}^n a_{ij}(t,x)\xi_i\xi_j \geq 0. \quad (8)$$

Theorem 3. *For the solutions to (7), it is possible to define a local energy function $E(t,u;\Omega)$, for each open set $\Omega \subset \mathbb{R}^n$, and to prove an a priori estimate leading to the wellposedness in the Gevrey classes γ^s with $s < 2$, on every determinacy cone.*

Remark 1. The equations

$$u_{tt} + u_x = 0, \qquad u_{tt} - 2tu_{xt} + t^2 u_{xx} = 0, \qquad u_{tt} - 2xu_{xt} + x^2 u_{xx} = 0,$$

are examples of hyperbolic equations for which the Cauchy problem is γ^s well posed only for $s < 2$.

Remark 2. By using the energy estimate stated in Theorem 3, we can prove the *local* analytic propagation for the solutions of $\mathcal{P}u = f(u, \partial_t u, \partial_{x_1} u, \ldots, \partial_{x_n} u)$ which are bounded in some γ^s with $s < 2$.

Remark 3. The *finite speed of propagation* for a hyperbolic system of the type (6), means that, if $U(t,x)$ and $\widetilde{U}(t,x)$ are two solutions coinciding on the base Γ_0 of a determinacy cone Γ, then $U = \widetilde{U}$ on Γ. Under the assumption that U and $\widetilde{U}$ are Gevrey function in x of order $s < \overline{m}/(\overline{m}-1)$, such a property was stated by various authors (see Ohya and Tarama [OT], Rauch [R], Colombini and Rauch [CR]). However, the finite speed of propagation does not imply the existence of a local (in space) energy estimate.

2.1. Sketch of proof of Theorem 3

For the sake of simplicity, we shall consider the case of an homogeneous operator with coefficients independent of time. The general case requires minor additional computations. We write our operator in the form

$$\mathcal{P}u = \partial_t^2 u - 2\sum_{j=1}^n b_j(x)\partial_j\partial_t u - \sum_{i,j=1}^n \partial_i(a_{ij}(x)\partial_j u)$$
$$= u_{tt} - 2\langle b(x), \nabla u_t\rangle - \operatorname{div}(A(x)\nabla u)$$

where $b(x) = (b_1(x),\ldots,b_n(x))^t$, $a_{ij} = a_{ji}$, $A(x) = [a_{ij}(x)]$, and $\nabla = \nabla_x$.

The hyperbolicity assumption reads

$$\widetilde{A}(x) \equiv b(x)\,b(x)^t + A(x) \geq 0.$$

2.2. A transformation

Introducing the *derivation*

$$\partial_b = \partial_t - b(x)\cdot\nabla \equiv \partial_t - \sum_{j=1}^n b_j(x)\partial_j,$$

our operator becomes (up to lower-order terms)

$$\mathcal{P}u = \partial_b^2 u - \operatorname{div}(\widetilde{A}(x)\nabla u),$$

where $\operatorname{div} = \operatorname{div}_x$. Note that the operators ∂_b and ∂_t are not commuting each other. In the following we shall use the shortened notation

$$u_b = \partial_b u, \qquad u_{bb} = \partial_b^2 u.$$

2.3. The energy density

For any $u(t,x)$ we define the non-negative function

$$e(t,x) = \frac{1}{2}\left[u_b{}^2 + \langle\widetilde{A}(x)\nabla u, \nabla u\rangle\right]$$

$$\equiv \frac{1}{2}\left[\Big(u_t - \sum_{j=1}^n b_j(x)u_{x_j}\Big)^2 + \sum_{i,j=1}^n \big(b_i(x)b_j(x) + a_{ij}(x)\big)u_{x_i}u_{x_j}\right],$$

and the vector

$$\theta(t,x) = -\Big\{e(t,x)\,b(x) + u_b\cdot\widetilde{A}(x)\nabla u\Big\}.$$

Remark 4. The symbol of our operator $\mathcal{P} \equiv \mathcal{P}(x;\partial_t,\partial_x)$ is the polynomial

$$P(x;\tau,\xi) = \tau^2 - 2\langle b(x),\xi\rangle\,\tau + \langle\widetilde{A}(x)\xi,\xi\rangle.$$

Thus, denoting by $\mathcal{P}_\tau$ the operator with symbol $\partial_\tau P \equiv 2\tau - 2\langle b(x),\xi\rangle$, we have

$$u_b \equiv u_t - \langle b(x),\nabla u\rangle = \frac{1}{2}\,\mathcal{P}_\tau u.$$

2.4. The crucial estimate

Lemma 1. *For any function* $u(t,x)$ *it holds the identity*

$$\partial_t\,e(t,x) + \operatorname{div}\theta(t,x) = \frac{1}{2}\mathcal{P}u(t,x)\cdot\mathcal{P}_\tau u(t,x) + \Phi(t,x),$$

where Φ *is a function satisfying the estimate*

$$|\Phi(t,x)| \leq C\,|\nabla u(t,x)|\,\sqrt{e(t,x)} + C\,e(t,x),$$

for some constant C *independent on* u.

Proof. By a direct computation we have

$$\partial_t e + \operatorname{div} \theta = u_b \cdot \mathcal{P}u + \Phi_1 + \Phi_2 + \Phi_3,$$

where

$$\Phi_1(t,x) = \sum_{h=1}^{n} \partial_h u \, \langle \tilde{A} \nabla u, \nabla b_h \rangle,$$

$$\Phi_2(t,x) = -\frac{1}{2} \sum_{h=1}^{n} b_h \, \langle (\partial_h \tilde{A}) \nabla u, \nabla u \rangle,$$

$$\Phi_3(t,x) = e \cdot \operatorname{div} b.$$

Therefore we find, by the Cauchy-Schwartz inequality,

$$\left| \Phi_1(t,x) \right| \leq \sum_{h=1}^{n} |\partial_h u| \, \langle \tilde{A} \nabla u, \nabla u \rangle^{1/2} \langle \tilde{A} \nabla b_h, \nabla b_h \rangle^{1/2} \leq C_1 |\nabla u(t,x)| \sqrt{e(t,x)}.$$

To estimate the second term we apply to the Glaeser inequality,

$$\left| \partial_h \varphi(x) \right| \leq \sqrt{\varphi(x)} \, \sqrt{2 \, \|\partial_h^2 \varphi\|_{L^\infty}},$$

to the nonnegative function $\varphi(x) = \langle \tilde{A}(x)\xi, \xi \rangle$, with ξ fixed, and then we take $\xi = \nabla u(x)$. Thus, we get

$$\left| \Phi_2(t,x) \right| \leq C \, |b| \, |\nabla u| \, \langle \tilde{A} \nabla u, \nabla u \rangle^{1/2} \leq C_2 \, |\nabla u(t,x)| \, \sqrt{e(t,x)}.$$

The third term can be directly estimated as

$$\left| \Phi_3(t,x) \right| \leq C_3 \, e(t,x).$$

Combining the above inequalities, we get the wished estimate. $\qquad\square$

2.5. The local energy

Taking into account that $e(t,x) \geq 0$, for each $u(t,x)$ and each open set $\Omega \subset \mathbb{R}^n$ we can define the local energy

$$E(t,u;\Omega) = \int_{\Omega} e(t,x) \, dx.$$

In particular, having fixed a solution $u(t,x)$ of $\mathcal{P}u = f(t,x)$, we put

$$E(t) \equiv E(t,u;B_t) = \int_{B_t} e(t,x) \, dx, \quad \text{where } B_t = B(x_0, r_0 - \Lambda t), \quad \Lambda > 0.$$

Therefore we have, by Lemma 1 and equation (7),

$$E'(t) = \int_{B_t} \partial_t e(t,x) \, dx - \Lambda \int_{\partial B_t} e(t,y) \, dy$$

$$= \int_{B_t} \left[u_b \cdot f + \Phi(t,x) \right] dx - \int_{\partial B_t} \left[\langle \theta(t,y), N(y) \rangle + \Lambda \, e(t,y) \right] dy,$$

where N is the outer normal at y. By the definition of the vector θ, it follows

$$\langle \theta, N \rangle = -e \, \langle b, N \rangle - u_b \langle \tilde{A} \nabla u, N \rangle,$$

so

$$|\langle \theta, N \rangle| \leq e\,|\langle b, N \rangle| + |u_b|\,\langle \widetilde{A}\,\nabla u, \nabla u \rangle^{1/2}\,\langle \widetilde{A}\,N, N \rangle^{1/2}$$
$$\leq e\,|\langle b, N \rangle| + e\,\langle \widetilde{A}\,N, N \rangle^{1/2}.$$

Note that the characteristic roots of our equation are

$$\tau_{1,2}(x, \xi) \;=\; \langle b(x), \xi \rangle \mp \sqrt{\langle A(x)\xi, \xi \rangle}\,,$$

thus, for $\Lambda_0 = \max\limits_{x,\xi}\{|\tau_1(x, \xi/|\xi|)|,\ |\tau_2(x, \xi/|\xi|)|\}$, we have

$$|\langle \theta(t, x), N \rangle| \;\leq\; e\left\{|\langle b, N \rangle| + \langle \widetilde{A}\,N, N \rangle^{1/2}\right\} \;\leq\; \Lambda_0\,e(t, x)\,.$$

Hence, if $\Lambda \geq \Lambda_0$ (determinacy cones), the boundary integrand function in the expression of $E'(t)$ is negative, and we get the estimate

$$E'(t) \leq \int_{B_t}\left\{u_b \cdot f + \Phi\right\}dx \leq C\int_{B_t}\left\{|f|\sqrt{e} + |\nabla u|\sqrt{e} + e\right\}dx, \qquad (9)$$

whence

$$\partial_t\sqrt{E(t)} \leq C\sqrt{E(t)} + C\left\{\int_{B_t}|\nabla u(t, x)|^2 dx\right\}^{1/2} + C\left\{\int_{B_t}|f(t, x)|^2 dx\right\}^{1/2}. \quad (10)$$

2.6. Approximate energies

In the *strictly hyperbolic* case (10) yields an energy estimate. Indeed, in such a case, the matrix $\widetilde{A}(x)$ is coercive, so $|\nabla u(t, x)|^2 \leq C\,e(t, x)$. Hence (10) gives

$$\partial_t\sqrt{E(t)} \leq 2\,C\sqrt{E(t)} + C\left\{\int_{B_t}|f(t, x)|^2 dx\right\}^{1/2}.$$

In the *weakly* hyperbolic case, we define

$$\mathcal{P}^\varepsilon = \mathcal{P} + \varepsilon^2\Delta, \qquad \widetilde{A}_\varepsilon(t, x) = \widetilde{A}(t, x) + \varepsilon^2\,I,$$
$$e_\varepsilon(t, x) = e(t, x) + \varepsilon^2\,|\nabla u(t, x)|^2$$
$$\theta_\varepsilon(t, x) = -\left\{e_\varepsilon(t, x)\,b(x) + u_b(t, x)\,\widetilde{A}_\varepsilon(t, x)\nabla u(t, x)\right\},$$
$$E_\varepsilon(t) \equiv E_\varepsilon(t, u; B_t) = \int_{B_t} e_\varepsilon(t, x)\,dx.$$

Using Lemma 1 we find the identity

$$\partial_t\,e_\varepsilon + \operatorname{div}\theta_\varepsilon = \frac{1}{2}\,\mathcal{P}^\varepsilon u \cdot \mathcal{P}_\tau^\varepsilon u + \Phi_\varepsilon,$$

where now

$$|\Phi_\varepsilon(t, x)| \leq C\,|\nabla u(t, x)|\,\sqrt{e_\varepsilon(t, x)} + C\,e_\varepsilon(t, x) \leq \frac{C}{\varepsilon}\cdot e_\varepsilon(t, x).$$

Since $\mathcal{P}^\varepsilon u = f + \varepsilon^2 \Delta u$, we cannot estimate $\partial_t e_\varepsilon$ in terms of e_ε; indeed, instead of (9), we get (noticing that, as above, the boundary integrand in $E'_\varepsilon(t)$ is negative):

$$E'_\varepsilon(t) \leq \int_{B_t} \left\{ u_b \cdot f + \varepsilon^2 u_b \Delta u + \Phi_\varepsilon \right\} dx$$

$$\leq C \int_{B_t} \left\{ |f| \sqrt{e} + \varepsilon^2 |\Delta u| \sqrt{e} + \varepsilon^{-1} e \right\} dx.$$

Thus, we are forced to introduce the *energy of order one*, i.e.,

$$E_{1,\varepsilon}(t) = \sum_{h=1}^n E_\varepsilon(t, \partial_h u; B_t)$$

$$= \sum_{h=1}^n \int_{B_t} \left\{ |\partial_b(\partial_h u)|^2 + \langle \tilde{A} \nabla(\partial_h u), \nabla(\partial_h u) \rangle \right\} dx + \varepsilon^2 \sum_{|\alpha|=2} \int_{B_t} |\partial^\alpha u|^2 \, dx,$$

to get

$$\partial_t \sqrt{E_\varepsilon} \leq C \left\{ \varepsilon^{-1} \sqrt{E_\varepsilon} + \varepsilon \sqrt{E_{1,\varepsilon}} + \|f\|_{L^2(B_t)} \right\}.$$

Going on, we consider the sequence $\varepsilon_j = j^{-1}$, where $j = 1, 2, \ldots$, and we introduce the *jth-order energy*

$$E_{j,\varepsilon}(t) = \sum_{|\alpha|=j} E_\varepsilon(t, \partial^\alpha u; B_t)$$

$$= \sum_{|\alpha|=j} \int_{B_t} \left\{ |\partial_b(\partial^\alpha u)|^2 + \langle \tilde{A} \nabla(\partial^\alpha u), \nabla(\partial^\alpha u) \rangle \right\} dx + \varepsilon_j^2 \sum_{|\alpha|=j+1} \int_{B_t} |\partial^\alpha u|^2 \, dx.$$

Proceeding as above, we find

$$\partial_t \sqrt{E_j} \leq C \left\{ j \sqrt{E_j} + j^{-1} \sqrt{E_{j+1}} + \|f\|_{H^j(B_t)} \right\}.$$

2.7. The Gevrey energy

We define the local Gevrey energy of order $s = 2$, as

$$\mathcal{E}(t, u) = \sum_{j=0}^\infty \sqrt{E_j(t, u)} \, \frac{\rho^j}{j!^2} .$$

Proceeding as in [ST] or [S], we find the a priori estimate

$$\mathcal{E}'(t, u) \leq \sum_{j=0}^\infty \sqrt{E_j(t, u)} \, \frac{\rho^{j-1}}{j! \, (j-1)!} \left\{ \rho' + C_1 \rho + C_2 \right\} + C \mathcal{E}(t, f) \leq C \mathcal{E}(t, f),$$

for a suitable choice of the radius-function $\rho \equiv \rho(t)$. This estimate leads to the local γ^s wellposedness, for $s < 2$, on the determinacy cones.

Finally, proceeding as in [ST], we can prove the analytic propagation on the determinacy cones for the semilinear equation $\mathcal{P}u = f(u, \partial_t u, \partial_{x_1} u, \ldots, \partial_{x_n} u)$.

2.8. Gårding energy forms

We recall the classical techniques developed by Leray ([L]) and Gårding ([G]) in order to construct an energy function for a *strictly hyperbolic* equation. Let

$$P(t,x;\tau,\xi) = \tau^m + \sum_{j=1}^m p_j(t,x;\xi)\,\tau^{m-j}, \qquad \tau \in \mathbb{C}, \ \xi \in \mathbb{R}^n \quad (m \geq 2),$$

be a strictly hyperbolic polynomial in τ, where the $p_j(t,x;\xi)$ are real homogeneous polynomials in ξ of order j. Consider the differential operators $\mathcal{P} = P(t,x;\partial_t,\partial_x)$ and $\mathcal{P}_\tau = (\partial P/\partial \tau)(t,x;\partial_t,\partial_x)$. Then, for all real function $u = u(t,x)$, one has

$$\frac{1}{m}\mathcal{P}u \cdot \mathcal{P}_\tau u = \partial_t\{A^0(t,x;u)\} + \sum_{j=1}^n \partial_j\{A^j(t,x;u)\} + \Phi(t,x;u),$$

where $A_0, A_1, \ldots, A_n$ are Hermitian forms in the derivatives of order $m-1$ of u, i.e., putting $D = (\partial_t, \partial_x)$,

$$A^j(t,x;u) = \sum_{|\alpha|=|\beta|=m-1} a^j_{\alpha,\beta}(t,x)\,D^\alpha u\, D^\beta u, \quad j = 0,1,\ldots,n.$$

Moreover, there is some constant $c_0 > 0$ such that the form A^0, frozen at any point $(\bar{t},\bar{x})$, satisfies the *positivity condition*

$$\int_{\mathbb{R}^n} A^0(\bar{t},\bar{x};u)\,dx \geq c_0 \int_{\mathbb{R}^n} |D^{m-1}u|^2\,dx.$$

From this it follows, by a partition of the unity, the classical Gårding inequality

$$\int_{\mathbb{R}^n} A^0(t,x;u)\,dx \geq C^{-1}\int_{\mathbb{R}^n} |D^{m-1}u|^2\,dx - C\int_{\mathbb{R}^n} |D^{m-2}u|^2\,dx, \quad \text{for some } C > 0.$$

In conclusion, putting

$$e(t,x) = A^0(t,x;u), \qquad \theta(t,x) = (A^1(t,x;u),\ldots,A^n(t,x;u))^t,$$

we can write

$$\frac{1}{m}\mathcal{P}u \cdot \mathcal{P}_\tau u = \partial_t e(t,x) + \operatorname{div} \theta(t,x) + \Phi,$$

and a *global* energy for the operator $\mathcal{P}$ is provided by

$$E(t,u) = \int_{\mathbb{R}^n} \widetilde{e}(t,x)\,dx, \qquad \text{where} \qquad \widetilde{e}(t,x) = e(t,x) + C\,|D^{m-2}u(t,x)|^2.$$

Remark 5. In order to define a *local* energy, we need that the energy density $\widetilde{e}(t,x)$ be a *nonnegative* function, so that we can put

$$E(t) = \int_{B_t} \widetilde{e}(t,x)\,dx, \qquad \text{with} \quad B_t = B(x_0, r_0 - \Lambda t).$$

Moreover, in the computation of $E'(t)$ it appears the integral on ∂B_t of the quantity $\langle \theta(t,x), N(x)\rangle$, where $N(x) = x/|x|$ is the outer normal. So, to get the energy estimate in the weakly hyperbolic case, we need a *pointwise* estimate such as

$$|\langle\theta(t,x), N(x)\rangle| \leq C\,\widetilde{e}(t,x).$$

To this regard we recall that, as soon $m \geq 3$, a form

$$\sum_{|\alpha|=|\beta|=m-1} a_{\alpha,\beta}\, \partial^\alpha u\, \partial^\beta u$$

with symbol $\sum a_{\alpha,\beta}\, \xi^\alpha\, \xi^\beta \geq 0$, is not necessarily positive. This partly explains why the local theory for the equations of order $m > 2$ presents more difficulties than for the case $m = 2$.

Example (Third-order equations). Consider the *traceless* equation

$$\partial_t^3 u - 3 \sum_{i,j=1}^n a_{ij}(x)\, \partial_j \partial_j \partial_t u - 2 \sum_{i,j,k=1}^n b_{ijk}(x)\, \partial_i \partial_j \partial_k u = 0,$$

under the hyperbolicity condition

$$\Delta(x,\xi) \equiv \left\{ \sum_{i,j=1}^n a_{ij}(x)\xi_i\xi_j \right\}^3 - \left\{ \sum_{i,j,k=1}^n b_{ijk}(x)\, \xi_i\xi_j\xi_k \right\}^2 \geq 0.$$

Following Leray, we put $U = (u, u_t, u_{tt})^t$ to transform the equation into a *differential* system

$$U_t = A(x, \partial_x)U, \quad A = \begin{pmatrix} 0 & 1 & 0 \\ 0 & 0 & 1 \\ 2\sum b_{ijk}(x)\, \partial_i\partial_j\partial_k & 3\sum a_{ij}(x)\, \partial_j\partial_j & 0. \end{pmatrix}.$$

The natural symmetrizer for the matrix $A(x,\xi)$ is the matrix

$$Q(x,\xi) = \begin{pmatrix} 3\left\{\sum a_{ij}(x)\xi_i\xi_j\right\}^2 & 2\sum b_{ijk}(x)\,\xi_i\xi_j\xi_k & -\sum a_{ij}(x)\xi_i\xi_j \\ 2\sum b_{ijk}(x)\,\xi_i\xi_j\xi_k & 2\sum a_{ij}(x)\xi_i\xi_j & 0 \\ -\sum a_{ij}(x)\xi_i\xi_j & 0 & 1 \end{pmatrix}.$$

In view of a *local* positive energy, a good candidate for a density energy is the function

$$e(t,x) = \frac{1}{2}\left| u_{tt} - \sum_{i,j=1}^n a_{ij}(x)\partial_i\partial_j u \right|^2$$

$$\times \left\{ \left| \sum_{i,j=1}^n a_{ij}(x)\partial_i\partial_j u \right|^2 + 2 \sum_{i,j=1}^n a_{ij}(x)\partial_i u_t \partial_j u_t \right.$$

$$\left. - 2 \sum_{i,j,k=1}^n b_{ijk}(x)\partial_i u_t \partial_j \partial_k u \right\}.$$

The corresponding symbol is positive, but it is not clear whether $e(t,x) \geq 0$.

2.9. One space dimension

The one-dimensional case is very special, since each hyperbolic scalar equation of order m can be converted into a first-order system of the form

$$U_t = A(t,x)U_x + B(t,x)U,$$

where A is a hyperbolic matrix of Sylvester type. Therefore (see [DS]) it is possible to find a *quasi-symmetrizer* $Q_\varepsilon(t,x)$, i.e., a real symmetric matrix with the following properties:

$$C^{-1}\varepsilon^{2(m-1)}I \leq Q_\varepsilon \leq C\,I,$$

$$Q_\varepsilon A = T_\varepsilon + R_\varepsilon \quad \text{where:}$$

$$T_\varepsilon = (T_\varepsilon)^* \leq \Lambda_0\,Q_\varepsilon, \quad |T_x| \leq C\,\varepsilon^{1-m}Q_\varepsilon, \quad |R_\varepsilon| \leq C\,\varepsilon\,Q_\varepsilon,$$

with $\Lambda_0 =$ maximum of the eigenvalues of $A(t,x)$. Thus, we simply take

$$e_\varepsilon(t,x) = \langle Q_\varepsilon U,U\rangle, \qquad \theta_\varepsilon(t,x) = \langle T_\varepsilon U,U\rangle.$$

References

[AM] S. Alinhac and G. Métivier, *Propagation de l'analyticité des solutions de systèmes hyperboliques non-linéaires*, Invent. Math. **75** (1984), 189–204.

[BS] J.-M. Bony and P. Schapira, *Existence et prolongement des solutions holomorphes des équations aux dérivées partielles*, Invent. Math. **17** (1972), 95–105.

[B] M.D. Bronšteĭn, *The Cauchy problem for hyperbolic operators with characteristics of variable multiplicity*, Trudy Moskov. Mat. Obshch. **41** (1980), 83–99.

[CR] F. Colombini and J. Rauch, *Sharp finite speed for hyperbolic problems well posed in Gevrey classes*, Comm. P.D.E. **36** (2011), 1–9.

[CZ] M. Cicognani and L. Zanghirati, *Analytic regularity for solutions of nonlinear weakly hyperbolic equations*, Boll. Un. Mat. Ital. B (7) **11** (1997), 643–679.

[DS] P. D'Ancona and S. Spagnolo, *Quasi-symmetrization of hyperbolic systems and propagation of the analytic regularity*, Boll. Un. Mat. Ital. Sez. B Artic. Ric. Mat. (8) **1** (1998), 169–185.

[G] L. Gårding, *Solution directe du problème de Cauchy pour les équations hyperboliques*, Coll. Int. CNRS Nancy, 1956, 71–89.

[J1] E. Jannelli, *On the symmetrization of the principal symbol of hyperbolic equations*, Comm. P.D.E. **14** (1989), 1617–1634.

[J2] ———, *The hyperbolic symmetrizer: theory and applications*, Advances in phase space analysis of partial differential equations (A. Bove, D. Del Santo, and M.K.V. Murthy, eds.), Progress in Nonlinear Differential Equations and Their Applications, vol. 78, Birkhäuser Boston, MA, 2009, 113–139.

[L] J. Leray, *Hyperbolic Differential Equations*, The Institute for Advances Studies, Princeton, N.J., 1953, 238 pp., reprinted November 1955.

[KY] K. Kajitani and K. Yamaguti, *Propagation of analyticity of the solutions to the Cauchy problem for nonlinear symmetrizable systems*, Ann. Scuola Norm. Sup. Pisa Cl. Sci. (4) **28** (1999), 471–487.

[L]   P.D. Lax, *Non linear hyperbolic equations*, Comm. on Pure Appl. Math. **6** (1953), 231–258.

[Ma] R. Manfrin, *Analytic regularity for a class of semi-linear weakly hyperbolic equations of second order*, NoDEA (Nonlin. Diff. Equat. Appl.) **3** (1996), 371–394.

[Mi] S. Mizohata, *Analyticity of solutions of hyperbolic systems with analytic coefficients*, Comm. on Pure Appl. Math. **14** (1961), 547–559.

[N]   T. Nishitani, *Sur les équations hyperboliques a coefficients höldériens en t et de classe de Gevrey en x*, Bull. Sc. Math. **107** (1983), 113–138.

[OT] Y. Ohya and S. Tarama, *Le problème de Cauchy à caractéristiques multiples dans la classe de Gevrey. Coefficients höldériens en t*, Hyperbolic equations and related topics, Proc. Taniguchi Int. Symp., Katata and Kyoto/Jap. 1984 (1986), 273–306.

[R]   J. Rauch, *Precise finite speed and uniqueness in the Cauchy problem for symmetrizable hyperbolic systems*, Trans. Amer. Math. Soc. **363** (2011), 1161–1182.

[RY] M. Reissig and K. Yagdjian, *Levi conditions and global Gevrey regularity for the solutions of quasilinear weakly hyperbolic equations*, Math. Nachr. **178** (1996), 285–307.

[S1] S. Spagnolo, *Some results of analytic regularity for the semilinear weakly hyperbolic equations of the second order*, Rend. Sem. Mat. Torino Fasc. speciale (1988), 203–229.

[S2] _____, *Propagation of analyticity for a class of nonlinear hyperbolic equations*, Rend. Mat. Acc. Lincei **22** (2011), 135–149.

[ST] S. Spagnolo and G. Taglialatela, *Analytic propagation for nonlinear weakly hyperbolic systems*, Comm. P.D.E. **35** (2010), 2123–2163.

Sergio Spagnolo
Dipartimento di Matematica "L. Tonelli"
Università di Pisa
Largo B. Pontecorvo 5
I-56127 Pisa, Italy
e-mail: `spagnolo@dm.unipi.it`

Giovanni Taglialatela
Dipartimento di Scienze Economiche e Metodi Matematici
Università di Bari "Aldo Moro"
Via C. Rosalba 53
I-70124 Bari, Italy
e-mail: `taglia@dse.uniba.it`

Progress in Mathematics, Vol. 301, 315–324
© 2012 Springer Basel

The Final Problem on the Optimality of the General Theory for Nonlinear Wave Equations

Hiroyuki Takamura and Kyouhei Wakasa

Abstract. The general theory of the initial value problem for fully nonlinear wave equations is to clarify lower bounds of the lifespan, the maximal existence time, of classical solutions in terms of the amplitude of small initial data according to the order of smooth nonlinear terms and space dimensions. All the results had been obtained till 1995. So we have been interested in the optimality of the lower bounds. This can be obtained by blow-up results for model equations. Among such several results, only the case of the quadratic semilinear term in 4 space dimensions has been remained open for more than 20 years.

This final problem on the optimality has been known to be the critical case of Strauss' conjecture on semilinear wave equations. The technical difficulty prevented us from proving even the blow-up of solutions in finite time. This was finally solved by Yordanov and Zhang [5] in 2006, or Zhou [6] in 2007 independently. But the upper bound of the lifespan was not clarified in both papers. Recently Takamura and Wakasa [4] have succeeded to obtain it including all the critical cases in higher dimensions than 4.

In this note, we present the result of [4] in the most interesting case, 4 space dimensions. It is much simpler than higher dimensions.

Mathematics Subject Classification. Primary 35L70; Secondary 35B05, 35E15.

Keywords. Lifespan, nonlinear wave equation, quadratic term, 4 space dimensions.

1. Introduction

Let us consider the initial value problem for fully nonlinear wave equations,

$$\begin{cases} u_{tt} - \Delta u = H(u, Du, D_x Du) & \text{in } \mathbf{R}^n \times [0, \infty), \\ u(x,0) = \varepsilon f(x), \ u_t(x,0) = \varepsilon g(x), \end{cases} \tag{1.1}$$

where $u = u(x,t)$ is a scalar unknown function of space-time variables,

$$Du = (u_{x_0}, u_{x_1}, \ldots, u_{x_n}), \ x_0 = t,$$
$$D_x Du = (u_{x_i x_j}, \ i, j = 0, 1, \ldots, n, \ i + j \geq 1),$$

$f, g \in C_0^\infty(\mathbf{R}^n)$ and $\varepsilon > 0$ is a "small" parameter. Let

$$\widehat{\lambda} = (\lambda; \ (\lambda_i), i = 0, 1, \ldots, n; \ (\lambda_{ij}), i, j = 0, 1, \ldots, n, \ i + j \geq 1).$$

Suppose that the nonlinear term $H = H(\widehat{\lambda})$ is a sufficiently smooth function with

$$H(\widehat{\lambda}) = O(|\widehat{\lambda}|^{1+\alpha})$$

in a neighbourhood of $\widehat{\lambda} = 0$, where $\alpha \geq 1$ is an integer. Let us define the lifespan $T(\varepsilon)$ by

$$T(\varepsilon) = \sup\{t > 0 \ : \ \exists \text{classical solution } u \text{ of } (1.1) \text{ for arbitrarily fixed } (f, g).\}.$$

In this note we assume $n \geq 2$ for the simplicity.

In Chapter 2 of Li and Chen [2], we have long histories on the estimate for $T(\varepsilon)$. All the lower bounds of $T(\varepsilon)$ are summarized in the following table.

$T(\varepsilon) \geq$	$\alpha = 1$	$\alpha = 2$	$\alpha \geq 3$
$n = 2$	$ca(\varepsilon)$ in general case, $c\varepsilon^{-1}$ if $\int_{\mathbf{R}^2} g(x)dx = 0$, $c\varepsilon^{-2}$ if $\partial_u^2 H(0) = 0$	$c\varepsilon^{-6}$ in general case, $\exp(c\varepsilon^{-2})$ if $\partial_u^b H(0) = 0 \ (b = 3, 4)$	∞
$n = 3$	$c\varepsilon^{-2}$ in general case, $\exp(c\varepsilon^{-1})$ if $\partial_u^2 H(0) = 0$	∞	∞
$n = 4$	$\exp(c\varepsilon^{-2})$ in general case, ∞ if $\partial_u^2 H(0) = 0$	∞	∞
$n \geq 5$	∞	∞	∞

Here c stands for a positive constant independent of ε and $a = a(\varepsilon)$ is a number satisfying $a^2\varepsilon^2 \log(a + 1) = 1$. We note that the lower bound in the case where $n = 4$ and $\alpha = 1$ is $\exp(c\varepsilon^{-1})$ in the general case in Li and Chen [2]. But later, Li and Zhou [3] improved this part. The remarkable fact is that **all these lower bounds are known to be sharp except for $(n, \alpha) = (4, 1)$** and $(2, 2)$. See Li and Chen [2] for references on the whole history, in which the open sharpness of $(n, \alpha) = (2, 2)$ is dropped. For this case, it seems that $b = 4$ is a technical condition which may be removed. See also a table in Li [1]. Recently, Zhou and Han [7] have obtained the sharpness for $b = 3$ in $(n, \alpha) = (2, 2)$ by studying $H = u_t^3$.

Our purpose in this note is to show this remained sharpness of the lower bound by giving a sharp blow-up theorem for $H = u^2$ in $n = 4$. This is a part of our previous paper [4] which establishes the result for model terms $H = |u|^{p_0(n)}$ in all high dimensions, $n \geq 4$. But the proof of this note is much simpler than that of [4]. Because we have an extra term in our iteration frame for $n \geq 5$. This exponent $p_0(n)$ is well known critical number of the famous Strauss' conjecture on model equations $H = |u|^p$. See Introduction in [4] for all the histories and the estimates of the lifespan on this problem. We note that $p_0(n)$ is a positive root of the quadratic equation $(n-1)p^2 - (n+1)p - 2 = 0$ and $p_0(4) = 2$.

Turning back to the problem (1.1) with $H = u^2$ and $n = 4$, we know two papers of Yordanov and Zhang [5] and Zhou [6]. They showed $T(\varepsilon) < \infty$ independently. In this note, we prove the following theorem.

Theorem 1.1. *Let $H = u^2$ and $n = 4$. Assume that both $f \in C_0^4(\mathbf{R}^4)$ and $g \in C_0^3(\mathbf{R}^4)$ are non-negative and do not vanish identically. Suppose that the problem (1.1) has a solution $u \in C^2(\mathbf{R}^4 \times [0, \infty))$ with*

$$\text{supp } u \subset \{(x,t) \in \mathbf{R}^4 \times [0, T(\varepsilon)) \ : \ |x| \leq t + R\} \ (R \geq 1/4). \quad (1.2)$$

Then, there exists a positive constant $\varepsilon_0 = \varepsilon_0(f, g, R)$ such that $T(\varepsilon)$ has to satisfy

$$T(\varepsilon) \leq \exp\left(C\varepsilon^{-2}\right) \quad \text{for } 0 < \varepsilon \leq \varepsilon_0, \quad (1.3)$$

where C is a positive constant independent of ε.

Our method is based on the one in [5]. Zhou and Han [8] have recently found another proof of this theorem which is based on the method in [6].

2. Blow-up lemma for ODI

We shall start with the following observation on our problem (1.1). Let us put

$$F(t) = \int_{\mathbf{R}^4} u(x, t) dx \in C^2([0, T(\varepsilon)).$$

Then it follows from the divergence theorem, the support property (1.2) and Hölder's inequality that

$$F''(t) = \int_{\mathbf{R}^4} u(x, t)^2 dx \geq |\mathbf{B}^4(0, 1)|^{-1}(t + R)^{-4} F(t)^2 \quad \text{for } t \geq 0, \quad (2.1)$$

where $|\mathbf{B}^4(0, 1)|$ is a volume of a unit ball in $\mathbf{R}^4$. Moreover, the assumption on the initial data in Theorem 1.1 yields that

$$F(0) = \varepsilon \int_{\mathbf{R}^4} f(x) dx > 0, \quad F'(0) = \varepsilon \int_{\mathbf{R}^4} g(x) dx > 0. \quad (2.2)$$

Therefore, as in Takamura and Wakasa [4], we shall make use of the following blow-up result for ordinary differential inequality.

Lemma 2.1. *Suppose that $F \in C^2([0, T))$ satisfies*

$$\begin{cases} F(t) \geq Kt^2 & \text{for } t \geq T_0, \\ F''(t) \geq B(t + R)^{-4} F(t)^2 & \text{for } t \geq 0, \\ F(0) > 0, \quad F'(0) > 0, \end{cases} \quad (2.3)$$

where B, K, R, T_0 are positive constants with $T_0 \geq R$. Then, T must satisfy that $T \leq 2T_1$ provided $K \geq K_0$, where

$$K_0 = \left\{ \frac{1}{2^3} \sqrt{\frac{B}{3}} \left(1 - \frac{1}{2^{2\delta}} \right) \right\}^{-2}, \quad T_1 = \max \left\{ T_0, \frac{F(0)}{F'(0)} \right\} \quad (2.4)$$

with an arbitrarily chosen δ satisfying $0 < \delta < 1/2$.

Remark 2.2. In view of (2.1) and (2.2), we already have the second and third inequalities in (2.3) with $B = |\mathbf{B}^4(0,1)|^{-1}$. So the key fact is how to obtain the first inequality in (2.3) with an appropriate choice of T_0 which should be almost like the desired lifespan. See Remark 3.4 below.

Proof. We prove this lemma by contradiction. Assume that $T > 2T_1$. First we note that the second and third inequalities in (2.3) yield that

$$F'(t) \geq F'(0) > 0, \quad F(t) \geq F'(0)t + F(0) \geq F(0) > 0 \quad \text{for } t \geq 0. \quad (2.5)$$

Multiplying the second inequality in (2.3) by $F'(t)$ and integrating it over $[0, t]$, we have that, for $t \geq 0$,

$$\frac{1}{2} F'(t)^2 \geq B \int_0^t (s+R)^{-4} F(s)^2 F'(s) ds + \frac{1}{2} F'(0)^2$$

$$> \frac{B}{3(t+R)^4} \left\{ F(t)^3 - F(0)^3 \right\} \geq \frac{B}{3(t+R)^4} F(t)^2 \left\{ F(t) - F(0) \right\}.$$

Restricting the time interval to $t \geq F(0)/F'(0)$ and making use of (2.5), we get

$$\frac{1}{2} F(t) - F(0) \geq \frac{1}{2} \left\{ F'(0)t - F(0) \right\} \geq 0.$$

Hence we obtain

$$F'(t) > \sqrt{\frac{B}{3}} \cdot \frac{F(t)^{3/2}}{(t+R)^2} \quad \text{for} \quad t \geq \frac{F(0)}{F'(0)}.$$

If $t \geq T_1 (\geq R)$, one can make use of the first inequality in (2.3) to obtain

$$\frac{F'(t)}{F(t)^{1+\delta}} > \sqrt{\frac{B}{3}} \cdot \frac{F(t)^{1/2-\delta}}{(t+R)^2} \geq \sqrt{\frac{B}{3}} \cdot \frac{K^{1/2-\delta}}{2^2 t^{1+2\delta}}$$

for any δ satisfying $0 < \delta < 1/2$. Integrating this inequality over $[T_1, t]$, we get

$$\frac{1}{\delta} \left(\frac{1}{F(T_1)^\delta} - \frac{1}{F(t)^\delta} \right) > \frac{1}{2^3 \delta} \sqrt{\frac{B}{3}} K^{1/2-\delta} \left(\frac{1}{T_1^{2\delta}} - \frac{1}{t^{2\delta}} \right).$$

Then, one can put $t = 2T_1$ because of $T > 2T_1$. Neglecting $1/F(t)^\delta > 0$ in the left-hand side and making use of the first inequality in (2.3) with $t = T_1$, we obtain

$$\frac{1}{K^\delta} \geq \left(\frac{T_1^2}{F(T_1)} \right)^\delta > \frac{1}{2^3} \sqrt{\frac{B}{3}} \left(1 - \frac{1}{2^{2\delta}} \right) K^{1/2-\delta}.$$

This inequality contradicts to the choice of $K \geq K_0$. Therefore we conclude that $T \leq 2T_1$. The lemma is now established. $\square$

3. Iteration of L^2 norm of the solution

In this section, we shall construct an iteration of estimates for L^2 norm of the solution to get the first inequality in (2.3). The final estimate will give us an enough growth of the norm for large time. To this end, we have to start with the following basic frame of the iteration for $F''(t) = \|u(\cdot,t)\|^2_{L^2(\mathbf{R}^4)}$.

Proposition 3.1. *Suppose that the assumption in Theorem 1.1 is fulfilled. Then, there exists a positive constant $C = C(f,g,R)$ such that F'' satisfies*

$$F''(t) \geq C \int_0^{t-R} \frac{d\rho}{(t-\rho+R)^3} \left(\int_0^{(t-\rho-R)/2} F''(s)ds \right)^2 \quad for\ t \geq R. \quad (3.1)$$

Proof. This proposition immediately follows from the combination of two estimates for Radon transformation, (2.14) and (2.21), in Yordanov and Zhang [5]. □

The next proposition is the basic estimate for the first step of our iteration.

Proposition 3.2. *Suppose that the assumption in Theorem 1.1 is fulfilled. Then, there exists a positive constant $C = C(f,g,R)$ such that F'' satisfies*

$$F''(t) \geq C\varepsilon^2 \quad for\ t \geq 0. \quad (3.2)$$

Proof. This is exactly (2.5') in Yordanov and Zhang [5]. □

The main estimate in our iteration is the following proposition.

Proposition 3.3. *Suppose that the assumption in Theorem 1.1 is fulfilled. Then, F'' satisfies that*

$$F''(t) \geq C_j \left(\log \frac{t-R}{(a_j-1)R} \right)^{2^j-1} \quad for\ t \geq a_j R. \quad (3.3)$$

Here we set $a_j = 2^{j+1} - 2$ ($j = 1, 2, 3, \dots$) and

$$C_j = \exp\left\{ 2^{j-1} \left(\log(C_0 C_1 8^{-S(j)}) \right) - \log C_0 \right\} \quad (j \geq 2),$$

$$C_1 = \frac{C^3}{2^2 \cdot 3^3}\varepsilon^4, \quad C_0 = \frac{C}{2^{10}}, \quad S(j) = \sum_{k=1}^{j-1} \frac{k}{2^k}, \quad (3.4)$$

where C is the one in Propositions 3.1, 3.2.

Remark 3.4. The case of $j = 1$ in this lemma is equivalent to the result of Yordanov and Zhang [5]. Such an estimate will give us a new estimate for F,

$$F(t) \geq \left(\frac{C^3}{2^6 \cdot 3^3}\varepsilon^4 \log \frac{t}{4R} \right) t^2 \quad for\ t \geq 4R.$$

See the proof of Proposition 4.1 below. Hence, making use of Lemma 2.1 with $T_0 = D\exp(\varepsilon^{-4})$ in which $D = 2^7 \cdot 3^3 K_0/C^3 > 0$ and ε is appropriately small, we have an estimate of the lifespan as $T(\varepsilon) \leq \exp(2D\varepsilon^{-4})$. But this is not the optimal one. See also the proof of Theorem 1.1 in the next section. The improvement of the upper bound of the lifespan is carried out by further iterations of this proposition.

Proof. First we shall show this proposition for $j = 1$. Replacing $F''(s)$ in the right-hand side of (3.1) by the lower bound of $F''(t)$ in (3.2), we have that

$$F''(t) \geq \frac{C^3}{2^2} \varepsilon^4 \int_0^{t-R} \frac{(t - \rho - R)^2}{(t - \rho + R)^3} d\rho \quad \text{for } t \geq R.$$

From now on, we restrict the time interval to $t \geq a_1 R = 2R$ and diminish the domain of the ρ-integral to $[0, t - 2R]$. Then we have $t - \rho \geq 2R$ in the ρ-integral. We now employ the following elementary lemma.

Lemma 3.5. *Let $M > 1$ and R be a positive constant. Then, $t - \rho \geq MR$ is equivalent to*

$$\frac{M+1}{M-1}(t - \rho - R) \geq t - \rho + R.$$

It is easy to prove this lemma. We omit the proof. Making use of Lemma 3.5 with $M = a_1 = 2$, we obtain that

$$F''(t) \geq \frac{C^3}{2^2 \cdot 3^3} \varepsilon^4 \int_0^{t-2R} \frac{d\rho}{t - \rho - R} = C_1 \log \frac{t - R}{(a_1 - 1)R} \quad \text{for } t \geq a_1 R.$$

Therefore (3.3) is true for $j = 1$.

Next we shall show (3.3) by induction. Assume that (3.3) holds but C_j is unknown except for $j = 1$. Later we look for the relation between C_j and C_{j+1} which yields (3.4). To this end, we restrict the time interval $t \geq a_j R$ to $t \geq (2a_j + 1)R$. Then it follows from (3.1) that

$$F''(t) \geq C \int_0^{t-(2a_j+1)R} \frac{d\rho}{(t - \rho + R)^3} \left(\int_{a_j R}^{(t-\rho-R)/2} F''(s) ds \right)^2 \quad \text{for } t \geq (2a_j+1)R.$$

Making use of (3.3), we have that

$$F''(t) \geq CC_j^2 \int_0^{t-(2a_j+1)R} \frac{I_j(t,\rho)^2}{(t - \rho + R)^3} d\rho \quad \text{for } t \geq (2a_j + 1)R,$$

where we set

$$I_j(t,\rho) = \int_{a_j R}^{(t-\rho-R)/2} \left(\log \frac{s - R}{(a_j - 1)R} \right)^{2^j - 1} ds.$$

Now we restrict the time interval further to $t \geq 2(a_j + 1)R$ and diminish the domain of the ρ-integral to $[0, t - 2(a_j + 1)R]$. Then we have $t - \rho \geq 2(a_j + 1)R$ in the ρ-integral. We note that one can diminish also the domain of the s-integral to $[a_j(t - \rho - R)/(2a_j + 1), (t - \rho - R)/2]$ because of

$$a_j R \leq \frac{a_j}{2a_j + 1}(t - \rho - R).$$

Thus we get

$$I_j(t,\rho) \geq \int_{(t-\rho-R)a_j/(2a_j+1)}^{(t-\rho-R)/2} \left(\log \frac{s - R}{(a_j - 1)R} \right)^{2^j - 1} ds.$$

It follows from $2a_j + 1 = a_{j+1} - 1$ that the variable in the logarithmic term is estimated from below by

$$
\frac{(t - \rho - R)a_j/(a_{j+1} - 1) - R}{(a_j - 1)R} = \frac{(t - \rho - R)a_j - (a_{j+1} - 1)R}{(a_{j+1} - 1)(a_j - 1)R}
$$
$$
= \frac{t - \rho - R}{(a_{j+1} - 1)R} + \frac{t - \rho - R - (a_{j+1} - 1)R}{(a_{j+1} - 1)(a_j - 1)R}.
$$

Hence, neglecting the last positive term in the above equality, we have

$$
I_j(t, \rho) \geq \frac{t - \rho - R}{2(a_{j+1} - 1)} \left(\log \frac{t - \rho - R}{(a_{j+1} - 1)R} \right)^{2^j - 1}
$$

which yields that

$$
F''(t) \geq \frac{CC_j^2}{4(a_{j+1} - 1)^2} \int_0^{t - a_{j+1}R} \frac{(t - \rho - R)^2}{(t - \rho + R)^3} \left(\log \frac{t - \rho - R}{(a_{j+1} - 1)R} \right)^{2^{j+1} - 2} d\rho
$$

for $t \geq a_{j+1}R$. Applying Lemma 3.5 with $M = a_{j+1}$, we have

$$
F''(t) \geq \frac{CC_j^2}{4(a_{j+1} - 1)^2} \left(\frac{a_{j+1} - 1}{a_{j+1} + 1} \right)^3 \int_0^{t - a_{j+1}R} \frac{d\rho}{t - \rho - R} \left(\log \frac{t - \rho - R}{(a_{j+1} - 1)R} \right)^{a_{j+1}}
$$

because of $2^{j+1} - 1 = a_{j+1} + 1$. Therefore we obtain

$$
F''(t) \geq CC_j^2 \frac{a_{j+1} - 1}{4(a_{j+1} + 1)^4} \left(\log \frac{t - R}{(a_{j+1} - 1)R} \right)^{a_{j+1} + 1} \quad \text{for } t \geq a_{j+1}R.
$$

It follows from $a_{j+1} - 1 \geq 2^j$ and $a_{j+1} + 1 \leq 2^{j+2}$ that

$$
F''(t) \geq \frac{CC_j^2}{2^{10} \cdot 8^j} \left(\log \frac{t - R}{(a_{j+1} - 1)R} \right)^{2^{j+1} - 1} \quad \text{for } t \geq a_{j+1}R.
$$

As a conclusion, if C_j is defined by $C_{j+1} = \dfrac{C_0 C_j^2}{8^j}$ $(j \geq 1)$ with C_0 in (3.4), then (3.3) is valid for all $j \geq 1$. This equality is rewritten as

$$
\log C_{j+1} = 2 \log C_j - j \log 8 + \log C_0.
$$

It is clear that C_2 defined by this equality is the one in (3.4). For $j \geq 2$, we have the following concrete expression of $\log C_{j+1}$ inductively.

$$
\log C_{j+1} = 2^j \log C_1 - \sum_{k=1}^{j} k 2^{j-k} \log 8 + \sum_{k=0}^{j-1} 2^k \log C_0
$$
$$
= 2^j \left\{ \log C_1 - S(j+1) \log 8 + \log C_0 \right\} - \log C_0.
$$

This is exactly (3.4). Therefore Proposition 3.3 is now established. $\qquad \square$

4. Proof of Theorem 1.1

We complete the proof of Theorem 1.1 here. In view of Lemma 2.1, our first step is to shift the estimate for $F''(t) = \|u(\cdot,t)\|^2_{L^2(\mathbf{R}^4)}$ to the one for $F(t) = \int_{\mathbf{R}^4} u(x,t)dx$.

Proposition 4.1. *Suppose that the assumption in Theorem 1.1 is fulfilled. Then, F satisfies*

$$F(t) \geq \frac{C_j}{4^{j+1}} \left(\frac{1}{2}\log t\right)^{2^j-1} t^2 \quad \text{for } t \geq \{(a_j+2)R\}^2, \tag{4.1}$$

where a_j and C_j are defined in Proposition 3.3.

Proof. Integrating (3.3) in Proposition 3.3 over $[a_jR,t]$, we have

$$F'(t) \geq C_j \int_{a_jR}^t \left(\log \frac{s-R}{(a_j-1)R}\right)^{2^j-1} ds \quad \text{for } t \geq a_jR.$$

Here we restrict the time interval to $t \geq (a_j+1)R$ and diminish the domain of the s-integral to $[a_jt/(a_j+1),t]$. Then the variable of the logarithmic term is estimated by

$$\frac{a_jt/(a_j+1)-R}{(a_j-1)R} = \frac{t}{(a_j+1)R} + \frac{t-(a_j+1)R}{(a_j-1)(a_j+1)R} \geq \frac{t}{(a_j+1)R}.$$

Hence we obtain

$$F'(t) \geq \frac{C_j}{a_j+1} t \left(\log \frac{t}{(a_j+1)R}\right)^{2^j-1} \quad \text{for } t \geq (a_j+1)R.$$

Integrating this inequality over $[(a_j+1)R,t]$, we have

$$F(t) \geq \frac{C_j}{a_j+1} \int_{(a_j+1)R}^t s \left(\log \frac{s}{(a_j+1)R}\right)^{2^j-1} ds \quad \text{for } t \geq (a_j+1)R.$$

Similarly, restricting the time interval to $t \geq (a_j+2)R$ and diminishing the domain of the s-integral to $[(a_j+1)t/(a_j+2),t]$, we get

$$F(t) \geq \frac{C_j}{(a_j+2)^2} t^2 \left(\log \frac{t}{(a_j+2)R}\right)^{2^j-1} \quad \text{for } t \geq (a_j+2)R.$$

Note that $(a_1+2)R = 4R \geq 1$. Therefore we finally obtain that

$$F(t) \geq \frac{C_j}{(a_j+2)^2} t^2 \left(\frac{1}{2}\log t\right)^{2^j-1} \quad \text{for } t \geq \{(a_j+2)R\}^2.$$

The proof is now ended by $a_j+2 = 2^{j+1}$. $\qquad\square$

Let us continue to prove Theorem 1.1. Assume $j \geq 2$ and define a sequence of time interval $\{I(j)\}$ by

$$I(j) = \left[\{(a_j+2)R\}^2, \{(a_{j+1}+2)R\}^2\right] \tag{4.2}$$

and set

$$K_j(t) = \frac{C_j}{4^{j+1}} \left(\frac{1}{2} \log t \right)^{2^j - 1}$$

which is the coefficient of t^2 in (4.1). Then it follows from the definition of C_j in (3.4) that

$$K_j(t) = \exp \left\{ 2^{j-1} \log L_j(t) - j \log 4 - \log(4C_0) - \log \left(\log \sqrt{t} \right) \right\},$$

where we set

$$L_j(t) = C_0 C_1 8^{-S(j)} \left(\frac{1}{2} \log t \right)^2.$$

In view of the definition of C_1 in (3.4), we have that $L_j(t) \geq e$ provided

$$\varepsilon^2 \log t \geq E \equiv 2 \left(\frac{2^2 \cdot 3^3 \cdot 8^{S(\infty)} e}{C_0 C^3} \right)^{1/2} > 0. \tag{4.3}$$

Because $S(j)$ is monotonously increasing in j, but converges to a positive constant $S(\infty)$.

From now on, we assume (4.3). Then it follows that

$$K_j(t) \geq M_j \quad \text{for } t \in I(j),$$

where we set

$$M_j = \exp \left\{ 2^{j-1} - j \log 4 - \log(4C_0) - \log \left(\log\{(a_{j+1} + 2)R\} \right) \right\}.$$

We have $\lim_{j \to \infty} M_j = \infty$ obviously. Hence, there exists an integer $J = J(f, g, R)$ such that $M_j \geq K_0$ for $j \geq J$, where K_0 is the one in (2.4) with $B = |\mathbf{B}^4(0, 1)|^{-1}$. This fact and Proposition 4.1 imply that

$$F(t) \geq K_0 t^2 \quad \text{for } t \in I(j) \text{ as far as } j \geq J.$$

Therefore the definition of $I(j)$ also implies that

$$F(t) \geq K_0 t^2 \quad \text{for } t \geq \{(a_J + 2)R\}^2.$$

Now we are in a position to apply Lemma 2.1 to our situation with $B = |\mathbf{B}^4(0, 1)|^{-1}$. First we set $T_0(\varepsilon) = \exp \left(E \varepsilon^{-2} \right)$, where E is the one in (4.3). Then there exists $\varepsilon_0 = \varepsilon_0(f, g, R)$ such that

$$T_0(\varepsilon) \geq \{(a_J + 2)R\}^2 \quad \text{and} \quad 2 \max \left\{ T_0(\varepsilon), \frac{F(0)}{F'(0)} \right\} \leq \exp \left(2E\varepsilon^{-2} \right)$$

hold for $0 < \varepsilon \leq \varepsilon_0$ because J and $F(0)/F'(0)$ are independent of ε as we see. If the lifespan $T(\varepsilon)$ satisfies $T(\varepsilon) > T_0(\varepsilon)$, then we have

$$F(t) \geq K_0 t^2 \quad \text{for } t \in [T_0(\varepsilon), T(\varepsilon))$$

by definition of $T_0(\varepsilon)$, because such a t satisfies $\varepsilon^2 \log t \geq E$. Lemma 2.1 says that this inequality implies that

$$t \leq 2 \max \left\{ T_0(\varepsilon), \frac{F(0)}{F'(0)} \right\} \leq \exp \left(2E\varepsilon^{-2} \right).$$

Taking a supremum over $t \in [T_0(\varepsilon), T(\varepsilon))$, we get

$$T(\varepsilon) \leq \exp\left(2E\varepsilon^{-2}\right) \quad \text{for } 0 < \varepsilon \leq \varepsilon_0. \tag{4.4}$$

The counter case $T(\varepsilon) \leq T_0(\varepsilon)$ is trivial. Therefore (4.4) holds for any cases. The proof of Theorem 1.1 is now completed. □

References

[1] T.-T. Li, *Lower bounds of the life-span of small classical solutions for nonlinear wave equations*, Microlocal Analysis and Nonlinear Waves (Minneapolis, MN, 1988–1989), The IMA Volumes in Mathematics and its Applications, vol. 30 (M. Beals, R.B. Melrose and J. Rauch, eds.), 125–136, Springer-Verlag New York, Inc., 1991.

[2] T.-T. Li and Y. Chen, "Global Classical Solutions for Nonlinear Evolution Equations", Pitman Monographs and Surveys in Pure and Applied Mathematics 45, Longman Scientific & Technical, 1992.

[3] T.-T. Li and Y. Zhou, *A note on the life-span of classical solutions to nonlinear wave equations in four space dimensions*, Indiana Univ. Math. J., **44** (1995), 1207–1248.

[4] H. Takamura and K. Wakasa, *The sharp upper bound of the lifespan of solutions to critical semilinear wave equations in high dimensions*, J. Differential Equations, **251** (2011), 1157–1171.

[5] B. Yordanov and Q.S. Zhang, *Finite time blow up for critical wave equations in high dimensions*, J. Funct. Anal., **231** (2006), 361–374.

[6] Y. Zhou, *Blow up of solutions to semilinear wave equations with critical exponent in high dimensions*, Chin. Ann. Math. Ser. B, **28** (2007), 205–212.

[7] Y. Zhou and W. Han, *Sharpness on the lower bound of the lifespan of solutions to nonlinear wave equations*, Chin. Ann. Math. Ser. B., 32B (4) (2011). (doi: 10.1007/s11401-011-0652-5)

[8] Y. Zhou and W. Han, *Life-span of solutions to critical semilinear wave equations*, arXiv:1103.3758 [math.AP] 19 Mar. 2011.

Hiroyuki Takamura
Department of Complex and Intelligent Systems
Faculty of Systems Information Science
Future University Hakodate
116-2 Kamedanakano-cho
Hakodate, Hokkaido 041-8655, Japan
e-mail: takamura@fun.ac.jp

Kyouhei Wakasa
The 2nd year of Graduate School of Systems Information Science
Future University Hakodate
116-2 Kamedanakano-cho
Hakodate, Hokkaido 041-8655, Japan
e-mail: g2111045@fun.ac.jp

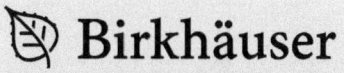 **Birkhäuser** | **birkhauser-science.com**

Progress in Mathematics (PM)

Edited by
Hyman Bass, University of Michigan, USA
Joseph Oesterlé, Institut Henri Poincaré, Université Paris VI, France
Alan Weinstein, University of California, Berkeley, USA
Yuri Tschinkel, Courant Institute of Mathematical Sciences, New York, USA

Progress in Mathematics is a series of books intended for professional mathematicians and scientists, encompassing all areas of pure mathematics. This distinguished series, which began in 1979, includes research level monographs, polished notes arising from seminars or lecture series, graduate level textbooks, and proceedings of focused and refereed conferences. It is designed as a vehicle for reporting ongoing research as well as expositions of particular subject areas.